IEE HISTORY OF TECHNOLOGY SERIES 14
Series Editor: Dr. B. Bowers

A SCIENTIST'S WAR

the war diary of Sir Clifford Paterson, 1939-45

Other volumes in this series:

Volume 1	**Measuring instruments—tools of knowledge** P. H. Sydenham
Volume 2	**Early radio wave detectors** V. J. Phillips
Volume 3	**A history of electric light and power** B. Bowers
Volume 4	**The history of electric wires and cables** R. M. Black
Volume 5	**An early history of electricity supply** J. D. Poulter
Volume 6	**Technical history of the beginnings of radar** S. S. Swords
Volume 7	**British television—the formative years** R. W. Burns
Volume 8	**Hertz and the Maxwellians** J. G. O'Hara and D. W. Pricha
Volume 9	**Vintage telephones of the world** P. Povey and R. A. J. Earl
Volume 10	**The GEC Research Laboratories 1919-84** R. J. Clayton and J. Algar
Volume 11	**Metres to microwaves** E. B. Callick
Volume 12	**A history of the world semiconductor industry** P. R. Morris
Volume 13	**Wireless: the crucial decade, 1924-1934** G. Bussey
Volume 14	**A scientist's war—war diary of Sir Clifford Paterson** ed. R. J. Clayton and J. Algar
Volume 15	**A period of change: electricity in mining** A. V. Jones and R. P. Tarkenter
Volume 16	**'Curiosity Perfectly Satisfyed:' Faraday's Travels in Europe, 1813-1815** ed. B. Bowers and L. Symons

A SCIENTIST'S WAR

the war diary of Sir Clifford Paterson, 1939-45

1st September 1939-9th May 1945

Edited by

Robert Clayton

& Joan Algar

Peter Peregrinus Ltd. in association with The Science Museum

Published by: Peter Peregrinus Ltd., London, United Kingdom

Peter Peregrinus Ltd.,
Michael Faraday House,
Six Hills Way, Stevenage,
Herts. SG1 2AY, United Kingdom

British Library Cataloguing in Publication Data
Paterson, Clifford 1879-1948
A scientist's war: the diary, from September 1st, 1939 to May 9th, 1945, of Sir Clifford Paterson FRS OBE.
1. Electric Equipment industries,: General Electric Company. Research laboratories, history
I. Title II. Clayton, Robert 1915-
III. Algar, Joan
621.31042072

ISBN 0 86341 218 1

Printed in England by Short Run Press Ltd., Exeter

Contents

Introduction

This book consists of a transcript of the diary kept by Sir Clifford Paterson between 1st September 1939, and 9th May 1945.

Clifford Paterson was born in 1879 and after attending Mill Hill School embarked on a comprehensive training in engineering at Finsbury Technical College and in two industrial companies. In 1901 he entered Faraday House as a special student-assistant, and shortly afterwards was joint author of his first paper. Subsequently he spent a year in Swiss industry.

In 1903 he joined the National Physical Laboratory in its newly established Electrotechnical and Photometric Division where he worked on photometry and the development of equipment for testing commercial electrical instruments. He was also responsible for much of the design and equipment of the new buildings for the Division — an experience which was to be of use in the future. Paterson became a recognised international authority on illumination; in 1913 he was appointed first secretary of the International Commission on Illumination, a post he held until 1928 when he was elected President of the Commission. A full account of his work at the National Physical Laboratory, including his activities during the 1914–18 War, as well as his subsequent career, is given in Obituary Notices of Fellows of the Royal Society, Volume 6, pp. 479–501, November 1949. For his wartime work on a height finder he was awarded the OBE.

In 1916 he was approached by the management of the Osram Lamp Company (a company in which The General Electric Company Limited had a large interest) who first asked him to suggest someone to organise a research department for them, and then proposed that he should undertake the task himself. After some initial doubts, dispelled by further discussions, Paterson accepted, but nothing could be done until the war ended in 1918 by which time the Osram Lamp Company had become a part of GEC and it was proposed that the new research laboratory should serve the whole of GEC. Paterson became Director of the GEC Research

Laboratories in January 1919, and held this post until his death in July 1948. He inspired and led its work, he conceived and was responsible for its philosophy, organisation and facilities. He also looked after relationships with the rest of GEC and the outside world. He was elected a Director of GEC in 1941. An account of his remarkable leadership is given in the book 'The GEC Research Laboratories 1919–1984' by Robert Clayton and Joan Algar.

Paterson maintained a substantial interest in professional affairs. He was President of the Institution of Electrical Engineers in 1930–31. He also delivered that Institution's Faraday Lecture, with many demonstrations at about fifteen centres in the British Isles, and in 1948 was awarded the Faraday Medal of the Institution. He was President of the Institute of Physics from 1937 to 1939. These Presidencies followed many years of service to the Institutions. He was also very active in the Royal Institution, regularly attending its Friday evening discourses (giving two himself) and serving on its Board of Managers for many years, several times as Vice-President. He served as President of the Illuminating Engineering Society and, after his period as President of the International Commission on Illumination, he again became its honorary secretary until July 1948 when he was elected Honorary President. Paterson was much in demand for service on government committees such as the Executive Committee of the National Physical Laboratory and Committees of the Department of Scientific and Industrial Research. He was closely associated with the work of the British Standards Institution and Chairman of the Committee of Management of *Science Abstracts*.

During the 1939–45 war, two of his principal public services were his Chairmanship of the Inter-Services Technical Valve Committee and his membership of the War Cabinet's Engineering Advisory Committee.

Paterson was elected a Fellow of the Royal Society in 1942 and knighted in 1946. His remarkable energy enabled him to carry out many public duties without neglecting the leadership of the Laboratories for which he was responsible, and his many contacts were to be of great value to the country and the Laboratories during the 1939–45 war.

An account of the organisation and work of the Laboratories is given in the book already referred to, but a brief account of their pre-war activities, and how they fitted the Laboratories for their wartime roles may help readers of the Diary.

The GEC Research Laboratories were originally conceived in 1916 to provide scientific support for the Osram Lamp Works at Hammersmith which, until the outbreak of war in 1914, had been dependent on technology from Germany. By the time it was possible to go ahead at the end of 1918 it had been decided that the new Laboratories should be the Research Laboratories to support substantialy all GEC activities. They started work at Hammersmith and, even before they moved to their purpose-built accommodation at Wembley in 1922, work on thermionic valves had been added to that on filament lamps. There were also supporting programmes on vacuum physics, glass, metallurgy and chemistry.

The 1920s saw continuation of this work and the start of work on communications materials, components and systems. Mechanical engineering skill, originally built up to support other programmes, was now employed on work for parts of GEC and for outside customers. Programmes were started on light from gas discharge lamps and on illumination, particularly street lighting. Large transmitting valves were developed in the Laboratories; they had cooled anodes and depended on glass-to-metal seals — a technique which was to be of vital importance in the 1939–45 war.

By the 1930s the Laboratories were skilled and inventive in vacuum and gas-discharge physics, in glass and metallurgy (particularly tungsten), in fluorescent materials, and lamps using these techniques. They were expert on photometry, the design of lighting fittings, visibility, and both internal and outdoor lighting installations. Their pre-eminence on large transmitting valves continued with the introduction of the CAT 14, which for at least ten years was the most powerful transmitting valve in the world. Work on sound broadcasting, acoustics and television, including cathode-ray tubes, was added to that on point-to-point communications. There were groups working on carbon, on batteries, on ceramics and on mineral dressing. All of these were supported by an integrated material characterisation team using chemistry, spectroscopy and X-ray diffraction, and by design engineers and highly skilled craftsmen in metal, glass and wood.

There were 'laboratory-factories' or 'try-out units' in which new products or processes could be tested before introduction into full-scale production, invaluable in the wartime work which was to come, and there were life-test facilities for assessing both new and production types of lamps and valves.

At the outset of war in 1939 the Laboratories were, by their experience and organisation, uniquely placed to make a major contribution to the war effort.

There were at that time research laboratories in other industries, such as the chemical industry. There were laboratories in some electrical, communications and radio companies; and there were development facilities in more such companies. The television industry in particular was well placed to pick up work on radar. However, there were few, if any, industrial research laboratories with the breadth of activities of the GEC Research Laboratories or with their multi-disciplinary approach, embracing mathematics, physics, chemistry, materials science, metallurgy and electrical and mechanical engineering. Nor were there many, if any, with their range, for example in communications – from materials and components to equipment and systems. They were thus able to undertake a wide variety of tasks in support of the war effort.

As the war progressed Government departments built up their own research teams, particularly by the transfer of staff from universities and government civil establishments, as well as by recruitment of new graduates. It was here, rather than with other parts of industry, that there were the biggest chances of overlap with the work at Wembley. Some of these are reflected in entries in the Diary.

The wartime work of the Laboratories is described in 'The GEC Research Laboratories, 1919 – 1984' (Ref. 1) but a summary may help the reader of the Diary to understand references to particular projects.

The war diverted the Laboratories for some six years from their prime task of serving the interests of GEC. In an address to members of the staff in 1944 Paterson described how, on the outbreak of war, Lord Hirst and Max Railing (the Chairman and Vice-Chairman and Managing Directors of GEC) instructed him to put the Laboratories at the disposal of the Government and to let no commercial considerations or industrial rivalry stand in the way of the war effort; nor did they wish the GEC to make any profit on research for the Services. Paterson would have found this instruction in line with his own views and it certainly appeared at working level that this is what happened. Work was undertaken because it seemed that the staff had the skills required to carry it out and not because it would lead to production in GEC factories or put the GEC in a strong technical or commercial position after the war. In fact there were occasions when this policy seemed to go against the interests of GEC; work started at Wembley being taken through to production in other organisations.

Paterson, whose standing and integrity were recognised by his appointment as Chairman of the Government Inter-Services Technical Valve Committee, had no respect or time for anyone whom he believed to be furthering their personal ends or playing at politics. His willingness to get on with a project without formality was to give headaches to Government administrators and accountants. At least one project is said to have been started on the basis of a discussion in the Athenaeum, the agreement being recorded on little if anything more than the backs of envelopes. It took a contracts branch two or three years to catch up with this one and accept that payment was due to GEC for work done.

The attitude of Paterson and GEC to profit did bring some benefit in that, in exchange for no profit, it was sometimes possible to protect GEC's interests in intellectual property rights to a greater extent than would otherwise have been possible.

There were a small number of discussions involving Paterson and occasionally one or two of his senior colleagues, together with colleagues from Head Office and Works, on what might be needed or proposed in the post-war period (such as television) and these discussions increased in frequency in the closing months of the war. It must also be noted that the provisional patent for what was probably the Laboratories' most important invention, the halophosphate fluorescent powders, carries a date of April 1941. But for almost every member of the staff the years 1939 to 1945 were devoted without distraction to work aimed at winning the war.

There were inevitably misunderstandings and occasionally friction between the Laboratories and others working in the same or related fields, including those who had been moved to new fields, for example from universities or Government civil establishments.

It was not unknown for high ranking officers in the Services to try to lure the Laboratories into projects (perhaps because they mistrusted their own technical organisations) with promises of facilities, such as a special squadron of aircraft, which might be more than were available to their own colleagues.

Paterson felt at the end of the war that less than fair recognition had been given to some of the work of the Laboratories, particularly on magnetrons and airborne radar. He moved quickly to put the record straight, using this Diary and the notebooks of senior colleagues and himself. All this having been said, the wartime years were a period of great collaboration between the Laboratories and other organisations, a collaboration which was fostered by Paterson.

The major contribution which the Laboratories made was their work on valves and cathode-ray tubes. Their own original researches led to a range of designs of disc-seal valves and associated transmission-line circuits for both pulse and CW operation at frequencies up to 3 GHz. The Laboratories also developed both vacuum and gas-filled modulator valves and small rugged valves for use in proximity fuses.

They were also asked to bring their expertise to bear on the development of devices originated by others, such as reflex klystrons and T–R cells. Their biggest contribution of this kind was that which they made to the cavity magnetron invented by Randall and Boot at Birmingham University. Members of the staff of the Laboratories first heard of this device in April 1940. Their report No. 8717, issued in August 1945, summarising Wembley contributions to high-power pulsed magnetron development, describes the Birmingham device as a CW oscillator using a large electromagnet; the valve being operated on a vacuum pump. However, the caption to an illustration of the Birmingham valve in the report (taken from Birmingham University CVD Report of May 1941) states: 'CW or pulse output order of half kilowatt'. Sir Mark Oliphant, who was Head of the Birmingham University Physics Department at the time has recently told us that 'the device was pulsed, yielding only low power when operated continuously, and was located in the magnetic field of a laboratory electro-magnet producing a steady field, which could be varied to find the optimum'.

The Wembly staff proposed and introduced developments which produced a high-power, pulsed, sealed valve, using a permanent magnet and with a new type of cathode, suitable for operational use. The first prototype of this design was tested on 29th June 1940. Quantities were produced urgently in the Laboratories, one of which was flown to America and demonstrated there. The valve which oscillated on 29 June 1940 is now in the IEE archives.

This was only one example of the needs of the Services being so urgent that there was no time to transfer a design to a factory, at least until early demands had been met. The Paterson concept of 'laboratory-factories' or 'try-out units' proved invaluable and pre-production units for many types were set up in the Laboratories.

Physically, the biggest effect on the Laboratories was the tremendous expansion of facilities for the pre-production and testing of valves, cathode-ray tubes and other devices developed for wartime purposes. More than 300 000 valves of 45 different types were produced in the Laboratories, including nearly 50 000 of one type. More than a thousand newcomers joined the staff, many of them part-time, and most of them with no experience of working in a laboratory or a factory.

A major radar project was undertaken by the pre-war Television Group. At the beginning of the war the RAF had only limited-performance radar in its night fighters, operating on a wavelength of 1.5 metres. There was an urgent need for a radar of higher definition and this implied operating at a much shorter wavelength and using a scanning aerial system. The Air Ministry asked Paterson to allocate one of his teams to such a project and the Television Group undertook this task.

There were to be complications when, with the transfer of many academics to defence establishments, AMRE (later to be known as TRE) gave the same task to one of its own new teams and for some months neither team seems to have been aware of the work of the other. The Wembley team nevertheless made a major contribution to AI Marks 7 and 8, particularly on their electronic circuits, on microwave components and on flight testing. A scanner developed at Wembley for AI was not used in this sytem but was the first scanner to be used in the development of the bombing and navigational aid H_2S. Later in the war the Wembley team contributed to the development of the centrimetric version of Oboe, the precision bombing system, to aerial design for other airborne radars, and developed an anti-aircraft radar for the Army and a compact range finder for Combined Operations.

An important development by this team was the use, in early 1940, of a silicon crystal frequency changer, in place of a thermionic valve, at microwave frequencies. The Laboratories certainly knew of no prior work when they did this.

A few tens of yards away from the AI team in the Wembley Laboratories, another team was developing, unbeknown to all in the Laboratories but themselves and the most senior people at Wembley, electronic countermeasures equipment for detecting enemy transmissions and for confusing and jamming enemy navigational and radar systems. The security on this work was so complete that an account of it, by its leader, was only recently produced. A description of the projects undertaken is given in Reference 1 and entries in the Diary record both progress and the close relationships of the Wembley team with the RAF and the other services.

By-products of the work on valves, radar and countermeasures were substantial contributions to measurement techniques at decimetre and centimetre wavelengths. These were the subject of an IEE paper shortly after the war.

Another major contribution was the extension of earlier work on dispersion of light by turbid media to predictions of the extent of scattering of centimetric radio waves by fog, rain, hail and dust.

After tackling a number of other projects the Telecommunications Group was joined by a team from RAF Technical Training Command to explore possibilities of torpedo guidance using detection by ultrasound. Before the end of the war models of a homing torpedo had been completed and had homed successfully on an artificial target in Loch Goil. This activity was to be the foundation of work on torpedoes which has continued ever since in GEC.

The Radio Group worked on development of VHF transmitter – receivers for aircraft use and invented a system for automatic selection of radio frequencies. Equipment incorporating this invention was flown to the United States for demonstration, accompanied by a member of the staff of the Laboratories. The group also contributed to Army mobile communications equipment, demonstrated the possibilities of narrow band VHF/FM mobile systems for use on land and in the air, and invented a low level radio altimeter.

When war broke out, the Illumination Group was probably the most experienced team in the world in handling light, and this expertise was to be of vital importance. Problems arose for all three Services, for Civil Defence, for war production and for the safety of the civilian population, particularly as it moved about in the black-out.

Lighting of streets was developed so that it was possible to move about in them without clues being given to enemy aircraft, The view from the air from bombing height was simulated in the laboratory, complete with moonlight and mist. Street lighting was also simulated and the results checked with night flights.

It was recognised nationally that mist was a vital factor in a distant view but there was no data on it. The Wembley team was asked to organise measurements of the properties of the atmosphere. The method used involved measuring the distribution of light scattered from about one cubic inch of atmosphere, in an aircraft at various heights, in the blackout. An ingenious piece of equipment, small enough to be carried in an aircraft, was devised and made ready for trials in two months, and shortly afterwards the first flights were successfully, though hazardously, made in an unpressurised Wellington bomber at heights up to 30 000 ft.

This data was used in work on the action of aircraft reconnaissance flares (mostly done in the laboratory in a small tank of turbid water but related back to the real atmosphere); on a proposal for, and the testing of, screened flares; on the working of searchlights, and the light distribution of a bomber illuminated by them; and on visual conditions in anti-submarine attacks by Coastal Command.

The laboratory collaborated with the Admiralty in setting up a range for the study of naval camouflage in which model ships could be observed at simulated long distances with controlled sunlight, skylight and mist. Some results were to be overtaken by the development of radar; other work such as the treatment of naval craft with disruptive patterns was of more lasting use. When a skilled observer was needed at sea, a member of the Group was seconded to the Royal Navy and was killed when the submarine in which he was on passage was torpedoed. He had made for this work an elegant, compact, telephotometer; after his death the laboratory staff made an improved version with which his work was completed. It was later manufactured by a GEC subsidiary and used for many years in a number of laboratories.

The Lamp Groups not only undertook a number of developments, including compact-source high-intensity lamps, but also provided staff for other activities such as valve development, production and testing.

The team which had worked on carbon before the war turned to development of powerful searchlights for both airborne and ground use. They worked with Air Commodore Helmore who used his personal approach and contacts to put through a number of projects without recourse to the normal administrative machinery. He was a great advocate of the use of searchlights in night fighters and later of a very large ground-based light. His final work with the Laboratories was on his concept of a 36 in diameter command guided torpedo, propelled on the surface by a Rolls Royce Merlin engine and by battery driven motors when submerged.

Other projects outside the main stream of the war effort were those on airborne infra-red systems (some of them with Captain A. Frazer-Nash).

From entries in the Diary this work seems to have involved trials in very adverse conditions both on the ground and in the air.

A new venture for the Laboratories was the development and production of diamond dies for wire drawing. Although the use of diamond as a hard die material for drawing down of metals into wire had been established many years before, dies had never been made in quantity in the United Kingdom. Wire mills had been supplied largely by imports from other European countries, and very little was known about their manufacture. A severe shortage developed as the war progressed. GEC had a major interest in the drawing down of very fine wires for lamps and valves.

In 1940 the Ministry of Supply set up an authority to control diamond die production and Clifford Paterson accepted the chairmanship of its technical sub-committee.

Development work started in the Laboratories which was aimed at production of dies from 0.004 to 0.0015 in. in diameter. There was very little information for guidance other than diagrams of parts of machines contained in patent specifications and impressions of machinery and processes used on the Continent which had been gleaned by lamp engineers during visits to manufacturers from whom dies had been purchased.

Machines were designed and processes developed at Wembley so that, by 1941, it was possible for young, inexperienced, female operators to produce the finest dies after only a few months experience. Before the end of 1942, fifty drilling machines were installed in the Laboratories. Frequent entries in the Diary record progress in the Laboratories and in industry discussions.

Other projects in the Laboratories included proximity fuzes, fog dispersal on airfields, infra-red communication, stopping fuze clocks in unexploded bombs, and gun location. In support of the war effort there was work on synthetic sapphires, alternatives to natural quartz, and on education of many organisations on statistical techniques for control of quality in production. The last of these led to a GEC Research Laboratories Handbook which in turn became a British Standards Specification.

Two members of the staff were killed in the course of their work; one, who was killed at sea while working on camouflage with the Royal Navy has already been referred to; the other was killed in an air crash when returning from the United States where he had been discussing the Laboratories' work on sub-miniature valves for proximity fuzes.

While the Laboratories' own organisational requirements remained simple and in accordance with Paterson's concepts during the war, there arose a major administrative load from Government requirements on manpower direction, security, salaries and taxation, contracts and accounting. For the first time Paterson relinquished responsibility for staff matters other than the graduate staff. Increasingly Mr. O.W. (later Sir Olliver) Humphreys assumed responsibility for the administrative management of the Laboratories as opposed to their scientific direction. He was to become Director of the Laboratories after Sir Clifford's death in 1948.

The Diary

There is no evidence that Paterson kept a diary for most of the time when he was Director of the Laboratories but he started a daily diary in September 1939, just as war broke out, and made entries regularly until he closed the diary when victory in Europe was being celebrated in May 1945.

In the Diary Paterson records the progress of work in the Laboratories and his own wider role in the planning and organisation of the scientific war effort against the background of, on the one hand, the progress of the war and, on the other, the personal problems of the members of his staff. There are also references to the administration and housekeeping which were necessary for the smooth running of the Laboratories.

There are accounts of visits from, and discussions with, senior Service officers, Ministers, and officials in many Government departments and research establishments. There are comments on the contributions that such people were making to progress and, on occasion, references to the delaying effects of some rivalries and bureaucracy. Paterson also expresses, from time to time, opinions on the proper roles of Government, the universities and industry in both war and peace.

In early entries his account deals with the transfer of the Laboratories from their peacetime to their wartime roles; subsequent entries deal with their sustained efforts over a number of years, including preproduction on thermionic valves and other devices in addition to research and development. Later entries deal with the planning of the return to peacetime activities.

To some extent the Diary is a 'puzzle book'. It has been possible to identify many projects and people but there are hard cores of each which remain enigmas. It is the Editors' hope that a number of readers will be able to supplement their efforts from personal knowledge and reduce the number of unknowns still further.

The original Diary, which is to be deposited in the archives of The Institution of Electrical Engineers, was typed on foolscap paper and filed in buff folders. There is no indication of whether it was typed from notes or from dictation, nor is there any indication of the typist. Such things as regularity of margins suggest that pages rather than days were typed at one session. Paterson is known to have used a dictating machine and it is possible that he dictated entries, usually at the end of a day, and that they were subsequently transcribed. Marginal manuscript notes indicate that the diary was used at the end of the war as a source of material for reports summarising the Laboratories' contributions in a number of fields. Two of these (on the cavity magnetron and on airborne radar) have been found and copies are in the IEE archives. The copy of the Diary deposited at the IEE appears to have been Paterson's master and possibly only copy because it includes his original manuscript marginal notes and the originals of the enclosures (such as letters) which he inserted. Most of these enclosures are reproduced as illustrations in the present book.

Obvious typing errors have been corrected where there is no ambiguity, as have some abbreviations which reduce readability. Where it is believed that there is a more significant typing error or other mistake, an intepretation has been suggested in square brackets. More speculative interpretations are indicated by a question mark at the end of the bracketed entry.

Notes have been provided for a number of entries to amplify, where possible, references to individuals, projects, and valve types. These are to be found after the main body of the diary, on pages 576–625.

Appendixes deal with the special relationships between the Laboratories and the CVD organisation, with valve types referred to in the diary, abbreviations used by Paterson, and the nomenclature of radar frequency bands. A brief chronology of outside events is provided in the following pages which may help to set diary entries in context.

Chronology of World Events

1.09.39	Germany invades Poland.
	Evacuation of children from British cities begins.
	Blackout starts.
3.09.39	Britain and France declare war on Germany.
17.09.39	Russia invades Poland.
30.09.39	National Registration in Britain.
	British Expeditionary Force lands in Europe.
14.10.39	H.M.S. Royal Oak sunk in Scapa Flow.
18.11.39	Magnetic mines sink 60 000 tons of shipping on English east coast in a week.
13.12.39	Battle of River Plate.
1.01.40	Bacon, butter and sugar rationed in Britain.
9.04.40	Germany invades Norway and Denmark.
10.05.40	Germany invades Holland, Luxembourg and Belgium.
	Neville Chamberlain resigns; Churchill forms National Government.
	L.D.V. (Home Guard) formed in Britain.
14.05.40	German troops enter France.
	Dutch army surrenders.
	British bomb the Ruhr.
28.05.40	Belgium capitulates.
28.05.40 to 3.06.40	British Expeditionary Force evacuated from Dunkirk.
10.06.40	Italy declares war on France and Britain.
14.06.40	German troops enter Paris.
22.06.40	France concludes armistice with Germany.
10.07.40	Battle of Britain (in the air) begins.
11.08.40 to 18.08.40	Battle of Britain at its peak.
23.08.40	All-night raid on London begins the 'Blitz'.
25.08.40	British begin night bombing of Germany.
15.09.40	Heavy daylight raid on London (Battle of Britain Day).
5.11.40	Roosevelt re-elected President for third term.

14.11.40	Heavy night air raid on Coventry.
2.12.40	Bristol heavily bombed.
29.12.40	Air raid with incendiary bombs on City of London.
22.01.41	Tobruk captured by Australian troops.
7.02.41	Benghazi captured.
11.03.41	Lease-Lend Bill signed by Roosevelt.
19.03.41	German air raids on London resumed.
28.03.41	Royal Navy victory over Italians at battle of Cape Matapan.
6.04.41	Greece and Yugoslavia invaded by Germany. Britain sends troops to Greece.
20.04.41	Rommel attacks Tobruk.
22.04.41 to 2.05.41	British withdraw from Greece.
6.05.41	Stalin becomes head of Soviet government.
10.05.41	House of Commons destroyed in London's heaviest air raid. Rudolf Hess lands in Scotland.
20.05.41	Germans invade Crete.
24.05.41	H.M.S. *Hood* sunk by *Bismarck.*
27.05.41	*Bismarck* sunk by Royal Navy. British withdraw from Crete.
22.06.41	Germany invades Russia.
25.07.41	Fighting round Smolensk.
11.08.41	Churchill and Roosevelt meet and sign Atlantic Charter.
19.09.41	Germans enter Kiev.
16.10.41	Germans 60 miles from Moscow. Odessa falls to Germans.
14.11.41	H.M.S. *Ark Royal* sunk near Gibraltar.
29.11.41	Russian counter-offensive.
7.12.41	Japanese bomb Pearl Harbour, Hawaii, and British Malaya.
8.12.41	U.S.A. and Britain declare war on Japan,
10.12.41	H.M.S. *Prince of Wales* and H.M.S. *Repulse* sunk by Japanese aircraft.
11.12.41	United States declares war on Germany and Italy.
25.12.41	Hong Kong surrenders to Japanese.
2.01.42	Manila taken by Japanese.
19.01.42	Japanese invade Burma.
9.02.42	Soap rationed.
15.02.42	Singapore surrenders.
9.03.42	Java surrenders to Japanese.
28.03.42	British Commandos raid St. Nazaire.
23.04.42 to 30.04.42	'Baedeker' air raids on Exeter, Bath, etc.
4.05.42 to 8.05.42	Battle of the Coral Sea
26.05.42	Rommel resumes offensive in North Africa.
30.05.42	British convoy reaches Russia despite heavy air attacks. R.A.F. 1000 bomber raid on Cologne.

3.06.42 Battle of Midway Islands between Japanese and American Navies.
21.06.42 Germans under Rommel take Tobruk.
Heavy British losses on convoy to Malta.
25.06.42 R.A.F. 1000 bomber raid on Bremen.
1.07.42 Vote of censure debate in Commons on direction of war.
3.07.42 Germans take Sebastopol.
26.07.42 R.A.F. raid on Hamburg.
28.07.42 Germans take Rostov.
7.08.42 Americans land in Guadalcanal.
15.08.42 H.M.S. *Eagle* and H.M.S. *Manchester* lost on Malta convoy.
19.08.42 Raid on Dieppe.
Montgomery given command of Eighth Army.
26.08.42 Germans reach Stalingrad.
23.10.42 Battle of El Alamein commences.
4.11.42 Rommel's forces in full retreat.
7.11.42 Allied landings in French North Africa under Eisenhower.
9.11.42 Germans move into unoccupied France.
13.11.42 British retake Tobruk.
19.11.42 German Army at Stalingrad surrounded.
2.12.42 Fermi initiates controlled chain reaction in the first nuclear reactor.
21.12.42 Eighth Army re-occupy Benghazi.
2.01.43 German withdrawal from Caucasus begins.
14.01.43 to 24.01.43 Churchill and Roosevelt confer at Casablanca.
15.01.43 Japanese driven from Guadalcanal.
18.01.43 German air raids on London resumed.
16 month siege of Leningrad ended.
23.01.43 Eighth Army enters Tripoli.
30.01.43 Russians destroy German army south-west of Stalingrad.
28.02.43 Heavy R.A.F. raid on Berlin.
1.03.43 R.A.F. begin systematic bombing of European railway system.
6.04.43 Allied armies in North Africa link up.
26.04.43 Heavy air raids on Ruhr.
7.05.43 Allies take Tunis and Bizerta.
12.05.43 German army in Tunisia surrenders.
17.05.43 Ruhr dams breached by R.A.F.
10.07.43 Allies land in Sicily.
25.07.43 Mussolini falls from power.
17.08.43 Allies take Sicily.
3.09.43 Allies invade Italy.
7.09.43 Italy surrenders.
10.09.43 Germans seize Rome.
23.09.43 Tirpitz damaged.
25.09.43 Russians take Smolensk.
30.09.43 Fifth Army takes Naples.

6.11.43	Russians capture Kiev.
28.11.43 to 1.12.43	Churchill, Roosevelt and Stalin meet at Teheran.
2.12.43	Direction to employment in mining industry introduced.
26.12.43	*Scharnhorst* sunk.
22.01.44	Allied landings at Nettuno and Anzio.
27.01.44	Leningrad completely relieved.
4.02.44	Allied troops reach Monte Cassino.
7.02.44 to 29.02.44	German assaults on Anzio Bridgehead.
15.03.44	Cassino destroyed by Allied bombers.
9.05.44	Sebastopol liberated.
18.05.44	Monte Cassino taken.
4.06.44	Fifth Army enters Rome.
6.06.44	Allied invasion of Normandy begins ('D Day').
12.06.44	Flying Bomb (V1) attacks on London commence.
27.06.44	Allies capture Cherbourg.
3.07.44	Russians take Minsk.
9.07.44	Caen (Normandy) captured.
20.07.44	Attempted assassination of Hitler.
15.08.44	Allies land in southern France.
19.08.44	Eighth Army takes Florence.
25.08.44	Paris entered by Allies.
4.09.44	Allies capture Antwerp and destroy flying bomb sites in Pas de Calais.
5.09.44	Brussels liberated.
8.09.44	First V2 rocket lands in Britain.
11.09.44	Americans cross German frontier near Trier.
17.09.44	British airborne landing at Arnhem and Eindhoven; subsequently withdrawn.
28.09.44	Canadians liberate Calais.
3.10.44	Warsaw rising crushed by Germans.
5.10.44	British troops land on Greek mainland.
14.10.44	Athens captured by the Allies.
7.11.44	Roosevelt re-elected for fourth term.
12.11.44	*Tirpitz* sunk.
16.12.44 to 5.01.45	German offensive in Ardennes; 'Battle of the Bulge'.
1.01.45	Fourteenth Army opens offensive in Burma.
17.01.45	Russians capture Warsaw.
23.01.45	Russians reach Oder.
4.02.45 to 11.02.45	Churchill, Roosevelt and Stalin meet at Yalta.
5.02.45	Allied troops enter Manila.
15.02.45	British troops reach Rhine.
16.02.45	Heavy American air raids on Tokyo begin.
7.03.45	Cologne taken by Allies.

23.03.45	Allied forces cross the Rhine.
12.04.45	Roosevelt dies; succeeded by Truman.
20.04.45	Russians reach Berlin.
23.04.45	Russian and American forces meet.
28.04.45	Mussolini killed.
30.04.45	Hitler commits suicide.
2.05.45	Berlin falls to Russians.
7.05.44	Germans surrender to Eisenhower.
8.05.45	Victory in Europe (VE) Day.
6.08.45	Atomic bomb dropped on Hiroshima.
2.09.45	Japan signs capitulation.

Acknowledgments

The Editors are grateful to many people who have made possible the publication of this Diary with its explanatory notes. Some of them are listed here. Apologies as well as thanks are offered to any who have been omitted.

Particular thanks are due to Mr. and Mrs. F.G. Brown (Sir Clifford Paterson's daughter and son-in-law) and to Sir Clifford's sons for their ready agreement to publication and their encouragement.

Marian Creasey, Jean Peacock, and Pauline White contributed by their word-processing of the original typescript.

The staff of the GEC Hirst Research Centre Library have patiently resolved many queries and sought information.

Our thanks are due to Lady (Joy) Clayton and Leslie Algar for help, advice and patience during the many stages of editing the Diary.

Many friends have helped to identify people, projects or events named in the diary. Substantial contributions have been made by Sir Charles Oatley, Professor J.F. Coales, Dr. D.N. Truscott, Mr. Brian Callick and Mr. W.R. Stevens. Among others who have helped in this way are:

Sir Geoffrey Allen
Mr. R. Bodwell, GEC – Marconi Publicity Unit
Mr. Paul Bookbinder, Archivist, Marks & Spencer plc.
Dr. L. Cohen, Executive Secretary, Institute of Physics
Ms. Patti Dossett
Mr. K. Hale
Mr. D.M. Heller
Sir Alan Hodgkin
Mr. J.A. Hyde, CBI
Dr. A.E. Pannenborg
Mr. J.M. Price
Mr. R.H. Searle, Chief Librarian, Royal Aerospace Establishment
Mr. R.A. Storey, Archivist, Modern Records Centre

Mr. D. Taylor, GEC – Alsthom Electromotors
Mrs. E. Vallis, Archives Assistant, The Library, Nuffield College
Mr. A. Wilkinson, Deputy Managing Director, Frazer – Nash Group
Mr. G.P. Wright

Illustrations

War Diary
C.C. Paterson

Clifford C. Paterson

1939

September 1st to 4th[1]

Considered and arranged many staff transfers to work of more urgent importance: klystrons, receiving valves, cathode ray tube screen efficiencies and radio work.

Ryde and Harris	to klystrons
Laycock	to receiving valves
H.G. Jenkins and Bowtell	to CRT screen efficiencies
Lewer	to radio work Coventry

Started organisation under R. Le Rossignol of E 1046 valve production at Laboratories.

September 4th to 8th[2-7]

Intensification of activity on valve and radio work for Bawdsey and Signal School 'ZZ'.

Despatch by car of special CRTs to S. Coast for Fogg and Atkinson's application—drivers Waldram and Stevenson. [Stephenson?]

Heavy Alloy bullets made for certain trials.

Negotiations with Mr. Sims for making Elmet Metal at Wembley.

September 11th

Discussed with Stevens their plans for ARP lighting particularly in regard to Illumination Engineering Society/British Standards Institution activity. Considerable activity on subject at Wembley.

September 12th[8]

CVD meeting with Signal School. It appears that with Air Ministry moved North and Admiralty [moved] West, the Research Laboratories is the most convenient fixed point for future liaison between all concerned. It was agreed that meetings must be held in future more frequently, i.e. a full meeting two-monthly and intermediate meeting sectionalised.

September 13th[9-20]

Attended meeting of National Register—dearth of Radio men.

Meeting with all the tonnage in C.S. Wright's office to decide programme of special research to be undertaken at Birmingham by Oliphant with treasury grant. Watson Watt, Gough, Bragg, Appleton, Ure, Shearing, Paris, Cockcroft, C.C.P. present. Oliphant to devote main attention to klystrons.

September 14th

ARP organisation at Labs. has been organised excellently, by Gardiner, Brown and others. We seem ready for action now.

Many of the girls living at a distance are returning home in order to live in more peaceful conditions.

September 15th[21]

At Carbon Works with Hawkins and Cameron. The demands on the Works are very heavy for more than their total output for searchlight carbons alone. We have urgent investigations in connection with cerium recovery and with the starch process.

September 18th[22-24]

Saw Max Railing and Colonel Disney at HO for culmination of discussions re. shadow factory which GEC is to get going in Lancashire. Must give Smith a good man if possible to work with Wade till new factory is going. Probably Hall.

September 19th[25]

Learned that DSR [Director of Scientific Research: C.S. Wright] and Admiral Somerville are visiting us tomorrow to see valve etc. developments.

Promised Good of BSI to join in work of IES Committees on consideration of black-out street 'seeing'. It appears there is a public clamour for "some street lighting" and I fear public clamour is a bad master.

Inspected with R. le Rossignol the details of production of E 1046 valve which should be established on a 100 per week basis in course of October.

September 20th[26-28]

Encouraging visit by C.S. Wright and Vice Admiral Somerville who becomes OC for 'ZZ' work (Fig.1). He is Gossling's old chief. Gossling is doing a fine job on liaison with service departments on 'ZZ'. Somerville saw scope and importance of our functions. Told them I am spending £60 000 a year on Valves, Circuits and CRTs etc.

Called by Max Railing again over shadow factory negotiations.

Claude General being shut down.

September 21st

Went into activities of IES and BSI re. ARP lighting and seeing with Good and Ackerley. Illumination Dept. (Stevens and Waldram) co-operating.

Had ring at home in evening from Tonking at Ely AA Command to express thanks of the Command for expeditious development and supply of 24 SW Sets for signalling between searchlight units. They had been made at Labs. in 14 days. Matter had been initiated by Tonking and Lewer. Matter may now extend to all commands (6000 sets).

September 22nd

Atkinson of Bawdsey here re. CRTs and filters. Went into development of filters for CRTs with Ryde. An excellent and intricate piece of development. It appears difficult to get the Air Ministry to appreciate what we are doing and to give me authorisation for our work for them.

22nd September, 1939.

Dear Dr. Paterson,

I have received a report from the Director of Scientific Research and Admiral Somerville on their visit to your Laboratories at Wembley on the 20th September and have been told of the remarkable progress which has recently been made in the investigations which you are carrying out on behalf of the Admiralty. I should like personally to congratulate you, and the members of your staff who are assisting you in this work of great national importance, on the success of these investigations, and the ability with which your researches have been conducted.

Yours sincerely,

Bruce Fraser

Controller.

Dr.C.C.Paterson, O.B.E., M.I.E.E., M.Inst.C.E.,
Research Laboratories of the General
Electric Company, Ltd.,
Wembley,
Middlesex.

Fig.1 Letter dated 22 September 1939 from Sir Bruce Fraser, Third Sea Lord and Controller of the Navy

September 26th

Discussions with B.P. Dudding about substitute alloys and other materials.

September 27th[29]

To M.R. and Head Office again re. shadow factory and difficulties with I.S. in view of Air Ministry insistence on GEC handling of the new factory.

September 28th

Uncertainty at Osram Advisory Committee whether British Thomson Houston are pressing on with mains voltage fluorescent lamps. As we have virtually shelved this development we may have to resuscitate it again—but I hope not. Ryde is very keen on the HV [high voltage] fluorescent tube.

September 29th[30-33]

A long CVD meeting. Present—Harley, Lewis, Pollard, Shearing, Ure, Coales and others. A general clearing house for all and sundry matters. Harley very much in favour of getting GEC Labs. linked up by arrangement with the Bawdsey organisation instead of just giving isolated development contracts to us. It remains to be seen if Air Ministry Contracts Dept. will agree. Have written up whole of our programme of work for Watson Watt.

September 30th[34-37]

Had approached Tizard about fogs and poison gas at the suggestion of Pirani and Friedlander. These two went to see Prof. Heilbron at Ministry of Supply and had a pleasant reception.

October 2nd

Learned that first Air Ministry outfits are not to use E 1046 valves. This is a disappointment—but amongst other considerations satisfactory performance at reduced pressures at 8000 volts is not yet assured.

October 3rd

Again to Head Office to meet Mr. Railing and Col. Disney re. shadow factory. Mr. Shoenberg is helpful and difficulties re. GEC handling of the factory using MOV staff have thereby been removed.

October 6th[38]

Discussed with Good the work on better seeing in streets under black-out conditions. With the dark adaptation principle as their background the IES and Home Office Committee are going on the right lines. Very active in trials which we are preparing.

Discussed our CRT programme with Chelioti.

October 7th[39-40]

Hill from Admiralty Research Lab. is coming to see Ryde re. short period light pulses and possibility of producing them. Colonel Elsdale (Royal Signals) here to discuss AA Searchlight high frequency sets.

Dr. Railing reported unseemly behaviour by General Fuller in unfriendly open criticism of GEC carbons in comparison with Morgan on the score of burning rate. Later tests, which Fuller perhaps knows nothing of, has taken away the main basis of criticism and I shall have to prepare the case for Dr. Railing to refute the charges. There will be another nasty smell—as there has been before vide graphitised carbons.

October 9th[41]

With Dr. Railing over the aspersions of GEC carbon quality—decided wisest policy to speak with Colonel Evans. Better seeing in black-out in streets going ahead. Disappointed and annoyed to learn that Dr. Bowen (Bawdsey) had been encouraging Cossor to develop a 45° CRT when [Group] Captain Leedham had told me that they were relying entirely on us for this.

October 10th[42-43]

Saw Wright and Admiral Somerville to put up to them the sponsoring of our work on Radio Direction Finding (RDF) for DSR and DCD i.e. £30 000. Somerville is to see Watson Watt but is personally favourable.

October 11th

Dr. Hill and Dr. Smith of Admiralty called to discuss certain high intensity arc and lamp problems.

October 12th[44]

Saw Frank Gill who wants to make representations in high quarters to get industrial research laboratories used more. Prepared letter to C.S.Wright re. collaboration with him and DCD for work which I hope may amount to £30 000.

October 16th

Shadow valve factory—management fee discussions—Lab. fee £10 000 suggested plus life test costs. Had to resist Mr. Sims urgent request to allow another foreigner here (Dr. Bano) to work on Elmet metal.

October 17th

Meeting to consider commercial development of discharge lamps on grounds that commercial aspects of our work must be kept as active as may be.

October 19th

Harley came to discuss future valve production demands for various types. Took opportunity of heart to heart talk with him to remove misapprehension that GEC Research Labs. are working only with intention to get orders for MOV. Explained that special position by CVD arrangement with Admiralty put obligation on Labs. to eliminate commercial rivalries and to assist Services to secure orders to others than GEC or MOV. Harley seemed very relieved to have this categorical statement of GEC policy in regard to its Research Labs. I pointed out that it is implicit in the plans and policy for the shadow factory.

October 22nd

Ring from Harley (Air Ministry) to say that DCD's influence is to be in direction of not undercutting the production facilities of existing valve factories by transferring everything to the shadow factory.

October 23rd

Prof. Oliphant came to discuss velocity modulated devices for short wave generation. He saw Ryde and Gossling and a very fruitful discussion followed. His investigation of energy needed to 'run light' does not make it look more hopeful as a generator or amplifier. Tizard was to have come but was prevented.

October 24th[45]

Learnt from Mr. Wells (Coventry) that order had been definitely placed for 4000 short wave sets for anti-aircraft (see September 21st).

October 25th

Had a depressing meeting at IEE and the Ministry of Labour on National Register. They seem to want to eliminate any 'priority list' of firms—for no reason apparently than to save the face of the Register which cannot supply the expert telecommunication staff desired by the Services.

October 26th

Arrived to find there had been a fire in valve life test section in the night. Fortunately it was not extensive and does not affect the urgent service work.

October 27th

Learned that our CRTs for Air Ministry are of high quality from focus point of view, and we are congratulated by Cossor on the quality.

October 30th[46]

Whitehead from Leighton Buzzard called with Atkinson to discuss CRT activities and requirements for our tubes and filters. So much more of an important nature than at first appeared seems to be wrapped up with the filter technique.

Coales from Signal School came and is disappointed with the performance of his SW set. Wants our version hastened and our reception programme expanded.

October 31st

CVD meeting—very active discussions. Gave service people opportunity to meet alone—a welcome idea. Harley still anxious (but sceptical) to arrange running contract for CRT work. Again reassured him on GEC disinterestedness in their research lab. collaboration with the services.

November 1st

IES night tests near Tate Gallery on 0.005 f.c. [foot – candle] 'street lighting'. Found it of real value for pedestrians and not without some advantage for cars. F.C. Smith had arranged the units so that in very limited areas only was the source visible and therefore glaring.

November 2nd

Called on Mr. Dale Bussel re. a message he had sent re. submission of our CVD payment account. He was keen to get DSR to agree to shoulder under CVD £22 500 per annum—leaving £7 500 for DCD Air Ministry to pay for our CRT development.

November 6th

Had visit from Sir Henry Tizard who wanted to satisfy himself that all resources of industry were being efficiently focussed on radio and RDF. He is sure that 'this is a radio war' and radio will win or lose it. He came feeling that some of the Air Ministry problems would go quicker if the centre of responsibility for development were in industrial rather than Service establishment as at present. I partially supported him. He inspected our facilities and looked over our valve and CRT work. He is seeing me again after he has seen Watson Watt.

November 7th

Pollard (Christchurch) came to Lamp Works at our request as George Chelioti and his people were worried that they had no direct contact with those who would test and inspect their product. The meeting served the purpose I think.

November 8th

The National Register and assistance to Services with personnel has fairly boiled up. The main demand is by Admiralty stimulated probably by Adm. Somerville. Special meeting with Appleton presiding and all the Industrial Research tonnage to consider situation. In view of our relation with Admiralty, got Wells to attend which he did well. All industrialists of course wanted 'farming out'. I with others am writing up a memo best stating arrangement for industrial co-operation on RDF.

November 9th

Job came along from Admiralty 10 days ago involving heating of certain components on which O.W. Humphreys has been very busy and dodging from Portsmouth to Bath to Wembley. Cattell and Dando are handling with him—reporting to Dr. Railing.

November 10th[47]

Production of E 1046 has been subject to careful review with Dudding and Pearce. 200 a week at least wanted. All processes being changed from mechanic to girl handling.

Marris and Espley being approved for RDF has brought them at last into active collaboration. Espley has been making the round of Service establishments.

November 13th[48]

Report from Dundee (Dr. Lewis) that the E 1024 valve is more than satisfactory so presumably the factory can now have an order for these—which I urged the Ministry to arrange 4 months ago to save delay in getting going with production. All they had to do was to accept our guarantee that if the first samples were not right the second ones could certainly be made so. But they would not forego their formal trials!

Ekco repeat our tests that E 1046 is satisfactory at high altitudes. Good.

November 14th

There has been a fine lot of voluntary ARP work—fire watching at night—by 20 or 30 members of the staff, for 2½ months. This will now be regularised and paid for. A most excellent spirit shown by all concerned.

November 15th[49]

Jesty on return from Christchurch reports Works-made CRT for War Office passes in 1st grade. R. Le Rossignol and Bell go to Ekco to see about mounting and cooling E 1046 for Bowen's job.

November 16th

Busy with Elmet and Heavy Alloy and discussion with Mr. Rouse. Elmet is still not a going concern for lack of tungsten powder with suitable grain size. A.R. [Dr. Railing] has sanctioned extension of Metallurgical Lab. which gives us 40 ft for this and for other process work.

November 17th[50]

Long discussions about the place members of television group can take in circuit investigations for 50 to 150 cm receivers for Service purposes. Dudding and Partridge and other groups are heavily occupied in 101 things for Lamp Works and Osram Dept.—too numerous to mention.

Humphreys and his special job for the Admiralty gets on only so so. The Navy would get on better if they took him really into their counsels.

Farnboro [RAE Farnborough] with their large order to Radio Works, Coventry, insist on one of our Research men being at Coventry to ensure that no troubles occur in passing the Farnboro No. 2 model into a production article.

November 20th[51]

Had unexpected and extraordinary visit from Major Snape, a recent addition to staff of Air Marshal Joubert. He was after the blood of DCD and staff whom he regarded as unsuitable for their job. By visiting contractors and examining them he was obtaining such evidence he could find for his purpose. He kept me the whole morning.

November 21st

Lunched with Darwin (Director NPL) re. tests being made in the building of our shadow factory at Wolverhampton to see whether building operations can be carried on after dark with 0.005 [f.c.?] illumination (and the like).

November 23rd and 24th[52]

To Harrogate (Air Ministry) with M.R., F. Smith and Giggins to try to settle the basis of payment which GEC shall receive for equipping and running the shadow valve factory. Discussion 10.30 till 1.30. Found the Air Ministry officials woolly in their ideas and, on the whole, in their thinking; for not one of those who spoke seemed to know anything of the conduct of industry. Their expectations and comments were childish. They heard what GEC ideas were and one must hope that these will germinate and provide a more helpful background for the next conference.

Humphreys has been getting on better with his Admiralty heating job.

Heard from C.S. Wright that Admiralty consent to an increased payment to GEC for CVD up to £15 000 but cannot make arrangements with the other Services for additional joint payment of £7 500 for circuit work by the remains of our television group. Have still to hear from DCD as to what he proposes to do for the extension of our CRT development programme.

November 27th

Learned from Jesty's visit to Dr. Bowen that our CRTs are earning a very good reputation. The 45° deflection tube is giving place to two separate tubes.

November 28th

A day on searchlight carbons with the Service people—new methods of test and criteria. Our people demonstrated a compact source of 2 kW with

24 in projector giving 90 million candles intensity, and daylight photometry across the sports ground.

November 29th

Lunched with Watson Watt. He referred to Tizard's visit to us and now wants us to take on research and development for AI—breaking away if possible from orthodoxy—if we can get expectation of better results therefrom. He will visit us here in a few days to discuss and finalise an arrangement. We shall need to incorporate some service people in our group.

Party from Home Office and flying officers to examine and comment on Waldram's method for using two projectors and an aerial photograph to reproduce black-out street lighting possibilities from the point of view of visibility from overhead.

December 1st

Main CVD meeting with all the services represented. 50 cm receiving is still the chief problem. Followed meeting with one on CRTs in which Harley was having pressed on him the claims of CRT and filter development as being a continuous research and so requiring a continuous research contract. His contracts people seem very sticky on the point.

Some difficulty seems brewing in reported displeasure of DSR over our CRT meetings. Rang up DSR who claimed pressure of work but would meet me sometime. So we carry on for the present.

Magnetic mines and their frustration are all the talk but we have not been brought into the matter.

New class of men being called up but nearly all of ours come under reserved occupation.

December 4th to 9th

A week of activity but nothing special to show for it.

Humphreys has visited the Admiralty re. catapults. Jesty and Gavin have both visited Christchurch but with rather disappointing experimental results. The conditions at Christchurch were too difficult to settle whether the deflected beam valve was up to its job—and it had had to be brought to Wembley to be investigated. Watson Watt who was coming on the 8th has postponed his visit till next week—to discuss his new development contract with us.

December 11th

Visit from Watson Watt for three hour discussion on AI and proposed lines of work and number of people to be engaged. Next meeting to include Bowen to be here on December 22nd. Learnt much from Watson Watt of probable changes in direction of all this work, and found no sympathy with the view that our CVD meetings were exceeding their functions. Mentioned C.S. Wright's apparent attitude.

December 12th[53]

Fowler came for a short period and left us with impression that Tizard and his Committee regard it as most important to bring all industrial labs. into active work because the Service establishments are so far behind in ability to tackle the experimental work. Gavin reports that Christchurch conditions are not fit for carrying out this sort of work; Dundee is out of it, and Signal School is the only one which can be regarded as a going concern. But we know they are very lacking in adequate staff.

December 13th

Great activity in ARP lighting as the authorities have approved 0.0002 to 0.0004 f.c. for the streets, i.e. somewhat above the level of good starlight. It appears from conversations with Good that all fittings must be approved under a 'mark' scheme—and that we may have to be one of the approval bodies.

December 14th

Certain anxiety over our responsibility to supply glass for silica seals. I shall be glad when we have really got 12 months supply in stock.

Estimates of our expenses in producing E 1046 valves show a much lower cost than had been indicated.

Humphreys has been occupied further with heating jobs for the Admiralty and a letter has come from DEE [Director of Electrical Engineering] expressing appreciation for the GEC handling of the effort.

December 15th[54]

Lunched with Appleton to see how matters stood in regard to his support of our proposed AI work. He is chairman of an RDF sub-committee of

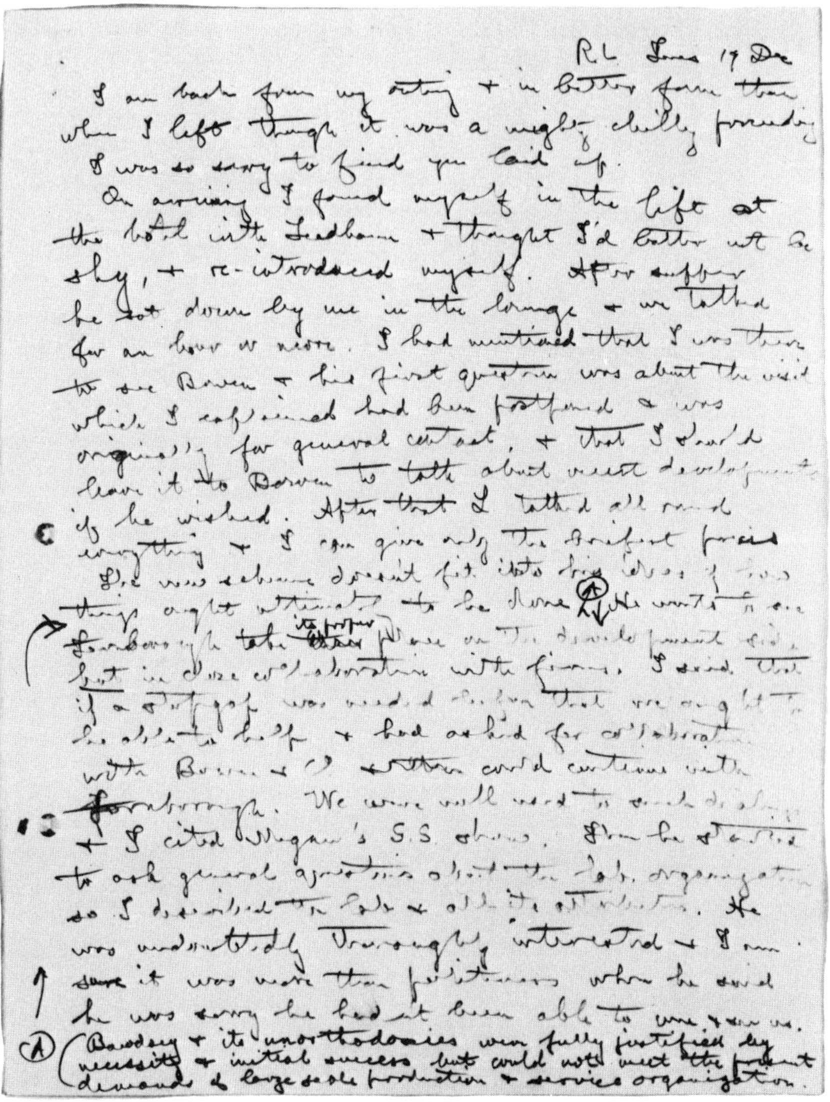

R.L. Tues 19 Dec

I am back from my outing + in better form than when I left though it was a mighty chilly proceeding. I was so sorry to find you laid up.

On arriving I found myself in the lift at the hotel with Leadbetter + thought I'd better not be shy, + re-introduced myself. After supper he sat down by me in the lounge + we talked for an hour or more. I had mentioned that I was there to see Bowen + his first question was about the visit which I explained had been postponed + was originally for general contact, + that I should leave it to Bowen to talk about recent developments if he wished. After that I talked all round everything + I can give only the briefest precis.

His own scheme doesn't fit into his ideas of how things ought ultimately to be done. (A) He wants to see Farnborough take its proper place on the service user side but in close collaboration with firms. I said that if a stopgap was needed before that we ought to be able to help + had asked for collaboration with Bowen + others could continue with Farnborough. We were well used to such dealings + I cited Wigan's S.S. show. Then he started to ask general questions about the Lab. organization so I described the Lab + all its activities. He was undoubtedly thoroughly interested + I am sure it was more than politeness when he said he was sorry he had not been able to come + see us.

(A) Bawdsey + its unorthodoxies were fully justified by necessity + initial success but could not meet the present demands of large scale production + service organization.

Fig.2 Manuscript note from B.S. Gossling to C.C. Paterson dated 19 December 1939

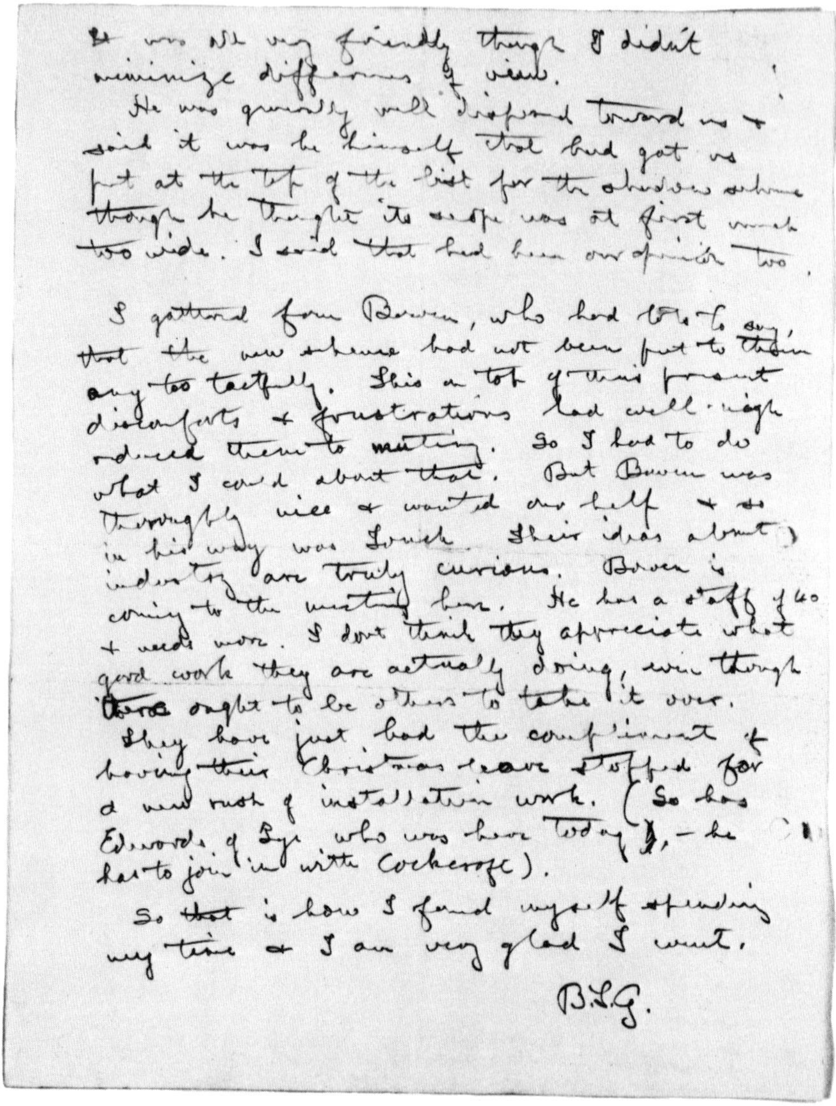

It was all very friendly though I didn't minimize differences of view.

He was generally well disposed toward us & said it was he himself that had got us put at the top of the list for the shadow scheme though he thought its scope was at first much too wide. I said that had been our opinion too.

I gathered from Bowen, who had lots to say, that the new scheme had not been put to them any too tactfully. This on top of their present discomforts & frustrations had well-nigh reduced them to mutiny. So I had to do what I could about that. But Bowen was thoroughly nice & wanted our help & so in his way was Lovell. Their ideas about industry are truly curious. Bowen is coming to the meeting here. He has a staff of 40 & needs more. I don't think they appreciate what good work they are actually doing, even though there ought to be others to take it over.

They have just had the compliment of having their Christmas leave stopped for a new rush of installation work. (So has Edwards of Pye who was here today, — he has to join in with Cockcroft).

So that is how I found myself spending my time & I am very glad I went.

B.L.G.

Fig.2 continued.

Tizard's main committee and is determined to foster the 'farming out' of service problems whenever possible. Furthermore he is concerned at the ineffectiveness of the effort of the Service establishments. I said I thought their morale had been undermined since Rowe and the Bawdsey group had been dispersed. They now were no longer an organism. Christchurch were trying to work under such bad conditions that people there lost heart. I told him I thought the main difficulty would lie with getting Air Ministry contracts department to agree a simple research contract.

December 18th to 20th

I have been laid up with 'flu and not closely in touch with all matters. Gossling had a valuable visit to Bowen at his Welsh Aerodrome where he also had a thorough discussion with Leedham. Gossling's typical report is attached (Fig.2). The bringing of us into AI may not have an entirely easy passage and Leedham clearly needs a lot of educating as to what industrial concerns are really like in place of what he seems to think they are like. The idea of bringing RDF into Farnborough—that place of bad tradition—is depressing.

Partridge reports that the new furnace is at last producing satisfactory melts of quartz seal glass and the emergency is not so acute. We have now three weeks stock which should increase further before Christmas.

December 22nd[55-56]

Have had an exasperating letter from C.S. Wright cancelling CVD meetings and talking of controlling his valve service to the other Services in some other way. Showed letter to Watson Watt today who found it incredible and he made arrangements to see Wright on January 1st before I see him. Lucky that Watson Watt was here today. He, Dr. Bowen, Dr. Touch and [Hanbury] Brown came to talk over arrangements for AI now to be known as 25 I. Had a good meeting and fixed up regular meetings with Bowen. Cavendish Lab. Cambridge are collaborating on the theoretical side of the aerial work but no additional establishment is to be brought in to do experimental aerial work. I told Watson Watt that I want an instruction to proceed for I had lost most of my confidence in the readiness of the Air Ministry Contracts Dept. to implement the actions of their technical staffs.

Came in on December 22nd to meet Watson Watt, Bowen, Touch and [Hanbury] Brown re. AI. Very agreeable discussion and found Bowen's attitude most helpful. We must get down to 20 or 30 cm which means a pioneer effort. The proposal is that when the gear is developed Bowen joins in and sees about installation (with us sharing). Agreed the group which is to be the main focus of the work, i.e. Marris, Espley, Edwards, Clayton, Cherry, Walter with mechanician help. Indicated about £10 000 annual expenditure.

December 29th

Air Marshal Joubert sent for me to satisfy himself that we have our marching orders right on AI from Watson Watt. Said he must insist on our using Northolt aerodrome for practical tests so as to be independent of people who may prevent our having full freedom of judgment and trials. There is thus a clash of view which I hope will iron itself out before our application work starts. Found A.M. Joubert most helpful.

There is a good deal of illness and absence from the Lab. I hope the New Year will start off with a good push.

December 30th

The Air Ministry still fail to come down with any definite contracts for our various pieces of research. I must write Wimperis again about payment for our men loaned to Farnborough.

1940

January 2nd

Went to Northolt aerodrome with Marris, Espley, Gossling and Megaw to examine aeroplanes with a view to AI and the aerials possible. Had useful discussion with officers from various establishments. The task we have taken on does not become easier as we examine it.

January 4th

Had Brigadier Fuller here with senior officers from Ministry of Supply (Fordham, Simpson, Paris). They spent five hours examining work and facilities. A very useful opportunity of letting them see what the Laboratories were doing and could do.

January 5th

A CVD meeting in depressing circumstances—for Wright of Admiralty is scuttling (or trying to do so) all that is most valuable in the meeting, i.e. the regular contact with the users of valves and those who are developing apparatus using them. I am seeing him tomorrow and shall try to salve something useful of all the work and effort we have successfully put in to secure the service which the development people need.

January 6th[57]

A change has come over the scene and all is bright again. Two hours talk with Wright and Brundrett (B.S.G. present) seemed to elicit the view that our meeting should continue—and Signal School would continue the secretariat. (Watson Watt told me that he had seen Wright on January 1st and told him that the meetings were the only thing in CVD he valued). There is to be a co-ordinating committee at the Admiralty for policy matters. It seems inconsistent with Wright's recent letter to me but all is well if it ends well.

January 8th and 9th[58]

Visited Salford Predictor, Quartz Crystal and Jewel factories with Heather. All very enterprising and important.

January 10th

Searchlight Carbon panel meeting (the last of a series) and programme now to be prosecuted instead of having its roots pulled up again.

January 11th

Meeting with M. Railing, F. Smith and Giggins re. shadow factory. To everyone's surprise Giggins has obtained a reasonable settlement with the Air [Ministry] Contracts Dept. and apparently things can now go on. A good job of work on his part. Apparently if Air Ministry Contracts Dept. is touched at the right point the figure suddenly works. I must try to apply the same trick to my long delayed development contracts with the Air Ministry.

January 15th[59]

Had visit at J.Y.F.'s instigation from KLG plug people who were in difficulty in rendering their aircraft plugs airtight. Brought Le Rossignol, Dudding, Partridge and Ryde into consultation, and it seems as if our technique for metallising refractories and brazing on metal would exactly fit what they want. We are making trials.

January 16th[60-62]

A day with Bowen on our new AI (or A25) job. Watson Watt came after lunch. Wing Commander Hart from Fighter HQ also present with

Hibberd and Robinson. GEC group described five or six methods on trial for 25 cm reception and Bowen put forward considered plan for signal utilisation. Excellent discussion. A heart to heart with Bowen (who was inclined to place a dividing line between generation and utilisation) ended in a similar talk with Watson Watt, Bowen, Marris and self in which it was agreed that for utilisation [system or display aspects?] GEC and Bowen's team combined. Bowen is a fine chap and I think our usual collaboration policy will pull things right with the rest of his group.

January 17th[63]

GEC television and radio meeting and a paper at Royal Society of Arts by Rooksby were a reminder that peace-time activities must not be overlooked.

January 19th

Met C.S. Wright at club. He is having to collect a large group for certain phase of the magnetic mine countermeasures—and asked advice. Had an hour with him.

January 20th

Had from Rowe [a] private letter re. St. Athan staff working with Bowen and possibility of our housing some of them here where research is possible under better conditions.

There is a very heavy pressure on all our facilities in valves and radio and somehow we must strengthen them.

January 26th

A week of ups and downs but nothing going wrong and general progress along the line both of directly Service things and also GEC requirements.

We are studying methods to strengthen up somehow in staff—but I find both workshop and scientific leaders find it difficult to visualise anything but already expert people, and these are not to be obtained, but we progress slowly.

Having started a number of lines in AI we have to await the gear wanted before really getting into the experimental work as actively as we should like. Ryde points out that the velocity modulated tubes give such outstanding expectations at very short wavelengths (15[cm] and under) that we must explore the possibilities fully.

There was a flutter about the beam valve failing again—but it just cleared its reputation again. We have had Robinson from Dundee here for three days to study short wave possibilities.

Usual rumours and sensations that Contracts Dept. at Harrogate might be going to do something regarding one and another of the many outstanding and delayed matters.

January 30th[64-65]

The worst traffic disorganisation on record—through ice and snow—imprisoned me in Hove where I was visiting R.W. Paul.

The resuscitation of tubular mains discharge lamps (3 ft and 5 ft) has thrown a heavy load on B.P.D., Jennett and others since Ryde and Harris are practically not available.

There is much sickness on the staff and work is going slow in places.

January 31st

Espley tells me that we may have to go down to 5 cm for the AI work—anyway the velocity modulation work has to be hastened in case we may have to both generate and amplify at these wavelengths.

February 1st[66]

Goodeve of [HMS] *Vernon* called on me to talk in confidence over reorganisation at *Vernon* which he is advising should be on the group system. We went over our experience at Wembley for about two hours and looked over the place. I was pleased that he thought of coming here to compare notes.

February 2nd[67]

Signal School variations of micropup seem to be of interest. Coales has been here. B.S.G. tells me that Gavin finds the milli-micropup is doing what is expected of it so far, and the results agree with theory. No information yet as to power.

Experiments at Christchurch show that the deflected beam valve repeats itself as a valve and can now be judged on its merits in relation to the acorn, whose manufacture on a large scale is not practical.

Megaw's work with 74X sets is up to Admiralty requirements and Coventry have kept their date; installation now in progress on ships.

February 5th[68-69]

Bennett of North Met. came to talk over mains borne radio service which P.P. Eckersley is advocating—and which experiments on N. Met. networks indicate as promising. W.E. Bragg for the night.

February 6th

Bragg spent the day till 2.30. C.S. Wright is interesting him in valve things. He has seen Oliphant and is to be initiated with our things. We introduced him mainly to problems of materials and electron behaviour rather than circuits but he could only touch the fringe of things. Hope for more visits from him.

February 7th[70]

A red letter day (*a*) the Air Ministry Contract Dept. sent their draft for a running contract for our work on CRTs—four months overdue! It is not a perfect instrument but may be able to be amended so that it can be worked to—so I shall accept it. (*b*) Rowe has been here to discuss many things including the possibility of some of his staff working here. Had talk with him, Gossling and Marris—the latter seemed to see nothing but difficulties and I hope Rowe has not lost his good impression of the helpfulness of our staff. Went with him to Fighter Command and talked with Larnder on usefulness and otherwise of RDF on different Services. Most interesting.

February 8th

CVD meeting—very useful I think—followed by a long talk with Mr. Bradfield of Harrogate re. CRT programme. But he is harnessed by the yoke of his contracts dept. and has little initiative, or desire to have any, however great the need may be.

February 9th

Trouble due to disappearance of an Air Ministry blue print. Informed the Dept. concerned and must await results.

February 12th

I misjudged Bradfield—in fact further talks with him and with Hibberd (on AI) indicate that there is a real effort by both men and by the Ministry to fit their organisation to the urgent requirements of the situation—so I hope for action as sympathetic as is their attitude towards our effort and its recompense. The fact is these Laboratories are anomalous and I must see if we can't get a special ruling which will save us from being tied up completely in their minds with commercial trading. Mr. Rabbin from contracts dept. is coming to see me at the end of the month and I am hoping to settle things with him helpfully.

February 13th

Commander Bathurst came re. prevention of leakage of information precautions. His dept. decided to have a CID man here over the lost drawing and he questioned all concerned, giving it as his opinion finally that he was not sure in spite of protestations that the drawing was brought back from a Farnborough conference. A message came through from Foot at Farnborough this morning that the drawing had been found there! Comm. Bathurst was helpful and level headed.

February 15th

An AI meeting attended also by Sir Geo. Lee who is now Director of Communications Development, Watson Watt having moved to Scientific Advisor to Air Council. Lee ready to take decisions and give forceful leads, is a very good addition to the management of affairs. Details of the meeting are taboo—but things are crystallising although Marris is fuller of the difficulties than of the likely possibilities.

February 17th

I have never known so much sickness in the Laboratories. It is all feverish colds—'flu and the like.

Our legal action on BTH internal frosting patent looks doomed—Cornock always said it was but other advisers put our chances at 50 to 60%. We shall await the result without much enthusiasm.

The milli-micropup valve has turned out according to plan and we hope for effective performance at 25 cm. This is a great relief and a real achievement. But it has yet to be put on a production plane. The kernel of the short-wave problem lies in efficient reception.

February 19th

Obtained Mr. Railing's sanction for an extension to the workshop partly because we are crowded there and also because we must be able to segregate the AI workshop work (£4500).

February 20th

On reflection it is an extraordinary inconsistency for DCD to ask EMI also to tackle the AI development in the face of the trouble they took in the autumn to force the GEC to get rid of EMI from the shadow factory management. M.R. has asked me to see Lee about it but Dale Bussell advises taking no action which would dissuade Air Ministry from their plan.

February 22nd

Efficiency of short-wave reception is still our obsession but it seems to me that we are getting on with the problem along one or other of the lines we are tackling.

Saw Wright today who showed me one of Oliphant's impulsive letters suggesting that the GEC Labs. preferred secrecy to openness owing to some remarks of Ryde's to him. One of those things which Oliphant might have spoken to me about before writing to Wright. I must ask him what about it.

February 23rd[71]

At BTH Rugby where Warren showed me various confidential things he was doing for the Services. Enjoyed the visit very much. Saw Davis, Kinman, Lucas and others. Smithells joined them at lunch.

February 26th

Had Mr. Rabbin here (one of the more senior Air Ministry contracts people) to discuss my AI and CRT contracts. He was very tired with a great deal of 'irregular commitments' to clear up. Found him helpful and understanding and ready to adopt the principle of a 1.8 overhead on experimental staff salaries. He left me with the feeling that if he has the handling of matters they will go through smoothly. I put a good margin on our AI total (£30 000).

February 28th[72]

Ure and Hughes from Signal School to a meeting with us at MOV to settle giving of specification to Signal School of E 960, E 1024 and E 1046 valves. Arrangements readily agreed and I hope all suspicions removed that GEC want to keep back information. The position might have been delicate because all MOV's own experience and special techniques are involved, which is a part of this development which cannot be regarded as the property of CVD.

Went to see Watson Watt and Sir George Lee at their request—the inconsistent position re. EMI being the main subject. It appears that in deciding to ask EMI to accept a similar contract as GEC for AI development, Watson Watt had clean forgotten all the trouble Air Ministry made over the shadow factory and refusal of EMI participation! Lee had never even heard of it and the letter from Air Ministry to GEC in October when I called attention to it appeared to be a jerk to them. They were ready to write to GEC reversing the position but I suggested they had better hold their hand for the moment as matters were now all under way.

The more serious thing mentioned by Lee was that Farnborough development dept. would have no staff or facilities to put on AI at its later stages (if any) and that the 'Contractor' i.e. GEC Research Labs. would have to carry the full burden of preparing the complete prototype and its working drawings ready for contract. Saw Gardiner and others about facilities and to break the news gently.

March 1st

To Witton re. carbon developments with Hawkins and Cameron. Tar and slow burners were the chief subjects but interesting results on 'flow' in the extrusion nozzles were described by Hawkins—work done by Thomas.

March 4th[73]

A very interesting account by Ryde and Harris of velocity modulation and its graphic representation at Physics staff meeting. Lockspeiser from AM Harrogate called to discuss with Gardiner, Ryde, Cornock and self the special problems of the balloon barrage. We shall be put in touch with their research station at Exeter. We made some suggestions.

March 5th

Much discussion as to how to handle 'GEC war time allowance' to *weekly* paid staff—but in the end such exceptions as there were it was found possible to deal with. The allowance will be appreciated.

March 6th[74-75]

A very interesting visit to Labs. by French Naval experts under wing of Signal School to see our SW radio developments for RDF etc. Captain Willett and Shearing came too and also two from Wireless Telegraph Board. Jervis Smith covered himself with glory as French interpreter—and Ryde's velocity modulation diagrams as an industrial mathematical effort seemed to surprise them as a phenomenon unknown in France. An intimate dinner at the Savoy which M.R. attended (and spoke charmingly) finished a very successful occasion. Megaw's and Jervis-Smith's French was the main factor contributing to success. The French were extraordinarily keen and live and it was a pleasure to have them.

March 8th

There has been great difficulty in fitting in the necessary development work on the 5 ft luminescent mains burning tube. Dudding, Harris, Jennett and others have had to squeeze in work when it could be managed—Claudgen are making and it seems to be going as well as a premature birth could be expected. We have suddenly met a snag in photoelectric effects in the 'glow switch' used with these tubes just as they are going on the market. Hope our people will find a way out.

March 12th[76]

There has been strong pressure from Signal School to produce an autodyne receiver for their short-wave reception for trials at Portsmouth this week. Bligh, Brittain and Falconer have been working day and night on it and we hear today that the set is showing up favourably compared with alternatives.

March 13th

Schnetzler is released from internment and back at work. The work on high-intensity compact-source lamps has become much more promising during recent months. We seem safe against bursts, and sizes up to 5 kW seem possible. The Services are interested for searchlight possibilities.

The provisional solution for the glow switch is to put a small mercury indicator lamp in the box containing the switch so as to give a source of weak ionisation.

March 14th

AI meeting all day. Bowen, Flight [Group] Captain McMillan and Barton (from Farnborough). Watson Watt was coming to examine Ryde's velocity modulation work, but failed. The meeting was lively with Bowen somewhat ruthlessly sweeping aside the guidance given by the last meeting when DCD attended. Wants us to concentrate on 10 to 15 cm and look upon the present 25 cm as only an experimental step for feeling our way. But he told me after that the 25 cm work would be invaluable for several other purposes. We showed the meeting our first lash-up on which Espley and all concerned had been working very hard—and Bowen admitted to me that if 25 cm was to be used our efforts are on sound lines.

March 15th

Very warm appreciation from Signal School and the French Naval Mission for their reception.

March 16th[77]

Wing Commander Hart turned up to tell us that he had seen the C-in-C (Stuffy) to report what they had been shown on Thursday and C-in-C was encouraging. But we must see what our electrical results are going to achieve. They may prove to be off the map.

March 18th

Message from Willett urging us to agree to develop our autodyne design to prototype stage as it is definitely successful. We will try and do it by drawing Bligh, Brittain and Foreman with Archer Thompson into the work.

March 19th

All day conference with Bearford of Signal School on details of autodyne. They gave us date of May 6th! Have told Willett we are game to try and DSR has agreed the addition of the work to our USW programme.

People are still going down with 'flu and measles.

S.V. Williams is wrestling with costs of Elmet metal which technically seems now to be on its feet. The vaunted 'Reutte' experience and knowledge has, as we thought, proved a bogey and the progress has had to be mainly the result of the efforts of our own metallurgists.

March 20th

Day before Easter. Am encouraging every one to take holiday. Am glad I set my face against overtime and curtailed holidays. The opposite policy is the paying one and many concerns who, for the sake of public opinion or because of mistaken ideas, insisted on long hours have I am sure discovered their mistake.

March 21st to 26th

Easter.

March 27th

Several key men away ill after Easter—Gossling, Bell, Megaw, Ramsay and others.

Had a ring from Air Marshal Joubert on March 21st asking if there is any hope on AI. After I had stated the position and asked him to get a report from Wing Commander Hart he asked me if money was any factor in the position as it would be worth a quarter of a million or more to have a practical result. He pronounced it the highest priority job of any, and urged me to appreciate our responsibility.

March 28th

Had long discussion with Marris and Espley and agreed that in addition to stiffening our facilities on 25 cm, we must make 12 cm an equally urgent objective and organise accordingly.

March 30th[78]

Had Picken to see me with a tragic story of the relations of his management to himself. Gave him advice re. his chairmanship of the IEE Wireless Section and such help as we can give him.

April 1st

Took further steps with all concerned to establish facilities for 12 cm transmission and reception. Proctor is to be brought in to velocity modulation. We are to get men from outside if possible.

April 2nd

Meeting re. shadow factory in M.R's office. The Ministry is at length giving permission for things to move! But only it seems with constant supervision at every point. *When* will they establish a technique for trusting and using honest men! Research Laboratories may have to help with personnel for CRT supervision at the shadow factory.

April 3rd[79]

DCD (Sir George Lee) is coming next week to see progress on AI and I will get the policy of 25 plus 12 cm programmes ventilated.

April 5th

CVD meeting all day. A full and fruitful meeting. Nine service members present. We showed our 25 cm milli-micropup oscillator with cooling of anodes in a copper block and inside cylindrical resonator. Also Ryde's theoretical work on velocity modulation.

The folly of the Air Ministry in delaying the E 1024 valve is now coming home to everyone. The valve appears to be extraordinarily better than the services expected and better than what they have been using so far. We told them in the summer what a good egg the valve would be—but apparently our views were not worth considering. They now cannot get any more valves for four months when they *might* have been in production by now.

April 9th

Visit from Lee (DCD) and Bowen to see progress on AI. Went over same ground again. Lee very much favouring 10 cm—as if there were no insuperable difficulties! Urged him not to throw cold water on a longer wavelength if there was a chance of making it practical. Gavin tells me that the milli-micropup operates well at 20 cm which will help.

April 10th to 12th[80-81]

Annual expenses and accounts under review. This is an interesting year. Receipts for sales and services rendered outside amount to £25 000. In consequence all allocations are less than last year. But I am having to add rapidly to the numbers of staff. We are about 550 now and constant demands to hasten on jobs.

April 13th

Am a bit worried over our inability to get the electrical part of AI settled quickly enough. Had a talk with Espley and must see if there is anything I can do to help. No sign of Air Ministry contracts yet.

Had word that Steel Controller had refused our permit for 14 tons of steel for workshop extension. Took action with Harrogate and Admiralty and hope for early reversal.

April 15th[82]

Had advice from Signal School that 1200 E 1046 valves would be wanted this year. We shall have to undertake them at Wembley I expect.

Went to see Lord Hirst who is slowly recovering from severe illness. Found him very glad to see me and interested to speak of the spiritual and psychological experiences in life. An affectionate and lovable man.

April 16th

Cockcroft came to talk over many things to do with RDF and also the use of light in powerful short duration bursts.

Coales from Signal School and Captain Perroux to see about arrangements for making E 1046B valves in France. A very busy and interesting day. Gavin has got the milli-micropup down [up] to 1 kW peak at 20 cm.

April 17th

Bristol Aeroplane Co. are inquiring for Heavy Alloy on a large scale which, if it eventuates, will mean five to ten times our present output. We have had many bites before, but this one looks more likely than others.

April 18th

R.H. Fowler's day. He had some encouraging news as to the value in recent operations of ASV (Aircraft to Surface Vessel), for which our new valves are used.

April 19th

Learnt that decision re. steel licence for workshop extension had been reversed—so our representations have done good. Strachan exerted himself to good purpose.

Had a good look over lab. buildings and progress with various extensions. We are clearing some more of the west end of gallery at Dudding's suggestion to accommodate staff facilities for starting on the E 1046B valve.

April 20th

Discussed Benjamin's Royal Society paper and the theories of his auto-electronic emission work for which, in spite of war activities, he has obtained some most attractive photos.

April 22nd and 23rd

Finalised our annual allocations to Works. Our overhead is *up* 10%—partly because of increased costs and partly because men are engaged on Service work which, for the the present does not take the full overhead. MOV is lucky to have so much pioneer work done on valves at other peoples' expense. Lamp Works allocation is down because various people have been taken off lamps for valve etc. work.

Gavin and Warren to the Clarendon Lab. Oxford with Signal School people to see a new HF receiving valve effect. Not much good as it is but should be followed up in case it can be improved. Offered Oxford people a visit here to study measurement.

April 24th

Air Ministry draft contract to hand at last but for CRT work only, so far. It is much on the lines I discussed with Mr. Rabbin in February except that it does not extend beyond a year from October 1st. The sum mentioned is £9000.

April 25th

A day on lamps and IEE matters. Young and I between us have got Jervis-Smith on to the IEE Council as an Associate Member. Excellent.

Giggins finds the Air Ministry contract on the whole sound and says Harrogate have great objections to 'running' contracts. They prefer these 'ITPs' (instructions to proceed) which involves agreeing a price later. I am accepting the contract with conditions as to payment as already in writing.

April 26th

An AI meeting. Very good I thought. De Burgh from Farnborough came and two from Dundee (to pick up all they could on short-wave receiving technique). Free and helpful discussion all morning—and examination of first preliminary working of 25 cm arrangement. This has increased enthusiasm for 10 cm and we have made arrangements for Proctor and Lamont to go on to VM [velocity modulation tubes] as particularly applied to AI reception.

Our milli-micropup generators seem to work excellently in their latest form. Gavin does not despair of working triodes well below 20 cm

April 27th

Spoke to B.S. Gossling (who sleeps at Lab. and has been taking a lot out of himself) about going easy.

April 29th[83-84]

D.S.R. anxious about Clarendon (Oxford) device as Lindemann and First Lord probably behind it. It has a long way to go to prove useful.

April 30th[85]

Special Lab. meeting to push velocity modulated tubes—Proctor joining in so as to ensure best chances of 10 cm reception.

Shadow factory meeting at Head Office with M.R. in the afternoon. There seems to be difficulty re. personnel for management as Wade finds conditions unsatisfactory and Grant goes over to Lamp Works as works manager under G.C.

May 1st[86]

ITC meeting all day. None too easy to get consensus of view of standardisation of 3 ft tube.

Randall from Birmingham came instead of Oliphant to get us to start quickly with replicas of three large magnetrons which Megaw thinks well of. Had to speak straight to Randall for Oliphant is an impatient person and expects first priority which we could not give him. I think Birmingham must have the main responsibility for the tubes or they will never be happy.

May 2nd

Shearing and Horton here—DSR rang to say that Clarendon reported great improvement in their device and would GEC take it on as a serious research. We showed willing, but report came through later that Clarendon staff denied knowledge of alleged improvement! Who started the story?

Horton doubtful, from his experience, of effectiveness of centimetre waves of [for] AI. We shall see.

May 3rd[87-88]

Clavier and colleague here from Paris Lab.—everyone found it a most stimulating visit. Marris hot on the idea of using Barkhausen circuit for reception.

Received AI contract from Harrogate. It looks all right.

Was notified of receipt of cheque for £1100 from Farnborough for the five men loaned to them. Wimperis must have pulled strings to get it settled with the Treasury. So our various long drawn out negotiations reach finality at last if we have enough patience.

May 6th

Valve meeting with Marconi. I hear that they are developing types of valves in their own labs. very similar to those we develop for MOV—we must see where this is leading, for it seems rather unsatisfactory.

May 7th[89]

Lunched with Mr. and Dr. Railing to advise them of the financial position at the Labs. We are now being paid some £30 000 a year from Service Depts. for our services and we are using mostly existing valve and radio staffs for this purpose—so that if and when the Service work falls away a heavy load of payment will fall somewhere and MOV and the Directors should realise this in good time. On the same subject I note rather carping attitude of MOV, because although their total allocation is no higher than the (reduced) total of last year they don't see why they should pay £4000 more on the receiving valve part of it. They don't realise that, were it not for the Services work we are doing, they would either be paying much more or they would be out of the picture. I shall raise this question at the Consultative Board.

May 8th

No end of a bother boiled up—not our fault I think. Dr. Ponte (of SFR) came at Signal School request to get full manufacturing particulars of E 1046B valve. Smith and I had asked S.S. for written instructions and sent copy to DSR. It turned out that this was the first DSR had heard about it and that these things can only be arranged at a level higher up and with Air Ministry. Watson Watt handled it with me and approval eventually came —after we had had to keep Dr. Ponte interested in other things for a day.

I made a brief call on Lord Hirst and was glad to find him better.

May 9th[90]

Had A.C. Bartlett here from Farnborough. Things don't seem very happy and certain technical jobs we know of are being delayed. The arrangement by which he joins Rowe at Swanage is not going to be frictionless.

Invasion of Holland and Belgium begun and everything is speeding up.

Ponte is also doubtful to what extent the AI problem is really going to benefit by going to lower wavelengths.

May 10th

Mr. Emmerson from Portland here, also men from HMS *Vernon* re. our daylight beam measuring set which we have been making for them and which seems to come out excellently.

Bank Holiday is to be cancelled and everyone I think ready to accept the sacrifice.

May 13th

Whit Monday but no holiday.

Autodyne receiving set design meeting with Signal School. Coventry are to make 200 sets. The special 'Services' committee room at NE corner of building is quite useful and secluded.

The thoughts [that] Cornock, Gardiner and Ryde have been giving to balloon problems are to result in a visit to Exeter next week.

Have a request from Appleton to make a return on all our researches to produce substitute materials to replace foreign importations. The review is quite impossible when one looks over it.

May 14th[91]

Grim reports from Holland. It looks as if she hasn't a chance.

The year's analysis of expenditure etc. is now complete. It has not any special features of note—oddly enough in spite of alleged rising prices expenditure under various heads has remained about constant.

Mr. Aldis is anxious to press on with the application of discharge lamps to his signalling devices for the Admiralty—so Francis and his group will be getting busy.

May 15th

People worrying whether they should not volunteer for the anti-parachute appeal of the War Minister. Got in touch with Capt. Bardwell who agreed that ordinarily this is not a service for such as we—that men in key establishments are wanted for production first and last and should concentrate on it.

May 16th

Everyone feeling very anxious on account of the issue at stake in Belgium battle and finding it difficult to concentrate undivided attention on their jobs. Saw Fletcher who told me of the straits of the Philips people through the occupation of Eindhoven and the efforts of their chief management to get over to England.

May 17th

Examined CRT position. We still have important developments for the Services. Ryde and Jesty have had tests at Christchurch showing that for Pollard's job the two colour lines can tolerate a wide diversity of screen colour—which will help manufacture. Signal School demands for CRTs become more important.

Elmet metal is settling down at last to a routine. Prices have been fixed with Sims for 12 months. Our success with these prices is dependent upon the rate of output being kept up. Development expenses appear to be about £1200, which I have managed to lose in our Heavy Alloy costs.

May 20th

The critical position of the battle in France is resulting in a general working of overtime wherever this will help the immediate future.

I have been canvassing the question of who shall act in my place if I get knocked out. An 'acting director' from the Lab. staff seems to be indicated without prejudice to a future decision.

Trouble with glass quartz seals seems to have arisen which must be rectified quickly but the inevitable result is the renewed demand for E 1024 valves somehow.

I understand the MOV will not urge us to relieve them.

May 21st[92]

Dundee (late Bawdsey) have now moved to Worth Matravers and members of their junior staff roll up from time to time. We discuss our work with them without reservation and are so, I hope, building up a lasting liaison.

There seems a danger of shortage of instrument pivots. Must send Leeds to Salford.

May 22nd

A difficult day for maintenance of gaiety and buoyancy. Morning news grave: Dudding down with temperature. People not sleeping enough.

Appleton spent the day with us and went over most of our RDF radio work. There seems to have been some very critical comment in higher circles of Bowen's AI work and I gather he may be affected. We discussed possibilities thoroughly with Appleton who is always helpful.

Had SOS from Harrogate re. supply of E 1024 valves and we must help Hammersmith with urgent hand-made samples. This is wicked in the light of our repeated warnings of a year ago to get on with orders to manufacture without delay for fear of the very consequences which now we are up against.

Have decided with workshop and other staff to work next two weekends on first priority jobs.

May 23rd[93-94]

Settled with Goldsmith that Sanderson must have an understudy on Battery work. He carries all the eggs in his own basket.

May 24th[95]

News still very bad and Channel ports seem untenable. Many arrangements to make re. overtime etc. to prevent wages and salary inconsistencies. We are now handed over body and soul to the Government and are to do and be paid etc. exactly what Ministers decide! What an opportunity for abuse of power! I was going to Worth Matravers on Sunday evening to spend the day with Rowe and discuss AI and the like, but Wright wants me at the Admiralty, so B.S.G., G.C.M. and D.C.E. are going.

May 25th

Ryde, Cornock and Gardiner returned re. balloon investigation from Exeter and I have got Lockspeiser to come on Monday. The principals at Exeter seemed to them to be 'one track' fellows. We seem to have got some interesting preliminary results.

Decided to have a brief Service in Library on Sunday morning for those who care to join in the National and Empire service of prayer. The first time I believe that a religious service has been held at the Labs. We may only have a very small attendance.

May 27th

Lockspeiser came and we had a useful talk with him and things will now, I think, move in the new investigation we are handling. He was distinctly interested.

DSR Admiralty had forced me to abandon my intended visit to Rowe at Worth Matravers in order to attend a meeting in his office with Standard Telephones (STC) and Mullard to secure 'co-ordination in valve research and development'. The matter did not seem to me at all urgent and I was in no very sweet mood. I had previously discussed matters with M.R. and agreed that I could take no decisions. A rotten meeting followed by some very straight talking to Wright: (*a*) for landing me without warning at such a meeting, (*b*) his risky friends. M.R., F. Smith and I had a private talk afterwards with Wright and Brundrett. I made up (*a*) with him, but (*b*) is his funeral.

May 28th

Searchlights—a good panel meeting. Belgium capitulated and everyone wondering what is to happen.

The weekend has been hectic with CID enquiries and preliminary organisation of anti-parachute squads. As I am chairman of Estate Committee, took action in getting in touch with the authorities. Humphreys of the greatest help. Jolley appointed OC and is pushing things with a swing. About 200 volunteers from Labs.

Shadow valve factory meeting in afternoon—has to be pushed hard now.

May 29th

More politics re. DSR and valve cooperation. Talk with Marris, Espley and Gossling on their first visit to Worth Matravers. They found an almost identical programme on AI as ours going on. A very frank and attractive visit.

May 30th

CVD meeting all day. We had a number of new advances to report; the chief [is] probably the large rise of sensitivity of the beam valve at 50 cm. The last member of the meeting left at 8.30 pm!

May 31st[96]

Agreed to give E.H. Rayner a job with Gardiner and wonder how the experiment will work. Attended a meeting with the Standard Co. [STC] at HO to agree cooperation at request of DSR Admiralty.

Sanderson tells me of his difficulties in helping Air Ministry to secure and specify one of the batteries they need. Another of the 'things that went wrong'.

Armstrong Siddeley seem really to want very large quantities of Heavy Alloy and we have to make serious enquiries.

June 1st and 2nd

Another weekend of work for nearly the whole staff. Arranged to increase Lab. hours to 6.0 (and 5.0 on Saturdays) and pay equivalent extra to weekly staff. Everyone very keen and stimulated by extraordinary withdrawal of BEF [British Expeditionary Force] from Dunkirk.

June 3rd

Urgent land mine job being handled by Jolley, Rayner and others.

Jolley appointed O.C. Local Defence Volunteers for the site and is doing his best to organise—but different views by the managements of the several concerns at Wembley doesn't make things too easy.

Gibson and Fremlin from the Standards Co. [STC] spent all day here and we make return visit on Thursday.

June 5th

An AI meeting. The full scheme is some way off and there was an urge to crystallise on an intermediate short-wave solution. Swanage has now come much more actively into the matter. A.C. Bartlett has now joined Swanage and I hope to spend next Tuesday there with Rowe and get my bearings. Bowen did not turn up at the meeting and he appears not to be quite so much as hitherto the focus of this subject.

June 6th[97]

Visit here from Cossor to interchange experience on CRTs for Air Ministry. Visit to Standards by B.S.G., J.W.R. and D.C.E. in return for theirs on Monday. Both appear to be useful exchanges.

June 7th[98-99]

Whiddington here all day. He is to help DSR Admiralty with valve development. There was too much to force into a single hot day and he will be coming again. His association with DSR may prove very helpful to CVD.

Saw Mr. Railing with T.W.H. to get his sympathy over the arbitrary cutting down of expenditure on valve development by MOV Board when any given year appears affected by financial stringency. M.R. left us to discuss the matter with C.W. and report again.

June 8th

We are working Saturdays but not Sundays and many others are inclined to do the same.

June 10th[100-101]

Our first CRT meeting at request of Harrogate. Walmsley, Bradfield, Atkinson and others present. Programme of work reviewed and confirmed. A very useful meeting run by Jesty on our side. Drove to Worth Matravers with Gavin at end of afternoon.

June 11th

Spent the day with Rowe at Swanage and discussed with him fuller amalgamation of our respective programmes on AI. Saw over the station and returned by evening.

There is renewed enthusiasm over the possibilities of the beam valve, the mutual conductance having been pushed up to double. If expectations are realised it will revolutionise reception at 50 cm even with existing valves—the group hopes to produce valves for below 50 cm.

June 12th

News is still grim—Paris will be taken by Germans—everyone keyed up to meet situation industrially and militarily. We are having to increase output of various products—and also to spread our risks by separation over the building. Defence volunteer movement tends to drag rather but Jolley and Castle will pull it through I fancy with patience.

June 13th and 14th

More AI consultations with Bowen—I am trying to settle some equitable arrangement for payment of lower monthly paid staff.

June 15th and 16th

Looked in on Sunday and found Espley excited over good sensitivity obtained with his 25 I. Also the beam valve folk have obtained still further improved results. Elliott from Christchurch also working at Wembley to get his beam valve circuits in order.

June 17th

French have caved in and are asking for 'honourable' terms. So we are to fight the war alone. More talks re. monthly paid staff rates. Am trying to get junior staff in the place to copy the seniors and put a little more confidence in myself and the GEC management—and not to squeal over 'injustices' before they materialise!

June 18th

Heavy Alloy plans for much increased production are now ripe. Saw Dr. Railing with Vaughan Williams and had his approval to proceed with discussions with Armstrong Siddeley. About £3500 in buildings is involved and £8000 plant for an initial order of about £75 000. He unloaded on me: (*a*) iron powder troubles, (*b*) diamonds for cable copper drawing, (*c*) substitute for sapphire jewels, (*d*) extrusion of aluminium alloys. All of them tricky jobs. Visited in the evening National Volunteer patrols on the site. First serious raids on England followed.

June 19th

Coales here from Signal School to discuss various things.

June 20th

Saw M.R. re. monthly paid staff and then settled whole matter to everyone's satisfaction, I hope. I think it is generous and fair.

June 21st[102]

Benjamin saw me about getters—we are about the only source of supply in this country and Admiralty and other manufacturers are now seriously interested. MOV have in the past given little encouragement to us. Whilst they were ready enough to see GEC carrying the baby, if they could buy from USA one shilling cheaper they were not very interested in our efforts. I am glad we persevered.

Dr. Bowen and Mr. Whittaker called re. a new and interesting application for the second stage of our AI work. It was a pleasure to link up with Whittaker again.

June 22nd

The conference on iron powder for Salford and a visit to ICI look as if they are supplying the clue to our difficulties. People will be working on it over the weekend.

June 24th

A visit by Good with Kodak and Professor Curters on camouflage research. Waldram and his group were up to the scratch as usual and provided demonstrations which seemed to settle the immediate problem of the use of models to discriminate between different hues of daylight and between sunlight and diffused light.

June 25th[103]

Consultations with Benjamin and Walters re. reinstallation of Azide plant and also the general consideration of the getter production problem—which the whole industry as well as the Admiralty are concerned about.

Learned that Christchurch experiments are giving the desired results with the beam valve at 50 cm. That will give everyone else courage and confidence to go on. Our hopes and fears have alternated.

June 26th

Ramsay's mercury, grid controlled, rectifier seems to be most promising for high-current supplies for RDF and is in great request inside and outside. Dunham deals with its application and we have established a regular group to concentrate on it. Winch has subscribed Machin to help.

June 27th

Had an Admiralty Lieutenant Tirrell (again with Kodak) to see me at home in the evening. 'Matter could not delay an hour'. Plenty of mumbo jumbo about secrecy and play acting on the part of the Lieutenant. He turned out to be a barrister in private life now attached to the intelligence service, and therefore accustomed to having confidence in his own opinion. I took on his job and must get going with it in the morning.

June 28th[104-105]

Asked Winch to be the focus of Tirrell's job and brought in Ryde, Waldram, Schuil and B.P.D. Demonstration for Friday afternoon in Illumination Lab.

Randall coming on Friday re. high power magnetron. Megaw and Boyland have been making splendid progress with these. There are two types—Randall's and Megaw's—the former water-cooled and the latter air-cooled. At the moment Megaw's is on circuit and giving some 2 kW at the desired wavelength—but Randall's will follow in the next few days. Boyland's gold wire joint seems to be one of the chief features giving the striking success.

The trials with Tirrell came off today and he was wheeled about the room in a 'scale' aeroplane. He is shedding the play acting and will probably be a good collaborator.

Another man from his group, Norway, has seen Ryde on our

experiments on the Balloon Barrage job which links up in an interesting way with the objectives of Tirrell and Norway. The head of this special group is our old friend Admiral Sir J. Sommerville.

June 29th[106]

Good reports today on all the work and I am glad Whiddington came yesterday to see things and to review valve matters with Gossling. E 1046 valves are touching 300 good valves per week. The group Le R., Pearce, B.P.D. and Leeds expect to start on the E 1161 (E 1046B) next week. We have 1500 to make.

This has altogether been an encouraging week.

July 1st

Output from Megaw's 10 cm magnetron has risen to 4 or 5 kw and is still rising.

Tirrell is tending to be a nuisance by trying to manage technical things himself instead of leaving us to do our own part of the job in the way we know how to do it. Styles came to a 5.30 pm meeting and I think Tirrell learnt where the shoe pinches. 10.30 pm trials fixed for Tuesday.

July 2nd[107]

Shaw shadow factory meeting. Discussions mostly about how to induce Air Ministry to let us get on with the job for which they have made us responsible.

Trials of Sashalite etc. with Goodeve, Tirrell and others in the evening. Results regarded as promising.

July 3rd

News of taking over by us of the French Fleet.

Visit from C-in-C Fighter Command and Wing Commander Hart who spent the whole afternoon inspecting and discussing our AI work. A stimulating and encouraging visit. Our responsibility is heavy. We have our AI meeting on Monday and hope some crucial reception results will be to hand by then.

July 4th

Talked with Castle re. Local Defence Volunteers and the lack of support given by certain leaders on the estate. We shall have an estate meeting on Saturday when the matter must be probed.

July 5th[108]

The Salford iron powder job goes slowly at Salford. ICI and RL are well on with their share. The essential cause of the trouble is now known but steps to improve it are so far not very successful.

DSR most interested in the 10 cm magnetron work.

July 8th

AI nearly all day. Watson Watt and Dr. Walmsley turned up as well as the others and there was a crowd in the Service [Services Conference] Room for a whole morning's report and discussion on our work. It was agreed that it is time to have a mutual disclosure on GEC and EMI efforts in the national interest—and a tentative date was fixed for the meeting. One part of the EMI effort has been the improvement of existing AI.

July 9th[109]

B.S.G.'s and J.B.'s report to DSR and CVD. This report is supposed to be monthly but has become two or three monthly! We have very good progress to report: (*a*) deflected beam valve amplification at 50 cm and below; (*b*) 10 cm magnetrons; (*c*) Ramsay's mercury grid controlled rectifier modulator.

July 10th to 13th

All busy days on the same work—progress still good on 10 cm but difficult to assess efficiency of reception. More people going onto it—Bligh's group; Megaw's on local circuit and valves therefore.

July 14th

Saw Fowler, Wright and others at the club—all very interested in AI and the policy as between 25 and 10 cm. Wright was very appreciative of the CVD work and our latest report upon it.

July 16th

Internal meetings on AI preparatory to our show-down with EMI, and conversations with the Services. Agreed with G.C.M. and D.C.E. to crystallise more definitely the two groups respectively on 25 and 10 cm so that one should not compete with the other for facilities. Joubert and also Watson Watt desire to talk over things with us.

July 17th

The whole day at our two-monthly CVD meeting. A rather memorable occasion for we recorded (*a*) beam valve amplification at 50 cm, (*b*) Megaw's air-cooled low field magnetron, (*c*) Ramsay's mercury modulator, (*d*) two velocity modulation receiving tubes—one with resonator for 10 cm. General demand for specimens.

July 18th

Our Air Ministry cathode ray tube meeting was arranged for today—but Air Marshal Sir Philip Joubert called to discuss our AI and we found him most direct and forceful in his comments etc. He made us feel that there is no room for anything more than 5 cm unless Yagis [Yagi aerials] are used and then we may as well use 50 or even 150 cm.

The CRT meeting went very well under Jesty's guidance with Walmsley presiding.

July 19th

All the morning with EMI and Walmsley in the chair for show-down of our respective AIs. I cannot say that EMI had got far but they appeared to agree with us as far as we had gone.

Am anxious because AI is losing its identity through our passing to Swanage everything we develop even in its embryo stage. Have taken it up with Walmsley and Rowe.

July 20th

Many matters to clear up. Tirrell is handled mainly by Winch and is constantly worrying us. Nash here urging me to push the Labs. forward in connection with technical service in making diamond dies. Shall write to F.E. Smith.

This has been a very full week. AI dominates everything.

July 22nd

Lunched with Watson Watt, Marris and Espley. W.W. is apprehensive lest the electrical part of the AI work which we are very suited for should be pushed less energetically than it should on account of the radiation [aerial] and representation [display]fields of work which we [it] seemed to him we were devoting too much time to. Discussed for a long time and showed him I think that there was more bark than bite as only two men were working at radiation etc.

Much effort to get an aeroplane wing for aerial experiments. Joubert had offered one and it has eventually come to hand, together with the attached unofficial communication from Air Marshal Joubert (Fig.3).

July 23rd[110-111]

Visits to B.P.D. by various people asking for help in technique—doctors and also Lamplough. The Air Sextant was among them. Wing Commander Hart turned up to lunch also anxious about AI as the C-in-C was bothering about it and what he had seen here. It appears that Joubert's point of view is gaining ground.

Poor Schoenberg (EMI) is now worrying about our liaison with Government Depts. on CVD, leading as it does to contacts with other valve makers—which he regards as dangerous. His motives are difficult to penetrate.

Various members of staff are at different establishments this week mainly trying to see what real gain in HF amplification is obtainable with the beam valve.

July 24th[112]

Met Sir F.E. Smith and Mr. Terry with B.P.D. re. means of fostering the production of diamond dies in this country. There emerged from the talk only a shadow committee with a proposed impartial chairman.

(See entry of July 22nd, 1940)

AIR MINISTRY,
Whitehall, S.W.1.

July 21st.1940

Dear Dr.Paterson,

Air Marshal Joubert thought you would be amused to see the enclosed cartoon which was sent to me after your "wing" had been obtained.

I hope it was found safely at the right garage !

Yours sincerely,

Dr.Paterson,
Research Laboratories ,
General Electric Co., Ltd.
Wembley.

Fig.3 Letter and cartoon from Air Marshal Sir Philip Joubert after Air Ministry had sent a damaged Hurricane wing to the Laboratories for aerial experiments. Letter dated 21 July 1940

July 25th

Lunch with C.S. Wright. Have learnt that Air Marshal Joubert has been discussing very seriously the non-utility of 10 cm AI and the matter may shortly be raised generally. He also ran over with me many other matters unofficially—a very helpful talk.

I fear evidence is growing that the use of the beam valve as a HF [High Frequency] amplifier does not result in an improved noise-to-signal ratio—but the noise seems to come from the associated circuit rather than from the valve.

Have had message that BTH are to be brought into a talk on 10 cm work.

July 26th

Ramsay seems to häve got on well at Swanage in fixing up his mercury vapour modulator. Warren and Bligh after returning from Christchurch and Signal School seem to confirm that beam valves in HF amplification do not improve signal-to-noise ratio. Warren's double tetrode seems to lead the field on 1½ metre amplification. Had helpful conversation with Rowe (Swanage) re. our relation with them over AI. They have only just realised that we have a contract from the Air Ministry for a complete prototype system.

July 27th

Arranged to go to Swanage with B.S.G. and G.C.M. and D.C.E. on Monday. Spent whole afternoon reviewing our memo on AI for single seaters.

July 29th

At Swanage all day with Marris, Espley and Gossling, to see what sort of marriage could be arranged between the two organisations on AI. Rowe made the courageous proposal that someone from Wembley might be transferred to Swanage to take charge of the work. A very attractive day with conferences on the downs in the open air. Our memo on AI for fighters was mentioned and sent for from Wembley, so Marris and Espley stayed over Tuesday to discuss it.

July 30th

We are beginning to realise that the beam valve is not promising for signal-to-noise ratio although it amplifies at 50 cm. We have not however given up hope. Had ring from Marris that he does not see the way clear to adopt yesterday's proposal on ground that the Swanage group on transmission and reception are not improving on our own work and are considerably behind it.

July 31st

Arranged with AI group to hasten everything so as to get all possible into the shop window for visit by Sir F. Smith and Sir George Lee on Saturday.

August 1st

Megaw attended (on Monday) a new liaison meeting on 10 cm reception. Walmsley presided and manufacturers with research facilities attended. It was a useful review culminating apparently in our own conclusion that the crystal mixer and local oscillator (probably magnetron type) is the best arrangement at present.

Diamonds loom large and Mr. Terry has practically turned down the proposals made at the discussion with Sir F. Smith. Anyway we here carry on as rapidly as may be, but the business will have to take the normal research course. No short cuts I fear.

August 3rd

Sir F. Smith and Sir George Lee's visit seemed to me very successful (Fig.4). It showed that we had got 10 cm responses and we had all our components for 10 and 25 cm in proper shipshape form for going into the air. They stayed three hours and afterwards Lee discussed with Marris and me the possible methods of liaison or partnership with Swanage. The idea at present is to take our present gear to Swanage and Christchurch and try in the air for strength of signal etc.

There are to be several more discussions next week.

August 5th

AI meeting—when we showed the various 'engineered' components. Barton very anxious there should be one place for GEC gear and one for

Visit of Sir F. Smith CTP
Sir George Lee DCD
Aug 3. '40

AI

SECRET.

DRAWING OFFICE ROOF.

Mr. Espley. Mr. Edwards. Mr. O'Kane. Mr. Knox.	New Hut.	Experimental 10 cm AI system expected to be in operation to give echoes from ground targets.
	Old Hut.	Engineered assembly of 10 cm system suitable for preliminary air trials in course of being tested out.

SPORTS FIELD.

Mr. Clayton.
Mr. Hawes.

Wing of Hurricane with aerial of Yagi type having polar diagram measured.

Mr. Gavin.
Mr. Cherry.

U.S.W. ROOM II.

Deflected beam valves as frequency changer. Other recently attempted triode and diode valves.
Spark pulse ~~generator~~ modulator

Mr. Megaw.
Mr. Willshaw.

U.S.W. ROOM I.

Magnetrons.

Mr. Ramsay.
Mr. Walter.

TELEVISION III.

New gas filled valve and circuits for pulse production under oscillographic examination.

Mr. Walter.
Mr. George.

TELEVISION I.

25 cm E1130 & E1190 2 KW & 20 KW transmitters dismantled for inspection.
Positive grid local oscillator for 10 cm reception under test for constancy of frequency, etc.

Mr. Warren.

VALVE DEVELOPMENT ROOM.

Double tetrode. Low noise valves.

Mr. Bligh.

RADIO I.

50 cm deflected beam amplifier experiments.

RADIO III/

-1-

Fig. 4 Programme for visit of Sir Frank Smith and Sir George Lee on 3 August 1940

-2-

Mr. Ryde. Mr. Harris. Mr. Proctor. Mr. Sloane.	RADIO III. ELEC. ROOM II. VAC. PHYSICS.	Velocity modulation amplifiers work at 10 and 25 cms, and local oscillator for 10 cms.

MAIN WORKSHOP.

Mr. Le Rossignol.
Mr. Bell.

Magnetrons.

E.1130	2 KW	25 cms.
E.1190	10 KW	25 cms.
E.1137	200 KW	1.5 metres.
E.1155	Pulsing tetrode.	

C. R. T.

Mr. Jesty. — Cathode Ray Tube development in general.

APP. DEVELOPMENT II.

Mr. Dunham.
Mr. Harvey.

Tank for imitation of high altitude voltage flashover and similar tests.

Fig.4 continued.

Swanage gear—fear it will be difficult to avoid the appearance and spirit of rivalry. The 10 cm signal behaved itself but is still uncertain.

August 6th[113-114]

A crowd of people here from many establishments including Oliphant and Randall and Ellis. It is difficult for people to get on with urgent work when visitors are so frequent but it cannot be helped. Megaw's improved 10 cm magnetron with eight chambers appears up to expectations. Randall with Duke's assistance (we introduced him to Oliphant) has made a good copy of the air-cooled magnetron for 5 cm.

Lord Hirst paid us a visit and was wheeled round to various rooms with myself, B.P.D. and R. Le R. looking after him—a delightful episode.

August 7th[115]

Schoenberg, Blumlein, Condliffe and Broadway with Bowen came for a thorough discussion of AI. It took a whole morning and until 3 o'clock. The general exchange of views and experience was helpful and the atmosphere of co-operation most encouraging.

August 8th

Had expected steps re. closer interworking with Swanage but learnt it was under discussion at higher levels. I hope nothing will be suggested which would sap our independence. Lewis and others are coming on Monday to examine our memo and to see our assembled components.

Heavy Alloy very slack all of a sudden. I wish one of the aircraft uses would really materialise.

Attended National Register committee at IEE. Nothing seems to satisfy the Services except bleeding industry of their own men. Offers to train men or suggestions that the Services should do likewise fall flat.

August 9th and 10th[116]

Put pressure on the diamond die production job—we must get started on the drilling some way or other.

Lt. Tirrell's job has resulted in the making of trial projector frames which ought now to be nearly ready for trial.

E 1046 production still going on and losses are reduced to almost negligible proportion. Costs are also down.

Glass ring seals now wanted badly and I put pressure on. Warren's tetrode seems a joint winner with Mullard EF 50 for 1 metre reception in place of the acorn.

We are trying to arrange two weeks holiday for everyone—it is really important.

August 12th[117-118]

What a day! Lewis, Dee, Skinner, Bartlett, Atkinson, Ward from Swanage to talk AI and means of co-operation. Except for Lewis there was no articulate idea on the subject from their end. They appear to me to be a group of individualists with little experience in team work—except Lewis. Pressed them to have fortnightly group meetings with us to review branches of work and detailed programme. Dee was non-committal. I thought it best to put our proposals in writing in a letter to Rowe. I don't believe they have yet got the idea of sharing responsibility with us.

August 13th

Further talk with Lewis—quite helpful and wrote Rowe. Marris off for a holiday and needs it.

Had stirring conversation with A.S. Gosling (Lamp Works) about money we are to receive for the first 1200 E 1046 valves. I did not see why MOV should expect to take a so-called 'profit'.

August 14th

Received final contract arrangements from Air Ministry re. balloon barrage investigation—up to £2000. I wish the course of the work looked as promising as it did earlier. We shall have to see Dr. Lockspeiser to find out if it is worth while carrying on.

August 15th

Consultation on cathode ray tube work which goes very smoothly now. Works have large contracts. Saw EAW film on good food campaign—part of which had been made at Research Labs.

One of the chief problems on valves now is to clear up the signal-to-noise situation on very short wavelength reception. Amplification seems to be an ineffective way of recovering dbs [decibels] lost in detection. The whole matter will need flattening out.

August 16th[119]

A day noted for the first two raid warnings which sent us all underground. In the second one Cockcroft was here and we continued our conference with him in K shelter. He is on a mission to USA and wanted introductions and the latest information on things in general.

August 17th

The diamond polishing and treatment is now starting actively and the first ground faced diamond has been produced.

D. Horton here from Signal School to learn the latest information about things in general and to assure himself about the supply of matched E 1046B valves. At these short wavelengths spot frequencies are the rule.

August 19th

Lunched with Gough (DSR Ministry of Supply) and heard many interesting things.

Walmsley came to tell me of proposals for executive committee for AI for which he will be chairman—a good arrangement.

August 20th

Tests are taking place this week at Christchurch to secure a signal-to-noise comparison of different receiving arrangements at 50 cm. B.S.G. [Gossling] is there for the week.

August 21st

A day with Farnborough and Air Ministry people on high intensity lamps and their measurement. Major Goad and Mr. Nye came. Farnborough has had some severe treatment from bombs in the recent raid.

August 22nd

Wired Superintendent, Farnborough, offering hospitality to any of his staff who can carry on their work here. Lamplough very pleased and will probably accept.

Hear from Gavin that Christchurch tests show that on noise-to-signal criterion the old diode mixer is second to none in spite of all our efforts to use more complicated and modern ideas for reception. The crystal is now being tried.

August 23rd

A day on diamonds with Dudding and Leeds at the De Beers head office. We were invited to an informal meeting with others to get a picture of the urgency of our problem of fine die drilling and the critical position of the Johnson & Matthey fine instrument wire production. The diamond industry is a most forthcoming and helpful group. Next week we hope to begin drilling.

August 24th

A day with two raids—8.30 am and 4 pm—but no signs of anything here.

August 26th

C-in-C Fighter Command here to enquire again into AI. Raid occurred whilst he was here but we would not take any notice of it. Bligh back from Signal School reporting that the autodyne sets not apparently so good on signal-to-noise as with the best diode or crystal mixers—which is a disappointment.

Letter from Dr. Walmsley advising me of an 'executive Committee' with 'advisory functions' to look after AI. Am appointing Marris with condition that I may attend. I expect it will be all right if Dee (from Swanage) will play.

August 27th

A seven-hour night raid made many people good for nothing. Shaw factory meeting showed limited progress—machine makers are the usual trouble.

August 28th

We get raids most days now and they are wasting much time—very heavy production congestion at MOV makes it difficult to visualise using newly developed valves on a large scale. Difficult to know whether all valve makers are in the same position. If they are, new production ought to be organised for.

August 29th[120]

CRT meeting with Ministries all the morning. More night raids, but people are beginning to 'weather' them. Shearing and Wright to see Megaw about USW programme and its application to Admiralty DF. Had useful talk.

August 30th[121-122]

Dr. Vick (Cockcroft's right hand man) from Min. of Supply called to discuss proximity devices and particularly valve developments by Pye. This will involve a considerable programme. Sanderson comes into it with special battery developments.

August 31st

Saturday—and in view of the trying week for everyone gave half day off. The first free Saturday afternoon since June 1st.

September 2nd

Lunched with M.R. and A.R. and obtained permission to extend Labs. at NW corner for diamond die work—£4500.

September 3rd[123-124]

CVD day with large attendance by scientific officers. They arrived in the middle of a raid at 11.15 and were greeted in a shelter. The afternoon's meeting was interrupted by a second (one hour) raid so we carried on with it in the shelter in sweltering heat.

At 7 pm Waldram, Stevens and I received committee appointed by the War Cabinet to reconsider street lighting. Mr. Rucker, Major Sandys, Air Marshal R.H. Peck and others including Good and Smith of the IES street lighting committee. They were well impressed by Waldram's aircraft night view of town under different street and background illuminations. They left at 8.30 pm for demonstration in town.

September 4th

Two short raids today. Have made arrangements so that half the staff remain at work till the last minute whilst the remainder go underground. Dr. Vick and Mr Edwards (Pye) came to discuss the special valves and prior to which Vick studied with Sanderson the special battery for the same device.

Dr Gordon brought interesting collection of diamonds of various crystal forms.

Two raids per day are now becoming the regular thing—followed by one or two at night. Espley, Megaw and Ryde are at Birmingham. Oliphant reports good signals by exact frequency identity of generator and local oscillator.

September 6th

To Carbon Works, Birmingham. In Birmingham, as usual, raid damage is made light of. I am planning to get away for a week from Sunday. On the whole members of staff are managing to get a fortnight in odd days—but for some it is difficult.

September 9th to 17th

My own escape to Devonshire was well worth while. Returned to find the violent attacks on London had caused some disorganisation at Wembley. Gas was not available for two days and posts are delayed several days. B.P.D. made preliminary arrangements to rent premises at Watford as a standby for gas for glass blowing.

Dudding has left for a fortnight's break.

AI is dragging. We cannot see how 10 cm can really afford a solution. Espley strongly favours 25 cm. Swanage plunges to shorter wavelengths still and wants to run everything as though they are cock of the walk. I see no special signs of real partnership. Walmsley struggles to keep things on the rails, but I can see that a sticky problem plus a lack of spirit of partnership on the part of Swanage is going to give rise to an unsatisfactory position for all concerned and particularly for the success of the job itself. It may very likely be that Watson Watt and others were in error in insisting on our tackling the extreme short-wave solution.

September 18th[125]

Saw Brundrett (DSR) with Megaw. USW contract was reviewed and Brundrett instructed me to allow for Bligh's group working on the autodyne to continue (or its equivalent) in addition to Megaw's group. I explained my costing to him.

September 19th

The valves for Dr. Vick have been much more promising so far than we expected. Brundrett advises that they are to come within CVD—which is a simplification. But an examination shows that the total annual cost of CVD is now well above the £15 000 per annum agreed a year ago. Have written DSR suggesting £25 000 from the New Year.

September 20th[126]

Leeds is doing well with diamond dies. The drilling machines appear to be operating effectively. Have a meeting at Ministry of Supply with Captain Lyttleton. M.R. has authorised building extension and application for licence for steel work has gone in. At the moment I am basing plans on output of 30 dies per week.

September 21st

Night raids this week have been very annoying in London area but day raids have amounted to very little. The first stick of bombs fell on North Wembley last night narrowly missing the Lamp Works. The members of staff who have suffered so far are Miss Grange (Bushey), Miss Rust (Fulham), Mr. Horrod (Richmond). Possibly others of whom I have not heard.

September 23rd

On diamond dies with Dr. Railing to Ministry of Supply. All the original government people have retired from the subject and left it with Raw Materials Dept. who are going to leave the industries to get out of their own troubles. Mr. Archer (Ministry of Supply) quite sympathetic to our work and will see about steel licence for the building extension.

September 24th

Roof watching during the Air Raid alert signals is now going well. I take from 10.30 to 1.00. There is more excitement than I expected.

September 25th

An afternoon on the Shaw factory. We shall all be thankful when we can feel that this is a going concern, for enemy action has been great and many bombs fall each night in the neighbourhoods round. Shaw suffers to some extent from delays caused by the enemy.

September 26th

Mitchell here from Farnborough reporting general success and steps to try his aerial arrays on short-wave RDF.

September 27th[127-128]

All the afternoon at the Air Ministry with Air Commodore Lywood on new job with Signals. R. Le Rossignol with me. We can really help I should think. A new development contract will be involved.

September 28th

Spent most of morning organising for the new job. Gavin to be segregated to take charge here, and Peters and his people to come in to help with a view to making the sets. Swanage has been asked to help also—I hope we are not to have more jealous overlapping.

AI executive meeting here all day with Walmsley in the chair. I did not attend. Barton remarked to me that the new arrangement seemed to have the only effect of slowing Wembley down to the pace of Swanage! But we are giving the new arrangement a good run for its money and shall back Dr. Walmsley.

September 30th

Peters and McAinsh here to be initiated into the Lywood job. Am glad to find they can and are anxious to co-operate wholeheartedly. Gavin is pushing on with the searching receiver and Bell (at BBC Research Labs.) in conjunction with Met. Vick., with adaptation of a GL transmitter. Peters is also considering making up an IF amplifier and modulator for

RDF which can be sold to Government experimental establishments. The absence of suitable modulators is hindering a good deal of work.

October 1st

Met Wright (Admiralty) and ran over CVD matters with him and found them proceeding on the whole effectively. He had a report from his radio people in the deputation in the USA who took samples of our RDF valves with them. Overheard by them: 'they've got us beat on toobes'—which pleased Wright very much.

We are beginning days with rain and overcast skies and German raiders come over singly and hang about, causing long raid warnings for little cause. Our watchers on the roof are proving effective and although a certain risk is inevitable we have had few breaks. The failure of gas supply has caused the main disorganisation so far, and we are to see if we can fix up a substitute gas gear. The Watford shop premises will soon be in our possession.

October 2nd

A number of *ad hoc* meetings with the Services have been started:

Subject	*Representatives*
10 cm reception	Marris
50 cm reception	Gossling and Bligh
10 cm measurement	Marris and Proctor
Modulators	Gossling and Dunham
Proximity devices (Dr. Vick)	Benjamin and Gossling

The last of these held today was very effective in clearing the air and making objectives definite.

October 3rd

Attended at Ministry of Aircraft Production to see Walmsley and Hecht re. development contract for Lywood's job (now designated AB). Fixed a maximum figure of £11 000 for 12 months. I think the arrangements proposed will be in order—but there are usually troublesome requirements stuck in at the end which have little connection with the course of the work.

October 4th

MOV have had some damage due to a nearby bomb and one building where specials are made cannot be used. Have offered help from Labs. and Benjamin is probably taking on some acorn valve pumping and Stabilovolt work. Saw him about his programme and arranged extra help so that the point discharge phenomenon could be tried for HF reception. This is an idea of Ryde's and, although speculative, if it works it would be very important. Benjamin's valves for AB seem to promise very well. He carries a very heavy load.

October 5th

These long Saturdays have certainly helped to keep work up to date. A group meeting on velocity modulation and fixing its priorities. Also long discussion about how to incorporate 3 and 5 cm VM oscillator generators in the programme. It seems to me that at these wavelengths magnetrons and VM tubes may tend to become very much the same thing.

Stevens reported on street lighting situation. Plan for more light has much weight behind it and a large scale trial has been made this week—Good in command. Stevens tells me that the authorities are more and more treating the laboratories as an impartial authority—which is very satisfactory. This is the tendency also in CVD. If this status can only be really established we can fulfil a unique function.

I have encouraged our people to issue brief reports of a strictly practical character on such subjects as those above (see October 2nd). They seem to be appreciated and are circulated to all the experimental establishments.

October 7th

Coventry is noticing a falling off in our assistance and Wells writes me to restore the *status quo*! Bligh has now one or two extra good men who should soon be able to pull their weight.

October 8th

Had criticism to face from Estate managers through my insistence on retaining responsibility for sending RL [Research Laboratories] staff underground. I don't agree with the procedure of the 'Site' and their devolution of responsibility to a spotter.

October 9th

Gossling and others back from meeting on 'modulators' at Haslemere. It probably did good by way of general exchange.

Called with [S.V.] Williams on Mr. Sims (Elmet Metal) to discuss outstanding points. Unfortunately situation overshadowed by complete lack of orders and we are nearly shut down. A pleasant talk but recovery of development expenses and other matters shelved.

October 10th

A Cathode Ray Tube Meeting with Air Ministry and others. Dr. Walmsley in the chair. An orderly and model meeting largely due to Mr. Bradfield's minutes and records.

Shearing and Wright from Signal School called to study what we are doing in the CM wave searching work of Gavin's.

October 11th

Arrived to find a bomb had dropped in the night at the entrance of Union Works and several houses had been demolished. Research Laboratories are unscathed and have done a full day's work. Gas and water services to the rest of the site are disturbed. We are actively canvassing the possibility of making sufficient gas on a small scale with 'auto' producers.

October 12th

Bligh has made arrangements with Clark at Coventry for a more orderly machinery of liaison. A nice letter from H.P.W.

October 14th

Had a ring yesterday (Sunday) from Samson and Barsdorf re. magnetic treatment of timing mechanism of enemy delay-action bombs. Got Macfadyen to take matter in hand—which is interesting and useful. Attached Humphreys to the job with Macfadyen.

Mitchell from RAE called. He appears keen and resourceful as usual.

October 15th

DSR Admiralty asked me to meet him and Mr. Banks of Canada House re. transfer of special valve technique to the Americas [Americans?]. On reaching the Athenaeum we found the glass of all windows blown out by bombs the other side of Waterloo Place. We lunched with open windows—and fortunately it was not too cold. There seem various toes not to be trodden on.

London has had it very hot the last two nights.

October 16th

Gavin's centimetre short-wave searching set is making good progress. The contracts have arrived from Contracts Dept. Air Ministry but not in the form I had discussed with Walmsley on October 3rd. Shall have to try to get extra provisions inserted for the development part of the programme.

Learned from Chelioti of the enormous crater against their north wall. Some five houses completely demolished and one of the Robertson buildings partly so.

October 17th[129]

Bell finds that he can double the output of the E 1166 valve for Coales by circuit adjustment—which is good work. Gavin's new edition of the milli-micropup is also showing some 30 kW at 25 cm.

Conference at Canada House re. valves in USA to which I got Smith and Fletcher. It turns out that we had better deal direct with the USA. Tizard whom I saw at the Club reports laudatory remarks regarding our valves from experts in USA.

October 18th

We are bothered about our own making of Kemet getter and licence from the American Co. We want a compulsory licence—and Mr. Burrell here today to tell us it is not so simple. Our quality is the best yet.

October 19th

Winch tells me of extraordinary sensitiveness he has achieved in detection of infra-red—a one inch disc of black velvet at 100 feet! Have put him in touch with Espley to discuss possible uses and must then see the 'DSR's'.

Sanderson's compact battery seems to be going well and to be more promising than a secondary cell.

October 21st[130]

Went with Gavin to conference on AB at Air Ministry. Found Group Captain Blandy there and in charge of enemy wireless matters. It was agreed that we attend meetings on future Mondays. With Cox (HO) to Ministry of Aircraft Production re. our contracts for AB. Found them somewhat sticky about following the precedents of previous (AI and CRT) contracts.

Patmore, now at Ministry of Aircraft Production, approached us some time ago re. substituting glass for extruded aluminium for control sheaths. Partridge is trying it out.

Party from Ministry of Supply to examine our bomb disposal experiments. They seem to be on right lines.

October 22nd

To Coventry by road with Bligh to see Wells and Clark who want to be assured that they can lean on us for a continuous effort in their radio programme and particularly that Bligh can be at their service all the time. Found that the enemy had been paying them a good deal of attention at night. Had helpful talk with them and must keep up the strength of our group on Coventry work.

October 23rd

Brundrett rang me to say that GL sets wanted more expert attention in service and would I give him a man for three months—Cabinet priority. Fixed up with Lewer to go.

October 24th

Learned from Sir George Lee that he had turned in his hand as DCD—said he could not carry on with the interference with his job of Smith and Watson Watt. Had a confidential talk with Walmsley.

Le Rossignol and Bell at Swanage for general contact and were asked to stay for a special meeting.

October 25th

Found Le Rossignol and Bell had had none too easy a time with the Swanage people and Harley who wanted (for improvement of GL) to have small production of 10 cm magnetron and hard valve modulators. He wanted to pass the Sutton velocity modulator tubes to Ekco although the technique was all ours. Rang him up and told him that we wanted to do any laboratory making up there was to do. Confirmed in writing.

October 26th[131]

J.Y.F. and others here to discuss passing manufacturing information to Bell Labs. on special valves.

Horton and Moullin here: (*a*) to be initiated into 10 cm position, (*b*) to try to persuade us to make a large modulator using Ramsay's valve. Had to say that we could only co-operate but not take overall responsibility.

Mitchell from Farnborough to get us to make a spark gap under pressure—Le Rossignol handled.

There is great pressure on our shops for all sorts of work. I hope they will be equal to the occasion. Our new workshop extension is now ready to be inhabited.

Mr. Carling here from Labour Exchange to collaborate with us on our training scheme for mechanics. George Myers is to be in charge of it.

October 28th

Discussions with M.R. and others re. transfer of information on RDF valves to USA (Bell Labs.) It includes GEC circuit work and GEC velocity modulation tubes—so the whole matter does not rest with MOV.

Had Heather to lunch. The GEC has not so far suffered at all seriously from enemy activity.

October 29th[132]

To Haslemere (to which one of the Service research centres has been evacuated) for GEC CVD meeting. Drove over with six others and had good meeting. The chief points are: (*a*) absence of any large enough modulator for testing out our new 50 cm valves; (*b*) phenomenal output of our first oxide-coated RDF triode—in this case the milli-micropup, which gives 30 to 40 kW.

October 30th

Leeds learnt one or two useful points re. diamond drilling at Philips, the chief being to dilute the diamond dust olive oil mixture to prevent 'packing'. A simple thing but it seems to make all the difference to him. Had meeting to review the machine design for diamond drilling in the light of latest experience.

October 31st[133]

Had visit from Captain Frazer-Nash with Mr. Whittaker re. applications for AI and the like to their problems. There seems a possibility of the use of our compact-source lamp suitably modulated and Ryde and Francis will work at it.

November 1st

Bradfield very frank in questioning relationship between ourselves and Swanage—enquiring if all goes smoothly. I gather that Walmsley is probably keeping on.

Internal meeting on modulators preparatory to a Service meeting next week.

November 2nd

Saturday, we leave off at 12.00 for the first time for six months. Lab. starting time in the mornings has been put back half an hour for everyone. Mechanics shops start at 7.30. It is likely therefore that we shall not lose so much as one feared. Raids start very soon after dark and we don't seem to have mastered the night raider yet—so staff has to be well away before dark.

November 4th

Letter from Oliphant. He tries us all by treating a superficial experiment of his own as if its conclusions could be set against thoroughly substantiated work here. Met C.S. Wright who tells me Oliphant is complaining of lack of candour on our part! I had to tell Wright that a man earns candour to the extent that he inspires confidence and a sense of good faith with others. Prain, of raw materials department of Ministry of Supply, lunched with me to ask me how I would like GEC diamond drilling effort to be regarded in relation to the effort of the other workers they have got going on the job. I am inclined to his plan of a general committee for circulation of information and consultation.

November 5th

Whiddington had been here yesterday to impress on us the urgency of 3 cm generation. 10 cm and below seem to be very much more in demand than 50 cm in various important uses. This probably means some easing off of pressure for higher power valves of the E 1046 *d* and *f* classes.

November 6th

Had long talk with B.S.G., G.C.M. and E.M. over complaint that Oliphant is not being treated frankly by us—a gross piece of maligning. I must see Oliphant alone and tell him that if a man asks for more than ordinary openness he must earn it by inspiring confidence in the research workers concerned. Conversation got onto subject of the 'dash for glory' type of research favoured by Swanage. A.C. Bartlett from Swanage came in the afternoon and opened out on exactly the same subject. No background of radio engineering to establish their apparatus, measurements etc. on a sound reliable basis.

November 7th

Dr. Railing asked for advice on a neutral chairman for a meeting on valve standardisation. Suggested: (*a*) Bradfield, (*b*) Picken or (*c*) Lee.

November 8th[134]

A CB meeting when I urged the building up of a financial reserve for MOV to maintain the level of valve research when the present support by the Services falls away.

Had Wing Commander Helmore to lunch, over the new job which Jolley is handling for the Labs. under his direction. It means a new contract of sorts I expect.

Attended the 10 cm meeting in the afternoon at Ministry of Aircraft Production. A noticeable inclination to emphasise everyone else's work but that of GEC!! I have no doubt that GEC does dominate the field on account of the thoroughness of its work.

November 9th

Am off to Swanage tomorrow to see what I shall see and hear what I shall hear.

Gavin's tests with Air Ministry people of his searching receiver for 10 to 40 cm were highly successful yesterday—more than up to expectation. As usual the heavy-weight representations of other establishments (at policy meetings) were found to be virtually without substance when the magnitude of their effort comes to be appraised!

November 11th[135]

At Swanage. Met Captain Willett and Horton and also F.O. Brown (new DCD) at the hotel on Sunday night. Willett explained he was there for a trial next day on submarine detection. Was able to speak to him re. so called Sutton tubes (velocity modulation) and Gavin's AB work.

Had long talk with Rowe on relation of Swanage to development establishments and places like ours. Found Rowe's views sensible but fear that his people think differently. Dee and Skinner seem to be 'Swanage uber alles' men, and the function of Wembley to supply them with components and no more. I had to be frank with Rowe and told him that I thought Dee seriously intolerant and entirely satisfied with his own ideas which are often narrow and unsound. Skinner may be better, but has no conscience in appropriating credit to himself and Swanage which should go to Wembley. I hope Rowe will manage to keep his wild men from doing too much harm. He told me of complaints from them that Marris keeps them from consulting Megaw. I told him that Megaw had now been fully warned of the sort of fellows they are and could safely talk with them, for he would not be indiscreet enough to say more than they could be trusted to hear.

I saw submarine tests which so far looked interesting—but they will be safe in Willett's and Horton's hands.

A pouring wet drive home.

November 12th

Drove over with Ryde to Nash and Thompson to talk over their infra-red ideas. Had very interesting look over their power turrets and saw their efforts to meet the desires of Swanage in regard to scanning of radiators.

November 13th[136]

The bomb disposal work seems to have got absorbed in Kramer's establishment, and Humphreys tells me that there are rivalries and the like amongst the official people—who do not however seem anxious to compete for the honour of handling the bombs personally. I believe our ideas have worked out well but the view is based on the circumstances that these days people don't refer to us unless things are working out badly.

November 14th

Dr. Martin and Dr. Evans here on Sanderson's compact battery and the conditions of contract. The terms of suggested contract were simple and sensible and it remains to be seen whether in its final form the financial people spoil it.

Helmore rang up and talked much. I told him roughly of Frazer-Nash's proposals with which we are helping, and it seems up to me to bring them together if there is need for exchange of information.

November 15th

Coventry badly bombed last night and GEC Works damaged. Saw Joubert at his request about our team, and hints he had heard that a certain amount of obstructiveness might be dealt with by me. I had to tell him I had been consulted by the individual in question and he had acted on my advice under difficulties brought about by Rowe's 'wild men'. Joubert was concerned at the manner Jerry is getting away with his night bombing, and talked over prospects with me.

Lywood lunched with me and explained various matters concerning our AB work. The emphasis at present is nearer 100 cm than 10 cm.

November 16th

We have been worried this week over shortage of applications from girls for our various kinds of work, so we are putting up notices and writing round generally. Diamonds are going slowly owing to this.

Magnetron local oscillator is not looking very healthy according to Megaw but Gavin's work on AB prospers and Captain Willett wants some receivers.

November 18th[137]

At Air Ministry at Lywood's Committee on RCM. Gavin's searching set welcomed except at Swanage whence the usual sort of criticism. The wavelength measuring device seems to work well and we are asked to extend it to 4 metres.

Group Captain Blandy has a 'Y' group working with Eckersley at Baddow and we are asked if willing to collaborate. They are Marconi engineers.

November 19th

Oliphant came with Randall—and had a serious talk with him—asking him to speak to me if he has a grouse, instead of belly aching to people in Westminster about our secretiveness etc. I fear Oliphant's mental processes are one-way concerns for matters of this kind.

Had hour's talk with workshop staff over overtime payment during lunch hour. It is difficult for the men to assimilate the idea of 'free overtime' to qualify for staff status. But they made a case which causes me to modify our previous proposal to them. I am afraid there was a set of irritating circumstances which gave rise to the incident.

November 20th

Wasted more time writing letters re. Oliphant's silly complaints. Ryde reports good progress with Nash & Thompson experiments; and Francis by improvement of electrodes is over 5 kW with his compact source lamps. We are glad to be helping in at least four schemes for tackling the night bomber. Coventry was damaged last week and Witton this—so we have no need for persuasion to do our best.

Willett rang up from Signal School very interested in Gavin's set and urging extreme haste in helping Wright to build one himself. Had a most appreciative note from Ellis over the freedom of the help we are giving him—such a pleasant and striking antidote to Oliphant's captious complaints. I have sent it to DSR to show that everyone does not think the same.

November 21st

Walmsley lunched with me and we discussed all the above matters again. He is a helpful person and a good influence and was very anxious that nothing should stand in the way of effective contact between us and Oliphant.

November 22nd

CRT meeting. Nothing special emerged because things mostly going smoothly. Had meeting also with outside people on AB. It is most difficult to contact with people who really know about this. They are all active operational officers without a research establishment and too busy to attend to development of effective apparatus. Gavin is carrying the job well.

Showed Bradfield our velocity modulation work and expect he will place an order with us to make an equivalent number of tubes to Ekco. I think he agreed my arguments.

November 25th[138]

Lywood's Committee on AB—atmosphere very friendly and helpful. Gavin able to report satisfactory design for 70 cm wavemeter. His searching centimetre receiver apparently working well in service hands.

Increasing pressure on 'Nash and Thompson's' and 'Helmore's' jobs as supreme stop gaps against night bombers. Service rivalry seems to exist but fortunately we have two entirely separate groups on them without clash of priority.

Marris sends a more satisfactory report from Swanage (AI) than one had feared and the components, both those made here and at Swanage, at least seem to work.

Had charming letter from Oliphant confessing and regretting recent erroneous attitude towards this place and giving generous tribute to our work. I hope this buries hatchets on both sides.

Clifford is giving a course of lectures on spotting for roof watchers.

Raids are much lighter in daytime now and I seldom have to take a roof duty. The night raids in the provinces are more serious and Coventry and Birmingham have had it heavily. It is all causing a slowing down of work and at the moment scarcity of girl labour is very embarrassing.

November 26th

Lunched with Lywood (with Cornock) to discuss Lywood's device for locating enemy aircraft.

Terry's committee on diamond dies in afternoon and discussed policy under consideration for the new Ministry of Supply Diamond Control Advisory Committee to be held tomorrow and which Dr. Railing prefers to attend instead of me. This work of fooling about for the sake of personal prominence disgusts me when there is serious work to be done.

November 27th[139]

Spent morning on IEE National Register Committee. The pressure for personnel on radio for the services is tremendous. Lively discussion with the new official, Mr. Guy, to assert the status of the IEE Advisory Panel.

Had annoying letter from Air Commodore Leedham refusing my request for consideration of our position on velocity modulation tubes. Better and juster counsels may prevail.

Marris's report from Swanage still encouraging but no evidence as yet that the new AI is markedly better than the old.

November 28th

Two hours in the afternoon with C.S. Wright and Whiddington on subjects of AB searching—possible alternative laboratory accommodation away from London—personal relationships and the like, also VM tubes and Leedham's letter. A very helpful talk.

Dr. Railing wants to see me re. the Advisory Committee on diamond dies. Something seems to have been decided but I have lost touch. It appears they want me to be chairman of the research committee.

November 29th

Trouble over no shadow factory for valves. MOV supply to BBC through Marconi. J.Y.F. rightly making a thorough fuss about it and wants Research Labs. brought in to see if they can help. Saw Smith and suggested asking Kodak to organise manufacture; at same time agreed that if MOV bombed we could accommodate their personnel and carry on production at Research Labs. *pro tem*. Probably seeing Davis of Kodak tomorrow.

December 2nd

A busy CVD day at Haslemere. Took party of five by car and returned with six. Meeting from 10.30 to 3.0 with half hour for sandwiches. Ure told me of enemy use of RDF for channel ranging. Oliphant came to meeting for first time. Many useful items of progress. Probably 'perverted' triode down to 21.5 cm was the newest feature.

Letter from Rowe referring to happy relationship between all on AI work at Swanage. Marris and Espley now back having decided on most of the components as a result of trials on a plane with Dee, Skinner and Co. I must now forget our unfavourable judgments of the Swanage individuals and hope that the hints they have had from Walmsley and others will bring out all the best in them.

December 3rd[140]

The serious situation re. supply of large valves to BBC discussed with M.R. and others prior to meeting Ashbridge tomorrow. Lunched with Kodak management (Messrs. Bent, Macmaster, Webb and Davis) but found we were 10 days late in our request to them. They have committed themselves to a large shadow manufacturing scheme.

December 4th

Examined diamond die progress with Chelioti, Dudding and Leeds. There seems little now to settle up technically except how to get production going on a large enough scale.

To MOV to conference with Ashbridge on how to secure production of BBC valves. It looks as if in an emergency they will have to count on the Research Labs.—but MOV must increase their present rate of production.

DSR is taking an interest in wide angle searchlights.

December 5th[141]

Some comment from outside about our not sending staff to shelters in last Tuesday morning's bomb dropping incident near the site. As our people were sent to emergency stations instead of into the open they were better off than the rest of the site some of whom were in the open when the bombs fell.

With B.P.D. to Dr. Railing who reported to me the Advisory Committee meeting on diamond dies. I am invited to act as chairman of the technical panel to secure interchange of information and arrange any joint experimental work required.

Gossling reported useful visits to Government research establishments over three or four days—in which he discussed current RDF applications with the officers concerned.

Lord Beaverbrook through Helmore is showing keenness on the job Jolley, Cameron and Hawkins are handling.

The bomb disposal job has come back to us in spite of its practical success as they want to know why the clock is stopped! Laudable curiosity.

December 6th

A quiet day for settling many internal matters. Fire watchers' staff training intensified in view of disastrous effects of large numbers of incendiary bombs on other factories.

To add to our congestion of work the GPO are insisting on complete screening of all radio radiation!!

December 7th

Had letter from Air Commodore Leedham agreeing to giving the Labs. an order for 60 velocity modulation reflection valves of the Sutton pattern. So we have won our point without leaving, I hope, any ill feeling. Moore, who is taking charge of the production of these tubes, showed me one today made with his own hands. A bright fellow.

Arranged with B.S.G. and others to make Machin in charge of modulator station for testing etc. our high power RDF oscillators. Machin is making a very good show at the radio art for a photometrist.

December 9th[142]

Viewed, with others, film of valve making processes produced by Kodak Co. A most excellent film which now may never go to USA on account of Bell Labs. not 'biting' particularly on to the idea of collaboration. To Duxford to visit De Bruyne and talk over laboratory organisation etc.

December 10th

Visit of Ministry of Supply experts on bomb precautions with magnets. Also of permanent magnet experts from Darwins and Jessops. There seems a good deal to do before the thing is out of the wood.

December 11th

Mr. Prain here to discuss details for the diamond die technical committee. I asked (*a*) for B.P.D. to be a member (*b*) for the secretariat to be at the Diamond Die Control. All seems agreed in principle. Lunch with Heather to discuss action in event of damage—air raid experience at Coventry, Witton and Southampton Works and conditions under which Wembley laboratories might move to a more isolated site. I have heard from DSR unofficially on the subject and shall have to consider it.

I hear Mazda valve works at Ponders End have had a hit.

December 12th

I gather the production position re. CRTs is critical and GEC is likely to go in for a special CRT factory in addition to getting Shaw into full working order. Air Ministry rang up to ask for Jesty's services at the Mullard works to help with CRT difficulties there. I have other evidence that our CRT work under Jesty has been very much appreciated by the Air Ministry. I could wish our own Works could be as whole hearted in their comment and resulting action.

December 13th

A day when birds come home to roost. Walmsley called and asked: (*a*) as the 'manufacture' of velocity modulation tubes by Sutton at Signal School was not going very happily (they have been visited by the enemy amongst other things) could Wembley push on with samples which the Services can use!! Perhaps it was a good thing that we insisted, against opposition and to the point of making ourselves disagreeable, on making these valves here; (*b*) can Wembley give general service on behalf of DCD department to contractors and others making CRTs. The answer to both is of course in the affirmative—and we may be able really to help in these ways if people will only regard us as out to do disinterested work for the nation irrespective of commercial consideration.

Heavy night bombing of industrial towns goes on—Coventry, Bristol, Birmingham, Southampton, Sheffield so far, not to mention Liverpool and others.

December 14th

Le Rossignol reports a request from Signal School for us to make and deliver by March, 200 E 1191 (Ramsay's gas triode for modulation).

Spent time trying to arrange with Ramsay for Price to take on the group for this work—and I must find someone extra to help Ramsay.

December 16th

These are busy days at Wembley with the canteen at lunch time full of visitors. Wing Commander Helmore here today with Jolley and Waldram, Cameron and Hawkins to discuss and carry through experiments. As usual the illumination group demonstrated what is possible without the necessity of carrying out difficult flights. Helmore expresses himself as extraordinarily appreciative of the Wembley ways of doing things.

Lywood's RCM Committee in the morning—all going to plan with the 80 wavemeters. Mr. Hecht of DCD is most helpful with priorities.

Edwards and O'Kane back from Swanage. They report the magnetron local oscillator performance as effective as Sutton's velocity modulator—as we expected.

December 17th

The victories over the Italians in Egypt and Albania are cheering everyone. Did further work on memorandum re. alternative site for some of the more vital laboratory work. Peters here from Coventry re. AB and AI work. He has got his place going remarkably quickly since the air bombardment three weeks ago.

December 18th

10 centimetre meeting most of the day in town where various rabbits were produced from hats including several velocity modulation tubes by Ryde and Broadway (of EMI). There seems to be some enigma about magnetrons—critical points of sensitivity which are rather disturbing but which Megaw does not appear to regard so seriously as the rest of the experts. I shall be glad when this matter is cleared up. Oliphant and Lewis think it is variation of the strength of the magnetic field.

December 19th

Ramsay back from Signal School concerned that his gas triode is not doing as well as he expects on life. As a number must be made by March,

I must arrange more help—particularly on the production side. Dudding suggests Douglas, who appears now to have ironed out the silica seal with Partridge.

OAC meeting—5 ft fluorescent tubes are marketing well and production at Claudgen is being pushed up. Jennett, Jenkins and B.P.D. have done good work under difficult conditions to achieve success.

December 20th

A day crammed with details of *all* the jobs in hand! Jolley and Cameron tell me that Witton people are working all out on the gear for Helmore's job and in consequence it is making very good progress. Winch told me two days ago that the N and T [Nash & Thompson] equivalent is not far off the trial stage.

December 23rd

This is going to be a week of clearing up details and trying to digest some of what we have swallowed. Signal School are inclined to press us to produce their small requirements of a new CRT—about 1000 in all. If we agree we shall have to organise yet another group.

December 24th

A day of Helmore and Nash. They are still in separate compartments but both going strong. The Nash gear will be in the air on trial first and looks (from the trial made at 6 pm this Christmas Eve in the little projection room) to be promising.

We are only breaking off for one day this Christmas and shall be in on Boxing Day.

December 26th

Everyone came in willingly today although a good many establishments closed. Had telegram from Harrogate instructing us to make 60 VM oscillator tubes and requiring us to give assistance to Ekco in their manufacture—one more little problem for MOV directors.

December 27th

Sat with Cockcroft at lunch [he] just back from USA. Bell Labs. prefer for present to co-operate by correspondence.

Chelioti agrees that if we can, RL had better do any CRT production in its power—so it now remains to see if we can organise it.

December 30th

Enemy tried to set central London alight last night and destroyed many old buildings. Dr. Railing's house in town was demolished. So it goes on.

Lywood's meeting on AB this morning at which we were able to report several 60 to 100 cm 'wavemeters' finished. They look as if they would be very satisfactory in use.

Helmore is receiving strong pressure from his sponsors to get his gear in the air. As the gear he is asking us to make involves entirely new carbons and lamps there is no assurance when, if ever, a satisfactory result will come. We wish he would send up what he can get and leave us reasonable time to develop the next stage.

December 31st

Oxford party with Dr. Keeley to study our short-wave (cm) work. Ryde's VM 3 cm tube gave signs of oscillating before something went wrong with it—ditto his 4.5 cm tube. It is clearly going to be a job to implement our policy of starting production groups on account of pressure on mechanics.

1941

January 1st[143-144]

Once again review of workshop facilities to see how to economise them in newly swelling programmes and demands.

Helmore has decided to send up his gear into the air next Monday. He intends to pilot himself.

An AB meeting with Reeves (TRE), Buckwell (Blandy) and Shearing and Wright (SS). Some wrangling over who shall have the first 10 cm searching sets. Signal School very anxious to get some to sea. Also Warren's little SW triode (E 1231) likely to be popular. Shearing talking of possible demand for 100 searching sets—if so we shall have to make the E 1231 in quantities of 1000 or so.

January 2nd

Internal meeting (B.P.D., H.B.W.G., R.le R., Jesty, Pearce and self) to discuss ways and means of launching all our various valve production efforts—the latest being Admiralty CRTs. We have arranged regular progress meetings with Gardiner, Pearce being OC production groups.

Saw Dr. Railing re. letter to Admiralty on alternative laboratory accommodation, and settled terms of reply to Admiralty. He strongly supports our proposed CRT building section—cost £4000. It remains to see if licence can be obtained.

January 3rd

Took part in conference with Fletcher, Gamage, Chelioti and others to review GEC's position vis-a-vis cathode ray tubes. 300 a week is maximum for Hammersmith for this war and GEC must make its position plain to the Government. I suggested the only possible course is to introduce entirely new firms into the industry who will bring fresh drive and resources entirely independent of GEC existing people and plant.

Dr. Hill here with Ryde on subject of infra-red. He had most interesting development to report which will help in Frazer-Nash device.

January 4th

The F & N job, of which Winch is the chief hero, went up in the air last night and he and Captain Frazer-Nash with it. The result appears to have been all that we could have expected although the conditions for the trial were not all we should have liked and a more thorough test must be made. All however were very pleased that an interception even under these conditions was effected. The cold was intense both on the ground and in the air.

January 6th

Lywood's meeting—all going well according to plan.

Most interesting meeting at Admiralty with Wright, Pye, Frazer-Nash, Dr. Hill on use and tactics of the gear Winch has made with Nash. The latter full of enthusiasm over his Friday night's interception—and also struck with possibilities of Dr. Hill's IR device. Frazer-Nash, in addition to getting a move on in fitting gear to planes, combines in himself pilot, designer, fighter, technician, and is *persona grata* in all the aerodromes with the essential personnel. So contacts with all the usual people which are so difficult to harmonise is no longer needed. Helmore combines some of these qualities. Such men and particularly Frazer-Nash are the only means of getting certain essential things done quickly.

If AI had had someone like him, *one* good experimental establishment to serve him would meet the case. But alas such establishments as Farnborough and Swanage must handle things too and, although they mean so well, their status in the effort and their views must be satisfied in the early stages and speed and enthusiasm are damped down. The policy of putting *certain* things in the hands of people like Frazer-Nash is right.

January 7th

To CVD meeting at Haslemere—but alas glassy roads stopped further advance much beyond Esher and we were glad to get back, although missing the CVD meeting was a sad event and impedes progress, for it is a general clearance meeting for so many matters.

January 8th[145]

The CVD meeting yesterday with us absent was a thin business and a proper meeting is to be held on Monday. Whiddington tells me that the panic over centimetre magnetron idiosyncracies grips all concerned, so much so that they are bringing in ST & C plus Oliphant to investigate. People our end look upon it as just the sort of 'dreadful happening' which often arises with a new device and probably has its origin in non-expert handling at Swanage. Megaw goes there this week and will discuss it—but has thoroughly studied the 10 cm magnetron itself to ascertain the extent of its discontinuities—which are small. CRT meeting with Walmsley presiding went smoothly. He tells me that our suggestion regarding the loan of Norman to help Bradfield at MAP is likely to go through.

January 9th

Saw Norman and found him agreeable to the arrangement.

Eckersley here from Marconi's to exchange our respective experience on AB and Lywood's job. Both sides felt the exchange had been well worth while. Marconi can help us on aerials and we can help them on the sets. But we shall neither make a radical change in our present programmes.

Examined the magnetron production position with Pearce in view of postponed CVD meeting on Monday.

January 10th

The days become more and more hectic—to a considerable extent due to so many overhead people having fingers in pies. The alleged anomalies in magnetrons seems to have stirred up a regular panic and half of Westminster appears to be boiling. Willett of Signal School rang up to ask the truth and I had to suggest that people should show a certain amount of balance, confidence, and a little patience in face of what appears to be the usual sort of happening and rumours with any new device.

January 13th

Postponed CVD meeting at Haslemere. This time we arrived in time in spite of fog and wintery conditions. It was a useful meeting on the whole, the main item of excitement being the before mentioned and alleged vagaries of the resonant cavity magnetron now in great demand for several prospective purposes. Megaw and Willshaw had summarised their data to date on uniformity and the like and it seems to go far to reassure those with cold feet (*a*) that there is serious work being done to establish a proper groundwork of knowledge about this valve, (*b*) that the extent and cause of possible discontinuities are known.

I have a letter from Air Commodore Leedham asking me to accept the chairmanship of a new Inter-Service Technical Valve Committee to standardise valve types with the object of checking the haphazard introduction of new types of valves. The membership of the Committee (four plus chairman) is not very strong and to put me as chairman is to set a thief to catch thieves! Apparently I am wanted by Smith, Fuller and Wright which is gratifying, and I may be able to do some good. I understand the committee is known as the 'anti valve committee'!

January 14th

Messrs. Abraham[s] and Doyle came to see how our work on diamond dies is progressing. B.P.D. and Leeds went into things with them. Leeds seems to be gradually finding how to avoid a condition of things producing a flat bottom to the hole which becomes polished and refuses to dull. We are submitting to a hiatus just now whilst the first 25 machines to our design by Bryant Simons are going through. We have only two of our machines to drill with so that opportunity is a suitable one for investigating reasons of slow progress in fine holes.

January 15th

Visits are beginning of delegates on RDF from abroad—Canada, New Zealand. USA are asking for many particulars. Have been setting up a 'proving' group for the new products we are making, expanding the work which Holland has so far been doing. Holland, Falconer, Willmott, Cheeseman—no easy job for them.

January 16th

Meeting most of morning with all concerned on launching the manufacturing groups and consequent workshop priorities.

January 17th

The storm over the alleged vagaries of magnetrons seems to be rising higher and gathering amongst the heavy tonnage up at Olympus. Westminster is a marvellous whispering gallery. I hear that a 'serious storm is brewing'. But no one comes to us to get their ideas cleared! The same sort of thing to a lesser degree applies to VM tubes where some pulling of strings is causing the CVD policy committee to issue rather silly edicts.

Spent two hours with B.P.D. and Miss Heath planning the launching of the Diamond Die Technical Committee.

The getter and the aluminium—magnesium wire work go very well under Benjamin, Ransley and Britten. Considerable quantities of wire are to be drawn. Orders for filters are also growing and Cooper will soon have four girls working at them.

January 18th

Had message from Winch (who had spent half Friday night in freezing cold and snow on the roof of Nash & Thompson factory) that his device looked promising with its new CRT attachment and that its sensitivity was up to expectations—500 yards anyway, not under the best conditions. The telescope of Dr. Hill did not have a comparable sensitivity and looks unpromising for use under these conditions. It is bad luck on him having to work under such arctic conditions—but success makes up for a lot.

January 20th

An interesting day for CVD. Heard various peoples' views about the 'Anti Valve' technical committee, Leedham (who lunched with me) C.S. Wright, Whiddington, Walmsley. Both B.S.G. and I had been working on the scheme of things over the weekend and I was glad to find that Leedham found our proposals good. But there are to be protests over the veto on new valve proposals given to my committee. Walmsley is joining it, I am glad to find. Wright spoke about the 'coming row' over the alleged failings of the 10 cm magnetron. Willshaw's report ought to calm it down!

Dudding doing fine work in sorting out and balancing the women and boy personnel of the place.

January 21st

Francis reports good progress with compact source lamps up to 5 kW. Reduction of blackening, 100 hours life, and tests to 40 atmospheres to prevent bursting are real advances plus the rotation of the lamp to facilitate burning in any position.

Jolley tells me the Helmore scheme should be in the air in a week's time.

January 22nd

Oliphant and Randall here to discuss the vagaries of magnetrons but the conversation concerned as much the vagaries of some people at Swanage. It appears that the latter have sold themselves and their institution very effectively to their chiefs, and whatever they say, however immature, goes in MAP circles. It looks as if F.E.S. [Sir F. Smith] was about to take some action or other of a rash kind—but we shall see. No one in upper circles has yet had any direct contact with Wembley. The real trouble is that when we sent our first raw sample of 10 cm magnetron to Swanage they chose to regard it as a thoroughly investigated and established valve, instead of an article whose main development lay ahead of it.

January 23rd[146]

Lunched with Cockcroft re. dispatch of information to USA and gave him a picture of the magnetron position. Also Willett on same subject. Glad to meet Brown (DCD) and Watson Watt both of whom assured me that DCD and Sir F. Smith quite appreciated the position and placed confidence in us.

Coales, Ratsey and Moullin called and amongst other things urged us to prepare for making the E 1046D (E 1166) with oxide-coated filament—deliveries to begin in August. This is one more type to be made in considerable numbers. We hear that the E 1061 is performing excellently in the Navy.

Ryde reports very good results with his VM mixer which is much appreciated by Marris and Co. I gather it is an improvement from – 6d*b* to + 3 or 5.

January 24th

Hill (on Winch's work) spent another night on the roof of Nash and Thompson's place having several accidents (due to no fault of theirs) with electrical components. But the gear worked eventually at a 450 yard distance under bad conditions of mist, so under better conditions it ought to have ample range.

Whiddington, speaking for Wright, warned me that the latter would have to protest at Leedham's terms of reference to my Anti-Valve Committee and that I might not have a smooth passage with the committee. So we must see what will happen! Hughes (secretary) came to see me and we discussed the whole subject, finding ourselves in entire agreement.

January 25th[147]

A Saturday afternoon meeting of the London branch of the Institute of Physics. Upwards of 80 present, some from long distances. Subject radiography for industrial purposes. Rooksby and Miss Jackman read an excellent paper.

Winch somewhat disturbed that his device developed with Frazer-Nash is being called by the staff there as the Nash interceptor. Rang Frazer-Nash to ask (if it is having a parentage assigned) if he will call it the Nash GEC interceptor. He spoke very warmly about Winch's work and I think the latter felt better on the subject after hearing F.N. But the question of names is always difficult.

January 27th

Lywood (AB) Meeting, Admiralty most anxious to have our 10 to 40 cm searching set. I persuaded Lywood to facilitate their having the first lab. model this week. Peters will be three weeks finishing theirs off.

Signal School are asking for 1000 velocity modulation tubes! or rather a quotation for them—which means magnetrons and modulators too.

January 28th

Marris and Espley back from Swanage. Marris gives me the impression that on the whole luck has been against them in selecting and making up the components for AI trials. He is hopeful that Ryde's mixer (VM) may be a real help in giving range and protection when the same aerial is used for transmission and reception.

Discussion at Labs. on the lives of magnetrons and milli-micropups which are problematical at present—and on the need for putting the 25 cm solution (instead of the 10 cm) before the Services for some uses, e.g. GL.

January 29th

10 cm meeting all the morning following a symposium with the experts yesterday at which some cooling down of the hot air took place. The conclusion some people reach is that magnets were at the root of the troubles but personally I cannot see where is the evidence for this. Anyway Westminster seems to be going through the motions of coming out of the fog it and Swanage have created, and these meetings are a stage of the process. Lunched with Schoenberg who had attended the meeting.

Manpower (engineers) meeting in the afternoon at Ministry of Labour.

January 30th

A hectic day with first diamond die technical committee from 10 to 3.30—Cockcroft and Whiddington also. Whiddington had disturbing rumour of proposed action by F.E.S. on reports of an extraordinarily false description from Swanage. Hope he will see me with a view to putting some balance into things.

January 31st[148-149]

An AB meeting here when Gavin showed their 10 to 40 cm receiver set (Lab. finished model to Foreman's mechanical design). This one is going to G.M. Wright (Admiralty) who is in urgent need of it. I left meeting before trials in this field were started.

Dr. Truscott to lunch and after, to familiarise himself with our filter and valve work.

To Hammersmith for first MOV-Shaw-Research meeting. We reviewed the Shaw valve programme and decided to meet monthly alternately at the three establishments.

February 1st

Learned from Whiddington that F.E.S. may view matters less drastically than we feared—but this remains to be seen. We have written DSR about the combined supply and testing of magnetrons and magnets and have given way in large measure to their desire that this should be done.

February 3rd[150]

More consideration of valve production, and demand for E 1155 hard modulator valve.

Interesting visit from Staff Sergeant Moore who, profiting from his experience on silhouette in street lighting, had applied some very sound principles to aircraft chasing. We helped him with some hurried printing and photography.

The Nash and Helmore jobs are getting ready to take the air (Fig.5).

Balloon barrage job is being thrown up as no longer sufficiently promising.

The bomb job is so far progressing well but each time we are in sight of finality something new turns up.

February 4th

My first [Inter-Services] Valve Technical (Anti Valve) Committee in Walmsley's office at MAP. Leedham came to launch the committee and Brundrett, Lycett, Truscott, Walmsley, McGinnety and Merren were there. Hughes was ill. The committee has very considerable powers and has agreed to take over the job of weeding out all redundant valves from the 350 types now used by the Services. My memo introducing the 'tentative' status for valves in development was welcomed—also Gossling's chart. A very satisfactory beginning I think.

February 5th[151-152]

Another day on valves. To Ministry of Supply Committee (Dr. Vick) with Benjamin who has made a very good job of the development so far and has arranged things so that two types of valve only are needed for this development.

Called for Air Commodore Leedham (DRP) and brought him to Wembley to lunch and took him with B.S.G. and R. Le Rossignol over the Labs., particularly valve production and radio. He was specially concerned with the embryo production groups and CRT developments.

N & T having another trial with Winch's gear this evening.

February 6th

The general condition of cold feet in Westminster and elsewhere over 10 cm magnetrons was not improved 10 days ago by news from USA that Bell Labs. had raised our figure of 10% efficiency to 50% and an increase of power of three or four times! Megaw made some special valves to test the point (feeling it nevertheless a waste of time) and yesterday actual samples came from USA by 'the Ferry' which Megaw has tested. It is all

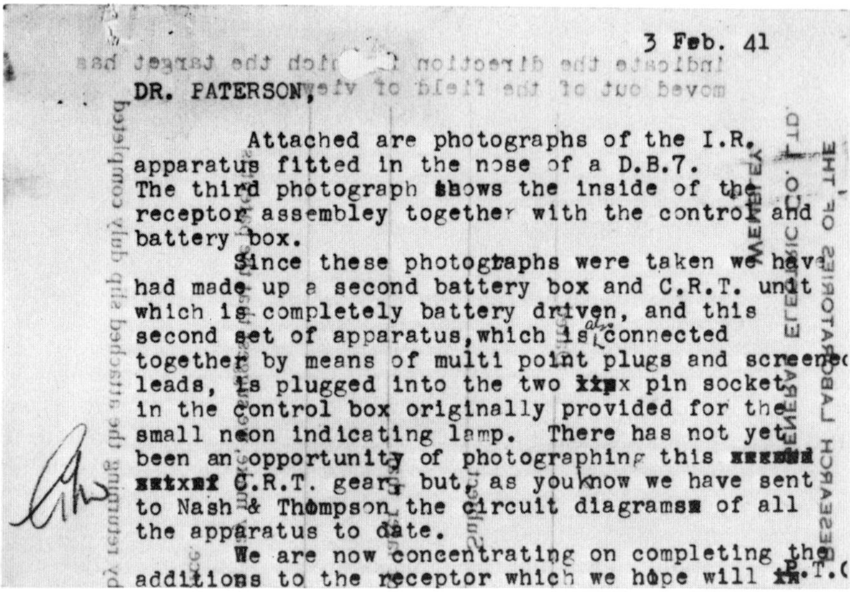

3 Feb. 41

DR. PATERSON,

Attached are photographs of the I.R. apparatus fitted in the nose of a D.B.7. The third photograph shows the inside of the receptor assembley together with the control and battery box.

Since these photographs were taken we have had made up a second battery box and C.R.T. unit which is completely battery driven, and this second set of apparatus, which is also connected together by means of multi point plugs and screened leads, is plugged into the two six pin sockets in the control box originally provided for the small neon indicating lamp. There has not yet been an opportunity of photographing this xxxxxx xxtxxx C.R.T. gear, but, as you know we have sent to Nash & Thompson the circuit diagrams of all the apparatus to date.

We are now concentrating on completing the additions to the receptor which we hope will ...

Fig.5 Note dated 3 February 1941 from G.T. Winch together with photographs of Nash & Thompson – GEC infra-red equipment

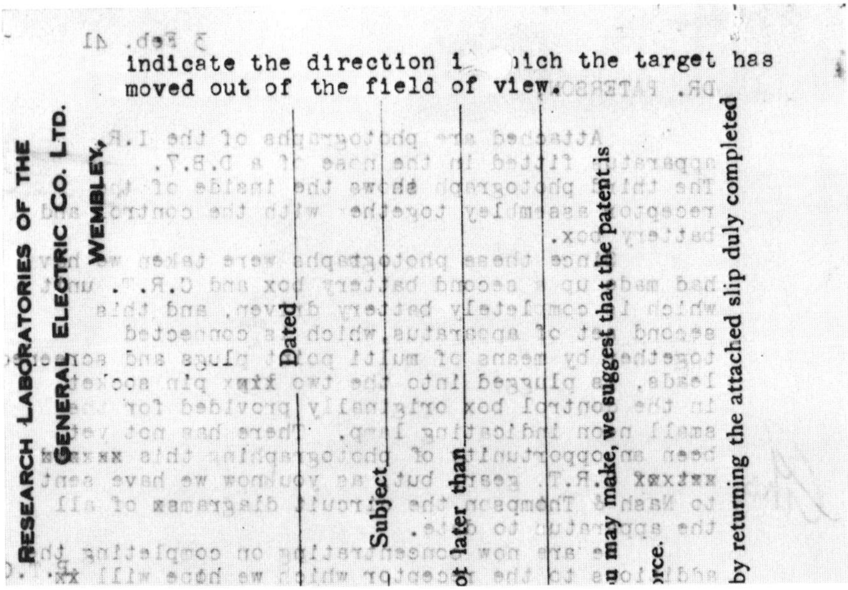

indicate the direction i[...]ich the target has moved out of the field of view.

Fig.5 continued.

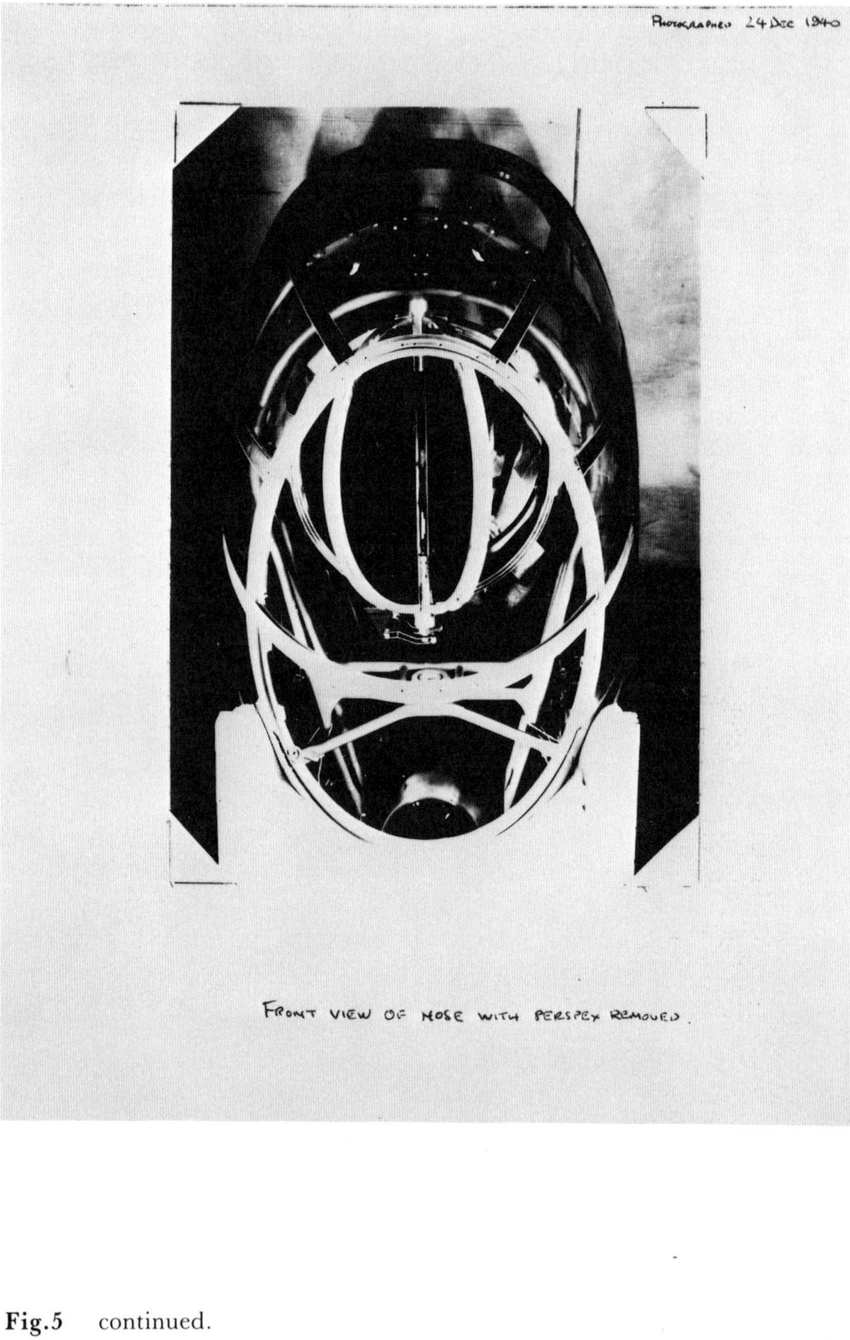

Fig.5 continued.

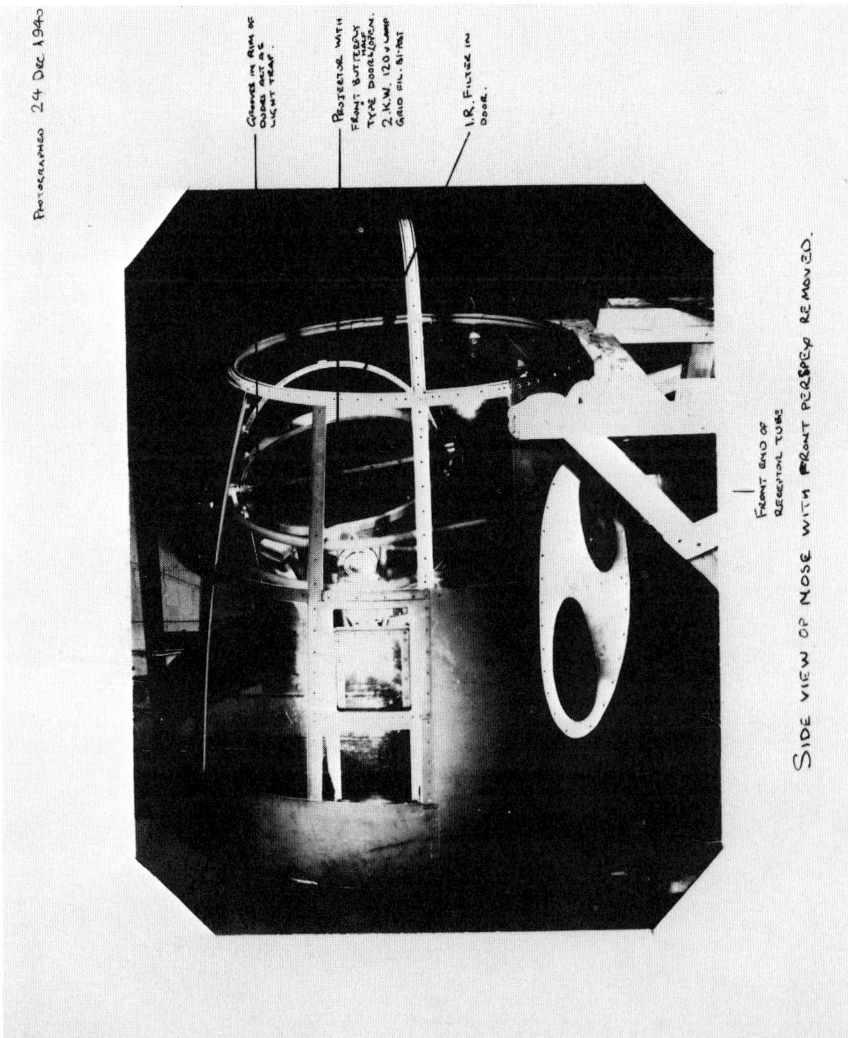

Fig.5 continued.

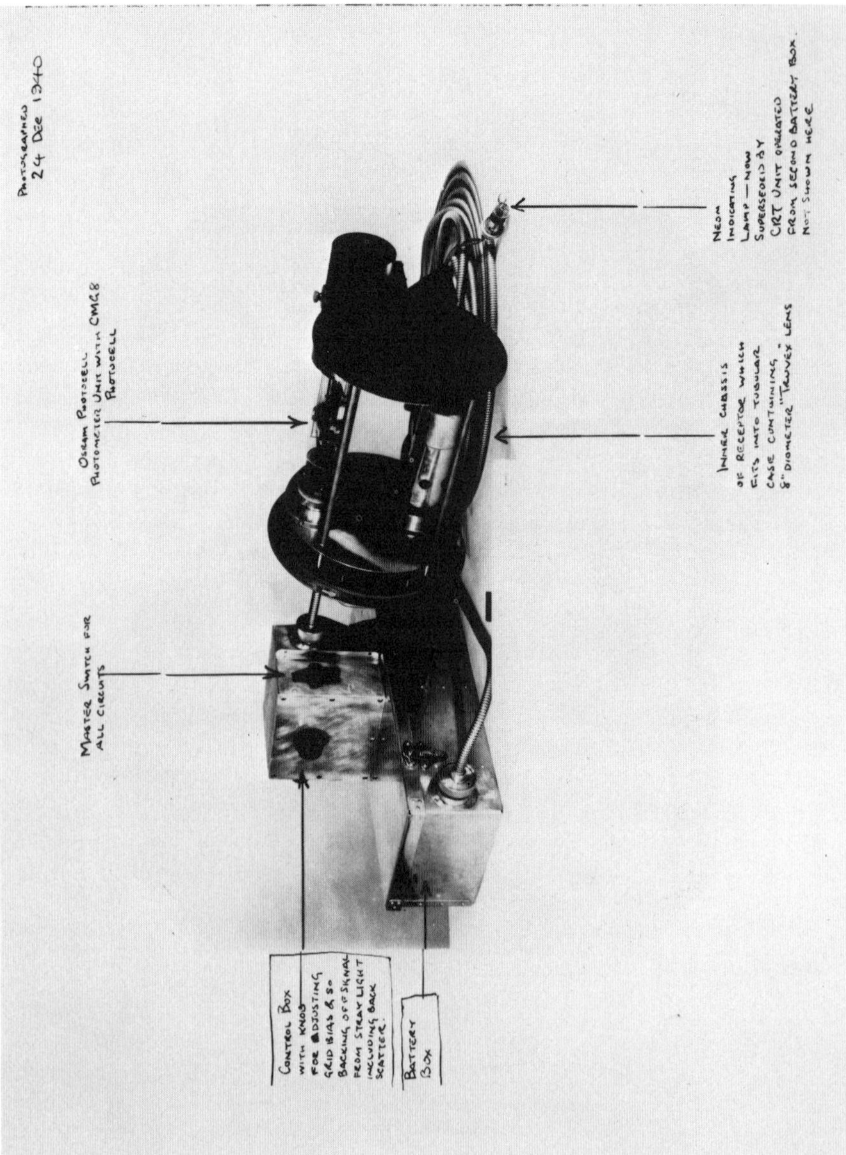

Fig.5 continued.

a mare's nest. Their valves are the same as ours. I fancy the cold feet have been warming up somewhat, as the result of our efforts to get people to keep their balance—and I hear today that Robinson working at Rugby has had consistent service from his magnetron. I fear that the real authors at Swanage are too self satisfied to learn the lesson.

February 7th

Staff Sergeant Moore and his Major (Clarkson) here at my suggestion to see how their scheme can be studied and demonstrated using the Waldram projector technique. Very hopeful.

Mr. Paterson and Mr. Stevenson from London central labour group to see our training unit and incidentally the Labs. They were appreciative and helpful. I gathered we shall be wise to get in as much of the labour we are likely to need as soon as possible, in view of the tremendous efforts and shortages ahead.

February 8th

Suggestion from Oliphant that Megaw should go to America for a quick visit. Nice of him but I shall have to speak out!

February 10th

Lywood's meeting—at which there seems to be a good deal more for us to do. People seem very pleased with the 10 to 45 cm searching set, the Admiralty having the first one now at sea.

Lunched with Capt. Willett (Signal School) who wanted advice on a staff matter.

February 11th

A quiet day when everyone could get on with their jobs. Glass for silica seals seems to be under the weather again and Partridge is having an anxious time—so are we all.

February 12th

Helmore meeting at Air Ministry. He is putting all his faith in 1800 A arc being worked on by Cameron with the Witton people. Lucas has an order for making the lamp to work these carbons. I fancy Witton feel they would prefer to make the lamp themselves. I am afraid Helmore is courting a long delay in getting his device properly tried.

February 13th

I saw the Helmore group working their 1800 A arc today. It is a fearsome instrument.

Most of the day on CVD meeting at Wembley—the morning in informal discussions, and the afternoon three and a quarter hours meeting. A hard but enjoyable time—and I find Oliphant agrees with my representation of the general position in my letter re. Megaw's visit and the cold feet constantly exhibited by Swanage whenever they find themselves up against an enigma.

February 14th

Lunch with Gant and B.P.D. The former has taken over Lamp Works management under Chelioti.

Air Ministry Signals are after a tunable transmitter which we thought several months ago was no longer wanted.

February 17th

Marris and party back from Swanage. Espley's handling of the 'hybrid' arrangement in the gear seems to have created a very good impression. We do not come out quite on top as regards the receiving unit, the TRE being some dBs better than all comers. The gear which is a mixture of Swanage and Wembley will be flying its trials shortly.

Have regular meetings on Monday mornings on progress with our valve and CRT production groups. We are getting on.

February 18th

CRT meeting today—all the morning. Walmsley in the chair and announced the change of status of the meeting which occupies the same relation to the anti-valve committee as CVD in valves. It is an excellent arrangement for all concerned but the irregularity of having it centred round GEC Research Lab. service may be called attention to sometime.

Spent afternoon with Humphreys going over definition of work of each member of scientific staff for the Ministry of Labour—afterwards designing with him the organisation to come into force when the Labs. are blitzed.

February 19th[153]

Winch's gear worked in the air again last night and repeatedly picked up the enemy at 300 to 400 yards. His CRT was giving trouble and the gear must be made more robust.

Urgent message from Ure of Signal School that order is coming for 500 VM tubes and 75 urgent by end of March. I ventured to ask if Ekco was being brought in but it was indicated that they preferred the GEC!! What a change since Signal School and Ekco could handle all that is necessary, and GEC were far too busy to take on anything more!

Megaw had a good time at Rugby introducing his large magnetron for their GL work. Machin goes down shortly on modulators—for BTH thyratrons and Wembley gas triodes are running together for first place and probably two types, unless fully interchangeable, will not be tolerable.

February 21st

Second anti-valve committee this morning in which the members very definitely intended the committee to exert its functions. Gossling's 'Magna Carta' is coming into its own and is to be kept as the running commentary of the valve development art. Lunched with Walmsley and I am afraid talked scandal, but some situations cannot be taken too seriously.

Jolley had a ring from Helmore that flying last night the equipment exceeded our highest expectations—so *that* is excellent. They spoke of a range of a mile.

February 22nd

There are more members of the families of the staff joining in our valve making—Mrs. Megaw and Miss M.Dudding. But it is very difficult to keep a stable staff of girls—they (or their parents) give no consideration to the importance of their sticking to their job in the national interest.

February 24th

Lywood's meeting. They are wanting generators, receivers and wavemeters for the whole range from 10 cm. to 5 metres and we have agreed to chart the whole wavelength range and show what is and is not available.

Whiddington lunched with me and I had to try to persuade him that centralisation of CVD priorities, orders, distribution of samples and the like is not likely to be of help. I wish Whiddington, who is an excellent chap, would master the valve job better by getting information direct from us, instead of living on his imagination and other exaggerations at meetings. I don't think he liked what I said.

February 25th

Helmore's job has gone into the air with excellent results albeit the 'enemy' was using his navigation lights. We are to go on with our 1800 A arc nevertheless.

Marris back from Swanage. The teams are working at Christchurch and meeting the expected difficulties in fitting things into planes.

We now have definite instructions to make not less than 600 of each of five types of valve in addition to the E 1061 which is now going well.

February 26th

An MOV service meeting in the afternoon when we displayed and reviewed all the live valves (some of them very live) which we have in hand. A most impressive range.

February 27th[154-155]

Saw Joubert at his club and pressed on him the promising nature of our IR (F—N) tracker arrangement and of Helmore's scheme. I think the talk did good.

Leeds and Eden are doing excellently with the diamond drilling after the former had proved that the stream of diamond dust in oil and its impingement into the bottom of the hole is the vital factor. He has got a die down to 0.0005 in.

O.W.H. has passed me his emergency plans (based on Brown Bros. scheme) for when we are blitzed.

February 28th

Whittaker from Nash & Thompson here to discuss actual position and next steps regarding the infra-red application. Whittaker is most sanguine and feels the device must be 'sold' to the pilots energetically so that best tactics can emerge from comments and experience.

The party who were at Signal School yesterday to see the new gear for which our latest valves are needed came back impressed with the seriousness of our responsibilities.

March 1st

The laboratory staff has now grown to 666, mostly by the increase of girls and student assistants. The increases in the scientific staff about balance out those who have left for the Services or for other reasons. Our plans for small scale production look as if they are on the low side for meeting requirements and we shall probably have to expand them by another 50 girls before we have done, but we are suffering badly for want of more workshop output which makes most things go slower than they otherwise would.

The battle of the magnets and magnetrons is going as we knew it must. The staff of the Establishments (SS and TRE) are saying that they have no use for this idea of supplying magnets with magnetrons and I am afraid Whiddington, who has been the mouthpiece for this scheme, must be feeling things rather. Magnetrons are being used and are giving good results and we hear nothing now of alleged vagaries—which is also what we predicted. But the Bell Labs. still claim 40 to 60% efficiency for their magnetrons against our 10 to 15%. At present this matter is a complete enigma.

Gavin in a state of fury having had a message that some inquisitive young gentleman (so far unnamed) at Swanage has opened Gavin's 10 to 45 cm receiving set (which he has struggled to calibrate and complete during the past fortnight) and ruined it. We will see how true the report is before making charges of amateur irresponsibility.

March 3rd

To Manchester for the second meeting of the diamond die control technical committee. Inspection of wire drawn dies and Frederick Smith's factories in afternoon and Committee meeting at Midland Hotel in the evening. A cheery and useful meeting followed by one to BI Prescott on Tuesday. Mr. Doyle joined in and is a great help owing to his inside knowledge of the diamond industry. There was throughout a continuance of the very frank exposé of techniques and practices by the individual members which was started at the first meeting.

March 4th

Afternoon at Shaw factory—a most excellent layout promising to be the best valve factory in the country. This is mostly due to Lindridge and Hall, under Wade's direction. Certain types of valve are beginning but it will be a long time before it achieves full output.

March 5th[156]

Went over some of Cobden Turner's new factories: (*a*) Warrington, (*b*) Haywood (Times Mill) (*c*) Sapphires. There was a meeting yesterday at Salford on powders, the factory process for which is not going well. Dudding, Marris, Chirnside, Smith, Polgreen and C.T. [Cobden Turner]. I believe a programme of work was arranged which is going to be stuck to this time.

March 6th

Had visit from Air Marshal Sholto Douglas, C-in-C Fighter Command, to see Helmore's lamp. I urged him to give equal attention to the IR solution being handled by Frazer-Nash. He said he at present had a distinct preference for Helmore's solution but I learn from Frazer-Nash that he has got his priority for an AI to be fitted to his machine—which is the main thing at the moment.

March 7th

Arranged to pass over the CVD etc. radio secret work from Miss Burbridge who cannot any longer tackle its increasing volume. Miss Fildew is now managing all the orders for valves of which there are many from the various Service establishments. There was a considerable tangle to unravel before the whole thing was straight.

Gavin has most appreciative report from sea of the performance of his 10 to 60 cm receiver. Very encouraging.

March 8th

Visit from Whiddington. Our conversations with him these days are usually stormy. As B.S.G. [Gossling] remarked, if he would try to function as a liaison officer assisting collaboration instead of a co-ordinating executive, things would be much happier.

March 10th

A very active day with visitors and internal happenings. Ministry of Labour here to examine skilled craftsmen staff lists. Our effort will shrink to the extent that men are taken from us. Whiddington and Walton here from DSR to see how we are organising production without detracting from our experimental effort. We are going to have a job to stick to all our present staff as reasons for leaving and taking on other jobs are frequent and people don't always put the national claims before personal preferences.

March 11th[157]

'Anti valve committee' (TVC) from 11.0 to 5.0; Harley back from USA. The sub-committees on 'preferred lists' and 'standardisation' are progressing well. We spent most of the time on the Valve Development Chart using it to get all our CVD valves reviewed from the point of view of the committee. There is a rather clear antagonism between MAP and the CVD machinery or at least some of the individuals.

Silica seals are still not properly on their legs and Partridge is now making a zirconia-brick small tank to avoid some contamination of the glass from molybdenum pots.

March 12th

To Diamond Die Control Advisory Committee where I reported the technical position, urging that the only way to make any progress was to order some GEC drilling machines for general use and get them into use quickly by any who are wanting to drill. Saw Watson Watt at lunch who wants to come and see the Nash and Thompson GEC IR etc. gear.

Helmore says MAP is ordering 100 of the sets we have been developing for him.

March 13th

All day on visit to Marconi, Chelmsford, with Gossling, Megaw and Gavin. Admiral Grant, Managing Director, was very warm in his expression of a desire for GEC and Marconi to work in closer collaboration than in the past.

Heard account of yesterday's meeting from Megaw and Gavin. At the 10 cm meeting the myth of the magnets and magnetrons which worked up such a storm over Olympus was dispelled, to the discomforture of those who had put more weight upon those who ignored our advice than on those who had confidence in us. I fear it was shown that Whiddington had backed the wrong party again. On most matters things went satisfactorily for the Labs.

Gavin reported that Air Ministry want GEC to fill up all the gaps in transmitting, receiving and wavemeter sets—quite doubling the present programme of work in hand.

Jolley is getting very busy with the rapid application of his job to night bombers.

Marris back from Christchurch. The AI set has flown twice and operated satisfactorily. But range at present leaves something to be desired. Our various chickens seem to be coming home to roost!

March 14th

The Ministry of Labour examination of our staff was completed today. Humphreys and Gardiner have handled it very well and the two representatives seem very satisfied if not impressed.

Everyone very pleased that 17 bombers brought down during the past two (moonlight) nights.

Salford [iron] powders to the fore again and B.P.D. has been getting statistical results which are somewhat devastating to the scheme of observation and tests from which deductions have hitherto been made.

March 15th[158-160]

A Saturday afternoon visit from Sir John Salmond, Air Comm. Whitworth-Jones, Air Marshal Hill, the Parliamentary Sec. to Balfour, Helmore and others, to see the Helmore equipment. Had a most useful talk at tea time in conference on the tactics of the business in relation to AI.

Jolley and his team on this are doing a good job of work and will I hope keep critical of practical possibilities. Lockspeiser and Green from

Farnborough were there and did not seem over keen I thought on making the scheme go.

Dudding has made a rush visit to Salford about their powders. His statistical analysis of their records seems to show that little weight can be placed on the basis used so far for discriminating between good and bad production batches. Dudding is also very active in looking after the new women members engaged for our increasing production demands. My wife has now been working in one of the groups for four months (November 1940) with Joan to take three days a week. Mrs. Megaw has started as a pumper. Perhaps we shall get a large number of wives of members of the staff.

Partridge has not been having a smooth passage with glass for silica seals in which we make at present only a hand-to-mouth delivery to MOV and Shaw. But trials he has made this week with a small tank furnace of zircon bricks has made some good looking glass—and he is now much more hopeful.

March 17th

My internal production meeting in the morning had some disappointments—but the sort of teething troubles which always come. Pearce and all of them are working with a will.

Gavin's and Bligh's programme on AB, extended to fill all the gaps between 10 and 250 cm in wavemeters, receivers and transmitters, seems possible to fit in and we have received ITP for them.

Sent in CVD account for June to December amounting to about £1200.

March 18th

Talks with Wells, Heather and Clark re. greater friendliness with Marconi short of disclosure of non-war equipment.

Lunched with Harley of MAP just back from USA. He was somewhat damping as to possibilities of pulling together over there—an interesting comment on the present atmosphere of elation here and in the USA over Roosevelt's speech bringing USA definitely out of her neutrality.

March 19th[161-162]

Went over with all concerned the prospects of our making all the samples of experimental valves asked for by the different establishments. Found the position not so difficult as we feared and it looks as if our mechanic strength would meet the present needs within reasonable time.

Espley back from ADEE Christchurch where Dr. Black is taking a serious interest in our 25 cm valves and gear and is naturally struck by its relative simplicity compared with the elaboration of 10 cm with its spot wavelength characteristic.

Winch tells me that he has prospects of doubling his range on the GEC—Nash & Thompson device which is wonderfully good news.

I saw Leeds first 0.0004 in diamond die today and he seems to know how to make them like this every time. He has one or two more only awaiting breaking through. The first of the new machines for drilling is out of our shops but the 25 machines from Bryant Simons will take several months yet.

A good day when many things seem to go right.

March 20th

Dudding and Chirnside report the making of powdered alumina which appears at last exactly to repeat the Swiss material needed by British Jewels for making sapphire boules. It now remains to see how it can be repeated on a production scale.

Ryde reports success with his electronic arrester for TRE and it would seem that they can now be made in quantity by Hammersmith.

Air Marshal Joubert came with Helmore for yet another demonstration. I think he was pleased with it. He has definitely been driven to the conclusion that AI needs something to help at the end.

March 21st

Partridge's silica seal glass seems to be all right and he has just founded a second batch up to 2000°C—a striking achievement. This gives a 12 month's stock of glass.

Examined Winch's IR gear in projection room which is ready for expected visit from Watson Watt on Monday.

Spent much time trying to clear up difficulties caused by Swanage paralleling with Randall our request to Megaw for an intermediate magnetron. What a lot of trouble these people cause by doing things the wrong way.

March 22nd[163]

Had a nice note of congratulation from Rowe (Swanage) on the result of our joint AI effort (Fig.6).

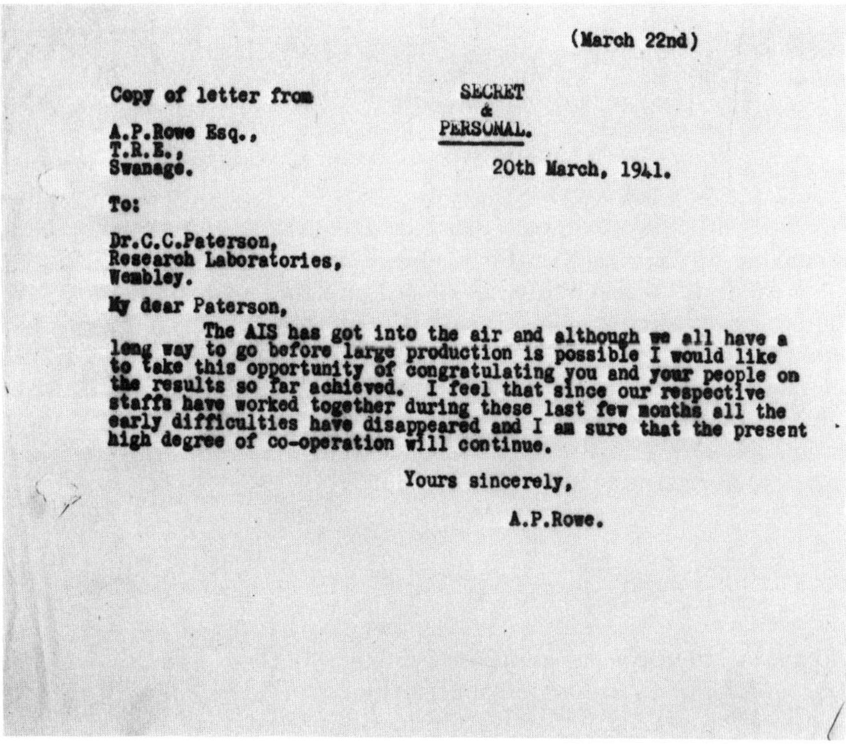

(March 22nd)

Copy of letter from

SECRET
&
PERSONAL.

A.P.Rowe Esq.,
T.R.E.,
Swanage.

20th March, 1941.

To:

Dr.C.C.Paterson,
Research Laboratories,
Wembley.

My dear Paterson,

The AIS has got into the air and although we all have a long way to go before large production is possible I would like to take this opportunity of congratulating you and your people on the results so far achieved. I feel that since our respective staffs have worked together during these last few months all the early difficulties have disappeared and I am sure that the present high degree of co-operation will continue.

Yours sincerely,

A.P.Rowe.

Fig.6 Copy of letter from A.P. Rowe, Chief Superintendent TRE, to Clifford Paterson, dated 20 March 1941

Ryde reports steady work on silicon crystals for reception but we must press forward production of our best quality material which requires the high frequency furnace. Various people are asking for it, and DSR authorises our work to be included under CVD. Detector crystals are needed for Dr. Dick's work.

March 24th

Lunched with Dr. Railing and talked over our work and the prospect of our taking up Hermsdorf insulator business and product.

Mr. Watson Watt came to examine the GEC.—Nash & Thompson equipment. Whittaker was present. I thought W.W. seemed somewhat subdued but he grasped the importance of the gear and said that it had been formally put into the programme of Dr. Pyke, DSR, Air Ministry.

Everyone anxious over what Yugoslavia will do. It looks as if she was going to buckle down to Germany like Bulgaria. Raids on ports and convoys have started on a heavier scale.

March 25th

A whole day on CVD meeting at Wembley. We had about 20 present and a useful meeting. The expedition to Wembley seems popular. Partridge's glass was admired but no one appreciates (but Hughes) what it means. The men have made 100 seals from it with no breakage.

March 26th[164-165]

To Lywood's meeting in the morning—but having thrown all their requirements onto us a brief indication of progress was all that was needed. The great desideratum is for the elimination of spot wave devices below 40 cm CW. Gavin and Bligh have organised the work between them and Coventry very well with a minimum of demand on our own mechanics.

I wish the authorities would hurry up and 'schedule' us so that our girls cannot get unsettled who are engaged on production jobs on valves.

Some of our 'experimental' valves like E 1231 (Warren's little valve), E 1190 (oxide milli-micropup), E 1209 (small gas triode) and others are being demanded in numbers which makes them also 'production' types.

March 27th[166]

A day of good news. Yugoslavia has thrown out her Quisling. Fall of Keren and Harrar. I hear from Walmsley that there is another 'flap' on the question of not enough investigatory work being done on magnetrons. I lunched with Warren of BTH today and had a very frank and pleasant talk ensuring that we hang together on our programmes of work on large magnetrons. Spent the afternoon with H.H. and Gamage going over the laboratory programme of work. The amount of our Government development contracts is of the order of £80 000 this year.

March 28th

Espley has passed me his memo on our 25 cm RDF for GL and such like purposes, showing that it has advantages apparently on most points over 10 cm. The trials have been made at ADEE Christchurch. It appears at first sight that 10 cm for GL may have been a rather hasty decision.

I gather from certain quarters that strings are being pulled to start a separate Service 'valve studying' establishment for 10 cm and below. How these fellows try to make jobs for themselves!

March 31st

The Epidiascope has revived again and Stevens is busy with a new scheme. I don't think it is for exactly the same purpose. A contract is to hand for a first model. Most of the CVD people seem to be travelling to the various establishments these days. These visits do much good and save many tragedies by catching up uninformed use of valves before it is too late.

The financial year ends today. It will reveal a curious financial situation, on account of the large amount of money earned from our charges for development contracts for the Government and for products made and sold at the Laboratories. I hope our overheads are sufficient on the Government work.

April 1st[167]

The news is magnificent of the successful action against the Italian fleet—six ships destroyed in night action. Saw H.G. Jenkins's group's work on 'settling' process for CRT screens. It appears simple and sure, and it appears possible to extend it to sulphides.

To lunch with Merren of the Admiralty and Mr. Carrell Wilson of USA Research Corporation (MIT) to see if we can get going on more rational and intimate contact between GEC Research and the Bell and

RCA Labs. in America. Merren mentioned one case in which RCA passed a question to their Research Board. The latter passed it to Dr. Bowen (of British purchasing commission); Bowen to Cockcroft, Cockcroft to Whiddington. Whiddington asked Merren to pass it to GEC Research who are the only people who can give the answer!

The question appears so distorted that Merren asked for the original telegram, which is being looked up. I understand it concerns means of getting sufficiently low in wavelength with the micropup.

Jolley tells me that the Helmore equipment was flown again on Sunday night and went as all expected it would. The 'enemy' plane crashed on landing but [the] Wing Commander, who was observer, escaped with a knock on forehead.

April 2nd

Brass hats are beginning to buzz round the ROF [Remotely Operated Fuse] job on account of its promise and want the tempo increased. Dr. Vick is coming to see me on Monday about the job generally. I feel the battery to be the more doubtful part, but Benjamin is keen, as usual, to come up to the scratch.

April 3rd

Megaw, Gossling and others back from visits to BTH (10 cm GL) and Oliphant. A useful time and discussion. Megaw and Randall are agreeing a joint memo on 'the next stage of magnetron development' covering higher voltages and fields which give expectations of higher efficiencies up to a max. (for the E 1198) of 15 000 volts and (?) 1200 gauss.

Both the magnetron job and Benjamin's ROF jobs are at present beyond the ability of their groups to tackle the task so have arranged for Donaldson to join up with Benjamin and Balls with Megaw and Willshaw.

I hear that a '10 cm and under' school is to be concentrated at Oxford—no details.

April 4th[168]

Anti-valve committee in Walmsley's office this morning and lists of preferred and black list valves approved. The work goes well and certain CVD valves are now passed to TVC.

Walmsley tells me that there are to be a dozen AIS sets made and the question arises can Spon Street make them.

I believe Coventry (Bradford) official attitude is unsympathetic but it seems very desirable.

With Dr. Railing and Cobden Turner over a proposal by the Ministry of Supply that TMC shall put down plant for making iron powders when Salford have all production capacity required.

April 5th

Saturday—Sir Henry Tizard and Helmore, to go into the equipment and its use. Tizard as usual very practical and to the point. Two lamps tested out (Lucas and GEC). Both stood up well—but it did not transpire how sensitive are the adjustments and carbon settings and I asked that this should be taken account of. I left the whole group with Helmore discussing the amalgamation of the best features of both lamps into an 'ATI' model.

Tizard very interested in our report of Winch's infra-red equipment and is coming again in a week's time to look round at what we have.

April 7th[169]

Visitors all day. Dr. Vick to discuss his work in its broader aspects. He seems most appreciative of all we are doing. Tells me that ADEE establishment is to become a research place for Ministry of Supply under Cockcroft as Senior Superintendent.

Couch (from Admiralty) came and also C.S. Wright (DSR) to see Waldram and Stevens demonstration of 'diffused lighting'. Very successful visit and fairly promising development. We shall probably have a development contract for it. Wright also mentioned real success of the Signal School submarine job and wants a still larger output of valves from us which we are already stretching ourselves to make. Simultaneously a message from Signal School on same lines but emphasising that 'anything DSR suggests is likely to be not ambitious enough'!

April 8th

To Swanage for an evening's talk with Rowe on things in general and to ensure that all is well between our two establishments. I found that all is as well as can be expected in this world of varying human nature—but I asked him to get his people to talk over difficulties with us before ventilating them before more august audiences. I can see it is difficult for them but I instanced the E 1198 magnetron and the magnets each of which had to be calibrated with its valve—and the waste of time which this folly caused. I heard in detail of the GCI [Ground Controlled Interception] and ELSI [Searchlight Control]. The former seems to be doing its stuff well—11 planes down last night.

April 9th

On to Parkeston [Parkstone?] (Bournemouth) for diamond die meeting at Simpson's works. A very pleasant time. The meeting recommended that the control should get a number (say ten) of our machines made and in use as soon as possible so that those could start drilling who were not yet doing so.

April 10th[170]

To ADEE Christchurch where Dr. Black spent the whole morning showing me around. I had a very useful talk with him on work in general. He is keener than ever on the 25 cm proposal for GL and says Cockcroft also wants it to go forward. So we are likely to have a development contract for it—but I presume ADEE will mainly handle. He is most anxious to change the crystal for a diode mixer.

April 11th

Coventry blitzed again last night and I hear that there will be delay in getting on with the AB and AIS jobs. Very sad if the delay is to be prolonged.

Examined the CRT group's testing station of which Jacob has been the centre figure. An extraordinarily good job and very elaborate—we must show it more often to interested visitors.

April 12th

Easter Saturday. We are only closing the Labs. on Easter Monday this year, at the Government request—but I doubt the wisdom of so short a holiday. Latest news from Coventry is that the damage to some of our 'AB' radio work for Lywood may not result in so long a delay as we feared. Peters will be able to tell us more in a few days time.

April 15th[171]

Dr. Bainbridge of USA visited us on Friday and Saturday. There was so much for him to see and discuss that he only covered a limited part of CVD. We got more out of his visit than out of all the contacts, indirect and direct, we have had since the war started. We are afraid information comes over which is deliberately withheld from Wembley—why, I don't know. Saw Dr. Railing and Cobden Turner re. proposal by TMC to make iron powder for radio cores with Government funds. GEC objects and we go to a meeting tomorrow with the Ministry.

April 16th

CRT meeting all the morning with Dr. Walmsley in the chair. It was arranged to have a standardisation sub-committee and to make provision for reference to the committee of experimenters who are seeking the most appropriate tube for any purpose. To meeting with Ministry of Supply, Salford, STC and others on iron powders. No decision on the vital issue of whether TMC should be subsidised to make iron powders.

April 17th

Severe blitz over London last night but we seem all to be prepared now for this sort of thing.

Rycroft, working under O.W.H. has done a good job in discovering why German bomb clocks can be stopped. The investigation was put up to several laboratories and we took the prize.

Saw Warren about the various new receiving valves he has in hand; 10 cm diode mixer, alternative for the USA grounded grid triode etc.

Have been going into the costs of various valves made at Wembley. We have made 10 000 VT 90 valves which may be worth £40 000 depending what figure is fixed by Hammersmith on the basis of their manufacturing costs.

April 18th

Helmore, with all MAP influence behind him does not seem to be able to get a machine fitted with AI for a month! He is appropriately angry.

Whiddington spent the afternoon here looking over our experimental valve production—he is being hounded on the accusation that the Admiralty always gets prior treatment by GEC Labs. Today's investigation did not show this to be the case. The 'grounded grid' triode or a GEC equivalent for AIS are being prominently pushed just now.

April 19th

Had call from Sir P. Joubert 'can we anyhow hasten the making of the 10 AIS sets it has been decided to proceed with'. Dee appears to be restive, not realising how long these things inevitably take. I am afraid with their recent blitz, Coventry will not be able to do better than stated.

April 21st

Have come to conclusion that we must give up pretending that we are keeping a patent department going. Chapple and Peachey must be absorbed into the technical work of the Labs.—leaving only Campbell and Cornock to run patent department.

Found review of work of our valve production groups full of disappointments—mainly due to difficulty of getting women trained to the more difficult operations. Very big outfalls at present and only a few good valves materialising.

Stevens and Turner to Glasgow re. 'diffused lighting'.

April 22nd

Campbell and Bligh on a curious mission to appraise rival tank radio equipments—at least so it appears from telephone information.

Meeting today on construction of modulators of various types needed for our valve testing. Every problem eventually boils down to 'how can we avoid having to ask the workshop to do anything'!

Strong pressure for the Helmore job and equipment of planes.

April 23rd[172]

Long 10 cm meeting at MAP which went better than might have been expected in view of our difficulties with the various magnetrons we are making.

Had Campbell's and Bligh's report resulting from a request from Sir Andrew Duncan that Campbell should help to adjudicate between a Pye tank radio set and a SEE (Signals Experimental Establishment) set. A somewhat delicate mission for NRC which I think he executed well.

April 24th

Learned from Whittaker that the IR gear in its plane is being used for trials and training of personnel in the tactics of the affair. I understand that Air Marshal Joubert and Air Chief Marshal Sholto Douglas have had a favourable conference with Captain Nash on the aerodrome.

Dudding has been holding an internal conference on glass breakage in valves and I hope it is going to lead to a measure of clearing up of causes of wastage in some of the valves we are trying to get into tentative production. B.P.D. is carrying a very big load and seems tireless in his maintenance of effective lab. and works collaboration where this is needed.

The news is none too good with the overrunning of Greece by Germany and the renewed threat to Egypt.

April 25th[173]

Mr. Railing (whom we are glad to see back in better health) held meeting to see what can be done to foster luminescent tube lighting in view of revolution in USA lighting. Something needs to be done and I hope we may induce George Wilson to go to America for the GEC on this subject.

Had SOS from MOV whose making of VT 90 (E 1046) valve is not going well. Delivery demand by MAP has become critical and Smith wants one of our people for a month!! Somewhat trying in the middle of our own troubles.

April 26th

It looks as if we can help MOV with their VT 90s by helping to train their rather raw supervisors. Pearce as usual is practical and helpful.

April 28th[174-175]

Review of our valve production groups shows a more hopeful outlook.

Hawkins reports that ten rush sets of Helmore's equipment are to be put through under his development contract.

Winch reports the possibility of a 350 watt IR equipment of about 300 to 400 yds range and he will work on this as his next compromise. This is very promising.

Heard from IEE (Beard) that I am nominated to an Engineering Advisory Committee of the Privy Council. I should not have sought this post but, as it has come, B.P.D. and also M.R. and A.R. advise acceptance.

Attended meeting at Ministry of Supply to consider compact source mercury lamps as substitutes for searchlights. BTH and Siemens were there and we showed our usual ability to collaborate together. Our loyalty is mainly to the Admiralty.

Churchill's speech at time of evacuation of Greece has made everyone realise what they are up against.

April 29th

Spent morning on annual Lab. accounts and allocations. It looks as if most would be smaller than usual—but until overheads are reckoned it is impossible to predict in a year such as this one.

Received extension of AI contract but the sum does not appear enough for this year (£20 000).

April 30th

Sir Henry Tizard called, at my request, to discuss with us Winch's infra-red equipment and the most useful lines of further development. He was helpful as usual and advised concentration on long range.

Helmore rolled up to talk over the progress of his gear which is having strong backing and makes us feel we are engaged on something well worth while.

To Bradford in the afternoon for a consultation on the RCM and AIS work which we want Peters to do at Coventry Spon Street Works. T.W.H. is joining Bligh, Gavin and myself.

May 1st[176]

A useful day at Bradford with Wells, Clark, Peters, T.W.H., Bligh and Gavin. Settled satisfactorily the policy of production and development particularly in relation to Peters's Spon Street sphere of work. The work they are doing with us for Air Ministry and Admiralty not seriously affected. The recovery they have made from their blitzes is an extraordinary achievement although the complete chaos in their small tools and jigs due to the fire is going to cause serious delay. Their second blitz was about three weeks ago.

May 2nd[177-178]

Went over the laboratory work with Mr. Railing and gave him a picture of our activities with the Services on our various development contracts.

Lunched with Lord Falmouth, Beard and Fleming about proposals to be put by us before Lord Hankey's Engineering Committee. We are not very definite—certain floating ideas, that's all.

May 5th

Heard indirectly that a meeting on infra-red at Wallop aerodrome (Lockspeiser being present) was in danger of turning down Winch's device. There is to be another meeting however, and I rang Lockspeiser to get him here on Wednesday to see the doings properly.

Waldram back from Antrim where at last, and not without adventures, he got his flights and his photometric measurements at 18 000 ft. The difficulties of photometering under these conditions seem to me almost insuperable. (Fig.7).

A long anti-valve committee [TVC] in the afternoon at which the meeting really faced up to the question of the tentative manufacture we are doing at Wembley and its consequences.

May 6th

A four hour's CVD meeting with 20 present and lively all through. Much insistence by Signal School that certain valves must be approved for their use and obvious tilting against TVC. The enigmas in the irregularities in resonant magnetrons are most intriguing. We seem to be hunting them down but are most interested in what they are going to turn out to be due to.

The velocity modulation tubes which Sutton, with Signal School support, declared to be such 'an easy manufacturing proposition' are turning out very troublesome although Signal School, EMI, GEC and Ekco all get the same wide spread.

May 7th

Had Lockspeiser and Hill here to study Winch's device. He had come in with a sad tale of previous night's trials when apparently vibration had set everything adrift and the gear failed to work. He was truly downhearted, and I confess to surprise unless the machine is a

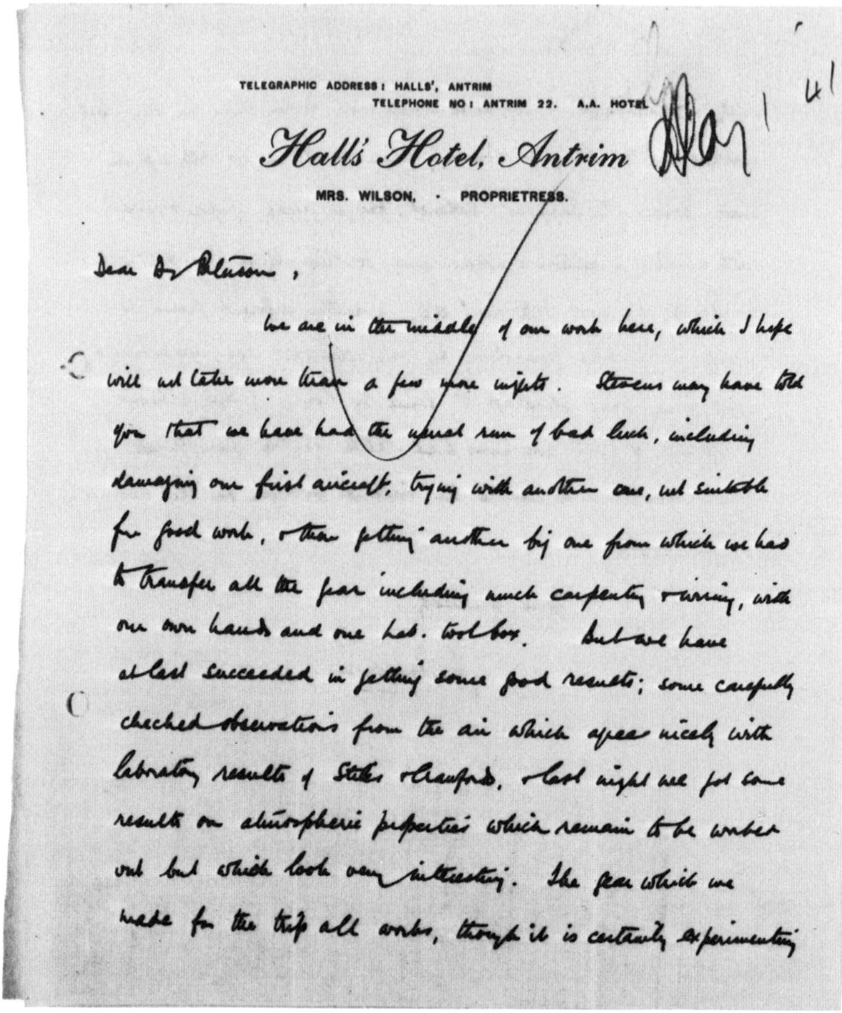

TELEGRAPHIC ADDRESS: HALLS', ANTRIM
TELEPHONE NO: ANTRIM 22. A.A. HOTEL

Hall's Hotel, Antrim

MRS. WILSON, · PROPRIETRESS.

May '41

Dear Dr Paterson,

We are in the middle of our work here, which I hope will not take more than a few more nights. Stevens may have told you that we have had the usual run of bad luck, including damaging our first aircraft, trying with another one, not suitable for good work, & then getting another big one from which we had to transfer all the gear including much carpentry & wiring, with our own hands and one lab. tool box. But we have at last succeeded in getting some good results; some carefully checked observations from the air which agree nicely with laboratory results of Stiles & Crawford, & last night we got some results on atmospheric properties which remain to be worked out but which look very interesting. The gear which we made for the trip all works, though it is certainly experimenting

Fig.7 Letter from J. M. Waldram to Clifford Paterson (May 1941) describing photometry flight trials.

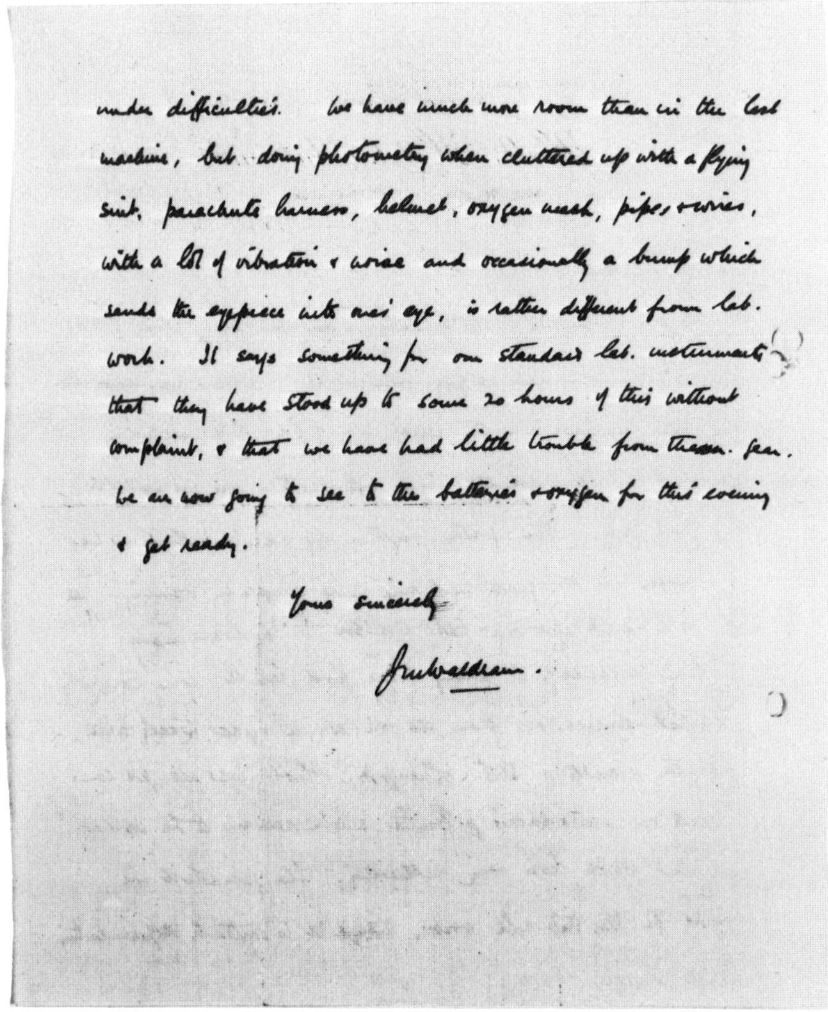

under difficulties. We have much more room than in the last machine, but doing photometry when cluttered up with a flying suit, parachute harness, helmet, oxygen mask, pipes & wires, with a lot of vibration & noise and occasionally a bump which sends the eyepiece into one's eye, is rather different from lab. work. It says something for our standard lab. instruments that they have stood up to some 20 hours of this without complaint, & that we have had little trouble from them. Gen. We are now going to see to the batteries & oxygen for this evening & get ready.

Yours sincerely

J.M. Waldram.

Fig.7 continued.

particularly vicious specimen. The visitors were very helpful and sympathetic and I think spent a useful time, and Winch handled his case very well indeed.

One of our hopes to explain the E 1189—98 mystery has gone squash!

Valve Works meeting in afternoon—very useful and helpful, but adequate production seems an impossible problem, so far as the new types are concerned.

May 8th

Preliminary look at laboratory accounts which are very different from the usual. Allocations to all Works and Departments are down—and overhead appears to be 180%. I hope this means that our overhead to Government departments has been about right this year but they will require careful scrutiny. Met Lord Hankey at IEE luncheon and found everyone expecting much of our labours.

Winch's IR job looks a bit sick from the policy point of view. Not sure whether Frazer-Nash is too down hearted to proceed! I trust the results will be judged on their merits, and not on the accident of the last trial.

May 9th

Annual accounts to the Board Meeting—very interesting. Gross sales of articles made £86 000. Overhead to GEC establishments drops from 2.2 to 1.8 about. Apart from this nearly all allocations are down.

First meeting of Hankey Committee which got down at once to bed rock on lack of engineering inside the Air Ministry establishments.

Went to see over a school building at Harrow as a possible alternative site for some of our work if we are blitzed. Whiddington tells me that formal instructions are on their way from the Admiralty.

May 10th

B.S.G. and Megaw report a good and lively modulator meeting at Rugby yesterday. Progress seems to be good and collaboration particularly so. Ramsay's small modulator E 1209 looks as if it would be needed in quantity. Another production job!

May 12th

Our valve production units are making gradual progress and getting their techniques more solidly established. The magnetron enigma is still unsolved.

I got Ryde to see a naval engineer who wants to use Winch's gear (modified) for a naval use. It looks as if we should have a further job.

Discussed with Gardiner how by some shift arrangement we can make our workshop machines go further.

May 13th

Winch reports a happier meeting with Lockspeiser, Frazer-Nash and others in which interest was shown in proceeding further with his equipment. The main trouble last time was slight lack of stability in the electronic tetrode valve. News from Salford that Polgreen has collapsed with serious breakdown and SOS that we will take responsibility for powder experiments. Must have conference tomorrow to settle what's to be done.

Searching for space in the laboratories where we can transfer production of VM valves. It is most difficult to find any other than in the gallery.

Large quantities of barium getter are being required by Shaw as a precaution and Britten is trying to plan an increase in production.

There is much speculation today at the news that Rudolf Hess has arrived in England as a refugee.

May 14th

Tremendous urge from MAP behind the Helmore job—three complete planes every three days asked for. The party at Heston aerodrome are having a job to ensure that the lamps always operate properly and I hear this evening that they have another day there tomorrow. Meanwhile Witton are not sure what to put in hand.

Had Shearing and G.M. Wright from Signal School at the Labs. to press for WT short-wave research since the RDF stuff is tending to crowd the other out. Fixed a programme with Megaw for which I hope we shall find staff and facilities for he and Willshaw are deeply involved with magnetron development.

Telephone communication is very troublesome as a result of Saturday night's blitz, and those are best off who can stay in their establishments and get on with their production or experimental jobs.

Decided with G.C.M. and B.P.D. to send C.G. Smith to Salford and to regularise the arrangement for helping Salford.

Mason is in for a long illness with his heart and is off to hospital. Must try to get another chauffeur for the time being.

May 15th

To DSR to talk over CVD policies generally. Suggested to him that I could help CVD policy committee if I attended—otherwise they had no one to help them on questions of valve making, samples supply etc. Found him in agreement with me that the proposed new valve research lab. at Oxford was a most unhappy idea. It seems to be fostered by MAP under an impulse to find a place to tuck away certain inconvenient people—but if founded it will certainly be fostered with some sort of going concern which will sap the strength from existing concerns.

Went to Heston Aerodrome where the Helmore group is struggling to make the equipment reliable—and listened to plans by Helmore for early equipment of 12 more planes.

May 16th

Saw Gossling after a tour by him of the southern establishments. Also had Walmsley to lunch at the Club, to talk over some of the things discussed with Wright yesterday. I have the impression from various sources that our academic-ridden radio effort is poisoned with intrigue, jealousy and uncharitableness. Industrial and commercial life seems clean in comparison. Had long talk with Walmsley asking for assurance as to future of AIS group. I told him frankly I have no confidence in MAP management if one or two straight men like himself were to leave.

May 19th

Went over our pre-production programme of valves with GEC Advisory Board, showing samples of valves. It formed an interesting story. Discussed with Warren and B.S.G. the grounded grid triode and the Clegg diode with a view to incorporating in our programme on receiving valves.

No news from Heston aerodrome and the Helmore party—so I hope things are going reasonably.

Am trying to get a closer tie up with BTH and Standards on valve development programmes so as to act as a more solid block upon which the Services can expend their waves of impulse and empirical effort. A regular meeting with BTH is indicated.

Fowler back from America will be here on Friday. Saw him at the Club.

May 20th[179-180]

Finished draft of a memo to the Hankey Committee and got B.S.G. to criticise. It emphasises the need to focus the research development design and production as much as possible in industry, for the sake of speed and reliability. Attended Hankey committee in afternoon when Tizard's memo was under review and the shadier spots in industry and the Services were probed. I hope my memo will be to the point.

Dudding tells me he has arranged with Warren for certain personnel to help with the new receiving valves. Lait with the grounded grid triode and Cutting with E 1232 and Boyland with the E 1248.

May 21st

More politics over the Helmore job—and had Walsh and NPL here over claims of Helmore lamp which are to be 'separately investigated by independent experts'. Jolley tells me that there are somewhat wild demands by operational people for the lamps.

Technical valve committee (anti-valve) all the afternoon and found confused thinking over preproduction of valves and responsibility therefore. Approvals seem to be going satisfactorily.

Benjamin and B.S.G. at ROF committee where Pye seem to have been lacking in collaboration by not informing us of important actuality trials. I understand they were quietly ticked off by the chair.

May 22nd[181]

Arranged with Jolley and Ryde that Francis should take on with Cameron some more fundamental work in Helmore's arc lamp so that we can know a little more about its background.

Bradfield here to worry about spark gaps which are not wholly satisfactory for all applications of T & R working. Arranged conference with B.S.G., J.W.R. and N.L.H.

Off to Stoke-on-Trent for meeting of diamond die technical committee. We have received the first two machines from Bryant Symons. Our drilling is going according to plan, and there appear to be several members of the Committee who can drill below 0.001 in. Conditions for optimum efficiency have to be determined.

The Laboratories War Weapons Week has so far realised well over £1000.

May 23rd

Meeting at Bolton's (Stoke-on-Trent) uneventful, but probably useful.

May 24th

Saturday. R.H. Fowler was here yesterday on his return from USA and had two hours at meeting with all concerned. They are getting on quickly over there but have not much to give us as yet. They seem to have discarded VM devices as competitors in almost any field. Oscillators and mixers appear all to be of triode variety. They seem to be in some difficulty in getting the micropup family to work down to 50 cm.

Winch has now a 'Mark II' infra-red equipment on paper and we are seeing Whittaker and Broughton about it on Tuesday. Must come in this afternoon to finish my paper work.

May 26th

Valve production meeting as usual. Some 1500 E 1232 valves required by SS [Signal School] and TRE.

Discussed the small resonant magnetron enigma with all concerned. They have been trying various changes but analysis by Falconer of 24 valves yields but slight clues. These are however being followed up and we hold another meeting in a fortnight to decide what to report formally.

Discussed arc lamp fundamental work with Cameron and functions of Witton and Wembley in connection therewith. I hope there will not be any rivalry.

To town to sub-committee of Lord Hankey's committee on engineering to meet BSI management for their views on present state of war standardisation.

May 27th[182]

Whittaker here with Broughton from Nash & Thompson to talk over Winch's proposal for Mark II of his gear. With modifications this was agreed—but Mark I with a 300 yard range must be kept a going concern.

Gossling, Benjamin and Sanderson were present at a difficult ROF meeting on Saturday when the chairman had to call Pye to task for not collaborating helpfully. Dr. Vick handled the matter tactfully. Pye have certainly wasted the time of everyone.

To Lord Hankey's committee where we questioned leading Admiralty engineers. I find their organisation good.

May 28th

CRT meeting with Services. There seems to be some outside criticism of the committee acting as a restraining influence to development. I think Walmsley has checked it, but I can quite understand those who fix the widest possible gulf between 'inside' and 'outside' the Services finding such a committee 'irregular'.

To Leedham about responsibility for preproduction and [who] urges that it should come under the planning committee for valve production. Whiddington who came this afternoon took the opposite view and wants it under CVD suggesting that if it is not, the Wembley Labs. would lose their independence and become a production controlled place.

Much trouble regarding the VT 90 production at MOV which is at present right down—but we think after studying the process and attention to grid size, it will rise again. Leedham full of threatenings about it, for AI and ASV sets are held up.

Very heavy fighting, and uphill, in Crete and Libya. Roosevelt's speech and sinking of *Bismarck* are the silver linings.

May 29th[183]

Dr. Kahn here to talk over Helmore matters. He supports the proposal for some fundamental work here. I mentioned it to Helmore who warmly backs the idea.

May 30th

Off from home at 6.00 am to Rugby to visit BTH at Warren's invitation to see their new 10 cm GL equipment. A very good job and most enjoyable visit.

May 31st

A trying incident with Whiddington who seemed to lose his equanimity because I could not free Megaw to attend a small meeting on magnetron fundamentals which he had called for next Wednesday. Megaw has an important series of tests next week and has to concentrate all his attention on it. I had promised Megaw I would not commit him for any other engagements that week, and had to stick to my guns. These fellows cannot do serious work and be called about to all sorts of conferences.

June 3rd

Evacuation of Crete has taken place. USA, Abyssinia and subjection of Iraq revolt are the bright spots. So far air power seems to mean more than sea power.

Had personal note from Whiddington which I hope means that the incident can be forgotten. Had note from Willett that Marconi Research Labs. are being brought into their valve research.

Hankey committee in afternoon. My memo on the incidence of responsibility between the Services and industry seems to be well received by other members of the Committee.

June 4th

Weather has improved and flights of Helmore and also Winch gear may now be expected. Meeting at Witton to discuss fundamental research on the former was a happy one.

Attended 10 cm meeting at which extraordinary statements were attributed to Dr. Black ADEE—about poor behaviour of large magnetrons and alleged 'suppression of an unfavourable report'. It was stated that the GL set at Rugby was being approved on the existence of only one satisfactory valve. Megaw spoke very well.

A sort of indignation meeting resulted with no one supporting the rumourists. I can't believe other than Black was a tool of some Quisling.

To Ministry of Supply (Cockcroft and Paris) with Espley to settle contract for 25 cm collaboration with ADEE. Probably £2000 per six months.

June 5th[184 - 185]

Have not yet heard from Mr. Craven who is touring the neighbourhood with an Admiralty man for alternative sites.

Had good discussion with MOV (Smith and Cosgrove) and colleagues from Wembley (B.P.D., R. Le R., A.G.P. and M.B.) on how we can help in the heavy job which MOV is tackling to expand its production (four new factories). I think we eventually agreed the process for passing over the processes of a preproduction group at Wembley to the factory: (*a*) by associating a given individual from Hammersmith staff with the preproduction from the beginning; (*b*) by contacting regularly with one of Davis's or H.[F.] Smith's men in consultation over tools and techniques. The man under (*a*) to go with the process to the factory.

June 6th

Talk with Lord Falmouth on how to make our proceedings on the Hankey committee more effective. We all feel that interviews with heads of the Services are leading us nowhere. I will see BEAMA and he will write FBI.

Lunch and heart to heart talk with Whiddington who told me of the criticism, mainly from TRE, that the service on valves is not ample and speedy enough and the difficulties he has to meet unfounded charges half way. He knew nothing of the action of Captain, Signal School, in bringing Marconi Co. into the valve picture.

June 7th[186]

Jesty reports very satisfactory visit to TRE on cathode ray tubes. He saw the alternative PPI on Scophony principles and found the liaison on CRTs very happy. Meeting on modulator interchangeability with BTH at Signal School yesterday went well.

Winch reported his gear satisfactorily installed and only awaiting change of engine of the plane. Lockspeiser is coming on Friday to go into Mark II with him.

Went to Haslemere in afternoon to see Shearing and Ure re. new Marconi valve activity. Under Bankes section at Baddow there will be only four men working so their programme will be very limited. Ure is confining them to start with 3 cm VM tubes. Baddow Labs. are being virtually taken over by Signal School.

June 9th[187-188]

At my usual preproduction valve meeting, the yield from the different groups was shown to have much improved. NC 11 and the VM tube are the only exceptions. Both of them are cathode ray devices and appear to have a difficulty all their own. Mr. Sutton from SS [Signal School] Bristol constantly affirms that it is an easy tube to produce but I am afraid this is an example of the amateur in the Government establishment dictating to the manufacturer and coming a crash.

Warren of BTH with Davis, Latham and another came over to discuss magnetrons and their manufacture—a very happy meeting and the first of a series. The 'discontinuity' enigma is as puzzling as ever for after making some 100 E 1189 valves with systematic discontinuities we unexpectedly turn out three perfect valves with none at all—and for no reason we can detect.

The Laboratories have now been 'scheduled' as a protected establishment —which means that no one may leave except under special conditions.

The Helmore job is going fast so far as equipment is concerned, but a long spell of most unfavourable weather is making trial flights impossible.

Ryde has a regular job on his hands with spark gaps for T & R working. They seem to be needed for several different applications.

June 10th to 12th

I have been laid up with a chill and out of touch with every day lab. matters. The regularisation by MAP contracts department of the Helmore expenditure still hangs fire and I am again urging a covering approval as an interim measure. The standing departments of Ministries do not like these red tape cutting adventures like Helmore's and I feel one must get cover whilst everyone is keen.

June 13th

Mr. Craven's activities with the Admiralty point to the 'Government Building' at the Empire Exhibition [site] as being the alternative site of the Labs. and on his visit to us today Brundrett was able to go into the matter with us and seemed to agree with it. He spent the whole afternoon on a first visit and discussed preproduction and its consequences. It was very useful to have him here for he is a helpful and well informed member of society. Gardiner has prepared a very useful 'flow chart' showing how the workshop organisation is utilised.

June 16th

The only production valve in the dumps is the NC 11. Something drastic will have to be done about it.

Winch reports that a flying trial will be made tonight. Jolley reports that all operational trials are suspended until a larger number of machines become available.

Hankey committee this afternoon at which we decided that we must probe in the industrial field if we want to find defects.

June 17th

CVD meeting again strongly attended, and there was plenty to talk about from 1 to 4.30 although no rabbits were produced from hats. The discussion showed the value of our sample making and preproduction of valves, but the pressure grows stronger and Signal School have a way of sending sample valves away to sea and then demanding production as if they were established types.

I have received an invitation to attend the CVD policy committee tomorrow through Whiddington, who implies that my attendance is desired in future as chairman of the TVC. This seems to me a very helpful move.

June 18th

I hear Winch's flight did not come off on account of engine trouble.

Announcement on the radio this morning on RDF spilling some of the beans. CVD policy committee at which there was a strong inclination to keep preproduction under CVD and TVC.

Peters to lunch at which he told me the inner history of the recent Coventry staff memorial [memorandum?] to the directors re. the management of Telephone Works.

To MOV re. new type valves for RDF and their production. I have MOV and Admiralty now conspiring to secure staff to learn technical supervision of the processes.

June 19th

Winch's apparatus went up in the air last night to check range, but one engine and other machine trouble after another prevented a proper flight and finally they crashed the machine and demolished the gear. So that is a sad conclusion for the time being of a very promising device ruined through continuous defective servicing of machines. There will now have to be a meeting to decide if they will take our range figures on trust and go ahead with the improved Mark II version.

Fowler came today and we had a talk about his American contacts.

The Lab. accounts for the year show some £78 000 of work paid for by Government departments and £118 000 by GEC works and branches. Our 'manufacture' of valves and other trifles amounts to about £74 000—a total of £270 000. A truly extraordinary result for a mere laboratory whose staff are supposed to be babes in such matters.

June 20th

Had Captain Ward to lunch to see if he can help with constructive criticism which will help the Hankey committee to probe weaknesses in the engineering effort. Spent an hour with Mr. Gamage going over the Lab. accounts and general financial background.

June 21st

Ryde tells me of interesting possible development with our compact source lamp pulsed in synchronism with GCI signals for giving intense illumination at the right point using the Scophony solution. Jesty also tells me of similar collaboration with TRE, also through Dr. Lewis, in which Hopkinson and his 'true representation' photography will probably come in to help with some of the night flying training equipment.

June 22nd

The Nazis have launched an attack on the Russians. What a situation!

June 23rd[189]

Had Admiral Grant and others of the Marconi Co. here for the day and, I hope, interested them in various things.

I hope I have got the man to take action who is responsible for the finance of the Helmore job. It looks as if it would cost nearly £100 000!

Hankey committee in afternoon at which Mr. Stannier [Stanier] gave evidence and we learned much about tanks and their production.

Am bothered about our failure to produce any satisfactory cathode ray tubes at the Labs. Production is down to one a week out of about 15 or 20 attempted!

June 24th

Long discussions on Helmore job with Jolley and others preparatory to meeting tomorrow with Dr. Railing on the general conduct of the work.

MOV management to meeting on our preproduction valves which are to be grouped under supervision of Davis and Smith. But these two are tremendously occupied with other responsibilities and I fear that their collaboration cannot for this reason be much more than a matter of form.

TVC all the afternoon. It is now probable that our preproduction will come under valve production Committee of MAP and that we shall have to face a serious expansion of it, but there is a meeting pending to confirm this policy after which we may expect lively times.

June 25th[190]

A well attended CRT meeting at Wembley at which EMI and Cossor research people were present for the first time. I had to leave after the opening for a meeting with Dr. Railing on the Helmore job. Upwards of 100 planes must be fitted and with Jolley as 'consulting engineer' with Helmore, the Witton organisation will function as with a normal contract.

I have to ask for the increase of our development contract to £82 000 to cover the experimental work plus the 'development' equipment of 15 machines for operational experience.

No sooner back at Labs. than a message from Sir F. Smith to see him at once re. VT 90 valves of which there is a critical shortage. Took R. Le Rossignol. There seems little doubt we must get back on to VT 90s at the sacrifice of NT 93 and 97.

June 26th

Dudding, Le Rossignol and Pearce had consultation re. VT 90 and have taken preliminary steps. We shall not shut down completely on NT 93.

Waldram is finding many demands for our expert assistance on visibility matters. The 'diffused lighting' goes according to plan and now 'flares' are coming up—probably an easily solved problem. It looks as though the Waldram—Styles combination is a monopoly one in this field.

June 27th[191]

At Diamond Die Technical Committee all day—meeting held at Rugby Control Offices and talk by Dr. Gordon. Progress is being made mainly by Leeds. He gets improved results by chemical clearing [cleaning] of the boart—so eliminating foreign matter which merely clogs. He is drilling so as to leave the hole perfect without special polishing. About seventeen 0.000x dies have been made by us so far, and machines are now arriving from Bryant Symonds and extra girls are in training.

June 30th[192]

Production meeting this morning showed more hopeful results with NC 11 and 12 CRTs for the Admiralty—11 good tubes.

Visit from Dr. Vick over supply of sufficient valves of the Benjamin type for Pye's large scale experiments. He also wanted collaboration with

Francis over use of high intensity sources for gas detection. So this device which we advised to all concerned about 12 months ago has suddenly blossomed into promising application.

Learned that the Wembley Exhibition site is not very much favoured in Westminster!

Major Schlesinger from Woolwich came to talk about carbons with all the exuberance of an army inspector.

Hankey committee in afternoon—and a waste of time. We cannot get hold of people who can make criticisms with some effective bite about them.

July 1st

Got out Lab. budget figure for current year for Mr. Railing. £245 000, of which we expect to receive back £60 000 at least from Government departments exclusive of receipts from sales. A considerable increase on previous years and a heavy amount to carry in normal times.

Had Chelioti and Dudding to lunch and talked with Chelioti over policy of 'try out' groups. He assured me of complete coincidence of view with us provided that the object of the groups is not to sell but to get experience—selling being only incidental.

Mr. Lockspeiser and Dr. Hill and assistants came to discuss the IR job with Nash & Thompson. It is to be shut down as a practical development for operational use, but measurements are to be carried further to see and agree what figure of sensitivity is obtainable and can be counted on for IR. Winch is quite reconciled to this outcome and Captain Nash, we are told, now withdraws from his contention in regard to the tactical operational applications—in other words the device is dropped on operational grounds in favour of the Helmore solution.

July 2nd[193]

Gossling reports a good (CVD) 50 cm meeting at Haslemere which is to embrace 25 cm in future. He heard from Horton (Signal School) that C-in-C in the *Bismarck* action had spoken very favourably of the behaviour of ASV, which is encouraging to those working on its components.

The considerable extension of CRT life testing seems inevitable.

Lunched with Dr. Dunsheath re. the policy of his IEE post war planning committee on research.

July 3rd

Had Mr. Parker from MOS here to examine our costing arrangements for development contracts. As usual he found them convincing.

Had long session at DSR particularly on preproduction. I begged for a joint Admiralty (CVD) and MAP preproduction committee to handle the matter with us—since I had to contend that it is bound to be a permanent condition up to say 50 per type per week—although for the next 18 months it is liable to be stretched to actual production of hundreds per week. It is left that I am to join both sides in a meeting. I am afraid it is true to say that if either Service were running the whole thing—experiment and production—there would be no difficulties in organising facilities to the best advantage.

July 4th

The 105 machines for the Helmore contract are, I understand, definitely coming to the GEC to put through, although it looked as if MAP contracts were going to queer the pitch by placing bits of it elsewhere.

Francis disturbed by statement that GEC wanted [are being asked] to supply 300 water cooled 1000 kW [1000 W?] lamps. This is another case of Ministry of Supply not taking us into their confidence at an early enough point.

Lunched with Dr. Railing who is coming to give evidence at Monday's meeting of the Hankey committee. It is a tall order to ask anyone to volunteer opinions of where, if anywhere, the whole engineering system may be at fault.

July 5th

The pressure on CRTs is heavy and Walmsley complains that we are not getting on fast enough with our programme.

Pressure is also heavy on Ramsay with his gas triodes which are being asked to function up to a recurrence frequency of 2500 [pulses per second].

July 8th[194]

A good day and the peak of a heat wave. Discussed compact source presentation for TRE with Ryde and later Cockcroft and Kirk came to smooth out the parallel efforts by ADRDE and TRE. Cockcroft was impressed by the new E 1239 valve, the 'largest ever' of the micropup series—60 to 100 times the power of VT 90, the original micropup.

Boyland showed me his latest 10 cm (hoped) triode for CW—an attractive design.

Marris reports a more cheery feeling about AIS, which is going according to plan. Sets are just arriving from Peters [Coventry].

Laycock and Warren described their latest diode of which they have had one exceptional good valve equalling the best crystal. We are quite excited about it.

Visited with Mr. Craven the British Empire Exhibition [building] which has been unofficially agreed as an alternative site. Mr. Craven is extraordinarily good, albeit he finds it difficult to get the Admiralty to function, as he did with MAP.

July 9th

Went over to Heston to see progress with the 17 planes—a most impressive effort. Helmore was there and showed me an enthusiastic report from Stevens, the crack night fighting pilot. It is difficult to get away from the Lab. but I was glad to have taken the opportunity to visualise all that was going on.

July 10th

Meeting at the Air Ministry (Helmore) of all concerned to finalise arrangements for the big contract. Helmore seems to have wiped aside all the usual procedure—no official inspection, no specification etc. Everything left in the hands of GEC and Helmore.

Hankey committee in afternoon at which we had Mr. Thomas of Morris Motors—mainly on question of tanks and tank production. I draw a veil over the revelations. Visited Lord Hirst at 5.30 to talk over various matters. He looked very well and was most interesting to talk with.

July 11th

A CVD modulator meeting lasting from 1.0 to nearly 6.0 with Moullin in the chair. He had rather a game of it in trying to get the meeting to start standardisation of gas filled modulators.

I received written approval in principle of the Empire Exhibition building as an alternative site.

July 12th[195]

I should think this is the climax of the heat wave which started six weeks ago. Some of the work positions, particularly where the girls are making valve components, have become almost unbearable. The girls are sticking it but work is inevitably affected.

Espley called me to see the new AIS helical scanner we have designed and made. It has surmounted several snags and all seemed very satisfied with it.

I saw also Ramsay and Price's transformer drive for the E 1191 gas triode at full load.

July 14th

Meeting at DSR Admiralty with Wright, Leedham, Brundrett and Harley on preproduction. A most valuable meeting at which I was able to debunk various ideas built on wrong conceptions of our preproduction; i.e. that it cannot be expanded rapidly and indefinitely—but that it can start early on a small slowly growing output which the factory can take on after an initial period with the expectation of a much more rapid production. My plan of a small guiding executive was agreed representing CVD and VPC (Valve Production Committee).

The members will be Harley and Brundrett—and will report direct to Wright and Leedham. Our production is to be increased at the Labs. in every way possible—using if necessary the British Empire Exhibition [building].

July 15th

Visited Lemington Glass Works with others and found it producing every bulb possible—but a dirty and most untidy looking place. Discussed the concentration of bulb making plants and Group Captain Leedham's alternative of United Glass Bottles. Newcastle showed no signs, that we could see, of enemy action.

July 16th

Regret I had to miss a 10 cm meeting today—as things move fairly fast in this field. I believe Megaw has a fairly good case to state on magnetrons generally and Marris, after flying his AIS yesterday, came back with considerable expectations of its adoption on a considerable scale.

I attended a diamond die advisory committee which had been well prepared by Prain—so was without incident.

Discussed dispersal plans with Dr. Railing and had some useful guidance.

15th July 1941

VISIBILITY EXPERIMENTS.

Contacts with Other Laboratories and Organisations

Camouflage Department; Dr. Meetham. Jan 17th 1941. Double Projection Method. Dr. Meetham is engaged at Compton Verney on Smoke Camouflage, and is installing a Double Projection gear duplicating ours, for the study of smoke problems and also as a standby. Explained the system and gave details of our apparatus.
Aslo saw Mr. Newsome, meteorologist attached to Compton Verney. June 13th 1941. Visited Compton Verney and saw gear partly installed.

Note. Dr. Meetham was formerly at Porton and later at Princes Risborough.

Dr. Sayce Saw the viewing apparatus, anddiscussed and advised on problems of photometry; simulation of haze; etc.

Telecommunications Research Establishment; Dr. Uttley. July 4th 1941. Dr. Uttley is working in conjunction with Dr. Goldie of Physiological Laboratory, Farnborough, under the Night Vision Sub-Committee, on night fighting problems. Is interested in atmospheric scatter. Schuil demonstrated double projection method and discussed work on scatter and haze. Referred him to Mr. Good for copies of reports.

Royal Aircraft Establishment, Farnborough; Dr. Piccard, Mr.Lamplough, and another. 4th July, 1941. Dr. Piccard is working on sights for night work, and is interested in providing an accurate simulation of the scene presented to the aimer. Demonstrated the Double Projection method and the work on haze and scatter, and referred to Mr. Good for copies of Reports.

Dr. Carpenter and Mr. Morris. (Armaments Dept) June 1941. Discussed with us problems in use of flares at night. We calculated the conditions which Mr. Morrissexplained, and shewed on the screen the effects to be expected. Later Schuil calculated further cases and discussed them at Farnborough with Dr. Carpenter (in charge of Armaments Dept.) We made suggestions for modifications to the flares which should give a much improved performance.

Physiological Laboratory, Farnborough. Dr. Matthews and Dr. Goldie. 14th July 1941 Dr. Matthews is in charge of the Physiological Laboratory andis Chairman of Night Vision Sub-Committee. Told him of work done under Visibility from Air Sub Committee, and suggested that he get into touch with Mr. Good with view to collaboration. Dr. Goldie has been working with Dr. Craik and Dr. Uttley on night fighting problems. We explained the work which we had been doing, discussed the effects of haze, and received from Dr, Goldie some figures of night sky brightnesses.

Physiological Laboratory, Cambridge. Dr. Craik. June 1941. Dr Craik is also working on night fighting problems and similar problems (vision from tanks in dark; effects of gun flashes; etc) under the Night Vision Sub-Committee. Showed him work going on here and discussed atmospheric scatter; suggested that he contacted Mr. Good. Discussed with Winch problems of measurement of gun flashes.

Admiralty Research Laboratory Mr. Cheshire. July 1941. Mr. Cheshire is also on Night Vision Sub-Committee, and is working on night glasses. Needed information on brightness of sky near horizon over sea at sea level. Gave him some data, and sent him copies of thre slides used in our D.L. experiments.

Admiralty, D.E.E. May 1941. Experiments of visibility with and without D.L. have been made on modification to double Projection gear.

A.D.R.D.E., Prof. Cockcroft; also Major Perowne, Fighter Command. Experiments on Visibility of Searchlight Beams etc.

Fig.8 Summary from J.M. Waldram of contacts on visibility problems for the Services, up to July 1941

Heard officially from Sir Frank Smith that we are to tackle 4000 VT 90 valves as soon as ever possible.

July 17th

The 10 cm meeting yesterday mainly discussed magnetrons and production of various valves required for the extended 10 cm programme. I am afraid there are some serious disappointments ahead if production on a large scale is wanted before the factory can get properly going.

Hankey Committee this afternoon discussed disorganisation in the radio industry.

July 18th

Had Giggins, Craven, Strachan and others to arrange procedure for the alternative site at the Exhibition [Building]. I think it is going well so far.

A long TVC meeting this afternoon at MAP to discuss standardisation and the like. Took the opportunity to describe the limitations of preproduction as a means of getting really large numbers of valves.

Had summary from Waldram of the many contacts with Service problems relating to visibility. (Fig.8). I believe I have noted previously that we are very much the focus here for all this work for the services—a most interesting commentary upon the value of the peace and war time research by Waldram's group.

July 19th

Visit of Exeter balloon barrage people to talk over 'sliding contacts' between cables and planes in view of latest enemy developments.

Waldram had attended meetings at Christchurch and London on visibility problems and has made excellent contacts. He tells me that Good has secured the establishment of his inter-service visibility committee—so our work will secure more formal status.

July 21st

One of the many extremely hurried mornings with every consultation rushed. The valve production meeting showed some progress on all types but with most we are not even up to 20 per week. Glass workers and training of girls are the main retarding factors.

Had first Advisory Board meeting with Dr. Railing and Mr. Gamage in their new capacity of joint General Managers of the GEC. It was a good meeting at which Macqueen gave a demonstration of GEC sound reproduction outfits for aerodromes.

Helmore job going according to plan with many jolts to keep it on the lines.

July 22nd

Dr. Bowen fresh from nine months in the USA spent the day with us and told us a very interesting tale of RDF in the USA. His was a graphic story of the display and trial at the Bell Labs. on a Sunday morning of the first of our 10 cm resonant magnetrons. Some eight or ten of the staff of the Bell Labs., the magnetron produced from a suit case—connecting up and immediately showing a great display of power of 10 cm. It was apparently quite a good moment.

He studied the AIS helical scanner just finished and the combined product, under Marris, of Espley, Clayton, Craig, Foxon and other workshop personnel. It appears to be far ahead of the American counterpart which has come over here in a Boeing plane.

July 23rd

Dr. Walmsley (DCD) here for a CRT meeting—showed him the helical scanner and compact source lamp in connection with TRE work. The CRT meeting was a full one with representatives of Cossor and EMI present.

Waldram had another risky experience flying from Farnborough at 24 000 ft., when the oxygen supply failed and they had to make a very quick dive to a lower level. He is very plucky over his dangers.

July 24th

Visit from Whiddington re. magnetron programmes and gas filled resonator arresters and VM oscillators. I got him to talk over the arresters with Ryde, and Gossling contacted him with Jesty and Moore on the oscillators. What had been done on the latter seemed to be somewhat of a revelation to him, since the impression we had given him had led him to think there was lack of interest or enterprise. I hear Sharp [Sharpe] put up a good story. Lunched with Lord Falmouth re. Hankey Committee programme.

July 25th

I hear from Marris that AIS is needing the E 1209 valve and its modulator as soon as possible. Competition from the USA in AIS is proving a sharpener of action to TRE who do not mean to be beaten. We must try to make BTH BT 53 and E 1209 interchangeable.

Wheatcroft came (from Mertz & McLellan) to be initiated into spark gap arresters for GCI. If it is true (as Wheatcroft says) that they are to be used for co-ordinating the contract and superintending the GCI installations I should think it would be good.

July 28th[196-197]

Consultation at Head Office with M.R., A.R., L.G., J.Y.F. and others on cost of research and its influence on cost of lamps, in view of BOT enquiring on selling price of lamps. My analyses of Lab. expenditure since 1935 under 'subjects' instead of under 'works allocations' came in useful at last.

July 29th

CVD meeting, attended by influential people, Dr. Black, Whiddington, Oliphant (who is off to America), Dr. Lewis, Shearing and others. Advance seems to be rapid in many directions and there is a feeling of confidence because the 10 cm technique is bearing fruit and proving itself in practice. I hear it is being really instrumental in killing Germans. Request from MOS that Benjamin can fly to USA and back, staying there three or four days to investigate why USA valve makers can't make ROF valves to our design.

July 30th

A CVD policy committee at which our preproduction problems were the principal item. It is gratifying to be able to be present and to help in this committee's work, without embarrassments arising from alleged commercial jealousies.

July 31st

Arrangements for preparing the Government Building at the Exhibition for alternative lab. accommodation in the event of enemy damage is really starting. We must press on first with the heating—but coal is short and one has fears regarding supplies.

The AIS helical scanner has passed RAF inspection and is now to be put in a plane as soon as may be.

Peters came to talk over Coventry and Bradford and Spon Street organisation. I hope everything is going well. I have rather differing accounts from Bligh and Peters.

August 1st[198-199]

An important day's meetings. Mr. Brundett DSR (Admiralty) and Mr. Harley (MAP) came to study with us the preproduction of valve problems. We had prepared our production group diagrams and other data and had a very useful talk and walk round the groups. They wish for an estimate for building extension to accommodate expansion of production. The discussions and progress will be resumed each fortnight. We have six main groups each capable of handling two new valve types.

Hankey Committee in afternoon on radio industry organisation—Sir F. Smith, Leedham and many others present. My previous suggestion that they should set up for radio the same inter-service secretariat and control as they had instituted for valves was supported, and by the time Leedham made his statement he had fixed it up and everyone was happy.

Several more development contracts have been fixed up with us and the list (Fig.9) forms a useful record of our present activities.

August 5th[200]

There is a general feeling of tempered optimism over war progress since the Russians have held the Germans for six weeks and have given themselves and us time to consolidate defences and for USA to get going.

In my preproduction valve meeting this morning all except the glass – metal valves and gas triodes and magnetrons look sticky. It is the very short wave techniques which involve the troubles due to the very short pinches and to some extent the small clearances.

I received confirmation today from MAP that our Helmore Contract (now known as the Helmore GEC Turbinlite) has been increased from £10 000 to £82 000.

August 6th

Dr. Vick and others here on means for detecting shell bursts—to discuss with Ryde and Jenkins.

22nd August, 1941.

CONTRACTS.

	C.V.D. Valve Development.	4/59027/38/F227 Vote 60.
	A.I. Aircraft Interception on centimeter waves.	SB.3110/C.31(a).
A.B. (Radio Counter Measures)	C.W. Transmitters.	I.T.P. SB.11759/C.32b.
	Crystal Controller Wavemeters.	SB.20195/C.32b.
	Receivers. (a) 10 to 65 cm. searching receiver (b) 1 to 60 " " "	I.T.P.SB.11758/C.32b.Req.DCD/136/III
C	Prototype Magnetron Oscillator	SB.22039/C.31d.
	C.W. Transmitter 120-300 mc.	SB.20970/C.31a.
A.B.	Wavemeters " "	SB.20972/C.31a.
	Crystal check Wavemeter 120-300 mc.	SB.20971/C.31a.
	Valves Oscillator V.M.Type	SB.11925/C.31(c).
	Gas-filled Arresters for common T & R aerial working.	CTTS/INST/541/C.31D.
	C.R.T. (Cathode Ray Tube development)	SB.3022/C.31(a).
C	Nash & Thompson (Night Bomber chaser).	AM/SV.29147. Priority 1A.
	Balloon Barrage.	SB.9149/40/C.21b.
	H.E.L.(Group Captain Helmore Searchlight interceptor).	SB.20061/C.32(a).
	H.E.L. (equipping of Wellington bomber).	SB.18594/C4C.
	Bomb Disposal. Immunity by magnetic agencies.	294/7544 (Con.2A).
	Batteries.(Dr.Vick, M.O.S.) for Radio Operated Fuse.	294/7170/Con.2A.

Fig.9 List of development contracts held by the Laboratories at August 1941

-2-

U.S.W. Ultra Short Wave C.W. development for Signal School.	CP.39156/36.
Diamond Dies.	-
Diascopes for RDF display.	CTTS/INST/668/C.39b.
Diffused Lighting.(D.L.) (Ship Camouflage).	CP.4G/37387/F.2847.
Compact Source Presentation for Ground Control Interceptor.RDF.	-
25 cm. M.O.S. RDF.	294/8872 (Con.2A).
Wind Tunnel.(Farnborough).	SB.22740/SAS/C.28b.
Infra-Red. M.A.P. (substituting Nash & Thompson).	-
Gun Flash Location M.O.S.	
Compact Source development for Searchlights.	

Fig.9 continued.

Excitement on the part of himself and Benjamin on the first operation of the ROF from the ground in firing trials. This is good news and will undoubtedly arouse much interest. Benjamin's work has contributed substantially to this result, which has been achieved by Pye as senior partners and extraordinarily good co-ordination staff work by MOS under Dr. Vick's chairmanship.

Had long discussion with deputation from workshop on question of payment for holidays. Arranged both sides would think over matter for a few days.

August 7th

Hankey Committee today not very important. Decided with Dudding and others that craftsmen's shops should have the extra six days of holiday as other weekly staff throughout the labs. Very glad they are so jealous of their staff status.

August 8th[201]

Went over some of our work with Mr. Railing, particularly preproduction and the possible building extension to house it—amounting (these days) to about £16 000. Mr. Railing discussed with Giggins and myself and decided to try to recoup the amount by a 10 shilling addition to the cost of valves made at Research.

Had second interview with craftsmen's delegates from shops and gladly agreed to full staff status to be maintained and to apply to holidays being uniform for all weekly paid staff.

Sir John Madson (Australia) and Professor Whiddington to see over our RDF work and talk over special Australian requirements.

August 11th

A day of trying to make staff and facilities spread over more work than they can. Things are getting more and more difficult in this way and in the middle of it we have ridiculous requests through the Central Register to release men like O'Kane and Houldin.

Megaw tells me that his 150 kW magnetron which had been 'grounded' as a miscarriage is not a dud after all and will go to 250 kW. This was shown by BTH Research whose co-operation with us is the most loyal and helpful.

Jolley's report re. the Helmore GEC Turbinlite is that Witton do not seem to have got all they should on order and there has been some heavy talking and thinking. But it is not easy these days to impress on big concerns that some of these very vitally urgent jobs require a short-circuiting of their normal and sacrosanct machinery!

Our preproduction is making heavy weather of it. Training girls in glass blowing is the bottle neck now to expanded output.

August 12th

Ramsay and Gossling met BTH people at Rugby yesterday on gas triodes and modulators and although the meeting did not see eye to eye there was a strong desire not to differ in public. Standardisation is always difficult, particularly where one side does not even see that it is necessary.

Was with Max Railing for one and a half hours bringing him up to date in things. It looks as if the GEC would pay for the building extension needed for preproduction.

August 13th

Megaw gives an amusing account of the collapse of contentions of a certain influential experimentalist at a Service establishment who had been stating against Megaw's declared experience that phenomenal changes were taking place in permanent magnets. Megaw suggested that perhaps he was testing them in their sheet iron cases or in close proximity to metal. It turns out that this was the case. Unfortunately (for him) the gentleman in question had committed himself to writing before we could advise him. We hope this will be the end of the incident—but we know it will not end the habit of believing any man with academic traditions before one in these Labs.

We had a visit today from Prince Bernhardt of the Netherlands, Mr. Murray Payne and Mr. Gamage. It was a pleasure to show them round with Dudding. We had to show them only non-secret things.

Message to hand of Benjamin's safe arrival in the USA.

August 14th[202]

A day of announcing Churchill's meeting with Roosevelt at sea.

Modulators to the fore in connection with standardisation, and also Machin's plans for still larger sizes.

Meeting with J.Y.F., G.C. and W.H.W. to consider how GEC is to avoid unfair treatment by Govt. depts. in the matter of lamp orders.

Our various building extensions are nearing completion—cathode ray tube extension, diamond die ditto (finished), two local workshops at east end of Labs.—also an additional lavatory.

August 15th

A Brundrett and Harley meeting on preproduction. I think they have now a true impression of the rate of production and its growth which we *can* manage, and it is clearly below that required if other people keep up their deliveries of sets. One thing is quite clear—you cannot train new operators by existing operators at more than a certain speed. The new building extensions are to be pushed on at the maximum possible rate and MAP will put their machinery on to get the necessary licences.

I learn that all our 17 Helmore GEC Turblinite machines are completed and in action, and the group are working on some experiments as well as helping where they can with the large Witton contract.

August 18th[203]

After a more cheery preproduction meeting Capt. Willett from Signal School came in a state of emotion regarding his NT 98 magnetrons and his NT 100 modulators—also his E 1248 arresters. He says the 10 cm sets for submarine hunting are producing exciting results and are really 'the goods'. He wants our output pushed up hard of these valves—well over the 50 per week which has been our present aim. He would like 250! He went round to see the groups and thrilled Miss Rowe, Mr. Falconer and Miss Starbuck with pictures of sending submarines to the bottom. I got Dudding and Pearce to join in the talk.

What a difference it makes to have a senior officer like him to visit the place and show the flag.

Hankey Committee in the afternoon when Mr. Kipping gave evidence. A very busy day.

Dudding called a meeting of all engaged on oxide coating techniques to establish the processes for everyone concerned on a surer basis.

August 19th

Mr. Aldis here on secret signalling and possible help we can give him in his development.

Megaw reports helpful talks as usual with BTH on large magnetron policy. The good results with the 150 kW magnetron indicate it as the probable best survivor of all present types above the 10 – 15 kW size.

Gossling reports a satisfactory visit to Christchurch where Professor Ellis is taking in hand the ROF job there.

Marris showed me a complete set of the AIS equipment as delivered from Coventry.

August 20th

I hear Benjamin will fly back from USA next week and that he seems (as far as can be judged) to be succeeding in his mission. Pye are wanting 30 trials per week which means that number of batteries and valves per week. There is still no move to manufacture Sanderson's dry battery here although there may be supplies from USA. USA and England are running neck and neck, apparently, on ROF and interchangeability of components is very desirable.

Had a delightful letter from Willett, Captain, Signal School (Fig.10).

Helmore rang me today—the 'Helmore GEC Turblinite' is going forward more smoothly—but Jolley came to me with a long face over all the formalities of the AID inspection machinery for which he (with Helmore) is having to take responsibility.

August 21st

Hankey Committee at which we discussed AID inspection and the advantages of using a firm's own inspection and testing dept. I had to make a friendly assault upon the draft report which was before us—mainly in [on] its presentation. I must have a go at a better draft.

August 22nd[204]

A day of clashing priorities. AIS wants many hard modulators which will conflict with Signal School (Navy) demands for their 10 cm sets. At my TVC meeting at MAP this afternoon this, and also magnetron claims, were keenly discussed. Dr. Truscott urging the impossible production position with which he is faced with large numbers of new RDF valves and applications.In fact the situation is at last arising which I have sought to bring about by which the Services really realise what they are up against in the supply of RDF valves. There is to be a similar acute situation at a special meeting on VT 90 valves now required for ELSIE, AI, and ASV.

We had our first ten VT 90 valves off our group last week since our effort to restart.

Telegrams "Navalsignals Haslemere".
Telephone:- Haslemere 1040.

H.M.SIGNAL SCHOOL,
HASLEMERE.

18th August,1941.

Dear Dr.Paterson,

Thank you very much for the kind and sympathetic reception which you accorded me this morning.

I know I can rely upon your team to do everything that is humanly possible to increase the rate of output of the NT98, NT100 and E.1248 Valves. I have spoken to Admiral Grant and he will be only too willing to help you in any way he can. I told him that you would be getting in touch with him.

In order to make quite sure that there is no delay or hitch in delivery from Wembley to ourselves, I have suggested to our people that a light van from here should call at Wembley once a week to collect the output of these three precious tubes and bring them straight to us without any intermediate handling. I have told our Mr.Harris to get in touch with your representatives about this. - I hope it will be of some help.

There is one point in which I failed this morning. - I do not think I made clear to Megaw, Gosling, Le Rossignol and Company our appreciation of the wonderful contribution they have made to our armoury by bringing these valves into being.

Looking back, over R.D.F. progress, it stands out very clearly that the progress made is the progress of valve development. I am very aware that the Country must thank the staff of the G.E.C. Research Laboratories for much that we have to-day.

Looking forward, I know that this history will in large measure be repeated.

Yours very sincerely,

20 AUG 1941

Dr.C.C.Paterson,
Research Laboratories of the
General Electric Co.Ltd.,
Wembley, Middlesex.

Fig.10 Letter from Captain Willett, Signal School, to Clifford Paterson dated 18 August 1941

August 25th[205]

A much better preproduction meeting. Some 280 valves turned out in all and some 35 each of NT 100 and NT 98 which are the critical ones wanted by Signal School. Was able to wire Willett that he could have his 32 per week for the next few weeks.

Had Walton to see me on mission from Cabinet, asking me to name people who are really expert on radio for a reason disclosed to me. I am flattered that they chose me to advise!

Hankey Committee at which Lord Hankey adopted my draft of the 'Inspection' report to work upon. Two hours work finished the draft for circulation.

Marris and Espley are hard at work on the specification for AIS to which the present restricted production is proceeding. Ferranti are coming in to supplement Coventry's production.

August 26th[206-207]

Heather and Macqueen here to talk informally over the situation of Telephone Works where the machine tool capacity is only able to cope with about one-third of the works requirements. They stayed to inspect the small 'larger' [layer?] battery and the frequency modulation demonstration being rehearsed for an influential MOS visitation tomorrow.

Mr. Prain here to talk over matters and to inspect diamond die installation. B.P.D. and I discussed prices of dies with him.

August 27th

10 cm meeting. One of the chief items was the substitution of the '150 kW' (E 1255) type magnetron for the E 1233 type, the latter becoming obsolescent. The Birmingham group Randall (Oliphant being in USA) had much to say about 2000 kW design with which they are working. 3.3 cm operation was also to the fore. I gather from Ramsay that there is a good prospect of a satisfactory gas triode (E 1209).

Visit by General Butler, Col. Elsdale, Dr. Paris and the Superintendent of SEE [Signals Experimental Establishment] to take part in demonstration of frequency modulation. Bligh and F. Clark are handling and I saw off the expedition from Wembley in our two vans for field tests.

August 28th

Had progress from Francis to report to the OAC on 2.5 kW compact source lamp which is becoming a 'technically' solved problem. West Drayton (Admiralty Lab.) have been giving real assistance in designing and making housings which will adapt these lamps for searchlight use.

Had Lang Brown from DSR Admiralty to lunch with me asking for a man (preferably Stevens) who can be seconded from us to Admiralty and to work in the Camouflage Department's Labs. at Lemington [Leamington?]. Must consult again with Waldram.

August 29th

There is a general feeling in the Lab. that the pushing up of the new magnetron from the '100' to '150' kW will put our modulator out of step with the magnetron development since the present modulator is not capable of handling the magnetron at its full load. I don't think the position is serious, it only means that the modulator experts have to scratch their heads for a means of making a modulator which will go to higher powers.

A very emotional meeting on preproduction with Brundrett and Harley in the afternoon to see how to attempt to approach the demand for NT 100 and NT 98 valves needed by both Navy and Air Force. (The latter demand is for our 100 AIS sets). Our best efforts to get near the figure are inadequate to keep pace with the always rising demand. We have begun to turn out VT 90s again and shall approach our promised 100 for August. Permits for the lab. extensions for preproduction are not going too well as between the Government and the GEC.

September 1st[208-210]

A sad production meeting this morning for the NT 100 valves so badly needed by Signal School and others have developed a poor emission epidemic. Bell came up to the scratch and agreed to organise evening shifts and do all possible to get things on the lines again so as to yield the promised weekly quota of valves. Other valves, excepting VT 90, are none too good.

A meeting on Saturday between Shearing, Butement (MOS), Ross (?) Megaw, Willshaw and Gossling discussing the extension of our USW programme to include transmission tests over long periods at different centimetre wavelengths. There is to be another meeting to make more formal recommendations.

Professor Rankin (now attached to Dr. Walmsley, MAP) came for the afternoon to study various matters which we have in hand for DCD including centimetre waves. He seems to have been rather messed about in one job and another since war started, first with DSR Admiralty on 'inventions' then with Watson Watt pretending to do something but really wasting time, and so on.

September 2nd

A tragic day for the Laboratories. On my way back from town I bought an evening paper and learned that Benjamin was on board the crashed ferry plane which left Canada last evening—and there are no survivors. Mrs. Paterson was fortunately working at the Labs. and I found her in my office on my return and took her round to Mrs. Benjamin to whom she had the tragic task of breaking the news. I drove to Hampstead to bring over Mrs. Benjamin's parents and sister.

Benjamin was one of the best of our rising men and will be a sad loss to us personally, and as a brilliant research worker. He had to a unique extent the love of research for its own sake combined with a genius for handling factory applications and problems. MOV will miss him seriously and the Ministry of Supply will hardly fully recover from his loss.

CRT meeting all the morning, and this afternoon I spent with Dr. Railing, Mr. Gamage and Mr. Heather discussing confidential items in the research programme and particularly the Coventry (Peters) position in supplying AIS equipment. I hope the Managing Directors will now see Leedham and get the matter clear.

September 3rd

Consulted with members of Benjamin and Warren's group plus Donaldson re. arrangements for carrying on. Had many telephone messages of sympathy. Lord Hirst rang up for Mrs. Benjamin's address so he can write to her.

September 4th[211-214]

Had internal meeting to clear a remonstrance by the modulator group here that the emphasis recently placed by Megaw and myself on the potentially higher output of the E 1267 magnetron was undesirable since there was no modulator in view for running it. The protest, so far as this covers it, was found sound and there ought to have been a fuller consultation before the E 1267 magnetron was promulgated.

Many letters of sympathy arrive on Benjamin's death.

P. Good came with Beard and Haldane to see the triple projector technique for breaking down illumination and visibility problems. They have various schemes, one of which seems similar to the 'auxiliary searchlight'. They had not heard of it—which is the common complaint—which reminds me that I had never had a satisfactory answer as to why it was turned down.

September 6th

Called on Rowe at Swanage on the way to Dorset where I am spending my week's break. Rowe and I had a long talk on the subject of why the GEC (apart from its Research Labs.) were not all out on 10 cm equipment in view of its vital importance for various future developments apart from the established ones of anti-submarine and AIS. They are badly piqued that the whole GEC organisation is not taking AIS more seriously from a production standpoint, looking upon it as first fruits of a big harvest of successful developments. I had some difficulty in getting across to Rowe the actual position at Coventry and my own powerlessness to influence the Coventry management, in view of (*a*) F. Clark's hostility to anything less than one metre (*b*) Coventry's preference to use their facilities for lines of work which will be commercially attractive after the war.

I hope the promised meeting between the General Managers of the GEC, T.W.H. and myself and Sir Frank Smith will help to clear the position.

This diary will miss the next week.

September 15th

Benjamin's body was brought home this week and there were many representatives from the Labs. and from MOV at the funeral—there have been many letters of sympathy and Cockcroft, Fowler, Vick and Gossling have sent a note on Benjamin to *Nature* (Fig.11). The tragedy for his wife and for us all hardly bears thinking about.

There have been no other outstanding happenings. Megaw has made a tour of the establishments—long overdue.

Visited Sir Frank Smith with Dr. Railing, Mr. Gamage, Mr. Heather so that Sir Frank could tell the story of RDF. He also took the opportunity to emphasise the urgency of deliveries of VHF sets from Bradford and

AIS from Spon Street—the former definitely having priority. VT 90 valves came in also for pressure. We reassembled at Magnet House to talk things over and particularly the possible equipment of Spon Street for 10 cm manufacture.

September 16th

CVD (GEC) meeting at which the chief interest was Randall's (Birmingham) research on magnetron resonances apparently explaining the parasitic oscillations and means of avoiding them and putting up the efficiency from 25 to 50%. A good piece of work. Had conversations with Mrs. Benjamin's father re. her financial position.

September 17th

CVD policy committee at which our preproduction facilities were much to the fore. Our CVD valves are wanted in thousands now instead of hundreds.

Lord Hirst has promised to write to Lord Beaverbrook urging a Government pension for Mrs. Benjamin. He presided at today's RMA luncheon and looked very well.

Attended in Harley's office (MAP) with Giggins to fight against MAP suggestion that it will be quite good enough by way of provision for our valve preproduction facilities if they put down some Nissen huts for us against our east wing!!!

Was able to assure all concerned that Labs. could realise their expectation not to let down the Navy or TRE in the number of magnetrons and modulators needed for September.

September 18th

Gavin's work on RCM makes steady and good progress. He and Bligh are working very well on this programme together.

Have fixed up arrangements for Schuil's services to be loaned to DSR Admiralty for naval camouflage work at Leamington.

September 19th

Had talk with Gossling over the modulator position to see if a direct effort should be made to avoid competition of the bad old type between BTH and ourselves over our respective types of thyratron and gas triode. We shall wait 'til after the modulator meeting to be held here next Wednesday.

Decided with Ryde to cut our losses over the velocity modulator mixer work and to release Proctor and Sloane for other work. Am getting Sloane to join up with Megaw on magnetrons.

September 22nd

Valve production meeting showed a not very successful week. So many chances of things going wrong. We lost 60 VT 90s through bad filament wire. MAP are ordering 1000 E 1231 valves which means the organisation somehow of another preproduction group. The NT 99 is wanted in large quantities and we can hardly begin because of the VT 90. But Boyland with some girls and Miss Dimmock is doing excellently at 20 per week.

TVC meeting in afternoon. The importance of its work and functions are increasing—but the tremendous number of new valves with no chance of effective production is tragic. It does not seem possible for the valve production committee of MAP to allow any of the existing programme of valve production to be pushed aside even a little to make way for the new types. So we start a small production at Wembley and just whet the appetites of the Service Departments for more!

September 23rd

With Gavin to Lywood's meeting at Air Ministry on RCM. Gavin has kept our stock very high with the RCM people who seem to have got exceptional (for these times) deliveries and developments from Wembley and Spon Street. We seem to be supplying almost the complete range of receivers, transmitters and wavemeters from 10 cm to 10 m. A recent receiver requirement by the Admiralty from 3 to 10 cm will keep us busy—if they give us the development contract. BTH are working from 5 to 10 cm.

Studied possibilities of space for preproduction of E 1232 (NT 99) valves and agreed with O.W.H. and Le R. to clear out the H and C apparatus from the household appliances life test room. This should enable us to work up to 50 per week.

Gossling and Jenkins back from Christchurch where consultations have taken place with Ellis re. ROF.

September 24th

A large CVD modulator meeting at Wembley with long technical discussions. BTH having received a development modulator contract from MAP figures largely. I explained to Whiddington and Bradfield that, under CVD, Wembley is following a programme of equal magnitude and should be considered as co-equal with BTH in this field. The action of MAP independent of CVD has embarrassed Whiddington—but he is on the right lines in seeking reconciliation of the programmes on the CVD modulation committee. He also mooted the idea of BTH programme of valve development being discussed at the GEC CVD committee which BTH would attend. On discussion he felt doubtful about the idea.

September 25th

A new Helmore scheme for 'ground lights' has been put up to us with a prospective development contract—and also another infra-red job for Winch. Helmore's devices for keeping his contracts under his own hand are well planned and diverting. It remains to be seen how long this particular piece of work proceeds before ADRDE come into it.

The financial checks on these contracts etc. involve a good deal of work and have to be kept well up to date. Also Le Rossignol's check which he keeps on the costs of valve preproduction are a weariness to his soul. I shall have to try to get him some intelligent clerical help.

September 26th

Jesty's CRT group is now occupying their new premises for preproduction and life testing but they are going to be crowded in spite of the extension.

Truscott and Elton came in place of Brundrett and Harley to the preproduction meeting but they had a well prepared schedule of future requirements which, when compared with what we can hope to do here is not too bad, except in the case of one or two valves like the NT 99. Warren's E 1248 looks as if it is getting over its teething troubles at last and if so we are hoping to produce at 50 a week by the end of October. His E 1231 receiving valve is now to go into a production group to make 1000.

September 29th

Valve production review this morning was more encouraging. The numbers of the vital valves (NT 89 magnetrons and NT 100 modulators) are going up. The 1248 gap valve for Signal School seems to be getting over its teething troubles and may be expected in large numbers. The 1231 Warren triode is described as one of the most difficult of all to make and has yet to show itself a production possibility. The NC 12 CRT is the only project about which I feel disappointment. We seem to find no way of breaking the spell of continuously high wastage. It was a vain hope that we should be able to show the factory how to manufacture CRTs. 325 valves made last week—about £2000.

We have passed over full designs of our diamond die drilling machine to the USA and they have decided to make 200 machines for their own use. Meanwhile, whilst making a certain number of very fine dies, the factory is asking for various larger sizes which we are glad to be making.

September 30th

The main war interest is the Russian front and efforts of the Germans to attack towards the Caucasus after overrunning Kieff and the Ukraine. Leningrad and Odessa still hold against the enemy. Activity is expected to begin again in the Near East.

Had £8000 development contract for Helmore-Wellington bomber. That group will need strengthening.

October 1st

The getting of girl labour in sufficient numbers is still a difficulty—and when occasionally one or two leave we are badly landed. Have written Sir Frank Smith to ask for a letter from influential quarters emphasising the importance of the work of the Laboratories.

Had discussion on small magnetron with Randall's modification to increase efficiency, Boyland having been to Birmingham yesterday to get particulars. Programme started for urgent investigation in order to incorporate such change as may be warranted in valve types now in hand.

Had Dr. Willis here from Ordnance Staff [Board?] to seek advice from Dudding and Jennett re. application of statistical methods to selection and testing of fuses.

The ROF visit to Pye, of Donaldson and Jenkins, revealed some unpleasant snags in their present operation plans which it is not easy to see a simple way of surmounting.

October 2nd[215]

Hankey Committee at which we considered first draft of our main report. Tizard was very helpful and clear sighted as usual and the chairman with his patience will get a good report in the end but we shall have a lot of sifting to do before we are all agreed upon the relative efficiencies and causes thereof of the three main Services Supply Organisations.

The first of Randall's new magnetrons arrived today so we shall now have an opportunity of seeing what it does.

Jesty back from TRE where Sutton (Admiralty) was demonstrating the 'Skiograph' type of CRT which has a black line on an illuminated screen using sodium chlorate.

October 3rd

Meeting with Dr. Railing, Mr. Gamage and Mr. Heather to go over the activities of the Labs. and particularly the confidential ones. We spent two hours at it. This afternoon a special TVC meeting to discuss ROF valves with Services and valve makers together. It was an 'informative' meeting but the Services considered it useful. We learnt for the first time that the manufacture of ROF radio gear in this [country?] is out of the question on account of congestion. The production of sets and valves in England will therefore be confined to what is necessary for large scale trials in operation.

Randall's 'strapped' magnetron of the small NT 98 type appears to have twice the output of the unstrapped type, from our first tests of one of our own and one of his valves. But so far an increase has not been achieved with the 10-segment E 1256 (150 kW) magnetron according to reports from Rugby.

Suggestion from DSR Air Ministry to cut down Winch's programme on sensitivity of infra-red devices by eliminating the scanning feature from the programme. Have told Lockspeiser that we want the results and will do the measurements on our own.

Le Rossignol and Elton visited Marconi yesterday and fixed up their magnetron preproduction programme. They cannot tackle the NT 100 till the magnetron is going properly.

USA authorities are having 200 diamond die drilling machines made over there to our exact design. Prain, the Controller over here, writes very warmly.

October 6th[216]

Had note from Edwards who is at Christchurch with our AIS and the Beaufort to say that on its second test flight the Coventry (Spon St.) equipment has already given some very satisfactory results approximately

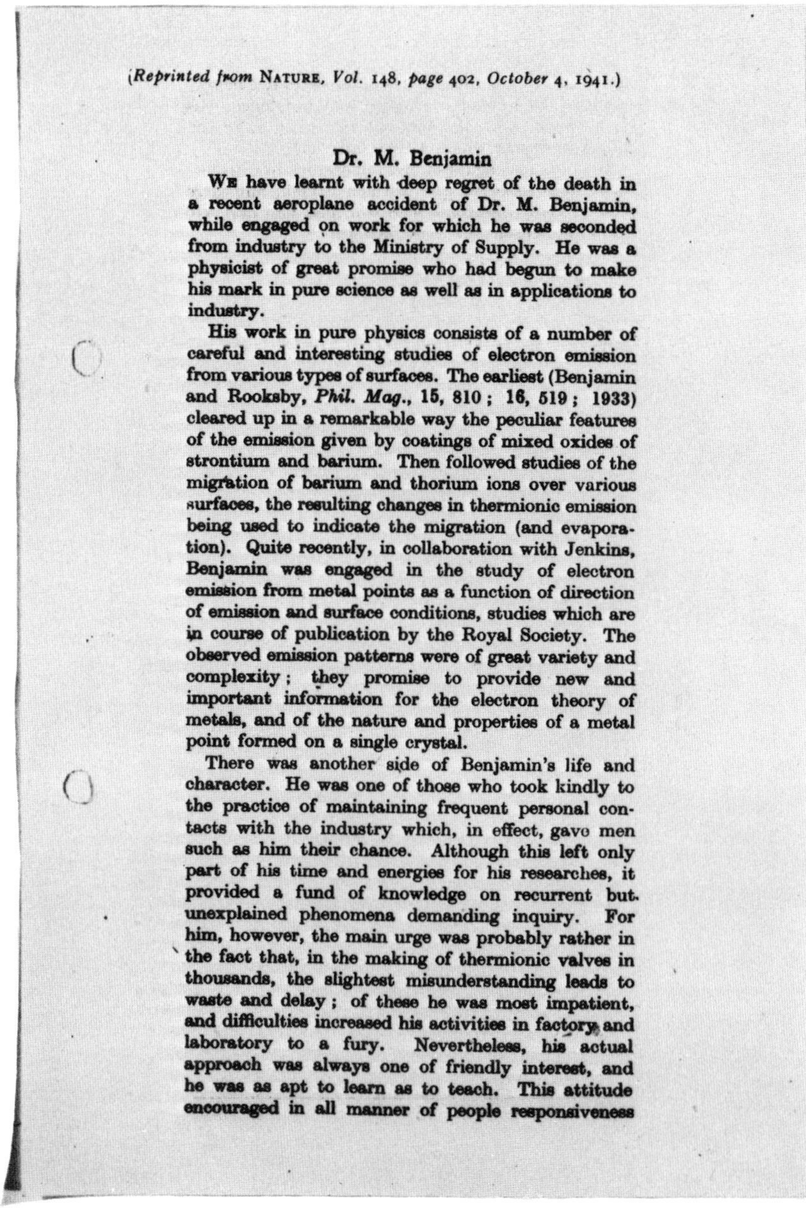

(*Reprinted from* NATURE, *Vol.* 148, *page* 402, *October* 4, 1941.)

Dr. M. Benjamin

WE have learnt with deep regret of the death in a recent aeroplane accident of Dr. M. Benjamin, while engaged on work for which he was seconded from industry to the Ministry of Supply. He was a physicist of great promise who had begun to make his mark in pure science as well as in applications to industry.

His work in pure physics consists of a number of careful and interesting studies of electron emission from various types of surfaces. The earliest (Benjamin and Rooksby, *Phil. Mag.*, **15**, 810 ; **16**, 519 ; 1933) cleared up in a remarkable way the peculiar features of the emission given by coatings of mixed oxides of strontium and barium. Then followed studies of the migration of barium and thorium ions over various surfaces, the resulting changes in thermionic emission being used to indicate the migration (and evaporation). Quite recently, in collaboration with Jenkins, Benjamin was engaged in the study of electron emission from metal points as a function of direction of emission and surface conditions, studies which are in course of publication by the Royal Society. The observed emission patterns were of great variety and complexity ; they promise to provide new and important information for the electron theory of metals, and of the nature and properties of a metal point formed on a single crystal.

There was another side of Benjamin's life and character. He was one of those who took kindly to the practice of maintaining frequent personal contacts with the industry which, in effect, gave men such as him their chance. Although this left only part of his time and energies for his researches, it provided a fund of knowledge on recurrent but unexplained phenomena demanding inquiry. For him, however, the main urge was probably rather in the fact that, in the making of thermionic valves in thousands, the slightest misunderstanding leads to waste and delay ; of these he was most impatient, and difficulties increased his activities in factory and laboratory to a fury. Nevertheless, his actual approach was always one of friendly interest, and he was as apt to learn as to teach. This attitude encouraged in all manner of people responsiveness

Fig.11 Note on Dr. M. Benjamin in *Nature*

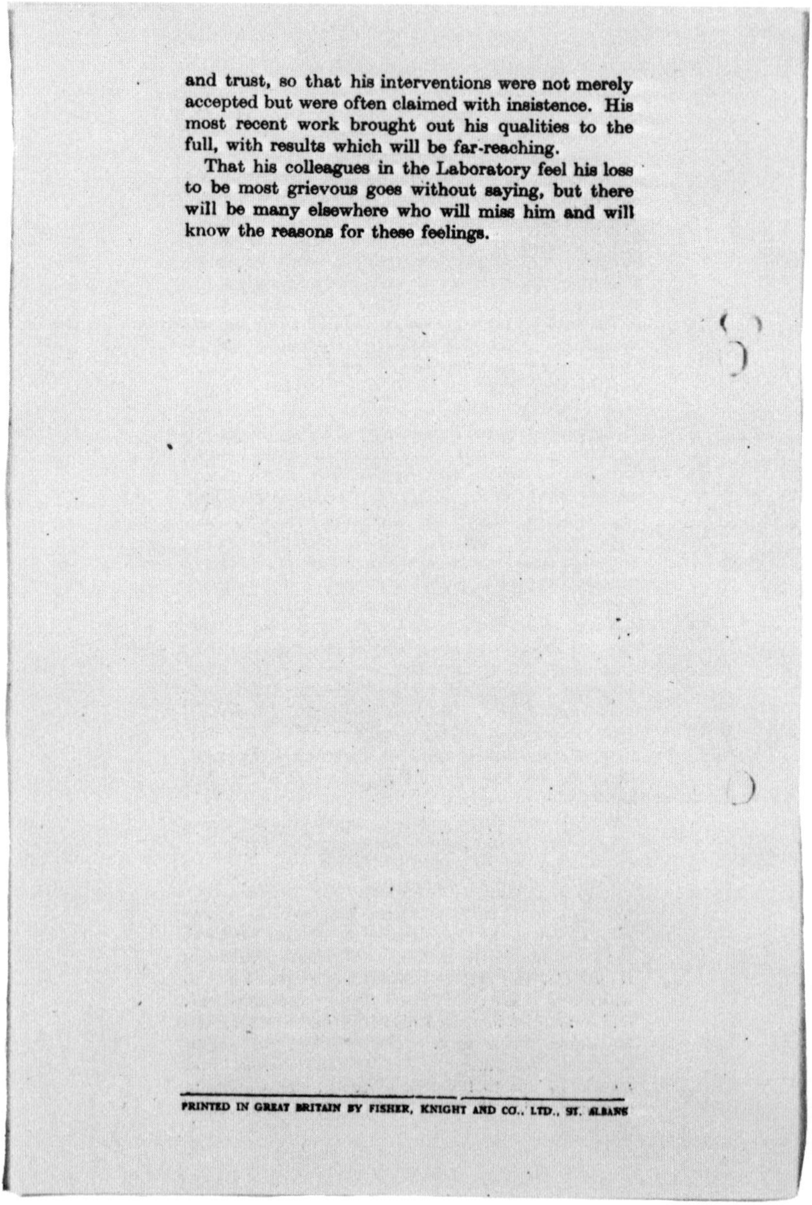

and trust, so that his interventions were not merely accepted but were often claimed with insistence. His most recent work brought out his qualities to the full, with results which will be far-reaching.

That his colleagues in the Laboratory feel his loss to be most grievous goes without saying, but there will be many elsewhere who will miss him and will know the reasons for these feelings.

PRINTED IN GREAT BRITAIN BY FISHER, KNIGHT AND CO., LTD., ST. ALBANS

Fig. 11 continued.

equal to those obtained from our experimental model. A very promising start.

Had a shock from Peters announcing that the cost of the AIS sets being made by him is fully £1000 each instead of the £500 estimated by us and on which the maximum sum on the contract was fixed. This is the first contract in which this has happened.

R.H. Fowler came today on one of his visits and we had a general talk over technical progress.

October 7th

10 cm CVD meeting at MAP all the morning and a talk by Oliphant on his USA visit in the afternoon. A useful and interesting day. Our people had made good progress in applying the 'strapped magnetron' advance to some of our valves. We paid a full tribute to the Birmingham group—and brought data and actual valves for Signal School to test out. The device has so far only shown definite advantage in the small 8-segment magnetron.

Many points in Oliphant's talk look interesting. He is coming to Wembley on Tuesday.

October 8th

Enquiry made from Bell as to an alleged USA Eimac Valve as a rival to the NT 100. There seems doubt about this but Bell has made a glass equivalent of the NT 100 in case it interests CVD.

An interesting Hankey Committee meeting in the afternoon at which we considered Tizard's redraft of the portion of the report comparing Admiralty, MAP and MOS organisation for equipment design and supply. We decided that the essential differences lay in having a highly qualified corps of engineers in high places, and in ensuring that supply and operations are respectively represented at the Army Council and the Supply Board.

October 9th[217-218]

Visited Farnborough with Jolley to see the superintendent (Mr. Farren) to see if J.S. Thompson could be spared for three months to help us on the new Helmore job. Farren gave the warmest report of the work of the four men from Wembley (Erith) who are on loan to the Aerodynamics section at RAE. They appear to be engaged on 'power jet' research of real priority and we felt it was out of the question to ask for the loan of one of them.

October 10th[219-220]

Had ring from Elton—'NT 98 and NT 100 supply question has now reached very high levels and Brundrett has to attend a meeting this afternoon. The set in which they are used is "the goods" in the Battle of the Atlantic. We want you to anticipate your promised numbers by December and get them out in November or before'. As preproduction meeting was in session in No. 3 conference room I went down with the message. Sent promise of not less than 60 per week from now onwards rising to 75 by mid-November. Elton seemed very pleased. We have always kept conservative in our promises but this will use up much of our margin of safety. Letter from Brabazon, MAP (Fig.12).

October 13th

Preproduction valve meeting indicates progress provided an epidemic with VT 90s can be countered.

A long internal meeting on Salford iron powder production problems, Pitkin and Dauncey having spent the past three weeks there.

October 14th

Oliphant spent the day with us describing RDF matters in the USA. We could not see that either in valves or their application the Americans yet had anything useful to tell us unless it be in certain uses which were hinted at. They were doing extraordinary things in valve design but many of them seemed complicated ways of avoiding simple designs or processes. Their failure to make good in the use of micropups is staggering.

Oliphant volunteered the observation to me that after getting to know establishments here and on the other side he had come to have an immense admiration for Wembley.

Later Helmore came particularly to examine Winch's infra-red proposals for a friend or foe detector. He spoke to me in the same sense and said that he constantly waxed lyrical over the helpfulness and effectiveness of our group here who are working with him (Jolley, Hawkins, Cameron, King-Brewster, Thomas etc.). He wants to make a 'special mention' of them. It is very nice to hear these things. (Fig.13).

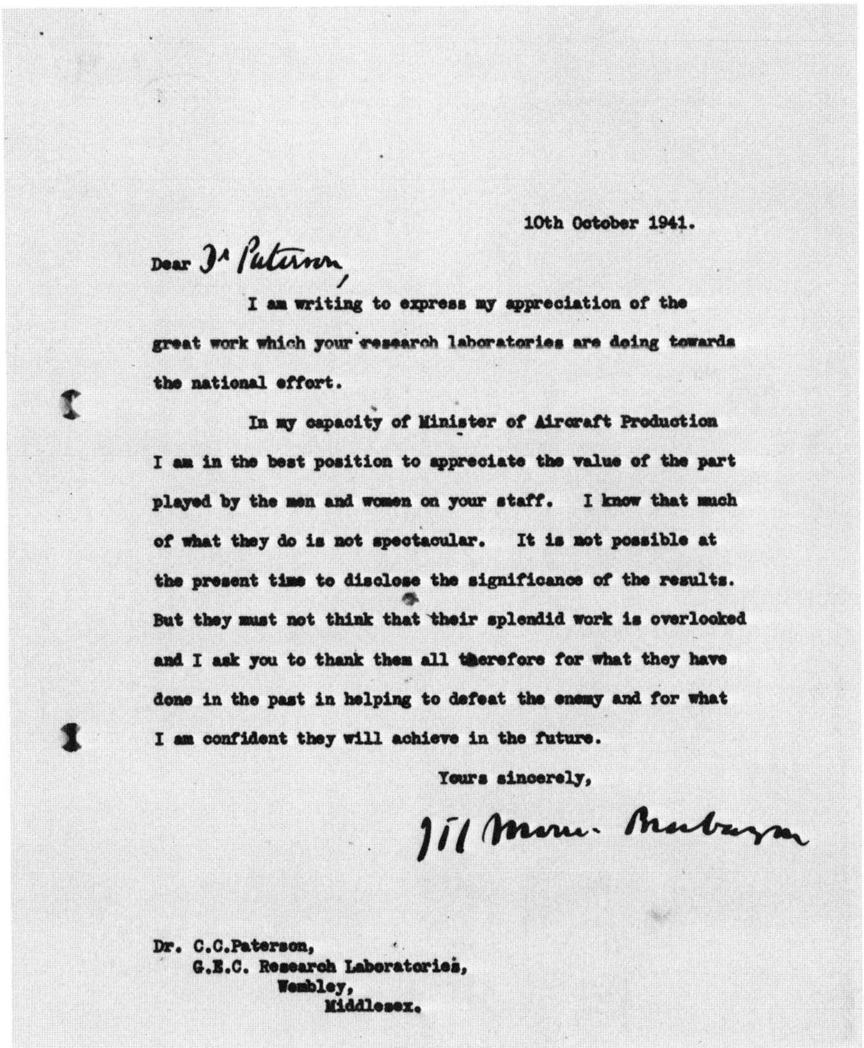

10th October 1941.

Dear Dr Paterson,

I am writing to express my appreciation of the great work which your research laboratories are doing towards the national effort.

In my capacity of Minister of Aircraft Production I am in the best position to appreciate the value of the part played by the men and women on your staff. I know that much of what they do is not spectacular. It is not possible at the present time to disclose the significance of the results. But they must not think that their splendid work is overlooked and I ask you to thank them all therefore for what they have done in the past in helping to defeat the enemy and for what I am confident they will achieve in the future.

Yours sincerely,

J T C Moore-Brabazon

Dr. C.C.Paterson,
G.E.C. Research Laboratories,
Wembley,
Middlesex.

Fig. 12 Letter from Moore Brabazon to Clifford Paterson dated 10 October 1941

Copy of
LETTER FROM

Group Captain Helmore,
Room 6094,
Ministry of Aircraft Production,
Millbank, S.W.1.

To -

16th October, 1941.

Dr.C.C.Paterson,
Research Laboratories,
The General Electric Co.Ltd.,
Wembley.

Dear Dr.Paterson,

As you know, the small team of research workers you so kindly provided for my searchlight experiments have now been working at high pressure for nearly a year, and have achieved very remarkable and useful results. The more I have worked with them, the more I have been impressed not only by their outstanding technical ability and efficiency but by their enthusiasm, determination and character which has enabled innumerable discouraging obstacles to be overcome.

Everyone of them has never hesitated to take on any job and they have frequently worked long hours night after night solely through personel keenness and desire to get on with the job. They have all, in order to get readings or research data, at various times personally flown under conditions of no little hazard in experimental aircraft.

I know you appreciate what they have done as fully as I do and now that the question of some recognition of their work arises I feel it might be of some assistance to you that I should put in writing my own views of the matter. As far as I am able I am endeavouring to represent my appreciation of the staff of the G.E.C. to the Service authorities and I gather from you that the Company itself may wish to consider the desirability of recognising their outstanding services.

It would be redundant in this letter for me to offer any personal thanks to you for all your cooperation and assistance as you already know how much I have appreciated your help but I very much share your desire that the staff actually engaged on the job should not be forgotten.

Yours sincerely,

W.Helmore.

Fig.13 Copy of letter from Group Captain Helmore to Clifford Paterson, dated 16 October 1941

October 15th[221]

Had discussion with Gavin and Warren about triodes and the use of their harmonics to obtain variable wavelengths between 3 and 10 cm. Mentioned matter to Whiddington who we believe is seeking information from all and sundry.

Our VT 90s (200) of last week's production seem to be safe, with a brief extra ageing period. Thank goodness.

Ryde's big report on the interference by mist and rain with 3 cm transmission and below will be ready for circulation this week end. Many people are wanting copies. Ryde tells me it is Part I and I am expecting it will be a sort of classic.

October 20th[222]

A plea from Goldsmith at Witton that Sanderson should devote more time again at Witton to help on factory problems. I think we can manage it. Lunch with B.P.D., Chelioti, Gant, Woodward and Strong to discuss long standing unsolved problems connected with lamp matters. A very useful and frank talk.

Presented shooting cup to Home Guard. Three out of four of the men in the winning team were from the Labs.

October 21st

Le Rossignol attended an MAP meeting—Air Commodore Leedham—to review the position of VT 90 production. STC had little contribution to make. MOV are making fine progress—over 500 a week. Research Labs. are up to 200 and Shaw is getting going. So GEC stands well and I hope all and sundry realise that the VT 90 is a sound egg for both manufacture and operation. It has been looked on very much askance in the USA.

Dr. Vick here on ROF valves and batteries explaining policy requiring about 500 valves for convincing operational trials before going to shorter wavelengths. Sanderson's batteries may be wanted at 100 a week for similar trials.

Constant evidence is coming in of the efficiency of Benjamin's mission to the USA—and a personal letter from Webster (British Science Office Washington) to Vick speaks in the warmest terms of him.

Megaw came up to tell me that their latest result with a small NT 98 magnetron was 550 kW output for 1000 kW input!

October 22nd

Meeting of Advisory Committee of diamond die control. I guaranteed our ability to drill 0.0004 dies whenever we want to. Polishing so as not to spoil the die is now solved with Leeds's new machine. Systematisation is the next move starting from diamonds faced down on both sides to uniform thickness. Breaking through with precision is the last major process to finalise. The Committee asked us to take up various jobs of specification and the like.

Meeting with MOV to discuss RDF valves and particularly the alternative hard modulators.

Attended demonstration with Macfadyen on Barn Hill in which he showed signalling with visible or infra-red over a mile by talking along the beam—using a small signalling lamp and modulating its filament current (only 1%) by superposing the microphone current on it. To me a most surprising possibility.

October 23rd

Discussed with Lord Hirst and Mr. Gamage pension arrangements for Mrs. Benjamin. The Treasury proposal is as mean as it could be.

OAC [Osram Advisory Committee] meeting and difficulty in producing water-cooled high intensity lamps for the Services precipitates a state of gloom—the GEC feeling its prestige waning in lamp matters. Francis at a meeting at MOS reported a demand for 150 water-cooled lamps by January 1st divided between GEC, BTH and Siemens. I believe there had been an intention to leave GEC out on account of their despondency regarding delivery.

Lunched with Warren, BTH, to clear up misunderstandings re. fluorescent powders and Levy and West. Also discussed magnetrons, crystals, non-centralisation of high intensity lamp correspondence with Services at BTH etc. etc. A very useful talk.

TVC meeting until 5.30—many new valves and some tendency to crab developments before they have had a proper run for their money.

October 24th

Diamond die technical committee at Rugby—voted a good meeting and discussion. We are being asked to draft specifications for diamonds and also dies, which always involves long discussions.

October 25th[223-224]

Various letters and messages from the Services. I gather from Ramsay that Signal School are favouring our E 1191 (CV 12) instead of the BT 9 which does not seem up to the steep wave fronts being handled. Whiddington sending sample order for 200 of Ryde's latest anti-burn crystals which have a good report from TRE. Meeting to be held re. Warren's diode (E 1273) which leads the field against Cosmos valve. Bell and Marris went yesterday to the general discussion on hard modulators with valve manufacturers and service representatives. Bell's new small modulator E 1271 seemed to win the day for a mass produced article for which it was considered that neither NT 100 or E 1209 would be suitable. The E 1271 can be used in pairs or threes.

October 27th[225]

VT 90 production suffered last week from filament on grid for some mysterious reason—otherwise we should have topped 300 for the week. Notwithstanding, our total of preproduction valves was about 450.

Letter from DCD confirming that Wembley do the development and Bristol the research on Skiatrons. Had serious talk with Warren re. continued backwardness of E 1248 and E 1231 valves.

Sir F. Gill and AID representatives come to consult Dudding and Jennett on use of production charts and the like for appraising the uniformity of a factory's product with a view to saving man power. Mr. Gamage here in the afternoon to get first hand knowledge of some of our work.

Letter from Stradling (Chief Advisor, Ministry of Home Security) asking if Waldram may serve on a select committee of experts on 'The photometric properties of the atmosphere'. He will be the only non-Government representative.

October 28th

CVD meeting—very full of interest and lasted three and a half hours. Magnetrons, diodes, modulator valves, crystals and a new proposal of Ryde's for a bulb of gas in the gas protective resonator chamber.

Extraordinarily high magnetron outputs are being found possible by USA, Birmingham and Wembley—but it will be some time before Signal School and others are ready to incorporate the result in their operational sets. TRE in fact, whilst agreeing that output is larger, cannot see any improvement in range in their apparatus.

October 29th

A CVD policy meeting and lunch with Whiddington took up most of the day. There has been for some time a tendency for the Services to look upon the GEC as so busy or having so much to do that they must be only used as a last resort. I am having to correct the impression wherever I can. The same impression has gained a hold at MOS re. water-cooled mercury lamps—but that is understandable because we always emphasise (at Lamp Works request) that developments will take a long time to get into production. I don't like the way that—entirely through chance—all the official requests for work in the 3 to 9 cm region so far pass us over.

October 30th

A day of settling many domestic points to help to keep a happy ship. The staff questions are innumerable, due to rapid changes of wages outside, opportunities to earn high rates, and reorganisations of work inside the Labs. B.S.G. returned from Christchurch where he had important talks on ROF with Professor Ellis who has taken charge of it there.

October 31st

The Labs. had a visit from the two General Managers of the GEC, Dr. Railing and Mr. Gamage, and Mr. Heather. They examined six of the special developments in hand at Wembley and I hope this is only the first of a series of such visits to get to know our work better. (Fig.14).

Jesty returned from a visit to TRE where various CRT matters were advanced a stage.

Regular meeting with Brundrett and Harley to review our preproduction. This is now in good running order. The necessary building here for expansion of our effort has not yet been approved by MAP. We have their building dept. representatives coming to 'examine us' on Monday. What a ghastly waste of valuable time and effort are caused by these departments, financial and otherwise, connected with these Ministries, which have the power to 'approve' and therefore to hold up vitally urgent schemes.

November 3rd

Visitation of MAP buildings and finance depts. on the subject of the character of the buildings for the extensions to be used for preproduction.

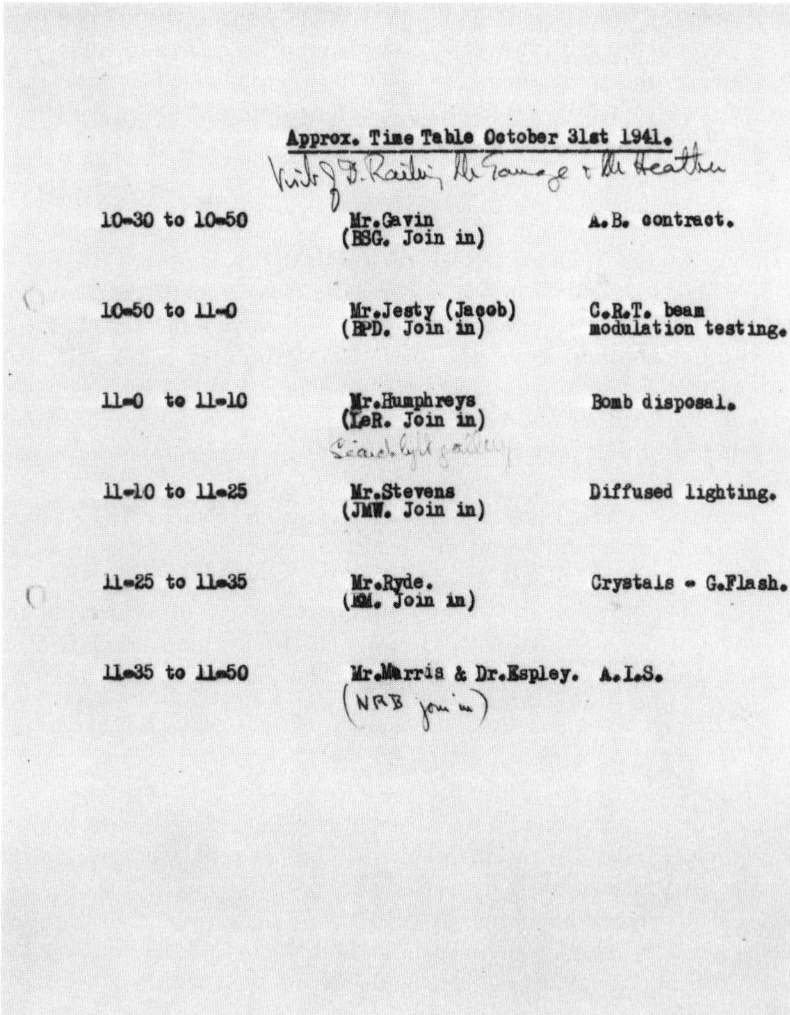

Approx. Time Table October 31st 1941.

Visit of Dr. Railing, Mr Gamage & Mr Heather

10-30 to 10-50	Mr.Gavin (BSG. Join in)	A.B. contract.
10-50 to 11-0	Mr.Jesty (Jacob) (BPD. Join in)	C.R.T. beam modulation testing.
11-0 to 11-10	Mr.Humphreys (LeR. Join in) Searchlight gallery	Bomb disposal.
11-10 to 11-25	Mr.Stevens (JMW. Join in)	Diffused lighting.
11-25 to 11-35	Mr.Ryde. (HM. Join in)	Crystals - G.Flash.
11-35 to 11-50	Mr.Marris & Dr.Espley. (NRB join in)	A.L.S.

Fig.14 Programme for visit of Dr Railing, Mr. Gamage and Mr. Heather on 31 October 1941

When the character of the work to be done was realised I don't think there was any further question on the subject. But we shall see.

First meeting of technical members of RCM committee at Wembley—now to be held monthly instead of our attending at Whitehall.

We have, with Lamp Works, to make between 50 and 100 water cooled quartz lamps for MOS—BTH and Siemens doing likewise.

November 4th

Dr. Bown of the Bell Labs. USA is on a visit to this country on RDF and such like. He has had a month visiting the Government establishments and the like. He is keenly interested and a most acceptable visitor. He will be spending another day. He was asking pointed questions as to Wembley's share in the total effort and when I told him he was not surprised, but said he had not been told any of it at the Establishments who seemed to think they had done it all! Truly the Professor type needs to learn how to give honour where honour is due.

This afternoon Mr. Lockspeiser (MAP) and Dr. Sykes (NPL) came to see if we had ideas and could do work on non-creep high-temperature steel alloys or their equivalent. Vaughan Williams had some ideas to be tried out with powder metallurgy and we agreed to accept a small research contract to start with. I am glad to be associated with the particular application for which these are needed.

November 5th

Our fire-watching arrangements have, I am pleased to record, reverted to a voluntary basis on an extended scale but everyone is ready to join in on the compulsory basis if necessary.

Our NT 98 preproduced magnetrons and their testing seem to require further attention. Megaw is concerned about them and all concerned are in a muddle about giving them a factor of safety.

Good reports of submarines being sunk and night raiders [being shot down] at the moment are scarce—but the Russians are losing the Crimea and the Germans are starting a heavy 'final' effort on Moscow.

Hankey Committee today trying to polish off our report.

November 6th[226]

Had copy of letter from Rogers Radio Tubes of Toronto which speaks well for the discussions Benjamin had there. It is a tribute to his ability and characteristic of his thoroughness. (Fig.15).

IS.

COPY.

ROGERS RADIO TUBES LIMITED
100 Sterling Road,
TORONTO.

SECRET

September 10, 1941.

Professor E.F. Burton,
Physics Department,
University of Toronto.

Subject: Abstract of the Discussion Benjamin-Parker
Monday, August 25, 1941

Dr. Benjamin and this writer spent several hours discussing the relative merits of several designs of projectile amplifier tubes. Benjamin apparently had studied my PK 117 construction and was familiar with the operation thereof., and admitted that it was robust for values of g higher than 20,000. I admitted that in view of the findings in Carnegie at Washington, that the glass tube was satisfactory up to 20,000 g. The discussion resolved itself to a comparison between the relative merit of the tension construction of the American tube and the compression construction of the British tube.

The American tendency, according to Benjamin, is to get into such small construction that it becomes difficult to produce these tubes in large quantities. He stated that the acorn type of tube gave him experience in England which showed that only by elaborate laboratory control it was possible to produce a few thousand of these tubes per year. I agreed that when the question of size leaves the realm of the naked eye and the fingers, and goes into the microscope and tweezers, the cost of the tube ascends rapidly and it enters the impracticable field. I agree heartily that the British tube, as exemplified by Benjamin model E1229-D pentode, is the most economic structure from the standpoint of uniformity and large production at low cost.

The E1228 thyratron, showed to me by Benjamin, is far in advance of the American type which is defective in the

Fig. 15 Copy of letter from Rogers Radio Tubes, Toronto, dated 10 September 1941, on their discussions with Dr. M. Benjamin

2.

region of operation around 45 volts. Benjamin stated that the British thyratron had a constant slope down to about 22 volts, whereas the American type loses its linearity below 60 volts, and is suitable, therefore, for only high voltage operation. There is no doubt in my mind that Benjamin has British quality in his thyratron construction and has an unusual foresight to large production which, curiously enough, has been overlooked in the American designs.

Going into the minutae of design pentode E1229-D, I asked a series of questions in regard to the details of construction. I find that the filament is tungsten, electrocataphoretically coated (a process which we have pioneered and used in Canada for several years). The filament suspension is by means of a specially designed spring hook which is designed for retardation as well as acceleration. Several important features are the anchoring of this spring hook, and the mica construction encasing the spring. One of the important reasons for using compression instead of tension is that the filament is not loaded by the weight of the spring during set back. The heart of the industrial secret in producing pentode E1229-D is the method of drawing the glass about the fingers of the lower mica so that the mica is held rigidly by the glass envelope. Dr. Benjamin explained that nitro-inert gas filling was used, on my questioning him upon this point, so that oxidation of the part did not take place. I suggested that additional support of G3 wires may be found necessary. He stated that in general, drape of G3 turns was not a serious matter when the tube was centrifuged at 20,000 g. I go into the greatest of detail and found Dr. Benjamin familiar with the whole field of construction and always had ready answers. He imparted to me a lot of experience covering years of endeavour in a very short time.

The thyratron E1228 has a nickel base filament because deflection of the filament during flight is not important on account of the large spacing. The hook tension likewise is different from the tungsten filament of the pentode E1229-D. The filling is 0.35 mm of argon fed through a capillary from an argon tank having a tolerance of 0.2 mm to 0.5 mm. The general glass construction applies to the

Fig.15 continued.

3.

lower mica of this E1228 thyratron as well as the pentode E1229-D. Benjamin pointed out that this thyratron had excellent linear characteristic down to 22 volts, and he stated that the associated circuits behaved in a more controllable fashion than in the case of the tiny American tube which is questionable for performance at low voltages. Benjamin gave me several references to look up covering some minutes of our conversation. He has a recent article in Faraday Soc., February, 1940, which covers the electrocaphoretic method that he is using in particular.

Benjamin asked me a series of questions to which I agreed:-

1. That the E1228 thyratron and the E1229-D pentode are reasonable and economical designs which lend themselves for mass production on a large scale.

2. Your company will accept an educational order for 500 tubes of each type.

3. If these experimental tubes are successful, your company would be willing to manufacture on a large scale, should they be required.

Accordingly, I confirmed this agreement to Dr. W.L. Webster at Washington. A copy of the letter which I sent to Dr. W.L. Webster was sent to Benjamin c/o the National Research Council, Ottawa, but was not delivered and was returned to me. This shows that NRC Mailing Department needs shaking up and at least they could have enquired to find out where Benjamin was in order to forward the letter to him. I am attaching Benjamin's copy of this letter herewith.

I have had a reply from Dr. W.L. Webster, copy of which is attached herewith.

In closing I wish to state that for the first time the War has hit home in a personal manner, and I deeply regret the untimely ending of Dr. Benjamin's career. He was very stimulating, and useful.

Fig. 15 continued.

4.

I must say that the compression type of construction, which Dr. Benjamin proposed, needs very serious consideration because it has very important merit.

Yours very truly,

ROGERS RADIO TUBES LIMITED.

(Sgd.) H.W. Parker.

HWP:MK

cc Professor Oliphant.

Fig.15 continued.

The 'Helmore turbinlites' have not yet gone up in anger for raids are still very slight.

Absences through sickness are seriously impeding work, especially the women.

I lunched with DSR Admiralty today—he is coming on Saturday to see Gavin's RCM work.

Polgreen called and spoke warmly of the iron powder situation at Salford, which at the moment is very encouraging enabling the Factory to be on top of all its demands.

November 7th

A good informal meeting on modulator transformers was arranged here by Macfadyen—well attended by all establishments concerned and Moullin kindly came. It was agreed that the transformer has a special problem to meet and Metvic were to be asked to attack it.

Warren attended meeting at Signal School re. use of diodes. Coales proved his case for their use and Cosmos agreed they could adopt our design and method of construction if necessary. Warren's and Laycock's diode seem to be the goods and is now going fairly well in first states of preproduction.

We are having to assert ourselves somewhat with ADRDE over mercury vapour lamps. BTH, due to their central position, have tended to be looked upon as the focus for this work. I spoke to Warren (BTH) about it and have now offered some lamps to ADRDE.

November 8th

At Salford to discuss powders, crystals and other things with Mr. Cobden Turner. He seems to have been expecting us to accept more responsibility in powders than in fact we have. Williams and Pitkin trials to secure a uniform product have been so far successful and Salford are pleased, but there is much to do before the processes can be established.

November 11th[227]

Trouble with NT 100 valves—leaks and flashings, 40% wastage. Most annoying after running for several months with only one epidemic of short duration. The NT 98 seems to be nearly out of its trouble. E 1148 and NC 12 are getting into the clear but VT 90 does not rise above 200 good valves per week.

Elton told Dudding an encouraging story of steps he hoped to materalise (internally) for bringing pressure from the Ministry of National Service onto their exchanges and officers to back us up. There is a real advantage in bringing the Brundrett and Harley combination into our confidence.

November 12th

A delightful second visit from Captain Willett of Signal School (the last visit was August 18th). He came to thank us for our successful efforts to meet his request for extra numbers of valves—and went to four of our preproduction groups to address the girls. After telling them that they had a big share in (*a*) sinking four submarines, (*b*) sinking the *Bismark*, (*c*) Matapan, (*d*) Captain Agnew's recent exploit in the Mediterranean, he thrilled to the thought that this had all been done by the irregular, almost illicit, manufacture of urgent valves in a research laboratory. It pleased him to think that the pompous German boasts were being outwitted by such an improvised set up! Altogether a happy occasion.

Willett spent two hours looking at some of our other work. We had a Hezekiah show of valves which had been prepared for Whiddington on Saturday and which would be a very useful permanent exhibit to set up. I spoke to Willett about our being 'out of' 3 cm developments.

Asked Walmsley to hasten decision re. our continuance of AIS work with TRE after our contract is out at the end of the year. Espley tells me that Mark II of AIS with his redesign will be ready this week. Also the 25 G for ADRDE. Both are going to Christchurch on Saturday for field trials.

November 13th

Had final meeting, for the present, of Hankey Committee and practically agreed our report. It seems to me rather innocuous in places but may do good.

November 14th

Mr. Kinman from BTH here yesterday to study our crystals with Ryde. The latter had had a pleasant day. Kinman was very appreciative of the design and took all particulars. Shortly after both Christchurch and Portsmouth rang him to report results which were in advance of anything yet in crystals. So the wheels are buzzing in that quarter.

Walmsley paid a chance visit in the afternoon to see Espley's redesign of AIS and also the 25 cm gear. It was a good day to come. The former of course is very impressive and the whole group is working long hours to have it ready by tomorrow—zero hour.

The snags with NT 100 valves are not yet cleared and this week's deliveries are less than 50 for both NT 98 and 100.

Waldram attended his first inter-service committee on 'photometric properties of the atmosphere' and was disappointed to find visibility question cut out of the terms of reference. Good may succeed in pulling it right. Young, Morse and Collins have gone for a high flight again today in connection with this research.

A morning with Dr. Vick (MOS) at the Lab. to see progress with our compact source lamps. Francis is talking of 10 kW and 15 kW lamps and feels fairly confident.

I also discussed with him:

Ever Ready collaboration on layer batteries
Morgan Crucible collaboration on carbons
Ferranti collaboration on special valves
Future of Searchlight Committee
Good's photometric properties of the atmosphere committee
Ryde's crystals etc. etc.

Dr. Walmsley rang up to say that our CRT contract was being extended normally for another year.

November 17th[228]

Preproduction panics are still worrying but I expect that is what we are here for and why we are doing it! Epidemics emerge from nowhere and give us a 50% loss at the least provocation! E 1248 and E 1231s however are more promising.

The AIS party have been working the whole weekend and are ready with the gear to go to Christchurch tomorrow with their Mark II version. This creates somewhat of a sensation and is leaving some people gaping.

I hear too from Cockcroft that the 25 S job is growing in importance and seems to have a very likely application.

Lunched with Heather and talked over many important matters of GEC policy which can hardly be recorded here. Advisory Board followed. A very good meeting, in spite of H.H., M.R. and A.R. being away.

November 18th[229]

Meeting with Chelioti and Levy and West for preliminary talk on subject of bringing them closer to GEC. It looked as if they were willing to fit in with a suitable arrangement.

I have been giving talks to groups of girls on the working and applications of valves. They seem anxious to understand more of what they are working on and this evening some 70 turned up in the library to hear an informal talk from me.

November 19th

The Libyan offensive was begun by the Imperial Army yesterday.

A 10 cm meeting at MAP with a good discussion. Ryde had a good story to tell re. his crystals and was congratulated. I gather from Gossling and Warren that Houldin has suggested that the coating between electrodes in the diode being a bad conductor affects the performance and that a thinner coating may save a few decibels. So the battle of the crystal *v*. the diode goes on each in turn taking a step. But the crystal has usually led, so far.

I met Mr. Gamage to settle with him Mrs. Benjamin's pension and feel that the proposal is a generous one.

November 20th

I heard last evening the MAP building directorate have again boggled over approving our extensions for preproduction. This time there appears to be some rumour at MAP that Dr. Paterson is obdurate or worse and is ready to defy the authorities of MAP if they don't agree with what he considers necessary. So we go on; allowing Jacks in office to hang up progress with the war. It is just three months since we asked for approval of the scheme and the best months for autumn building gone. I have to receive on Friday the highest official of the Directorate to enquire again into our little extension on the spot.

November 21st

Our MAP building directorate friends did not come after all—one was ill—but intend to do so as soon as may be. I understand they have been very severely restricted in the amount of building they are rationed down to.

A morning meeting at Wembley with Brundrett and Harley when we had to account for technical hitches with NT 100, NT 98 and E 1191 (CV 12) valves.

An afternoon meeting at Wembley of TVC which lasted three hours, and was well attended from the Services.

Helmore came to witness the first attempt at a '5 times' Turbinlite arc. It seemed to give promise of success. Helmore insisted to us on a suitable division of the work between Wembley and Witton. He had been to Witton last week and was very appreciative of the enthusiasm shown there.

November 22nd

Yesterday Humphreys attended an appeal before the National Service Board against one of our young men who insisted on leaving. We made a test case of it because we felt the Labour authorities were not using their powers to retain the services here of people who had been trained in our skilled jobs. They gave a good and sympathetic hearing and whatever the result we have confidence for the future that we shall have the support we need in all reasonable cases.

At yesterday's meeting Warren's new single ended grounded grid valve was welcomed by TVC members. Criticism of E 1231 termination arrangements—a useful talk on the subject.

November 24th[230]

Willett rang me up early to tell me (indirectly) that the sinking of the submarine which sank the *Ark Royal* (by the Corvette *Marigold*) was one in the bag for his gear and our valves.

Marris back from TRE and reports that the Mark II AIS is going according to plan.

Sir Phillip Bowyer-Smythe, new deputy Controller of Signals at the Admiralty, was brought by his predecessor Captain Brooking. I was very glad that they came to us specially for there is much to learn here. They were very insistent that we were the key to the present radio position.

Our preproduction meeting showed a better position with most valves—but output of the three main lines (except VT 90) has suffered from various troubles. CV 12 looks as if it had lost two complete weeks output.

O'Kane has made himself specially useful at TRE using our helical scanner on some very secret developments. Edwards is still there or at Christchurch on AIS flying trials.

Good came today to talk over recent changes in BSI senior organisation. He is now Director.

November 25th[231]

A visit from Mr. Ferranti and party from his works all day.

Capt. Signal School is arranging to use the E 1190 valve (millimicropup) instead of the ill fated E 1001 (which no one wants to make) in certain sets they have in use. It may be required in still larger quantities later. So one more of our babies is being adopted as the only possible.

November 26th[232]

A cathode ray tube meeting all morning—attended by all the usual people. To town afterwards with Walmsley and discussed with him the future of our AIS group and programme and other matters. E.K. Cole has been given the job of mass production of our new AIS set and we shall have to make one or two more prototype models in the Marris and Espley group with very little help from the workshop. But these days the staff is having to do what the exigencies of the war demand and 'making things' is often more important than inventing them.

Winch and Cameron have a serious job on, measuring the beam intensity of a 2 m searchlight in comparison with one of 1 m. There seems to be a dispute between a Mr. Savage and ADRDE, and so far our own measurements have favoured the contention of the former.

Davis and another from BTH here to study luminescent powder techniques after settling a silly misunderstanding with BTH over our Levy and West arrangements.

I spoke to Walmsley about Oboe and O'Kane's work and suggested we might take it up as a development.

We have passed over Ryde's crystal technique to BTH for adoption as soon as may be—and I think this action is appreciated by all concerned. It is very delightful being able to work in this non-grudging non-secret atmosphere.

November 27th

Had a delightful letter from Captain Brooking (Fig.16). These unsolicited testimonials are heartening.

OAC meeting at Hammersmith attended by both H.H. and M.R.. I was glad to hear that Heather was being invited to join.

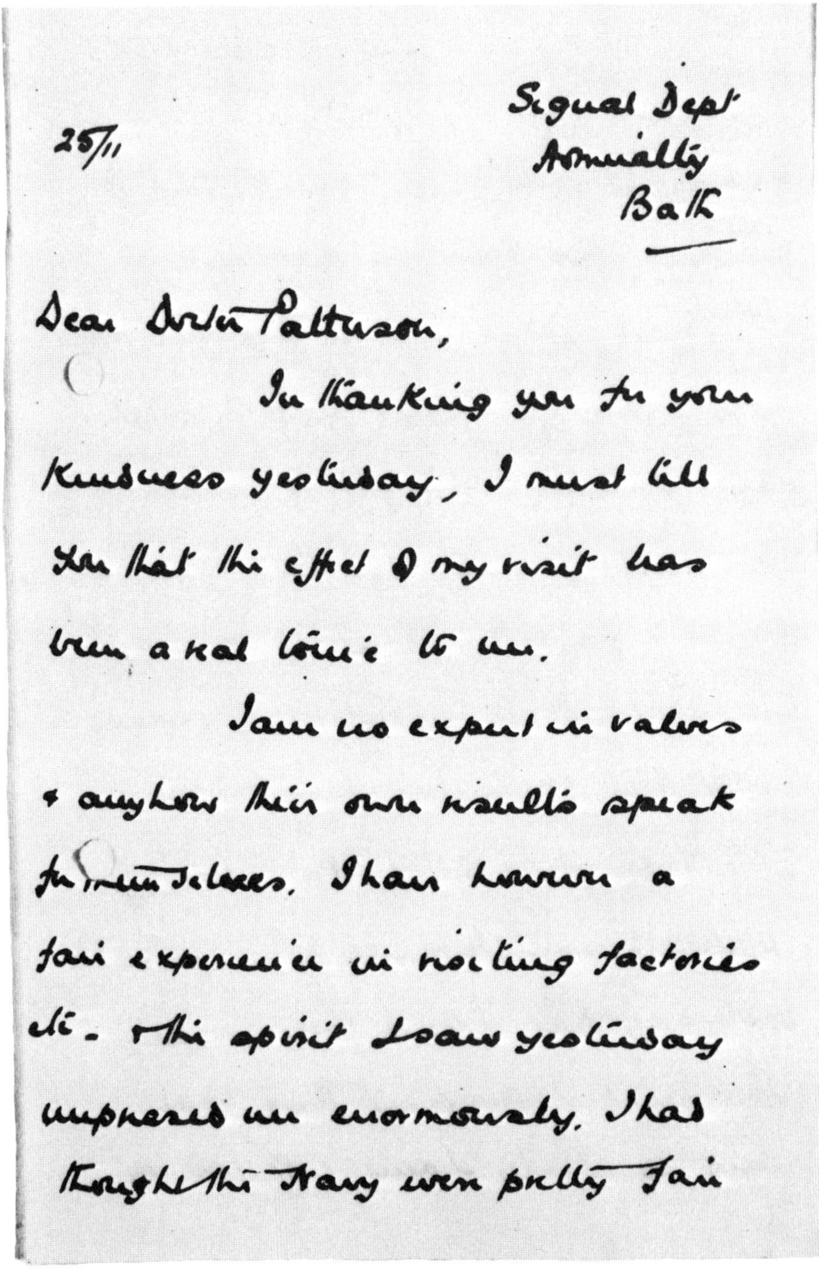

Signal Dept
Admiralty
Bath

25/11

Dear Doctor Patterson,

In thanking you for your kindness yesterday, I must tell you that the effect of my visit has been a real tonic to me.

I am no expert in valves & anyhow their own results speak for themselves. I have however a fair experience in visiting factories etc – & the spirit I saw yesterday impressed me enormously. I had thought the Navy were pretty fair

Fig. 16 Letter from Captain Brooking, RN, to Clifford Paterson dated 25 November 1941

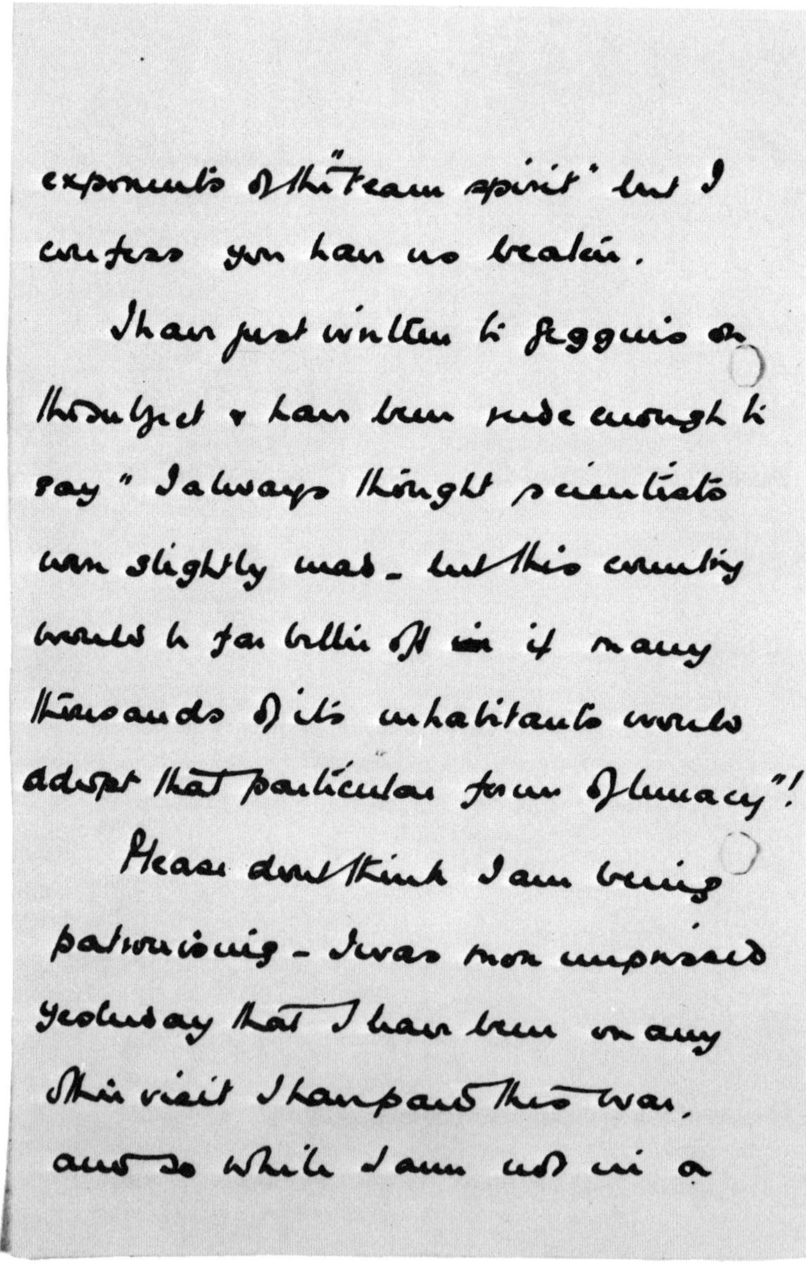

exponents of the "team spirit" but I
confess you have us beaten.

I have just written to Jeggins on
the subject & have been rude enough to
say "I always thought scientists
were slightly mad – but this country
would be far better off now if many
thousands of its inhabitants would
adopt that particular form of lunacy"!

Please don't think I am being
patronising – I was more impressed
yesterday than I have been on any
other visit I have paid this war,
and so while I am not in a

Fig.16 continued.

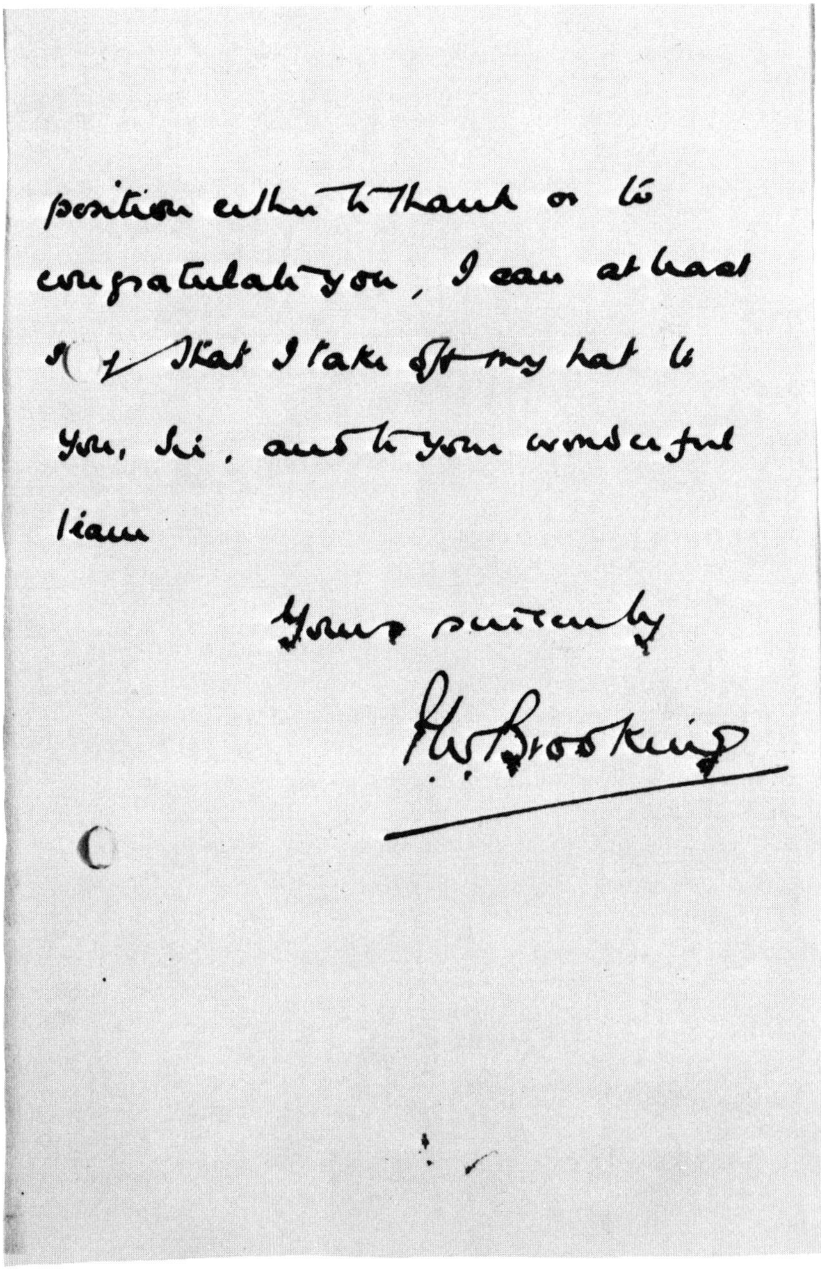

position either to thank or to congratulate you, I can at least say that I take off my hat to you, Sir, and to your wonderful team.

Yours sincerely

H.W. Brooking

Fig. 16 continued.

Had a letter from the Admiralty giving much guidance as to procedure in the event of invasion.

The AIS manufacture is going to cause some flap—GEC definitely desire to come in to it—and the making of further prototype sets is not going to be easy with the existing model to be cleaned up.

We managed sixty NT 98 magnetrons this week which is better than we feared.

November 28th[233]

Meeting with Railing and Gamage, T.W.H. being unwell. We went over the AIS question and I [also] pressed the importance of frequency modulation schemes with a view to ultimate manufacture.

Had O'Kane up from TRE to talk over his present very special work which at present has no name.

Hankey Committee (joining with the Scientific Committee) to consider the Ordnance Board and its raison d'etre.

November 29th

Examined the oscillator and resonator crystal work with Rivlin and Lewer, whose programme of work is expanding. Whiddington rang up to tell me that we were to be given a contract for 3 to 10 cm CW sets and Gavin (who had been with Megaw to a special meeting at Haslemere) tells me that Shearing was commissioning us under USW for much the same thing. Whiddington is contacting Shearing about it. The research will mean something new to be desired in the way of electronic devices.

December 1st

December opens with an anxious outlook on the hard fight in East Libya which has now lasted a fortnight—and possibly, very significant, the retreat of the Germans from Rostov and their being held at Tula south of Moscow and to the north west.

Our valve preproduction for last week reached a total of 625, which cannot be less than a rate of £150 000 gross per annum.

The tragi-comedy of our proposed preproduction building extension, as handled by the building department of MAP, drags on its deplorable course. After wasting another week for Mr. Colquhoun and Mr. 'Sam Brown' to come down to show still further distrust of mine and Harley's recommendation, they neither of them actually turn up but send

Mr. Millar (a very decent old Scot) and Mr. Bowley the type of civil servant who would lose any war. We again went over all the old points with them and they clearly had nothing to say but to agree with us but said they must convince the people to whom they are responsible. I left them arguing about a few hundred pounds with Strachan sitting round in the NT 98 preproduction space in the gallery! It was then 6 o'clock. So the precious days and weeks are squandered and the time and nervous energy of far more valuable people than they are, wasted!

December 2nd

Professor Ellis at ADRDE is making a lot of difference to ROF—and has found fundamental reasons for altering the form of Sanderson's battery, so things are somewhat in the melting pot again.

Winch and Cameron have been doing comparisons between the 150 and 200 cm searchlights here with ADRDE. There is some political background to this on account of the activities of Mr. Savage who presses the larger light but there is also a desire to press for a larger (250 A) carbon for which we are not perhaps so prepared as our competitors.

Visit from Sir Earle Page (Australian Cabinet representive) with Gamage to see what research in Great Britain is doing.

I hear from Marris (Bournemouth) and Peters that at the meeting at Swanage there was sympathy with the Coventry desire to take on from us the prototype of our new AIS set. But no move whatever to deviate from MAP decision to give it to E.K. Cole. So we shall have to co-operate with them in the matter. It is rough on our chaps, who have worked tremendously hard to make the existing model a prototype which Peters can turn into a production set, to find Coles turning it inside out to suit their particular designs.

December 3rd

Gavin at his AB meeting here yesterday and several new problems set to him in connection with RCM, the chief being the fitting of bombers with jamming gear against the enemy's anti-aircraft defences. It was a well attended meeting. The 'gap filler' receiver from 50 cm to 10 m has been in the nature of a competitive effort between TRE and Wembley. Our set is to go forward at once. The TRE set, which unfortunately seems ill designed, has not yet worked properly and Gavin seems disturbed that effort should these days be expended on what is not effective.

December 4th

Marris brought back a full report of the AIS meetings with TRE attended by Peters and others. There seems no chance of Peters coming in as leading manufacturer even if he eventually does so as shadower. All informal enquiries indicated that our AIS contract should be renewed in some form. The tussle between the design of Radio Dept. MAP, to use Peter's shop for model work and the GEC desire to use it for normal production, is a troublesome one. RCM is requiring all his facilities really.

I brought Cockcroft (Superintendent ADRDE) back to Wembley from town to discuss our searchlight tests which are to settle a political rather than a technical issue. He told me of the difficulties with GL Mark 3 and the possible use of our 25 cm gear of which he thinks highly.

We have had large new valve orders which Le Rossignol says bring the total value to something like £250 000.

December 5th

There is to be but one day's holiday at Christmas this year! But the establishment seems in good heart and moderate health. We strive to keep everyone who seeks to leave us—using the powers to the full. We are using many half-time workers—but still we are threatened by the withdrawal of key men for the army.

Hankey Committee this afternoon on the subject of the 'arsenal' facilities of the country.

December 6th

Had word (through Marris) that the flight tests with our AIS Mark I at Littlehampton are going well and that Edwards, who is handling this, is a very acceptable representative from us for 'selling' the set to the pilots. More talks C.C.P. and Dr. Walmsley—Marris and Dee (TRE) to try to get some concensus of view regarding the programme for our Marris-Espley group, in relation to the renewal of our AIS development contract. Walmsley sent me a draft of the specification for this which at first glance looks all right.

Re. our proposed extension for preproduction, I hear from Ackworth that Mr. Giggins was told at MAP that the financial arrangement agreed meant that the GEC would get an extension to their Labs. after the war at a knock down price, and therefore MAP have been wanting to cut down the buildings in all possible ways so as to frustrate this offensive scheme and render them useless to the GEC after the war. So that is what appears to be at the basis of all the delay and prevarication!

December 8th[234]

Japan declares war on USA and has got in some shrewd initial blows.

The preproduction meeting showed most types to be out of their main troubles—but VT 90 has had some recurrences of mechanical defects and wastage is considerable.

Had Dunsheath here to discuss his memo on post-war research and to see the work of the place. I strongly criticised the attitude shown in it to put in the shade the research done by the big industrial laboratories in favour of NPL, ERA and the like. He entirely agreed—the commercial competitive influence is the bugbear and causes secretiveness in utilising the research in ways which might help competitors. Dunsheath's big question mark was why were the GEC Labs. in such a different position in these ways than other concerns. The answer is that the Directors led by Lord Hirst have encouraged this attitude, particularly at the outbreak of war.

CVD meeting all the afternoon (3½ hours). Full of interest and of useful matter but nothing outstanding.

December 9th

A diamond die technical meeting—very dull. We were able to report dies between 0.0005 and 0.0006 of Leeds's make in use in the factory—but that diameters as low as 0.0004 could not be polished to size without opening them the additional ten-thousandth of an inch.

Gavin had a talk with Mr. Hecht (MAP) and Dr. Cockburn of TRE about RCM sets. It was necessary to tell Cockburn that his department's proposed receiver is not a very reputable design from the point of view of modern standards. He took our suggestion well and then discussed with us certain air-borne jamming equipment, which we agreed to push on with if desired.

Dudding is having his hands full in handling all the new female personnel for preproduction and like work—especially with all the changes in rates of pay which are going on. He is preparing for a lecture on the statistics of inspection for next Friday as a result of Sir F. Gill's activities.

Winch is progressing with his searchlight photometry with ADRDE. Fog and mist make it difficult.

The visit of Jesty and Sharp [Sharpe] to TRE to give a projection tube demonstration showed marked improvement owing to our new short-decay after-glow powder. But the colour trace tube ('Skiatron') was most in favour and it is expected that it will be chosen, rather than the new powder screen.

December 10th

A ragged day. CVD policy meeting till 2.30. I suggest that CW CVD meeting under Shearing's chairmanship should include industrial representatives.

News of sinking of *Prince of Wales* and *Repulse* by Japanese— a very serious matter.

We are to have interchange with BTH of information on 3 to 10 cm techniques in connection with Gavin and Megaw's extension of USW programme to include this. The number of receiving sets, transmitting sets and wavemeters being developed for RCM and the like is beyond most people to grasp. The job is to get the semi-skilled labour for them and also the support of the factories. There seems to be some prospect after all for Peters to have some share in AIS Mark II.

December 11th

A searchlight panel meeting—the first for two years—to discuss the doings in carbons since the war. Not much experimental work—mainly on higher current carrying capacity and greater resistance to shock. Mr. Couch here to discuss compact source lamps for our Admiralty contract therefore. He had a talk with the panel—and quickly agreed with me regarding the contract. He had heard nothing of the move to reorganise the searchlight carbon committee and I agreed to advise Dr. Vick as soon as possible to contact him.

G.W. Edwards looked in fresh from his adventures in installing and flying Mark I AIS. He has enthusiastic reports from pilots who have used it and permission has been given to use it against low flying mine laying planes.

Gossling reports much interest in crystal detectors for centimetric ROF—Ryde's capsules having stood up to the stresses.

Lord Hirst and Mr. Railing summoned me to Head Office and the latter asked me to join the Board of the GEC. I accepted the invitation, with no doubts in my mind as to the additional responsibilities involved in the step!

December 12th

Discussed with Heather his projected approach to DRP (Leedham) re. the use of Peters's facilities—the RCM work and possible AIS or other sets.

A CB [Consultative Board] meeting at which my colleagues were very charming. The Lab. finances are very abnormal and it is difficult to judge how they are shaping. Credits for about £50 000 this month for payment for valve sales and for development contracts.

Lecture by Dudding and Jennett to the Association of Production Engineers on inspection and statistics showing how 'inspection' should be part of a production system and not a system for testing and rejecting finished products. About 200 present. A good meeting promoted by Sir F. Gill.

December 13th

Humphreys tells me that we have some hard thinking to do in regard to the de-reservation of some of our men under the new rule, since 'Form 300' can only be used in acute cases.

December 15th[235]

Winch went to Christchurch aerodrome on Friday for flying tests with his IR detector, but experienced the same dangerous and incompetent experimental condition with the plane as has occurred with him and with Waldram in every test they have been on. On this occasion their radio went out of action and no communication with the ground was possible. This happened on a second machine—the first having a dud engine. His trial was a complete flop.

Had Chelioti here to look at some of the work in hand—he had difficulty in dragging himself away.

Had Dr. Milner here to advise us on the probable value to us of petrological techniques.

December 16th

Long talk with Marris on our function in AIS etc. As TRE have now equal or fuller facilities for engineering first models of equipment—and as the firm which is to manufacture *must* make the actual prototype to suit their shop methods, GEC Wembley have functions which are merely parallel with those at TRE unless their factory is selected for manufacture of the equipment. As Peters is not having support from DRP for mass production it is not obvious to the outsider why Wembley should be used by TRE. The answer of course is that unless they are used the originality and ability of the Wembley group is lost to the RAF war effort.

Lord Hirst saw me today to initiate me into the 'background' of the Board of the Company.

Attended a BSI Chairman's advisory committee on post-war planning in relation to standardisation.

Heard from Giggins that the Board had approved a 50/50 division of cost with MAP of cost of preproduction extension. He hopes that MAP will now agree to go ahead, so there we have one more wriggle forward.

December 17th

A post-war planning committee on research under Dunsheath's chairmanship (IEE). A very useful talk and consensus of feeling that the commercial competitive element needs to be eliminated from industrial research so that collaboration can take place for the service of the industry. It is sound to be considering post-war lines of thought, or the politicians will have it all their own way—but 'post-war' looks a long way off just yet.

Have been going into costs of our valve preproduction with Le Rossignol, later Brundrett. There are at present nine main types. We contend that a valve cannot be considered apart from the whole group of preproduction valves until it has reached about 2000. Up to this point the gains must help to balance out the losses. After 2000 a valve should be no longer regarded as in preproduction even though made at Wembley and should be subject to separate costing for the Government. Brundrett is most helpful—as always—and will speak with contracts department and then advise me how to act.

De-reservation is now coming up as a very active matter and Humphreys, Dudding and I are spending this evening considering all the cases of our own de-reserved men on January 1st. A strict enquiry into all cases seems most desirable provided the adjudicators are well informed. We may refer the tribunal to Brundrett on all our difficult cases, provided they are genuine.

Went to the English Speaking Union to take tea with overseas scientific representatives and made many useful contacts, and met Capon, Assistant DSR at MAP. He had been at Wembley with Waldram and Bomber Command officers to see about adopting our projection gear to representation of bomber targets. He was very appreciative.

December 18th[236]

I discussed yesterday with Heather and Robinson of Coventry the relationship of Bradford and Spon Street—Clark and Peters in regard to the various RCM and AIS and other sets which GEC is or is not asked to make. Particularly we discussed a recent letter from Leedham indicating that Spon Street was not going to be let into the production of airborne equipment like AIS because the Government had put down money at E.K. Cole and Ferranti's and would keep them full and efficient.

Coventry of course want to get Spon Street off mere model making work and to use their model making facilities for tooling up a regular production job. But the RCM and similar model and small scale work is insisted upon by Leedham. Marris is pessimistic about any solution to help Peters in these circumstances.

At Hankey Committee this morning, a sub-committee was appointed to help Ministry of Works and Buildings on question of improvement of war-time lighting. I was asked to be Chairman. We were also told informally of a committee for post-war lights which DSIR is appointing, sponsored by the same Ministry. I am told I am also asked to be chairman of it.

A lively TVC meeting at MAP this afternoon at which the desirability of adoption by the Navy of some of our valves now beginning preproduction (E 1231, E 1273) was opposed by Dr. Truscott on the ground that the number of types entering preproduction was much greater than the number passing out to factories—so that the position was heading for an impasse, seeing that the demand for every type in preproduction tended to grow beyond legitimate preproduction numbers.

December 19th

A lot of people are getting very tired and a feeling of things being on top of them. I think the limitation of the Christmas holiday to one day is a factor. Although news from Russia and Libya could hardly be better, that from the Far East is most disturbing—and this also has an effect. If we could work our social evenings again I believe it would make a difference.

The joint meeting at Wembley of the 'Valve Allocation Committee' and the preproduction committee on Friday under Dr. Truscott's chairmanship is reported as most successful. It gives our people first hand insight into the inter-Service needs and position. The Service people were taken round to see what it means to make valves like the E 1273 (diode) and the NT 99.

December 22nd

I was to have gone to Avonmouth today to see the diffused lighting work which Stevens and his group have been busy with for several months—but visit was cancelled, as so many are.

B.S.G. reports relatively successful ROF firing trials at 1½ metres (I think) which puts this 'hair-brained' scheme on the map.

Spent time trying to find 200 ft^2 more floor space for testing our preproduction valves. The diode, the millimicropup E 1190 and others are making embarrassing demands on an establishment already crowded to bursting. How we could do now with the extra building over which MAP have delayed so tragically.

December 23rd

Ramsay reported discussions at Signal School on the use of the CV 12 type of valve in various sets—but particularly a request that it should be perfected on the lines of the E 1209 to operate with shorter time limits [pulse lengths] for certain special operational use. This means a considerable piece of development for him. The CV 12 and also the E 1209 are being used in a number of ADRDE modulator models which they are making up in anticipation of requirements.

Megaw and Willshaw (on 3 cm magnetrons) and Ryde and Brittain (on crystals) visited BTH on Monday and had useful interchanges. The 3 cm magnetron is by no means finalised in its design because the weight of the magnet tends to become so great. Crystals have their ups and downs both here and at BTH!

December 24th

Heather attended a meeting with Leedham and Sir. F. Smith on AIS and reports, as a result of pressure from him, that Ekco had to admit October 1942 as the date for commencing delivery of Mark VIII sets and probably it would be January 1943. Peters's dates were appreciably more favourable than this and it now appears that he is to be given a contract after all and will make his own prototype! So he and the GEC have got their own way at last and Marris, who has been pressing matters from the TRE end, says that his and Espley's group will be very bucked.

I have an invitation from Appleton to be chairman of a DSIR cum NPL cum BRS [Building Research Station] committee on post-war lighting to advise the Department of Works and Buildings. It seems very like the old DSIR illumination committee revived again—but without research as its main function.

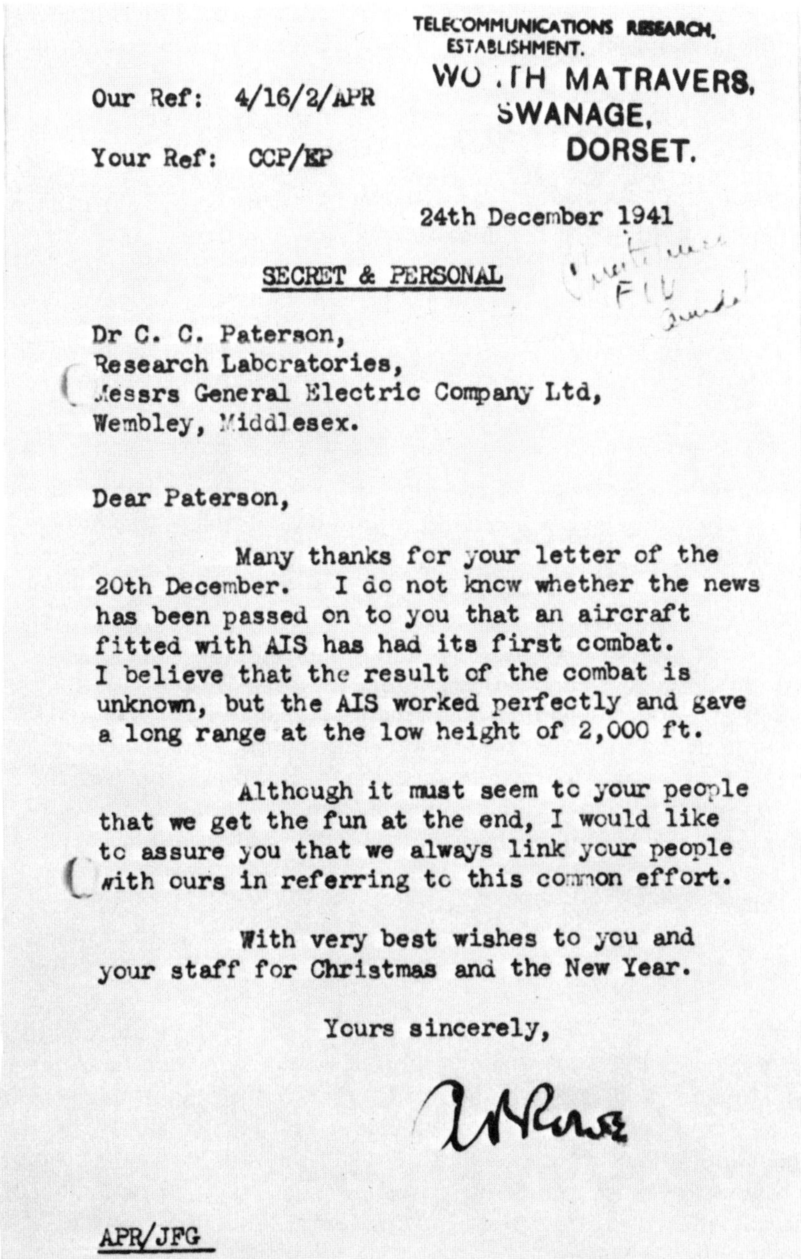

TELECOMMUNICATIONS RESEARCH ESTABLISHMENT.
WO .TH MATRAVERS,
SWANAGE,
DORSET.

Our Ref: 4/16/2/APR

Your Ref: CCP/EP

24th December 1941

SECRET & PERSONAL

Dr C. C. Paterson,
Research Laboratories,
Messrs General Electric Company Ltd,
Wembley, Middlesex.

Dear Paterson,

Many thanks for your letter of the 20th December. I do not know whether the news has been passed on to you that an aircraft fitted with AIS has had its first combat. I believe that the result of the combat is unknown, but the AIS worked perfectly and gave a long range at the low height of 2,000 ft.

Although it must seem to your people that we get the fun at the end, I would like to assure you that we always link your people with ours in referring to this common effort.

With very best wishes to you and your staff for Christmas and the New Year.

Yours sincerely,

APR/JFG

Fig.17 Letter from A.P. Rowe, Chief Superintendent, TRE, to Clifford Paterson, dated 24 December 1941

MAP have written to accept in principle the GEC offer of 50/50 share of expenditure on the new building extension! So until someone else puts a spanner in the works it looks as if the building could be started—four months late and all wasted time. Floor space is badly needed at once.

December 26th

The Lab. opens after a one day break only, for Christmas. I am afraid a lot of people will suffer from so short a break between August and Easter.

Took up with Le Rossignol and Gavin the question of pushing the concentric valve which has rather gone to sleep through everyone being too busy to keep it awake.

December 27th

Had note from Rowe of TRE telling us of the first combat in anger of AIS (Mark 7) and it seems to have acquitted itself well particularly at low altitudes (Fig. 17). Edwards happened to be up today and he gave us more details. Apparently if our man had not needed to draw level for identification purposes it would have been a sitting shot. As it was the enemy was hit and may have come down in the sea.

Major Sherlock and another from the MOS came to put some enterprise into the high-current Army searchlight problem. ADRDE appear to be conservative on the point—much more so than their opposite member (Dr. Paris for instance) at the MOS. Cameron had the two Majors in hand, with Winch and Waldram.

We hear that events in the East may affect our tungsten supplies for Heavy Alloy.

December 29th[237]

I am sure it was a mistake to shorten the Christmas holiday. Output on valves is down last week to about what it would have been had everyone had Thursday and Friday and Saturday off. And now everyone is feeling the effects of a holiday they never had. We have to take in hand the 'leadership' of our groups—and a sense of responsibility in their running on the part of those immediately in charge. We have to remember that many of them were raw when put in charge—but it is a sense of not allowing themselves to be let down that we want.

At last notification has come through from MAP instructing us to proceed with the extensions—four and a half months delay.

Gossling, from conversations with Moon, gets the idea that Service establishments do not sufficiently think of Wembley in terms of their new projects, because the idea gets round that Wembley is too busy.

A 10 cm meeting at MAP but nothing specially dramatic. Great emphasis on the 3 cm programme.

December 30th[238]

Settled details of new extension. Lunched with Ackerley to discuss advisory committee on lighting to the Ministry of Works and Buildings.

Discussed various plans for getting more esprit de corps into each of our preproduction groups.

Hankey Committee on the part of the Operational Staff in determining design. One would think that it should not be necessary to insist on this but it is as a matter of fact very important to do so.

The last day of 1941 and a feeling of thankfulness looking back that the Labs. have been able to build up and become increasingly strong to tackle its important tasks.

Iron powder for radio cores has been going well and the arrangements made for implementing our plans have been well backed and supported by all at Salford. Loading coil iron dust is now the chief remaining problem. Pitkin with Vaughan Williams's support has been making a first class job of his share—a main one—of the task.

Had an hour with Jesty's group on CRTs and the latest developments—colour trace screens, testing gear, poured, instead of sprayed, screens etc. A good show from a group which gets things done.

Afternoon meeting with EMI, Standards, BTH, MOV, Birmingham and Signal School on the vagaries of copper for copper-glass joints. All evidence pointed to traces of phosphorus being responsible for bad and difficult joints.

1942

January 1st[239-240]

Had Evans (Building Research Station) and Walsh here to discuss terms of reference and personnel for the new advisory lighting committee for the Ministry of Works and Buildings. Had some difficulty to move Evans that it would be sound policy to get Good to serve. Appleton seems to have some prejudice here.

Met (with Fleming) on behalf of Hankey Committee a small group of industrial people who urged the better use of experts from small industrial groups.

Whiddington here to see various valve matters, and he asked me if it were not possible to get more research weight with Ryde on crystals. I must look into this.

January 2nd

Lord Hankey told us at our meeting today that he had had a number of appreciative comments on the usefulness of our main report and he thought it was having indirect as well as direct value in strengthening certain tendencies where this was welcome.

A helpful letter from that good fellow Brundrett on the prices we charge for preproduction valves. He has seen his contracts people and made our position clear to them and eased the steps by which we can charge a price for the first 2000 which, taken over all our preproduction, should pay expenses fully. After 2000 a valve should be capable of price reduction and be able to be treated individually.

January 3rd

The holder for the E 1231 valve seems to be the subject of a major dispute—certain service people contending that the pins in the pinch recess are liable to get broken etc. The team here has now designed an excellent fool-proof socket which should put the pessimists to shame.

An amazing comedy to brighten the New Year. Ackerley rang me today to say that he had attended a hastily called meeting of the IES group under the sponsorship of the Ministry of Works and Buildings to make recommendations for an advisory committee to the Ministry on lighting with a view to post-war!! Perhaps the most intriguing touch is that both the Ministry representative (our friend G. Smith) and the IES group has independently decided to ask C.C.P. [i.e. Paterson] to be chairman! So we have the interesting situation of two committees in course of formation both sponsored by the same Ministry one through DSIR and the other through the IES for exactly the same purpose and each suggesting the same chairman.

January 5th

Preproduction has no serious set back but something always happens to prevent each group earning its desserts. The VT 90 quality shows troubles in operation which have been operative for two years but have only just come to our knowledge. Grid emission is one factor.

A long and cheerful CRT meeting with Walmsley presiding. Collaboration between all manufacturers is effective and cordial. Skiatrons were to the front and Sutton turned up from Bristol.

Spoke to G. Smith, Min. of Works and Buildings, on the duplicate lighting committee question. He seemed unperturbed.

January 6th

A quiet day with everyone getting on with their job. We suggested (on Jesty's proposal) that CRT manufacturers should get together to exchange experience re. cathode and heater designs for benefit of all and possibly to achieve a measure of standardisation. Jesty will call the meeting at Wembley and Dr. Walmsley is invited to preside.

January 7th

Visit from the management (Mr. Strong) of New Southgate section of STC about valves suitable for certain short wave applications of theirs.

They seemed to find what they wanted and were very acceptable visitors.

Meeting of the Modulator committee yesterday at Rugby was uneventful. The spark gap modulator seems to be coming to the fore in expectations for the future.

January 8th

My first GEC Board Meeting but clouded by Railing's very serious illness. Work of the Labs. was usual—no emergencies.

January 9th

Consultative Board Meeting at which the Lab. accounts, having received credits from outside of about £50 000, are once again level with the budget figure.

My general anticipation, overheads etc. seem to be working out about right.

Hankey Committee at which we saw representatives of Society of Chemical Engineers. Much hot air.

January 10th

Had word from ADRDE that GL Mark IV is adopting E 1190 and E 1209 provided we can see our way to preproducing sufficient of these types; must take up the point at the preproduction meeting on Monday.

Humphreys is very hard pressed with technical jobs on paint drying and other heat applications as well as having on his shoulders all the extra burden of handling de-reservation postponements. We are particularly anxious about those young fellows of 19 to 23 who combine craftsmanship with technical skill.

I received an extraordinarily nice letter of congratulation (on joining the GEC Board) from E.B. Moullin, Professor of [Reader in] Electrical Engineering at Oxford and temporarily attached to Signal School. As it really applied to the Lab. staff generally, I being merely the symbol, I attach a copy (Fig. 18).

January 12th

Preproduction is getting on pretty well—the chief anxiety being still with the magnetron group. The cure is psychological as much as physical. About 670 valves altogether in the week.

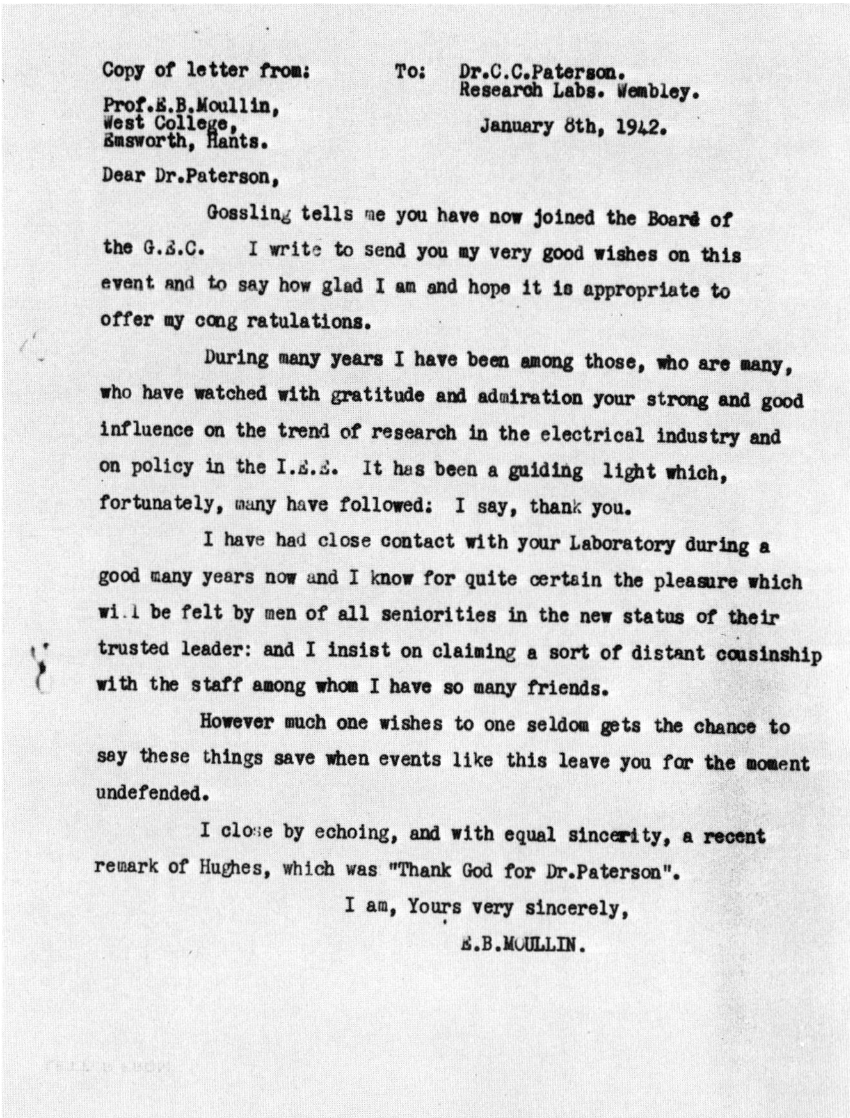

Copy of letter from: To: Dr.C.C.Paterson.
Research Labs. Wembley.

Prof.E.B.Moullin,
West College,
Emsworth, Hants.

January 8th, 1942.

Dear Dr.Paterson,

Gossling tells me you have now joined the Board of the G.E.C. I write to send you my very good wishes on this event and to say how glad I am and hope it is appropriate to offer my cong ratulations.

During many years I have been among those, who are many, who have watched with gratitude and admiration your strong and good influence on the trend of research in the electrical industry and on policy in the I.E.E. It has been a guiding light which, fortunately, many have followed: I say, thank you.

I have had close contact with your Laboratory during a good many years now and I know for quite certain the pleasure which will be felt by men of all seniorities in the new status of their trusted leader: and I insist on claiming a sort of distant cousinship with the staff among whom I have so many friends.

However much one wishes to one seldom gets the chance to say these things save when events like this leave you for the moment undefended.

I close by echoing, and with equal sincerity, a recent remark of Hughes, which was "Thank God for Dr.Paterson".

I am, Yours very sincerely,

E.B.MOULLIN.

Fig.18 Copy of letter, dated 8 January 1941, from E.B. Moullin to Clifford Paterson

Lunch with Chelioti and Gant and discussed: (*a*) Competition v. non-competition for post-war industry; everyone seems to favour a more co-operative type of industry, (*b*) The GEC – MOV – EMI triangle, and the suggestion that GEC may be blamed for all it has done for MOV in the valve field for the Services.

RCM (Gavin) meeting with the Services at Wembley. There seems an inclination to direct the 3 cm development our way in parallel with BTH! We angled for this six months ago but were turned down. It is very similar to the development we are starting under USW for the Admiralty.

January 13th

Serious consultation, myself, Le Rossignol, Dudding, Pearce, to rearrange magnetron preproduction. Falconer who has done yeoman service takes all testing, quality control and equipment. Pearce gives up all other valves and concentrates on the production part. D.A. Wright takes charge of lamp room valve group and Le Rossignol of NT 100 group. The group will have two days holiday at the end of the week and will start up anew and rapidly become, we hope, a happy ship.

January 14th[241]

A morning on Dunsheath's committee on post-war research organisation at which I felt I was doing most of the talking. We formulated some general principles for co-ordination.

Had the sad news that Mr. M.J. Railing had died last night. We shall greatly miss him although it had been apparent for some time that he was ailing very much.

Lunched with Heather and discussed GEC affairs.

Gossling and the ROF group had spent two days at Christchurch and done some useful contact work there. Our GL Mark IV (25 G) set seems to be going strong and Walter is following it with ADRDE into its operational stages.

January 15th

We are feeling hopeful that the de-reservation and calling up of key scientific and technical staff is not going to be applied to a disastrous extent—but in these matters we live from day to day.

The success of Japan in the Far East is affecting various products. Tungsten for Heavy Alloy is liable to be a difficulty and the Control requires special licences for purchases.

Had word that 50 more Helmore Turbinlite equipments are to be ordered. It is satisfactory to have so successful an issue of the joint efforts of Wembley and Witton.

A Hankey Committee this morning on the civil engineering work of the services. Lively and interesting.

Hope we have finally settled up with A.S.G. payment for our VT 90 valves made a year ago, MOV having 3 shillings per valve. It is very difficult to know whether Wembley has any real margin considering the amount of gear and subsidiary expense in which we are involved.

January 16th

A meeting took place here yesterday of Service people to consider special valve for Torpedo Factory, Greenock, very similar to one of the ROF types—wanted in a hurry. I think MOV will take it on probably.

A Shaw factory meeting showing good progress. Output about one-third possible but most of the floor area now occupied. It is a long time since Shaw has been referred to in this diary. It is becoming a valuable asset in the production programme of MAP and Admiralty and is well managed. It is curious to look back two years to our meeting at Harrogate when M.R. [Max Railing] came down and we attended a meeting at which the Government representatives seemed very uncertain what they wanted and when Leedham took up the attitude of having had a very unwanted project thrust upon him. From a purely commercial point of view we have yet to see what sort of an asset Shaw proves—but from a national and technical point of view M.R. had a sure instinct when he secured the management of the venture to GEC – MOV.

Had Signal School representative here to press claims of CV 8 (diode) valve and desiring a further expansion of numbers up to 150 per week. Had Algar along who gave hopes of doing it in spite of the need for establishing preproduction of the E 1190.

Marris has had long talks at TRE and with Dee. The position has to be pressed home that we are not an engineering and production establishment which TRE would have us be, but an experimental establishment like TRE itself and that we want difficult research and development tasks to handle from first principles.

January 17th

Our group working at Heston Aerodrome (Helmore) under Jolley seems to be taking on general functions of an advisory kind to Fighter Command on technical and scientific matters. This seems to me as it should be: it is another small step to break down the monopoly of the Government establishments in guiding the war effort.

January 19th[242]

Much worry today over the first indication of how the calling up and de-reservation arrangements are going to work out. There is rough work going on—and parallel and unconnected callings up (*a*) by a BEAMA committee stimulated by Lord Hankey, (*b*) by the Ministry of Labour de-reservation committee. The initial quota before March is nine for the RL and more to follow.

Saw Walter who had a cheery report on GL Mark IV and told me the reasons for its adoption instead of the AIS wavelength. Lobe size of the beam is the main reason.

The 'diffused lighting' party are off to Glasgow for trials. Stevens brought me photos of the gear and installation.

January 20th

Macqueen throws new light on the calling-up subject—apparently radio firms are not to be subject to call up and the technical commission's press gang should not operate on Wembley. I hope it may be true.

A very full and interesting CVD (GEC) meeting. How much more attractive Wembley CVD meetings seem than those held under strictly Services auspices! The most important contribution was a new triode valve by the receiving valve group (E 1344) which goes down to 9 cm CW, continuously variable on the fundamental [mode]. Ryde had interesting work on constructing spark modulators in glass envelopes under pressure. Gossling showed 12 months collection of valve chart monthly slips with the progress of each valve marked on the web of time.

January 21st

Jesty gave good report of the meeting at MAP of CRT manufacturers on cathodes and heaters which on his suggestion we had fostered. Also a satisfactory visit to TRE to go over their needs to try to reduce the number of bulb types.

Lunched with Fred Smith of MOV and talked over various difficult problems, personal and otherwise.

January 22nd

A CVD meeting at the Admiralty. The discussions were calculated to make me conceited in respect of Wembley work and standing. Great pressure applied for priority to be given to Megaw and Gavin's work on short wave CW set because, although various people are on it, Shearing feels that GEC is the only concern capable of producing a satisfactory solution on account of their combined valve and circuit facilities. Some amusement (to me) by Dr. Black's attack on 'preproduction committee' because they have not produced for him the one or two dozen EMI samples of velocity modulated oscillators needed for his first GL sets.

Preproduction (Brundrett) passed blame to Truscott (IVPC) and IVPC (Truscott) passed it to Whiddington (CVD) and twice the ball went round! The fact was of course that as compared with Wembley's preproduced valves EMI had not put themselves out to make sufficient hand-made samples to keep the establishments going—nor do Wembley boggle about who shall pay for tools—but pay for them themselves.

The meeting agreed my proposition to have industrial representatives present at the CVD CW meetings.

Espley gave me a report on his activities with Edwards and Hawes at Christchurch getting AIS Mark VIII into form. They have had very great troubles of a technical kind which could not have been dealt with apart from practical trials in the field. The troubles now seem to be surmounted.

Waldram showed me the addition we have added to Service epidiascopes to represent targets as they will be seen by pilots.

Schuil has been working here on a special photometer for his camouflage studies with the inter-Service camouflage establishments.

We have been offered a dozen young trainees!

January 23rd

OAC meeting—the first since Mr. Railing passed over. Heather attended for the first time. It was a good meeting because everyone was 'natural' instead of certain members play acting before the senior Directors. Called for Brundrett and brought him out to the, now, two monthly preproduction meeting.

We had to sit in sackcloth over the NT 98 magnetron on which we are so sadly down in deliveries—only half the number we led the Admiralty to expect. The rest of the valves were pretty well up to expectations—barring cathode ray tubes for AIS which are again behind.

We had very frank comments and warnings from the Service side on the blundering ignorance and ruthlessness of National Service press gang.

Men are apparently to be extracted without regard to the importance of the work they are doing.

January 24th

Bligh reports the first air trials of their frequency modulated sets are to take place on Tuesday and he speaks of exceeding expectations on sensitivity. Ministry of Supply tell us to get on with the work but are very slow in issuing contracts. We are collecting all our priority contracts numbers like scalps, to push to the attention of the press gang.

January 26th[243]

At preproduction we discussed the groups for E 1336 (a midget valve for torpedo school) and E 1344 (the new 10 cm triode). It looks as if the former would be handled by Newton and the latter by Cutting, both under Douglas's supervision. Knowing the torpedo school needs and the requirement to keep McLeod's shop free of new production jobs, everyone felt it best to organise independently of his shop.

Went with Cornwell to look over the standby accommodation for the laboratories at the Empire Exhibition. It is going on surprisingly fast. LSO and Strachan's office and Minters have done well to progress as they have.

January 27th

A meeting at Head Office with J.Y.F., L.G. and T.W.H. and G.C. to discuss the arrangement to conclude with Levy & West for the manufacture and control of luminescent powders. Substantially the scheme discussed by Chelioti and myself was agreed if L & W will also agree.

There is still a bad lot of absence amongst women through illness. It is puzzling to know why this should be. Pearce in his magnetron group had all four glass workers away.

January 28th

Humphreys and I lunched with Captain Bennett of the Fuel Utilisation Research Association and two of his people. They offer collaboration in the application of electrical equipment in the use of solid fuel in domestic use. We agreed at once because they are serious and also enterprising people.

January 29th

A meeting at Head Office to consider for a few hours the domestic appliances position of the GEC for the future so that the normal activities of the company and its interests are not completely overlooked. It was little enough we could do—but I think what we did encouraged us all. Gamage, Heather, Winstanley, Cattell, O.W.H., Ackworth and Duncombe were present, also Settle.

I gave a lecture to Polish engineers in the evening—with the help of Jenkins and Bowtell—Electric lighting brought up to date.

January 30th

A meeting with Dr. Railing, Leslie Gamage and T.W. Heather on our service contracts. Radio bulked largely particularly our openings in frequency modulation and RCM 10 cm applications using the E 1344 valve. Lunched with Shearing of Signal School who urged our speedy attention to the above 10 cm work.

TVC meeting at MAP and lively discussion on interdepartmental and standardisation matters.

January 31st

The air trials of the frequency modulation set prepared by Stenning and Bligh and their group were apparently very successful. The transmitter (at 70 watts) was in the plane and speech was carried on with Wembley as far as Exeter. I understand the Farnborough receiver was not so sensitive.

Ryde tells me that his crystals are apparently up on s/n ratio by some 4 dB—a considerable amount if true.

February 2nd

Preproduction reached a record of over 700 valves last week—and the NT 98 (magnetron) pulled back to about 50. But the absences of women through illness are unbelievable. They all give doctors' certificates, but either they are very unhealthy or doctors' certificates are easily come by.

I am afraid it is mainly the former. I believe half the people are just worried—no source of peace within themselves.

February 3rd

Battery meeting at Head Office (T.W.H. unfortunately away ill). Discussion on layer batteries in view of present and future demands. There will probably be a MOS request to develop a small battery both in layer and in cylindrical types.

Lunch with Helmore but little business talked—mostly about persons and his broadcasts. I always enjoy a talk with him. There is probably a contract coming for the general odds and ends which turn up at Heston wanting study.

Winch reports good progress with his various infra-red IFF detectors. He gets 1200 yards with a 75 watt lamp and 7 inch mirror scanning 20° and a plain photocell and two valves. There are several applications under study.

February 4th[244]

To diamond die control advisory committee. Was able to report Leeds's technique for drilling slowly so as not to damage the walls of the drilled hole. He finds that efforts to hurry by the use of coarse powder and heavy blows causes incipient defects which cause flaking of the surface.

I lunched with Prain who pressed the need for a specification for stones.

To H.G. Jenkins's paper on luminescence at the Royal Society of Arts. A full attendance with Kennedy presiding.

Report from Stevens (who is on the Clyde with his group) to say that their 'diffused lighting' is much less of a flop than they feared it would be—and some who came to scoff remained to pray [praise?].

Dudding, amongst his many other things, is much occupied with the monograph by himself and Jennett on statistical control and its application to production. It is in the first place for BSI.

February 5th

Saw Machin and Macfadyen to examine all the modulators their group has been building. An outstanding effort seeing that the gear has had to be built up by themselves in whatever corner of the lab. could be found. We have at last got life test capacity for Ramsay's valves and an opportunity of telling the Services where we and they stand as regards modulator life.

Waldram back from Arran where the diffused lighting trials are in progress. He is very pleased with his Nephelometer which he went there to test.

February 6th

Saw Gossling and Bell re. proper life test arrangements for all preproduction valves. We have to be sure whether we or the Services are going to be responsible for life test. So far as possible we should be.

Had ring from Espley who is at Christchurch to say that our 25 cm GL Mark IV is meeting with still more favour and GEC is asked to look into possibilities of manufacture as soon as possible.

A diamond die technical committee at which Dudding and I managed to get through a rough-out for a specification for stones for die making.

February 7th

Pearce with his magnetron production has had a much better week, and there is now a chance of making up for lost production. The E 1231 and 1273 are having their troubles, particularly the latter.

February 9th

A day at TRE Swanage where I went to talk over things generally with Rowe. Dr. [W.B.] Lewis and Dee joined in later. They were anxious that GEC should make itself (for its own sake) able to be still more useful to TRE by closer collaboration in its designs with Spon Street so that it would 'design into the Spon Street factory'. Rowe told me a remark of Dee's that of all the firms he dealt with he preferred his contracts with GEC. I must see what can be done in conjunction with Peters.

Various possible pieces of work were talked over for Wembley under our new contract with MAP to replace '25I'.

Rowe was red hot on the subject of men in industry who are not pulling their weight where they are and could be effectively used at TRE. He exonerated GEC—or at least Wembley.

I afterwards saw all their training devices for AI—most helpful in visualising things, particularly GCI.

February 10th[245]

Walter (GL Mark IV) back to make another prototype. He says everyone is pleased with progress and emphasises what Espley reported.

Last week most of the main preproduction valves went well and we had a message of thanks from Signal School for them.

Had Colonel Becher here from National [Central] Register, and a colleague to see something of the work of such a place as this. As he has a say in the calling up of technical staff it was a very useful opportunity. Humphreys and I did not by any means throw it away and Col. Becher seemed convinced that we were a model establishment in the employment of staff.

February 11th

The news these days is not bright, Singapore looks as if it would go the way of Hong Kong, and Libya is by no means a romping success. Only the Russians seem to be showing signs of successful operation.

Attended 10 cm meeting this morning—but by the time I left there was no particular excitement. Randall has a modified magnetron which is more elaborate to make and shows certain limited advantages which may or may not outweigh the disadvantages.

Back to Wembley to meet Robinson (of Coventry) who came to get general impressions of the radio work. He speaks of coming every three months—which I hope will result in a better liaison between all concerned.

February 12th[246]

ROF trials (firing) are reported satisfactory at high levels, so the general scheme is likely to go forward.

Warren's substitute for the RL 18 valve has shown promising results at 48 cm—which is what was required.

Bomber Command (Mr. Dickens) [Dr. Dickens] have written warmly of our assistance with their epidiascope representation of the 'target for tonight'. Illumination department have been doing this.

The Birmingham strapped version of the normal magnetron is reported to give slightly higher efficiency and a more favourable position of 'frequency jump' so there is a flap about it and we must at least get our drilling jig under way in case the valve is wanted as a type in preproduction.

February 13th

A Consultative Board Meeting at which I was able to indicate the following approximate financial position:

Amount outstanding to be allocated at March 31st . . .	£160 000	Receipts to come from Service contracts . . .	£70 000
Additional for Feb.—March . . .	£40 000	Receipts to come from sales . . .	£30 000
	£200 000	Balance to be allocated at March 31st . . .	£100 000
			£200 000

This assumes all stocks of materials written off and all equipment obtained for preproduction etc. also written off. But we shall see in due course what are the actual figures and whether this financing of our effort passes the auditors.

February 14th

A good deal of talk over the revised Birmingham design of magnetron for AIS. We think it a gross case of unnecessarily introducing a new component when an equipment is designed and just going into production. But from the CVD point of view we have to show willing and are putting in hand six or twelve samples but are not going to let them interfere entirely with our 3 cm magnetrons which we are just now getting going.

February 16th

Everyone is very angry and put out over the news. Singapore is lost and everywhere we seem to be 'facing overwhelming odds'. People ask why? Churchill asks for confidence and people are ready to give it to whoever will handle things with real foresight and sense. At present if anything happens the remedy is more 'controls'. However, if we look at things from the enemy's point of view it must look even worse to him, though Japan's successes will have bucked him up a bit.

Our valve making goes on reasonably well and on the whole we are meeting our obligations. Fowler and Whiddington were both here today to keep abreast of the work. Whiddington is specially welcome because we have found it difficult to interest him in our more fundamental work on receiving valves and the like.

An Advisory Board meeting at Head Office where discussions were brisk. Discussed with A.R. and T.W.H. a reply to a letter from Leedham suggesting that all radio development facilities are now 'saturated'.

February 17th[247]

A cathode ray tube meeting and very full discussion of many phases of the subject. DCD (O.F. Brown) visited us to study valves and infra-red (FF) [IFF?] devices and we found him a most receptive and helpful visitor.

Had an annoying invitation to Birmingham in glorification of Dr. H.T. Randall and his work on the production of luminescent materials. Good, whom I rang up, sympathised with our attitude. I have written a letter which I expect is too strong.

February 18th

Spent a morning at Dunsheath's committee on planning for post-war research and found myself doing most of the talking. The chairman showed some signs of cold feet but we pressed that it must be either pre-war conditions or an entire change. Some of the girls in the office are asking to help in the war effort by evening manipulative work—that's the spirit.

The new building extension is starting in spite of the continued cold and unfavourable weather.

Went into the cost of our diamond [die] making with Leeds. It ought not to leave us short if we can obtain 'control' prices from the Works.

February 19th

Gossling attended yesterday the first meeting of the CVD (CW Section)—it appeared to be uneventful.

Dudding and Jennett are drawing to the end of their monograph for BSI on control charts for quality of production. Frank Gill is pushing the business through the medium of a meeting of the three Engineering Societies—but I cannot see that it will do anything but stir things up, for them to settle back again unless some real missionary work is done.

I have heard that the proposed members of my lighting committee for the Ministry of Works and Building (through DSIR) have now been invited to serve. Post-war planning does not look a first priority matter at this juncture! I was glad to have Lord Hirst's views on post-war conditions particularly in relation to our deliberations on Dunsheath's committee of the IEE. He pressed for recognition by Government of interworking arrangements within industry.

February 20th[248 - 249]

A modulator meeting yesterday at which Gossling said the discussion was mostly about procedure. A clash with Mr. Yeoman Robinson on the small hard modulator and ensuring that his valve is interchangeable with Bell's E 1271. Ramsay reports promise of marked improvement in performance of his gas modulator by an 'ionisation' impulse before the actual pulse. This should stop local intense and unevenly distributed electron emissions.

Smithells came for an hour or so and to lunch. He looked round the place and it was delightful to have him about again. His sparking plug is a really good job.

Valve allocations meeting in the afternoon—with Truscott in the chair. Very businesslike and interesting. The Navy have great difficulty in deciding between the possible production of the NT 99 and the NT 93 (and 97). We finally decided on 65 of the latter and all possible of the NT 99s.

A large meeting with the Torpedo School at Wembley re. the E 1335 valve being made here and at Hammersmith. It appears to be going better than expected.

It looks as if the volume of our CRT work was outrunning the contract.

February 21st

Arranged with T.W. Heather for Mr. James of Chamberlain & Hookham and Mr. Southwell [Southworth] of Witton to go to Christchurch on Thursday to inspect our GL Mark IV and see whether either could take on the manufacture in spite of their not being radio manufacturers.

February 23rd

A record week for preproduced valves (745) notwithstanding absences through illness which were serious in most groups. Even the E 1231 is beginning to look up, and only the diode is still in the dumps.

Considerable Home Guard and immobilisation plans for machinery and the like for which we have to organise against invasion. It may have to be done at 15 minutes notice.

An RCM meeting at Wembley at which Gavin was able to give a good account of our activities. There is probably to be a meeting at which all contractors meet the Air Ministry officers, but Mr. Hecht tells me that they will probably decide to continue the GEC meeting at Wembley; I asked them to do so.

Helmore called, and we had two hours talk on contracts and particularly on patents and his share in any after-war activities on very high current arcs.

Just before 5.0 Mr. Rowe came from DRP to see about our plans to supply getters to the whole British valve industry. Britten's plans are progressing but he wants us to run double shifts till we get our increased production going. We need draw-benches and other things fixed up.

February 24th[250-251]

Gavin seems pleased with the visit of officers yesterday. The concentric triode is giving useful results down to 20 cm CW and one test went to 11 cm.

Francis reports progress with compact source lamps to 10 kW and 15 kW and West Drayton has arranged air cooling arrangements of much more efficient type. Searchlight and beam trials are to begin at Wembley shortly and should be exciting as the first time really high power compact sources have been run in searchlights.

Dudding has offered, to Sir F. Gill, talks on practical works control chart applications by way of missionary work as soon as he likes. He seemed to appreciate the gesture very much.

Peters and McAinsh came for a talk with Marris, Gavin and myself on the TRE desire that the GEC Research and Works team would act more as a unit. We came to the conclusion that very much was already done along these lines and that the solution was psychological rather than anything else.

February 25th

Had an estate meeting to review immobilisation plans in case of invasion—we must get the details of ours perfected. Humphreys has this in hand for the time being. Had long talk with Marris on subject of magnetrons for AIS and afterwards with Walmsley in order to make clear the extent of the gamble in going for the Birmingham version of the small magnetron. The gamble is (*a*) the uncertainty with a new thing that the unexpected may turn up, (*b*) the delay in getting samples is undetermined and may be considerable—which holds up the finalisation of the sets.

Gavin had news of satisfactory decision by the RCM special (Service) meeting yesterday. Two contracts are to come to us; (1) for airborne jammers, (2) for general collaboration with TRE on jamming. The meeting decided to continue the special GEC meeting. So all is as we hoped it would be—which is a tribute to the way Gavin has inspired confidence in all concerned.

Attended meeting of Home Guard in evening to present cup for shooting. The GEC unit seems efficient and keen in spite of complaints of poor facilities and discomfort. Major Castle spoke to me seriously about his continuing as OC of Home Guard.

February 26th

OAC [Osram Advisory Committee] meeting at Hammersmith—attended by Hugo Hirst. Chelioti exposed the extremely serious position re. tungsten wire and rod supplies—and molybdenum. Another case in which a need has been known long in advance but no one is sufficiently vigorous to push and apply the remedy against the dead hand of the controls—particularly the financial controls.

Hankey Committee in afternoon on civil engineering facilities. Arranged to see Hankey about the tungsten situation.

A new camouflage job for the Admiralty has come to Illumination Engineering Group who can use the diffused lighting technique for it.

February 27th

Searchlight panel meeting with the usual strong penchant of ADRDE for Morgans [Morgan Crucible] in evidence. Thomas had fixed up a cathode ray tube method of inspecting and probably measuring crater brightness of carbons scanning a photocell across various diameters. Very pretty.

To Hammersmith in afternoon to get particulars of the history and present position of tungsten supply for the rapidly expanding valve industry. If I see Lord Hankey on Monday it will be to urge a change in financial control whereby a Ministry on the technical side can give an ITP to proceed with a scheme up to a certain amount so as to lose no time in authorising a manufacturer to place his first orders.

Gave L.F. Britten a help with priority for his much needed draw bench for expanded getter production.

February 28th[252-254]

A Saturday, but full of news. Gossling returned from a round to TRE and Christchurch. He had a serious discussion with Professor Ellis at

Christchurch on ROF and other projectile borne devices. ADRDE are inclined to concentrate on the proximity fuse for which the key valve is the main uncertainty. Gossling gathered that Wembley were the only people they would wish to collaborate with on the valve side, and that we may be asked to take some responsibility in connection with its production. If this comes along it may mean an expansion of the ROF group.

The altered scanner for O'Kane's job at TRE is just finished and is being sent to TRE. I have arranged for O'Kane to work at and with TRE for 12 months on his very important job which at this stage must not be mentioned.

Megaw returned from several days' trial with his 9 cm transmission and reception of speech. His group have been making a set using a resonant cavity magnetron as transmitter and a E 1210 magnetron as local oscillator. He seemed very pleased with the results having got good speech reception at 50 miles. Probably the first time a speech transmission has been made at this wavelength. The Service Establishments (Admiralty & Army) seemed impressed.

Jolley, Hawkins and Buswell brought up their map and document destroying device which had just been demonstrated by Helmore to Sir Archibald Sinclair and Sir Charles Portal. They were so impressed with its effectiveness that they discussed the possibility of postponing a certain important operation until it was available.

Marris report of the visits of Mr. Southworth (Witton) and Mr. James (Chamberlain & Hookham) to ADRDE (to study our GL Mark 4 with a view to production)—did not give strong expectations of a satisfactory outcome—but there may be a chance with Witton, particularly if they would be willing to work with Peters. The testing of the gear might be a difficulty except by really expert radio people.

Dudding mentioned report from Salford of very favourable iron dust core characteristics as a result of substituting kaolin insulation for the present Bakelite materials. Partridge suggested kaolin as the result of Rooksby and Chirnside's diagnosis of strain produced in the iron particles by the insulating medium covering them.

March 2nd

Letter from Sir Frank Smith appointing Walmsley to co-ordinate research and development contracts on behalf of MAP. The reply gives me a chance to rub a few points in. Walmsley rang me today to say that he had no need to 'co-ordinate' us, but hoped we would be able to help him in advising some of his lame ducks.

Preproduction was 744 valves—constant illness keeps us from doing what we might and it is disappointing to those in charge who are pressing and scheming all they know to push the total up. The E 1273 (diode) is still a very lame duck, but Warren came to see me and to give me quite a hopeful picture.

Gossling reports that TRE are pressing for 100 E 1271 (hard modulator). We are making four to six per week by hand and that is all we can manage. I hope MOV are getting along with it.

Lunch with Chelioti and had interesting general talk. Wire quality is one of the things to be looked into.

Hankey Committee—strenuous conversation at which Gough was present about relative places of Army and Ministry of Supply in directing development. I think the chairman got tired of it. Had private talk with him and Tizard about inability of MAP to give an ITP to Lamp Works to order machinery for urgent tungsten wire and rod production. Two months have already been wasted and there is going to be a bad shortage.

March 3rd

At Head Office to see H.H. Saw T.W.H. and Mr. Wells when Heather and I urged the plan of a regular meeting on transmission and Service reception with Clark, Peters, Wembley, HO and Robinson. The constant new sets emerging from Wembley to be made by someone might just as well be manufactured in the GEC but were apt to pass us by due to lack of virile and concerted action by GEC establishments.

Much talk and uncertainty in Marris's tests of magnetrons. According to Megaw the upshot appears to be that the Birmingham magnetron has *not* an increased efficiency as claimed and is likely to be troublesome to make and not worth changing to compared with the E 1324. But the decision rests with TRE and MAP.

Dr. Vick is wanting us to take up two more jobs—magnetic testing of bullet cores and thermo-junctions mounted in bulbs.

March 4th

Discussed with Gardiner, Breadner, Dudding and Foxon means of linking up mechanics with the glass workers in the development of suitable tools for improved glass techniques.

Lunched with F. Smith, MOV, who told me that the valve making activities at Wembley are well known and recognised by Schoenberg and that Smith is reporting regularly to MOV Board on that matter.

At Hammersmith to Services Valves meeting. Our people disappointed to hear that E 1305 (grounded grid) valve of Warren's is probably being transferred to Mullard for extensive manufacture.

Heard from Rowe (TRE) agreeing my proposals re. O'Kane's attachment for work for 12 months at TRE.

March 5th

Attended with O.W.H. [Humphreys] a meeting at the IEE to present and foster the '6 months extensive training scheme' for young men in industry, the object stated being to 'up grade' them to professional engineering status so as to help to raise the general level of engineering ability in the country. It had a somewhat mixed reception—and in spite of certain lukewarm support the impression was left that it is an academic idea which will probably fizzle out.

The most poignant news is that Lord Hankey is no longer a member of the Government and I believe going almost into retirement! He is a fearless and helpful civil servant whom everyone respects and I am very sorry to hear the news (Fig. 19).

Professor Ellis of ADRDE comes this evening to spend the night with me and to talk over ROF and proximity fuse matters.

March 6th[255]

The upshot of my talk with Ellis was some tentative 'heads' of an understanding that we would be willing (subject to MOV) to act towards Chesham Hivac factory in the same way as to MOV in connection with proximity fuse developments for MOS and ADRDE. If this comes off (as Gossling discussed a week ago) we shall probably ask Donaldson to take the lead of a group.

Megaw tells me that the Birmingham magnetron instructions sent to us must be all wrong because the sample valves come out wide of the mark on wavelength. A considerable flap is working up on the gamble of assuming that this magnetron will yield a uniform product in manufacture.

March 7th[256]

A good deal of 'feeling' in some quarters over the E 1305 50 cm receiving valve whose Mullard form (RL 37) we have helped them to evolve. We are asked to discontinue our development, as manufacture will be at Mullards. A bit of tactlessness on Truscott's part by ignoring the recognition due to the Wembley team for originating the valve. Nobody minds passing over a design to any manufacturer provided the design is

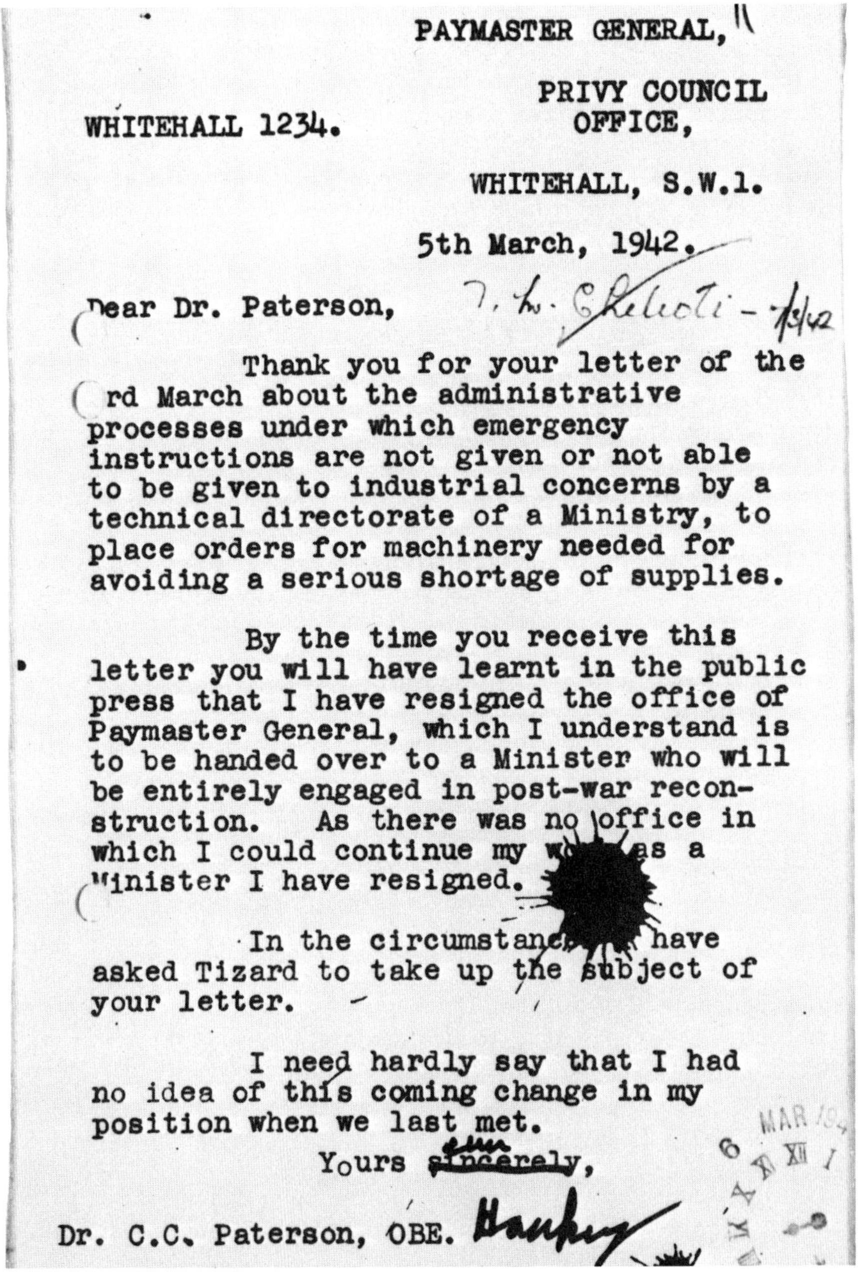

PAYMASTER GENERAL,

PRIVY COUNCIL OFFICE,

WHITEHALL 1234.

WHITEHALL, S.W.1.

5th March, 1942.

Dear Dr. Paterson,

Thank you for your letter of the 3rd March about the administrative processes under which emergency instructions are not given or not able to be given to industrial concerns by a technical directorate of a Ministry, to place orders for machinery needed for avoiding a serious shortage of supplies.

By the time you receive this letter you will have learnt in the public press that I have resigned the office of Paymaster General, which I understand is to be handed over to a Minister who will be entirely engaged in post-war reconstruction. As there was no office in which I could continue my [illegible] as a Minister I have resigned.

In the circumstan[illegible] have asked Tizard to take up the subject of your letter.

I need hardly say that I had no idea of this coming change in my position when we last met.

Yours ~~sincerely~~ ever,

Hankey

Dr. C.C. Paterson, OBE.

Fig. 19 Letter from Lord Hankey to Clifford Paterson dated 5 March 1942

not thereby ascribed to the manufacturer, instead of to its actual originator.

Had Pitkin to the Labs. (from Salford) to consult with Dudding and Jones as to the use which we can make of our tungsten making plant at Wembley, to enhance our own supply of tungsten rod for valves, in view of the approaching shortage. Can get no answer until they have looked into things more.

Got Walmsley's agreement to put the E 1342 magnetron in the same priority as the Birmingham 707, so that both can be developed further before choice between them is made.

March 9th

Preproduction was 830 valves last week. B.P.D. showed curves of rise, which is fairly steady. All goes well—considering. But the magnetron, and what is acceptable in wave jumps, is our main problem at the moment. The CV 8 is desired for a still higher current and will be difficult to handle at such a rating.

Letters are now in explaining that the E 1305 valve decision is purely on production grounds—and it appears to me that the same course of events is likely to be repeated so long as MOV are unable to take on new manufacture.

Couch came to settle up points re. the compact source contract and also the Morgan non-collaborator issue.

Attended meeting at the Royal Society on how to get better information and liaison with Russia. An interesting picture of things there—patchy but with possibilities and most difficult to establish contacts.

March 10th

An investigation of priorities in the workshop with their excellent system of allocations. We have long reached the position where the workshop is booked up months ahead on main essential jobs, and nearly all radio sets and many other items are made in local workshops and laboratories. Leeds and Gregory have decided to instal their own diamond machines and shafting.

Chirnside and Doxey [Dauncey?] showed me a delightful method for estimating minute traces of sulphur in nickel by burning in oxygen and treating the gases passing off.

A four hours CVD (GEC) meeting with long discussions on many subjects—the Birmingham 707 magnetron being the most contentious.

Further examination does not indicate any easement of the valve as a production proposition.

March 11th

Jolley has been asked by Helmore if we will arrange an experiment for sinking the *Tirpitz*. Like most of Helmore's schemes it is a big idea launched without a great deal of regard for our resources but with unlimited confidence in our ability to surmount difficulties!

MAP are after our Exhibition Building (alternative accommodation) and have applied to the Admiralty. They want it for tungsten manufacture and asked the management of Osram to have it inspected. Osram were told that MAP understood entry could be obtained through a man of the name of Paterson but added the warning that he might be found to be a difficult customer. Chelioti had had a good laugh over the incident. I take it as an unsolicited testimonial. I am afraid we shall have to give the building over.

Attended a CVD policy meeting which was full of interest. Had a very nice testimonial from Walmsley at the meeting, who reported the universal opinion from radio concerns round the country that the GEC was always helpful in collaboration and that BTH also had a good name but he could not say as much for the others.

March 12th[257]

Attended a tungsten wire meeting at No.7 factory with Chelioti and Dudding to dip into a rather difficult and troublesome story of loss of wire quality for coiled coil filaments. Split wires is the great trouble and whilst concentration is on uniformity of powder packing in the pressed bar it is not established that that is the real trouble.

Met Captain Margesson as new GEC Director at the GEC Board Meeting. He is coming to the Labs. next week.

Had talk with T.W.H. about the procedure and objects to a meeting at his and my instigation between the leading personnel of the GEC Radio and Factory departments. I hope this will develop into a general clearing house for all radio matters.

March 13th

Leeds advises considerable requests from Wire Dept. for diamond dies—mostly large sized. He and Godfrey are working late installing more drilling machines and the section is getting into good shape.

Meeting at MAP for all radio research industrial establishments under chairmanship of Walmsley the co-ordinator. There was a readiness to collaborate but the larger concerns pressed, and it was agreed, that interchanges should usually take place as the result of a specific action by Dr. Walmsley.

A TVC meeting this afternoon at which the final preferred list and Class II list of valves was up for passing. A vigorous protest from Dr. Truscott on behalf of IVPC led to a long discussion on whether the number of valves was greater than necessary. We compromised on a minute regretting that the list could not be shorter.

I was glad to note approval of E 1210 as oscillator at 6 and 7.5 cm for Megaw's CW transmitters which we gather Bradford are ready to make.

March 16th

815 valves preproduced last week.

Walter reports to me that the prototype GL Mark IV set on operational trial at ADRDE has been completely burnt up under conditions difficult to explain. It appears to have urged them to press us to strong efforts to replace.

A special magnetron meeting to try to see our way through wave jumps, frequency variations and many other troubles to which the type is heir. I hear it was very helpful.

A very long (adjourned) TVC meeting from 2.30 to 6.0. On the whole useful—but I get the idea that these young bloods in the new governing classes (Ministry controls) will soon grow into Nazi ways of intolerance if not kept on the rails.

March 17th[258]

Bell tells me that the main receiving valve of the German RDF set captured from the French coast is extraordinarily like Warren's E 1321 valve. The rest of the set appears to have little in it of startling interest but we are anxious to see the rest of it.

We are organising a campaign against slacking in the Labs.—wasting time and the like. There are a lot of new people who have no special tradition and who don't do a full day's work. We must get this mended.

I have word that Tizard and Whiddington are coming tomorrow to see something of CVD.

March 18th

A very nice letter from the Admiralty on Megaw's centimetre wave propagation over sea within the optical range (Fig. 20). Speech over 50 miles at under 10 cm.

Dudding has called a meeting of all concerned with oxide emitting cathodes—to pool difficulties and experiences. A lot seems to depend on the interlayer between the core and the oxide coating.

Tizard came with Whiddington and spent the whole afternoon with us getting a view of CVD valve matters. It was, I think, a useful overall view, and I hope he was impressed with the magnitude of the effort.

Gavin, James and Warren and Boyland are doing good work in pushing triodes down to centimetre wavelengths. Gavin and Boyland's all metal concentric valve (E 1374) is reaching below 15 cm and the 'gossamer grid' valve (E 1344) has touched 8.5 cm with nearly half a watt.

March 19th

A meeting of T.W.H., H.P.W., Macqueen, Clark, Peters, Robinson, Bligh and myself to review all the GEC radio activities with a view to behaving as a team rather than as a group of individuals. I instigated T.W. Heather to call the meeting and he ran it very tactfully. He and I hope it is the beginning of a breakaway from a disjointed past.

March 20th

A very full day—full of a mass of details and issues, one of them being congratulatory messages on my Fellowship of the Royal Society, all of which were very gratifying especially those from colleagues.

Brundrett and Harley to lunch at which Brundrett found out Dudding's keen interest in statistical handling of production problems. Dudding offered to give five lectures to his inspectors on quality control.

Examined roughly the costs of our diamond die making. Except for the costs of the finest dies (which are at least three times the others) it looks as if our costs would not come out too badly.

The Brundrett and Harley preproduction meeting went well so far as we are concerned for we are up to, and above, expectations on all types of valve. We were advised of the probable large use of E 1210 magnetrons of a wavelength shorter than 10cm for tank use and requiring a preproduction group. This results directly out of the South Coast speech trials referred to recently in this Diary.

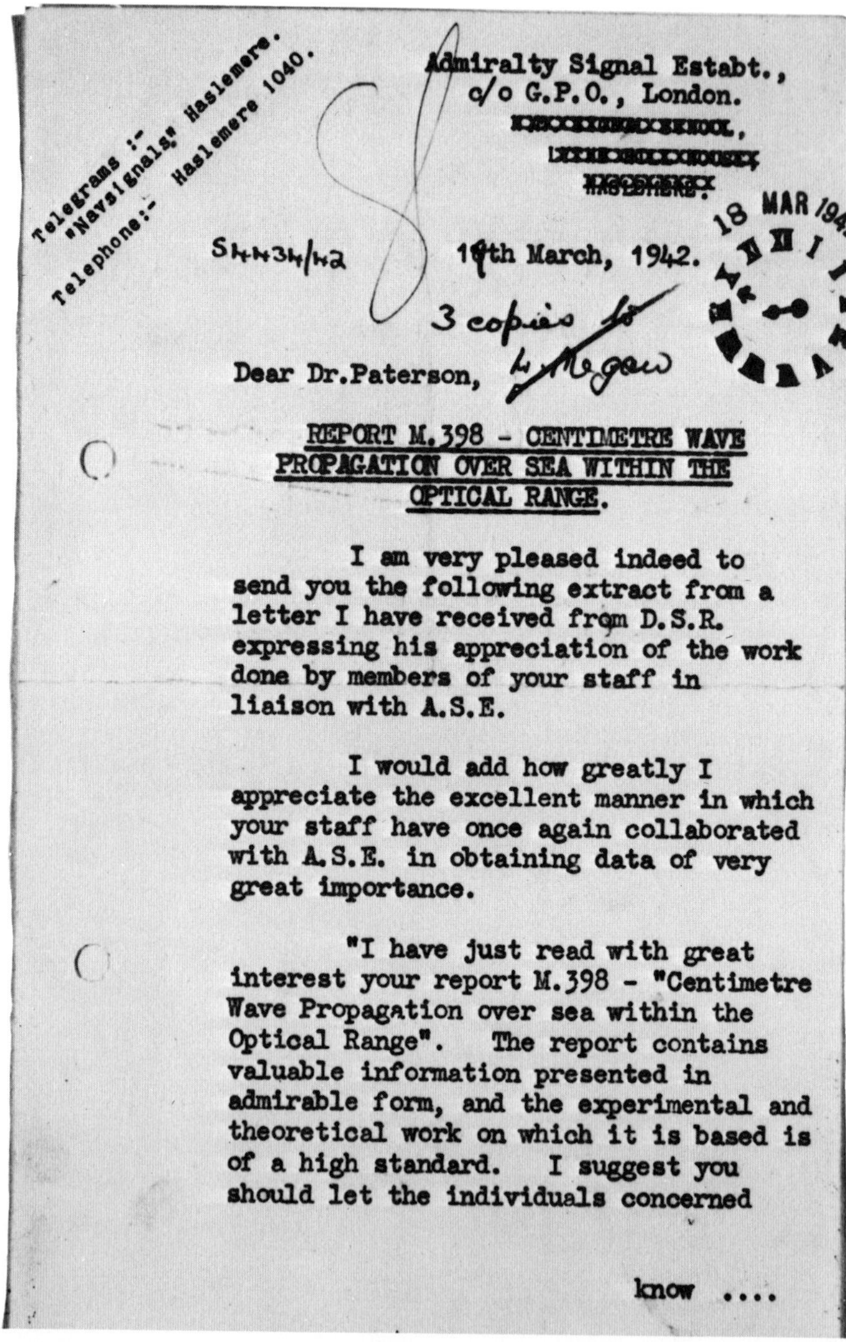

Telegrams :- "Navsignals" Haslemere.
Telephone:- Haslemere 1040.

Admiralty Signal Estabt.,
c/o G.P.O., London.

S4434/42

19th March, 1942.

Dear Dr.Paterson,

REPORT M.398 - CENTIMETRE WAVE PROPAGATION OVER SEA WITHIN THE OPTICAL RANGE.

I am very pleased indeed to send you the following extract from a letter I have received from D.S.R. expressing his appreciation of the work done by members of your staff in liaison with A.S.E.

I would add how greatly I appreciate the excellent manner in which your staff have once again collaborated with A.S.E. in obtaining data of very great importance.

"I have just read with great interest your report M.398 - "Centimetre Wave Propagation over sea within the Optical Range". The report contains valuable information presented in admirable form, and the experimental and theoretical work on which it is based is of a high standard. I suggest you should let the individuals concerned

know ....

Fig.20 Letter dated 19 March 1942, from Basil Lythill, ASE, to Clifford Paterson

know that the work they have put in on this investigation is highly appreciated."

Yours sincerely,

Dr. C.C. Paterson,
Director, Research Laboratories of
The General Electric Co. Ltd.,
Wembley, Middlesex.

Fig.20 continued.

March 23rd[259]

Preproduction meeting reported that we are suffering from shortage of testing people and facilities. Shall have to see Bligh and Jesty and try to get someone transferred. Otherwise all valves except small diodes seem to be going well.

Dr. Darwin, Director NPL, here all day to see over the place and have a good talk with Dudding and Jennett on statistical control of quality. I think Darwin liked the visit.

Admiralty want CV 55 (1190) valves in much larger quantities by July. I should think we ought to manage it.

Megaw back from 10 cm tests on S. Coast. I gather they had surprising fadings and intensifications of signal between Beachy Head and the Isle of Wight.

March 24th

'The Minister' MAP is meeting the 'Radio Manufacturers' for a talk about the new development and production programme and T.W. Heather called to talk over the issues. Mr. and Mrs. Rowe (TRE) came to spend the night at home and Rowe told me much of the background (Fig. 21). I fear that Leedham and DRP Dept. is defeatist regarding any expansion of production possibilities in the radio industry, but they put no romance behind their requests and their contracts. If they would inspire the production people with some of the operational information there would be an enthusiasm which would work miracles.

Rowe discussed GEE and H_2S and other plans and I reminded him that we wanted a difficult problem for Espley and others to get their teeth into.

March 25th

Discussed with Bligh the testing staff needed for all our preproduction valves. The testing and the proper plotting of results is a big job and at present we are not doing it adequately.

10 cm meeting at MAP at which the principal topic was the BM 717 magnetron and its possible difficulties in manufacture. Megaw reported several 717s made here by craftsmen which repeated in wavelength to ±1% so the meeting decided on the gamble of adopting 717 as first string and E 1342 as second string. BTH are to be given the order and have to get a move on quickly with their preproduction.

Dr. Vick came in the afternoon for an hour's talk on PF [Proximity

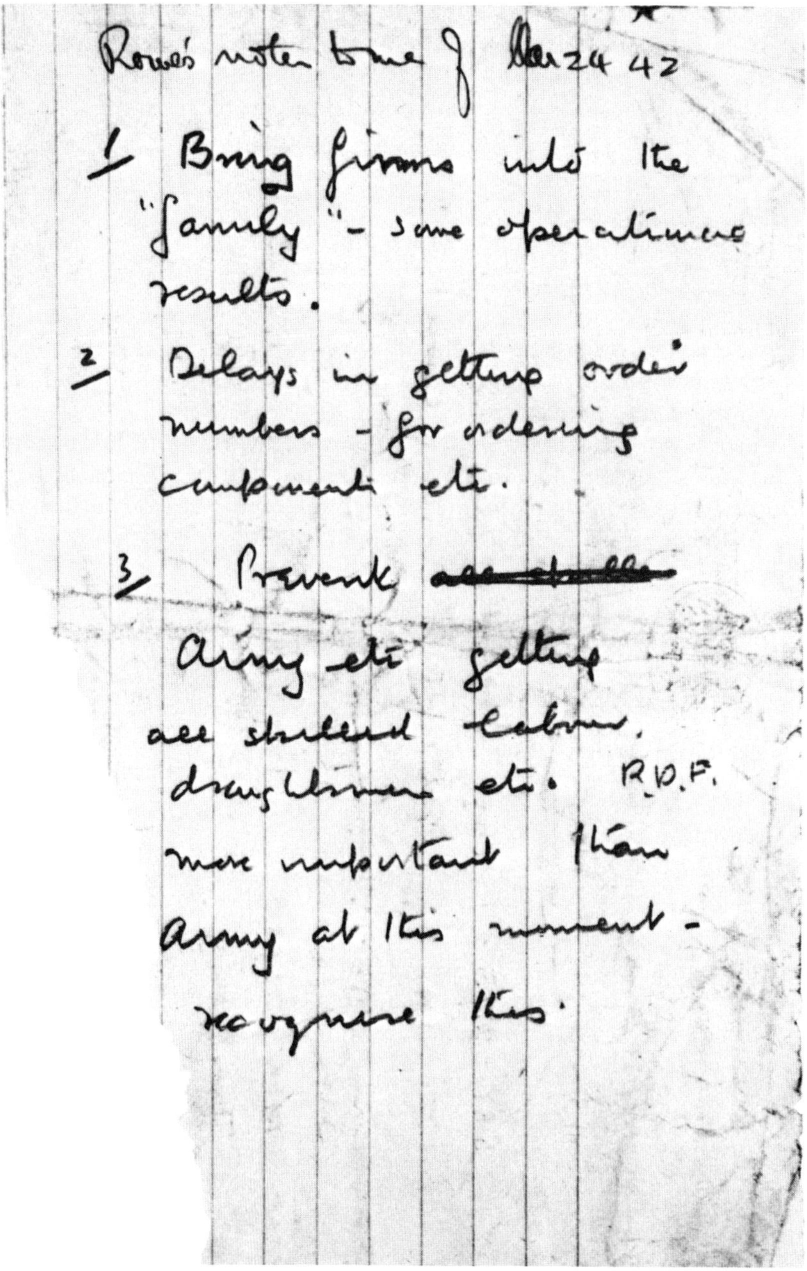

Rowe's notes to me of Mar 24 42

1/ Bring firms into the "family" – some operational results.

2/ Delays in getting order numbers – for ordering components etc.

3/ Prevent ~~[illegible]~~ Army etc getting all skilled labour, draughtsmen etc. R.D.F. more important than Army at this moment – recognise this.

Fig.21 Note of discussion between A.P. Rowe and Clifford Paterson on 24 March 1942

Fuse] valves, which are now to take urgent priority. There are three main applications Guns, Rockets, Bombs. The ROF job is to be very much second priority and is to take a changed form.

An iniquitous circular from Engineering Employers Federation trying to commend a bargain with AEU [Amalgamated Engineering Union] by which advisory committees on production are to consist only of trade union members.

March 26th[260]

Our compact source lamps seem to be not so bad up to 7.5 kW but above this, Francis reports that the temperature of the tungsten causes evaporation and blackening after a few hours. So unless they can be kept cool we are up against it.

Lord Hankey having resigned, Mr Butler, Minister of Education, has taken his place and has made a very favourable impression at the first meeting. He seems game to tackle anything within reason and expects to get us linked up to the new Minister of Production in such a way that he will use the committee where he wants an impartial body to make enquiries.

Our 'Warship Week' is on this week and Mrs. Paterson is hard at it again. We are well over £1000 with another day to run.

Heather tells me that the great meeting of radio manufacturers with the Minister did not amount to much. They want model room facilities increased by 50%.

March 27th[261-262]

Had visit from Macqueen and Rosen to help me in connection with my invited contribution to a discussion on radio education requirements on Sunday April 12th at which 30 Heads of University Physics departments are to be present for a conference in London. I have to present the needs of industry. Brought in Gavin who is very interested in the subject and he and I will now consult over the matter.

Lunched with Chemical Engineers and was glad to meet various people including Tizard whom I drove back to MAP. He told me how useful my awakener had been at MAP on the subject of tungsten supply and that the instruction had now gone out to put everything in hand—three months after it might have been done.

Dudding had his talk with the other openers of the coming great discussion on the application of control charts to mass production.

Dudding is much less interested in the showy propaganda than in the spade work of teaching production engineers.

Had a nice word of thanks from Walmsley for help given him in the chair at the 10 cm meeting on magnetrons.

Smythe [Smyth], from Northern Ireland, reports that the aerial tests in connection with diffused lighting came off last night and were successful.

March 28th

Our 'Warship Week' collection at the Labs. has practically reached £2500. Jolly good. There was pessimism about it amounting to defeatism except for Mrs. Paterson's optimism or wishful thinking. When the week once began Downing's posters were a great contribution to success followed by Cornwell's broadcasts and Leeds's and other's consistent spade work.

March 30th

Stevens is back from Greenock where the trials of diffused lighting of ships were discussed. Much to everyone's surprise the ship was less visible from aircraft when illuminated. This is rather confounding to those who have been prepared to condemn 'diffused lighting' as a foregone failure. It looks as if the report would be 'a useful and effective proposal but very troublesome to install on all ships, and almost impossible to find the necessary manpower'.

Preproduction went OK last week—about 800 valves. No special panics but some additions are upon us, i.e. 717 (Birmingham magnetron), E 1199 and E 1210 glass magnetron. Warren's E 1344, Gossling's [gossamer] grid valve, is also bound to come to preproduction. The E 1335 valve for torpedo establishment (most urgently needed) is already being made at 30-40 per week.

Prain here to settle diamond specification with Dudding.

Campbell's and Francis's typed book on 'noise in valves' is now complete and ready to go out to all (and they are limited in number) interested in the subject.

March 31st

A CRT meeting with 22 present. The Skiatron figured largely and it is up to us to justify making it in soft glass for easy production and Sharpe thinks he can do so.

To Head Office for a very useful meeting on MOV/GEC/EMI/MWT relationships to consider some rather surprising proposals from EMI.

April 1st

More talk on magnetrons (717) and the extent to which Research Labs. of GEC must carry on with them. We decided that we must do something somehow, and pending our drilling jig for the new block we would try to get Marconi to jig bore half a dozen blocks for us.

Winch reports all set for trials on Friday night of infra-red FF devices before a distinguished company of officers.

Lunched with Mr. (now Sir Frederick) Cook chief engineer of Ministry of War Transport. No honours or rewards would change his level and straightforward view of life.

Meeting with L.G., J.Y.F., G.C. and Levy and West re. a joint company for manufacture of luminescent powders.

April 2nd

This is the Thursday before Easter that the holidays used to begin! We are restricted to three Bank Holidays plus a week per annum, but I believe people can stand it if they keep cheery and buoyant. We have had much absentee trouble through sickness this winter but I believe there are many who have not learnt the way to be happy in spite of anxieties and irritations.

Megaw's work on his 6½ and 9 cm sets is requiring model shop facilities in the same way as Gavin's and all the rest. Have been trying to settle with him how to get together a new model shop nucleus for him.

A Butler (late Hankey) committee on searchlight production and co-ordination of design. It was a foregone conclusion that the Army and the Navy require radically different types of gear and we wasted our time talking about it.

Furthermore a number of the younger people (and some older) have not learnt the way to work hard. Attached is a memo circulated to all such (Fig. 22).

April 3rd

Had Vick and Whiddington here for a preliminary talk about Hivac collaboration and contract re. miniature valves for PF (shell) and other uses. I had prepared a paper suggesting how responsibility should be

23rd March 1942.

(A copy of this is being sent to every member of the laboratory staff).

WASTING TIME.

There is a large section - probably half - of the people in the Laboratory who work hard and pull their weight for the whole of the daily working hours. They are in to time and start up work at once; they are determined to get as much into the day as possible. At the end of the day they take the minimum possible time for clearing up.

But a very considerable number of people fall short of this. They are not working "all out" as the country requires of them and as the importance and urgency of the Laboratory work justifies.

People stroll about the place instead of proceeding briskly.

There are women and girls who, even if they arrive in time, slack about until ¼ or ½ past eight. This is obvious in some of the experimental laboratories but is not confined to them.

Youths are seen talking and hanging about their work when men are risking life and may be dying for lack of what it is their job to supply.

Girls seem to think that attention to their hair and personal appearance will justify long absence in cloakrooms during working hours. Their value to the country depends on the amount they do here - not the hours they spend here.

Others/

-2-

Others, both men and women, forget that it is up to everyone to help to maintain a cheerful atmosphere and to keep going in a brisk and buoyant manner - remembering that we are all working for the cause, and for the men and women whose lives and safety are in our hands, rather than for ourselves or the G.E.C.

The fighting spirit is an ungrudging one. It goes "all out", with unbounded loyalty and support to the teams we are with and their leaders.

In factories, slackness is watched and checked. Here we rely on people themselves maintaining a high standard of personal conscientiousness. If some individuals do not do so, we cannot allow the morale of the place to suffer by their example, and other means which we are reluctant to take will be necessary to bring these people into line.

At all costs these Laboratories must pull their fullest weight in the important work which has been entrusted to them.

This notice is for you who are now reading it - you know to which of the two classes of worker you belong - so please face the truth and make sure you are on the right side.

C C Paterson

BMO.

Fig.22 Note 'Wasting Time' sent by Clifford Paterson to all staff of the Laboratories

divided if we are to act toward Hivac as we do to MOV but was surprised to find considerable consternation about it—which was afterwards repeated when I spoke to Prof. Ellis at ADRDE on the phone. I had to be stiff and stick to my guns and insist the matter is too important to leave to a generalised co-operation requirement. The job is difficult and not a matter of brain waves, but of careful research towards a known objective and we feel we must be in charge of the team and not have Hivac complicating matters by going off on their own devices.

A full dress demonstration of Winch's infra-red FF [IFF] devices at Heston with Helmore—beginning at 6 pm in daylight and ending, I am told, at 2.30 am in a midnight conference. I left at 10.0 pm when demonstrations were going on under impossible conditions of extraneous noise—pitch darkness and sentries challenging the attendant experts all the time! In spite of this Winch did very well, and —with the help of Helmore's preparations with the Air Marshals—the decision made appeared to be sound. As usual the IR telescope of Admiralty Research Laboratory was trotted out again whether practicable or not.

April 4th[263]

A round table conference with Hivac, CVD, Admiralty Torpedo Factory and others—Whiddington presiding. We concentrated upon the Torpedo Factory valve, and were able to show willing and to discuss issues without reserve. I hope the meeting successfully broke the ice and will pave the way for the more closely regulated programme of work which will be necessary if the PF oscillator valve is to be successfully evolved.

Holland tells me that he has despatched 1023 valves this week—the first time the 1000 has been topped. Miss Parslow's efforts at packing have been largely responsible.

April 7th

Return to work after one day's break for Easter. There is a general suspense in the expectation of Spring offensives. Everyone hopes we are not to be on the defensive any longer.

Had meeting at Wembley with T. Smith and Evans (of BRS) over the work and approach of the new Illumination Committee to advise Ministry of Works and Buildings. We are to concentrate on small private residential house lighting. It does not seem to be a job for winning the war!

Lunch with G.C. at which we discussed Marconi/GEC/MOV affairs and then to RMA to settle my remarks on the training of radio graduates for Sunday's conference in town.

Some of the girls from the office are volunteering for valve work downstairs in the evening. They helped with the record despatch of valves last week.

A flap over statistical inspection and control of valve production with an emergency meeting called at MOV by Dudding.

A further 8.5 cm magnetron required.

April 8th

Helmore called unexpectedly to see the 4000 A. arc on its maiden run, a fearsome sight and noise. Witton have had one going on trial. I cannot say I was impressed by the steadiness or usefulness of the phenomenon, which was explosive, nor by the clarity of thinking on how it is to be used. But Helmore agrees that study of the matter by Waldram's group should be helpful.

Helmore and Jolley have been running their 'directed' motor boat on Staines reservoir not without adventures including, I believe, a ducking. But the visibility arrangements showed that the scheme was quite practicable and Helmore will try his red tape avoiding tactics to get the thing pushed with our assistance.

Edwards back from AIS Mark VII trials for a few days. The set is behaving reasonably, but wave jumps on the CV 38 magnetron may be making range variable.

Chelioti and F. Smith called to consult over the CV 8 and CV 55 valves, which they urge should be pushed forward by MOV if anyway possible. It looks as if Wembley would go far to fill Signal School needs for 1942.

Signals Experimental Establishment report 60 miles reception over sea over long periods in the recent centimetre trials.

S.V. Williams's thoria – nickel – chrome 'non creep' alloy is beginning to look interesting and tractable. Also Heavy Alloy bullet cores have been prepared and are being tried.

April 9th

Dudding had a further good internal meeting re. oxide cathodes and their techniques. Several points emerged which were worth following up and the research is being pursued on a broad front.

Had discussions with Gosling of Lamp Works and Mr. Gamage re. disposal of profits or margins on RL productions which this year must amount to a considerable sum. There seems no chance of building up a reserve against a rainy day.

April 10th

Arranged for Ramsay and Machin to tackle research on oxide cathode problems, in the ambit of Dudding's group meeting on this subject.

Laboratory party returned from Shaw valve factory meeting much impressed by the layout and general appearance of the factory.

A Butler committee on which I urged enquiry into effect of industrial rivalries upon the efficiency of the war effort.

News continues to be depressing. Much loss of shipping and naval vessels to the Japanese in the East.

April 11th

Schuil turned up from the Camouflage Department at Leamington. He has fixed up an extremely good model room where his doings can be demonstrated and studied.

Had a conference on the subject of increasing our model making capacity in line with a request to the industry from the Minister of Aircraft Production. It looks as if we could make a satisfactory arrangement with Claudgen to collaborate with us in the preproduction of prototypes made up at the Laboratories. I shall put this up to Heather.

April 13th

Preproduction keeps going well but research problems—particularly connected with oxide coatings—still keep turning up. Stevens tells me that gun blast tests on the ship fitted with DL [diffused lighting] have gone off with excellent results. So DL has not a stain on its character except trouble in installation.

Macqueen came to tell me of a long talk he had had with Leedham of an unusually frank kind on the subject of reorganisation and of demands pending by TRE for more first class people from industry. It looks as if the GEC may come in for a talking to, for certain misdeeds and delays.

Lunched with Good (BSI) to talk, amongst other things, over the special committee at which I am to preside for reducing the copper used in the electrical industry. I returned to Wembley to find that a meeting there with Hivac over the CV 61 (Torpedo Factory) valve had been going very well.

April 14th

Glass Advisory Committee at which wastage of glass under present conditions of pressure and of inadequate staff were much emphasised.

Everything conspires to make for carelessness over breakage and wastage.

Our returns show that apart from the 11 000 VT 90 valves made prior to 1941 our total output of valves made at Wembley is 20 000. The largest groups are:

Magnetrons e.g. NT 98	2200
Hard modulator e.g. NT 100	2500
Gas modulator e.g. NG T8	1500
Triodes e.g. VT 90	4100
Triodes e.g. NT 93	4800
Triodes e.g. NT 99	1400
Diode e.g. CV 8	1500

There are 21 types altogether. The VT 90 is so far the only type we have discontinued making and passed over to the factory. Orders against us are for 50 000, i.e. 1000 per week.

April 15th

Winch reports visit to an Army experimental station where sound ranging is studied and found that our 'Nash & Thompson' reception gear is just what they appear to want. It gives 10^{-11} sensitivity instead of 10^{-7}. Cockcroft and other inspecting officers were present and there may be a chance of this coming into its own again but on visible rather than on IR.

Gossling reports conversations with Ellis from which it appears that we shall be asked to proceed with the E 1335 oscillator without reference for the time being to Hivac.

The great public meeting on statistical control in industry took place this afternoon and appeared to be quite successful. Dudding spoke and did an experiment showing batches with 1%, 1.5% and 2% defectives and the chance of random selection.

April 16th

A day of internal talks and settlements—very useful. Our building extensions for preproduction are getting on but we shall not get into them for two months.

BTH paper on fluorescent lamps at IEE. Very distorted history I fear, which I endeavoured to correct.

April 17th

A diamond die meeting. Specification for diamond stones, for which Dudding has done most of the spade work, was passed. The expansion needed seems very great and most of the stones appear to pass through London.

A TVC meeting all afternoon with many valves to be approved. The dimensions of the 717 magnetron seem nearing settlement except that BTH want a new rating of cathode. Marris is taking the lead in these discussions.

April 18th

We are having to go thoroughly into the techniques connected with making boules for Salford sapphires. We have one of the boule making furnaces fixed up at Wembley and Dauncey is taking on the job under Chirnside. He has practically finished the development of nickel coated getters (in place of copper) and made a success of it. The nickel coated getter is much more gas free and is very much preferred.

Waldram is booked for a talk at Shrivenham to a large audience of searchlight instructors in May.

April 20th

Preproduction meeting yielded 800 valves made in the week. The diode (CV 52) has at last got out of its despondency and 60 were made. The USA are very interested in it and have asked for full particulars in order to pre-produce. Le R. went over cost figures with me up to January 31st. They show that counting all in, and writing off all losses and failures and purchase of gear for making the valves—and testing, we are likely (with 200% overheads) to come out square without any profit—which if true will be a good result.

Saturday visit to Hivac by B.S.G., Donaldson and Jenkins was a very friendly affair and all seems to be going well with the Torpedo Factory valve. Space in the Labs. for packing is fearfully curtailed and I have been trying to hasten transfer of radiant heating gear (O.W.H.) to the Empire [Exhibition] building.

Waldram demonstrated to me a method of using Helmore's ground light which we are developing so as to enable it to serve as background for silhouetting enemy aircraft. I am glad to see something more than guesswork coming into this project. We are still awaiting the enemy over in sufficient numbers to present him with the results of our work so far.

April 21st

A meeting at Head Office re the Marconi/GEC relationship. The subject is most interesting but not for record.

Marris and Espley are bothered over 'spark over' troubles in the T & R box of the AIS receiver above 25 000 ft. It has been primarily TRE responsibility but we have not objected to the common T & R box.

Espley reports conversations at ADRDE and now a formal meeting, coming down strongly on the side of 25 cm gear and its utility for offensive and defensive purposes. We are therefore pushing hard to get our replacement sets complete after the fire at Christchurch. So the discouragements poured upon our 25 cm effort of 1940 by TRE and DCD were soundly resisted by us.

April 22nd

A second GEC radio general staff meeting. It was a blessing to see the free and general discussion of all radio matters amongst all concerned. Next time we must come to some precise thinking. A meeting at BSI.

April 23rd

An OAC meeting, and a talk with Gamage after on the possibility of GEC getting ascendancy in any field apart from alliance with big concerns abroad. Met C.S. Wright and F.E. Smith after lunch and learned of prospective changes in Government experimental establishments.

Dudding had a good statistical propaganda meeting in Manchester and keen discussion. He is likely to be wanted there again.

April 24th[264]

Went through Lab. accounts with Kersey and O.W.H. preparatory to allocations. The separating of production from experiment and the separation of direct Service contracts from lab. work is not difficult but the pre-assessment of overheads upon production and service contracts leaves the result very much in doubt. We have taken 200% overheads which, if too high, will leave too large a credit and throw doubt upon the claim that we make no profit on our experimental work.

A valve allocations (preproduction) meeting in the afternoon showed that our efforts have been going well and that we have met our promises on all types.

April 25th

Warren reports a meeting with Signal School re. CV 8 valves at which the change in cathode design was pretty well finalised and trials with spark gaps were pressed in case it is possible to use them for some of the sets.

Ryde has had a request from TRE for a hundred 3 cm common T & R 'waveguide cells' by December 31st. The experimental ones have been made by Harris and Shaylor so far, and they will continue to do so. From January about 100 a month (?) are required. These use the flat glass to metal technique which we employed for the sextant bubble graticule—and look like no valve which has yet been seen!

A conference with Walton (DSR Admiralty) on the wording of an agreement for conditions of exchange of information with industrial concerns in the USA, i.e. GE, Westinghouse, RCA and Bell Labs. The proposals are all right in principle but can be improved from a practical point of view.

April 27th

Further message from Christchurch (ADRDE) in the nature of an SOS to take in hand the assembly and general getting going of GL Mark IV the first model of which was burnt. We are well on with the replacement and it will mean doing the whole assembly and testing it out at Wembley.

Much discussion of the merits of the USA lighthouse valve compared with our concentric, E 1344 or E 1231. We must decide whether our concentric is or is not better than the American. Gave some help to Britten to get his transfer of getter making to his new quarters. Peters and his colleagues here to examine [mixer] crystal making on a large scale.

It looks as if tomorrow's CVD meeting will be a lively one.

April 28th

An emergency request from Air Ministry. Mr. Grosse called on Gavin to request we put aside all our present work on RCM to take over making of 12 sets for special service. They are required in six weeks! We told him we were game!

A long and interesting GEC CVD meeting. Ryde gave a good account of crystal and waveguide T & R developments. The chief discussion was the USA lighthouse valve and our alternatives. We recommended that it would be hopeless to expect the lighthouse valve to be in production on a large scale in 12 months with all its special USA techniques and, if it is wanted in due course, it would be best to get the Americans to foster its manufacture here. Meanwhile we recommended our concentric (E 1274) all metal valve. It was agreed that we send samples to TRE as soon as possible.

April 29th[265]

A CVD policy meeting at which I had to speak more than I like. Hughes and Vick want to launch a preproduction establishment in our shadow accommodation at the Exhibition and DSR seemed convinced it is the right thing to do. I had to say that I was very doubtful of the wisdom of the scheme and whether it had been properly considered. I was asked to join an *ad hoc* committee to discuss the proposal.

The USA lighthouse valve also came up and our E 1274 alternative of which I passed round a sample. I found the MAP people against the scheme but reluctant to speak too openly.

April 30th[266]

A household appliances meeting at Head Office with L.G., F.W., T.W.H. present. Some lines of general policy were fixed to great advantage and will help to direct everyone's thoughts instead of dispersing them.

Winch reported message from Mr. Phillips at Larkhill sound ranging experimental station that MOS will ask us to co-operate with them in the use of the Winch/Nash & Thompson IR transmitter – receptor unit as forecast three weeks ago. A very happy development because we rather felt that the previous effort had been wasted. I believe there is not a single serious development here since the war which has not come into its own sooner or later although discarded at first—i.e.

> Winch/Frazer-Nash IR units
> Helical scanner for O'Kane's job GEE [not GEE – H_2S] with TRE
> 25 cm transmission and reception
> —and others I expect if one could think of them.

A TVC meeting all afternoon at which contacts with USA re. standardisation was the most interesting subject. I advocated getting places like Wembley, EMI etc. to keep (through official channel) their opposite numbers informed in USA and must write a memo on it.

The position re. Salford production of boules for sapphires is becoming serious and Wembley is being looked to to get it right quickly.

May 1st[267-268]

A useful Research Services meeting with A.R., L.G. and T.W.H. with the GEC 'radio activities chart' on the table.

Captain Margesson visited us with Leslie Gamage. Signed himself 'Margesson' (for he is now a Baron) for the first time in our visitors book. A most interested visitor and a pleasure to show round. The demonstrations by Ryde, Leeds and Stevens were very impressive.

May 2nd

A midget valve meeting yesterday at which inter-Service establishment rivalries stood out rather seriously. Unfortunately this tends to grow.

Had a formal request to schedule all our development contracts for MAP on radio and give monthly reports on them. I am afraid I blew up on the telephone to Walmsley on the subject. He toned matters down somewhat but an effort to keep progress charts of developments (which is what it looks like) seems to suggest the substitution of the letter for the spirit—and the impersonal handling or control of development as if it were a routine production matter.

We are also notified of the Inter-Services Radio Research and Development Contracts Committee.

Raids have begun again on a small scale—Bath, Norwich, Exeter, York and our Turbinlite has been in action and drawn first blood. It is of course dependent on the efficiency of AI and GCI and the teams on the latter appear to require a lot more practice. The performance of the Turbinlite is considered to be quite up to Helmore's hopes.

May 4th[269]

Nearly 1200 preproduced valves last week. The CV 55 (1190) is not yet 'going' but causes of wastage are understood. Craig's automatic grid winding machine which he has made himself is a distinct success.

Called on Sir F. Smith by request to answer for priorities in RCM in the light of the special request of emergency production of 12 sets in six weeks. Gavin has things partly shaped already.

To the Admiralty DSR *ad hoc* committee under Brundrett's chairmanship to consider a proposal by Hughes (Signal School) and Dr. Vick to use our shadow building for a training place for valve supervisors and operators. I had to point out a number of blemishes and was glad to find that MAP valve production department agreed with me. In any case it was ruled that the whole subject was one for Harley's department at MAP rather than for CVD.

A Royal Institution Manager's meeting to consider appointment of successor to Sir William Bragg—bless his memory.

Waldram gave his lecture to a conference of searchlight instructors on Saturday with great success, and it is to be repeated.

May 5th[270]

Have news from Marris that in recent small scale raids there have been three interceptions with AIS Mark VII and the planes were shot down. We are all bucked about it for there are not many sets in use as yet.

Waldram reports that his 'visibility in the air' experiment from a balloon was a wash out because the enemy came over and the heavens were otherwise occupied.

Was told by a special staff deputation led by Mackenzie of the desire of members to present me with a photograph enlargement of myself to mark my election to Fellowship of the Royal Society. A very gratifying incident indeed and one which I value more than anything else.

May 6th

A 10 cm meeting—not specially eventful. The CV 64 (Birmingham strapped magnetron) seems to be getting on its legs so far as samples are concerned. BTH will be making in larger number and we shall make about 200 at Wembley.

Heard during the meeting that Heather had been taken very ill—a serious shock which I can hardly face up to yet.

Had the first meeting of my Ministry of Works and Buildings Advisory Committee on Lighting. A very good general talk round the subject but next meeting we must decide courses of action.

The annual Lab. accounts are due to be finished by accounts department this week. Allowing 200% overheads on our valve preproduction and after paying Lamp Works and Head Office their expenses, the Labs. make a loss of about £3000 on a total turnover of £130 000 which is as I should have liked it. This will be taken forward into the coming year when there should be no difficulty in making it up.

May 7th

Dee from TRE reports that last Thursday night in a diffuse raid, out of seven enemy planes brought down six were caught and held by Mark VII AIS which was responsible for their destruction. Even when they came down quite low they were followed and held.

Had internal meeting on the use of Claudgen facilities as an extension of our model shop—the scheme will work up to a point but if pushed to serious quantities the testing and approving becomes a difficulty because Research Laboratories could not handle it.

The Research Laboratories annual accounts show alarming tendencies to bulge with money we can't, or are not permitted by the rules of the game to, dispose of. After the payment of overheads (200%) on Government contracts and on preproduction etc. there are not enough overheads left to spread over the work done for GEC Establishments. We shall charge therefore all GEC work with their usual (60%) overhead and lump the remaining debit balance into a single charge to Head Office.

May 8th

Consultative Board Meeting at which only Chris Wilson could be present. The above results by the annual accounts were taken. The accounts came out excellently, an overhead for the GEC Establishments of 180% (same as last year) being applicable. Expenditure payable by GEC is now only half that for the last peace year.

Looked over Levy & West's premises at Wembley and Shoot Up Hill with L.G. and G.C. and had lunch with them in town afterwards.

May 9th[271]

Heard that Espley and Walter have received an IEE premium and that Cherry has received the Kelvin premium.

Letter from Helmore about some of our minor jobs for him which makes me doubt whether he is able to carry MAP contracts department all the way with him in demands for a free hand!

May 11th

Jolley and Hawkins and party are fitting one of the latest Mosquito [air] craft with a Turbinlite. No further news of interception. The Hun seems shy of coming over.

Churchill made an invigorating speech last night. Told Jerry where he gets off if he uses poison gas—and what he is in for this summer.

Heather is reported on by the specialist more favourably but we must expect a six months absence from duty. Had Macqueen to lunch with me to talk over the steps we should take in connection with the GEC radio group. He will suggest to the others carrying it on with me presiding *pro tem* in Heather's absence.

May 12th

A CRT meeting—full. People however seemed acrimonious on the Services side particularly Mr. Sutton and Mr. Stewart from TRE, over Skiatrons. I must see Jesty and find out what is wrong. I think Walmsley was worried too.

Waldram reports that Morse (one of his group) has been down to Boscombe Down and had the apparatus fitted in a new aircraft and has sent back his first results up to 30 000 ft.

Smithells called to ask our help in making thoriated platinum wire by powder technique for sparking plugs to use on Rolls Royce aircraft engines. We are going to have here one of the men from Baker Platinum Co.

May 13th[272-274]

Had Mr. Enticknap here to talk over accounts and to see something of the place.

Espley reports growing urge on us from ADRDE for 25 cm equipment but applied to '4X' instead of to GL Mark IV. It is suggested that the GEC might take over the contract for the whole 4X gear and completely engineer it. We gather that a decision has to be reached quickly.

We also hear from Dr. Walmsley that TRE have a very hush hush and also difficult application for us to work upon in place of AIS now virtually completed as far as we are concerned.

I spoke by phone to Walmsley about TRE attitude to our CRT research and particularly Skiatron work.

A morning at the 'Chairmans [Chairmens] meeting' of advisory committees at the Ministry of Works and Buildings. It will probably have valuable co-ordination functions.

Had discussions at Head Office with Dr. Railing and Southworth and Espley on the outcome of GL Mark IV known as GR. Unless something goes awry at the last moment or Leedham makes difficulties in regard to GEC taking on the production job of GR we shall be in for it in conjunction with Witton and Spon Street. Went to the hut where assembly is taking place of the replica of the GL Mark IV prototype which was burnt up at ADRDE. A very good and well designed looking job.

May 15th

Gossling had a report on the modulator meeting this week which brought out advances in the spark modulator and particularly in the good work of BTH Labs. Espley may need to adopt this type for his GR job.

Have finally sent off our statement and financial review of our preproduction activities from the start until March 31st. They show a loss of about £6000 and no profit. But we expect to make this up during the first part of the new financial year. The total figure is £150 000 covering about 20 000 valves.

A TVC meeting—uneventful.

May 18th

Another preproduction week topping 1000 valves.

Had letter from Sutton of a very friendly character. Walmsley had evidently dropped a word in that quarter that all is not as it might be between Service and industrial teams. Jesty has prepared memoranda on our programme in regard to Skiatron showing that our work to make tubes with ordinary materials yields a product not inferior to that made at Bristol by Sutton. I am surprised however that we have not been able to improve on his tubes.

Mr Redhead from Kodak's came today to photograph me in my office for the presentation the members of the staff are making to commemorate the F.R.S. (for which they are mainly responsible)! It is a most delightful thought.

I finished off a popular broadcast on diamond dies for BBC—(except for the giving of it) which may come off on May 30th.

An Advisory Board meeting but several away unwell.

May 19th[275-277]

Confirmation from Y group through ADRDE that 25 cm is approved for RX and the word now lies with Leedham to approve GEC handling of the development contract. We are seeing about an Army shed to accommodate the extra personnel from Witton whom we shall need.

Megaw back from transmission trials off the Isle of Man. The results show strong signals beyond the optical range and therefore offer no advantage from the secrecy standpoint to the Admiralty on CW. (Their chief use would seem to be for RDF.)

Had Captain Hodgkinson and Mr. Phillips here on Winch's aid for sound ranging. We showed him Macfadyen's speech along beam gear which has been made by the boys for Home Guard use. He seemed to think well of it.

Have had formal notice of transfer of our Home Guard unit OC from Castle to Chapple. Castle has had great difficulties and prejudices to face since he took up the formation of the unit two years ago and he has done well.

May 20th

The complete schedule of the field of experimental work covered by the visibility (i.e. photometric properties of the atmosphere) committee is very wide. Waldram and his group are some of the chief actors and enter into more of the branches than any other single organisation.

I heard from Dr. Guy (secretary of the Mechanicals) that he had been appointed the independent chairman of a Service cum 'assessors' committee to enquire into the research and development resources of the Ordnance Board. I am to be asked to be one of the assessors. Affairs at the Lab. have on the whole been free from emergencies and generally busy. Half of our new building extension is nearing completion and the half adjoining engineering is well started.

Offensives alternately Russian and German take place on the Russian front. It looks as if it would be a ding dong business.

May 21st[278]

Gavin's special rush job is engaging the efforts of his group early and late and seems to promise all right—although it will be impossible to know until field tests can be made. There is no time to wait for these so it is a case of taking a gamble. It is known as Mandril [Mandrel], a magic word we find whenever difficulties arise with suppliers. Whisper the word and all avenues open!

The 'Ordnance Committee' invitation has come from the Minister of Supply and I have promised to help so far as I can.

We have to test our magnetrons on an actual distant object and there is nothing for it but a hut on the roof of Waldram's projector testing room. The speed with which Dr. Black's secretary got going with provision of the hut was exemplary.

May 22nd

A preproduction committee meeting at which some loose link was revealed in the means to be taken to get going with special valve requirements for ADRDE. Gossling at Hivac and a meeting afterwards with Whiddington on midget valves for PF developments. Humphreys did well as usual in a visit to the chief of the Willesden Labour Exchange for unofficial advice as to how to reserve our young fellows who are our mainstay for assistance on radio work and the like. O.W.H. is carrying the whole of this side of our worries with the authorities and is doing extraordinarily well.

May 23rd[279]

After last CRT meeting Dr. Walmsley must have had a word with Mr. Sutton and Mr. Stewart as I have had a note from each regretting any apparent antagonism. Jesty has drafted a memo restating our outlook and work on the Skiatron job.

Quick work by ADRDE in providing us with hut for Espley's RX job. It looks as if we should get it quickly.

Had long talk with Ryde as to our manufacturing policy on crystals. There is likely to be a number of types, but we are inclined for policy reasons to base them on the new GEC crystal technique, even though the surface is deliberately not always developed for high breakdown resistance when this is not required.

May 26th[280]

A wet Whitsun.

A luncheon with General MacNaughton GOC Canadian Forces—with Beard, Ricardo and Fleming to get ideas from him as to engineering tasks and organisation. The main points touched upon were:

(1) England as a huge maintenance unit
(2) Expert Engineering and Scientific Groups at GHQ in the field
(3) Master General of Ordnance to be an Army man and not to mix up the three Services in an executive body like the Ordnance Board
(4) Handicap caused by secrecy regulations must be broken down
(5) A financial floater to an organisation to bridge time between when a job ought to be started and when the finance department can put through all its required formalities
(6) The establishment of an Army Technical Development Board.

Discussed Lab. finances in light of past year's activities in order to judge correct overheads for current year.

Found Colonel Hemming and Mr. Phillips here from Larkhill on Winch's gear for sound ranging. He also saw Macfadyen's IR speaking down the beam gear in its last form and was very impressed. He took it away to try out.

May 27th

Peters here to discuss and settle policy on mixer crystal manufacture at Spon Street. All agreed.

Truscott wants us to make more E 1271 hard modulators—the fault being that MAP have not pressed manufacture at MOV sufficiently hard.

May 28th

These days are somewhat hectic—for several of our chickens have come home to roost all together. A great flap over RX—the ADRDE–GEC project (the outcome of GL Mark IV). Walmsley's small inter-Service committee to consider development contracts actually rejected the GEC offer and approached Murphy. Mr. [Dr.] Grey representing MOS on the committee did not know the facts and Walmsley guessed at them. This raised the fighting spirit in the GEC and particularly Espley's. He pulled Grey over yesterday and Cockcroft came up today to remonstrate and throw in his weight. If the GEC now shows an effective organisation for handling the job and reasonable prospects for production we shall probably have it. We then have Oboe to take up—Marris and the remains of his group to handle. I understand word has gone out that it is of greater urgency than previously and Dr. Walmsley wants to know whether we can organise to tackle both jobs.

The IR talking down the beam has also a rising stock.

Second post-war lighting committee meeting. A heavy but cheerful one.

May 29th

Organisation of the RX job to the fore. Had talk with Marris and with Walmsley who was worried about whether we could tackle the RX job for the MOS as well as the TRE Oboe job which he thought would necessarily come to operational fruition before the RX could so do. (I am not so sure.) He also mentioned another job which Dee from TRE wants us to do. So we shall have to expand our groups and particularly Espley's.

A diamond die meeting—but I feel there are more important matters afoot.

The Navy officers are showing distinct interest in the 'talking down the beam' set for use between vessels in convoy. Its complete secrecy is what they want.

May 30th[281]

Further talks with Espley and examination of the gear and predictor round which our electrical stuff is to be built. The general layout of the scheme was considered yesterday and on Monday there are to be internal meetings with all concerned including Dr. Railing.

There have been further troubles with the CV 55 (E 1190) valve—emission (which is explained), sagging heaters (not explained) accurate centering gear (always with us).

Had young Stewart from TRE to see me to talk over the recent incidents over our Skiatron co-operation, when he and Sutton seemed to be developing an antagonistic attitude. I hope our talk will help to keep things on the rails.

The Ministry of Labour are insisting on 'hours of work' of not less than 46—which apparently is asked to be formal. I wish they would realise that managements know far better than they how to run their jobs, and that this strain towards uniformity is purely a complex of the bureaucrat and has no benefits in industry—on the contrary. We shall have to add two hours a week to our present 44 and make after hours work, which comes in bursts, more difficult.

June 1st

Preproduction is down due to troubles with various valves—mainly emission. The NT 100 has its troubles too and CV 55 is not yet launched. We have responsibility suddenly thrown on to us for making the E 1271 hard modulator which we thought MOV could have taken. We are inclined to encroach on the efforts of the NT 100 group for this.

'Talking down the beam' (IR) was reported on by Chapple and Macfadyen who gave a demonstration at Larkhill, Salisbury Plain, to many Army officers under difficult conditions. The Army seemed very keen about its usefulness.

Consultation with A.H. Railing at HO with Peters, Southworth, Espley and Maskell, as to GEC set-up for handling GR. A fortnightly meeting with a central responsible man at HO to act for the Company. Dr. Railing agreed to come to next Friday's meeting at Wembley with the quality.

Sir John? Mr Abrahams and other[s?] from Diamond Corporation came to see how diamond dies were going on. As usual they were ready to help us with special supplies—this time boart.

I gave my BBC broadcast on Saturday evening on diamond dies—quite good fun.

The news is more cheering—a terrific raid over Cologne—Libya going not too badly—German failure in Ukraine. Nothing final but all good omens.

June 2nd

Gossling and Jennett attended meeting yesterday with Walmsley on the glass envelope thermocouple we have been developing with Elliott Bros. for TRE. Our part of the work (which has involved electrical characteristics) seems to have been all right.

Paint drying investigations and schemes under Humphreys and Connell have been growing in magnitude and importance for several months. Humphreys has been looking after the technical side of this as well as of the bomb disposal researches, in the intervals of administering the Ministry of Labour ordinances on reservation, calling up etc.—a very fine effort. Paint drying is now centred at the Empire Exhibition Research Lab. reserve building and it is soon to be followed by the bomb work [and] by a workshop section under Hobday for making valve components. This will give more space here for the actual making, inspection and testing of valves.

June 3rd[282-283]

Had visit from Miss Haslett, Mrs. Railing and Mrs. Gamage and showed them round; useful talks on girls' work, hours and Ministry of Labour requirements.

Ratcliffe from TRE paid emergency visit on AI Mark 7 which is more and more in favour with the operational people but is having constant pulse transformer breakdown. Discussed with him, Marris and Dunham the emergency alternatives of which there are several.

Had 'security men' here re. steps for keeping secret the resonant cavity magnetron.

June 4th[284]

Much preparation for tomorrow's meeting on GR and also for an FM meeting between manufacturers interested in FM.

Met Bragg to discuss his talk on physicists after the war and made some criticisms and suggestions. TVC meeting at which the members of DCD and DRP departments are very worried at their having to shift out of Thames House at short notice to go to Addison Road. With this upset at MAP and at TRE and ADRDE simultaneously we are being left fairly well alone for a few days.

June 5th

A very full day for the GR team. Dr. Railing attended at Wembley—and Dr. Grey presided. Cockcroft and Walmsley and others were here. Pollard presided at an afternoon meeting of the experts—and it was agreed that weekly or fortnightly meetings should be held at Wembley. We had to satisfy Walmsley that our other development would not be interfered with.

Waldram reports that Air Commodore Toomer having heard of our 'searchlight fan' tactics would like to come tomorrow to study it.

Bligh's party of competitors in FM added to the influx of visits and the canteen staff had to cater for some 80 people.

June 6th[285-286]

This has been a terribly hot week for people trying to work in galleries and other corners of the Labs. and I am thankful for the weekend break as a respite for them.

The war news is not so bad—terrific air raids on Germany—German offensive in Libya unsuccessful, a striking naval victory for the USA over Japan—and the much talked of German spring offensive in Russia not yet materialised.

Had a letter from Dee of TRE of which the following is an extract:

> 'You will be interested to know that one night last week there were three certain [kills] and one probable achieved by AI Mark VII. The 100 sets of Mark VII has clearly been a tremendous success and I am fully aware of the GEC's contribution.'

June 8th

Air Commodore Toomer and Lt. Colonel Bray—Fighter Command—came on Saturday and were very interested in Waldram's searchlight fan idea. There is to be a memo for the use of the C-in-C. Helmore did not seem specially interested when it was put to him for the large ground light.

The information re. the upper atmosphere has yielded an interesting set of curves from 15 flights 10 of which were over 25 000 ft.

Jolley reports a flagging of interest in the 'Helmore Turbinlite' for reasons entirely unexplained. But a new hare is 'the fish' a super torpedo and with which we are helping—when we receive a contract to carry out the work with him.

Had lunch with Lord Rothschild re. security matters and a half hour's talk with Watson Watt reviewing things in general. He had been analysing all the incidents in which AIS has been employed or attempted and was very pleased with the result.

June 9th[287]

This morning we lost our loyal and much loved friend Woodget. He had a heart attack in the Laboratories about 8 am and died in hospital within a couple of hours. Last evening when I spoke with him I thought how well he looked. A wonderful passing but we shall miss his cheery and courageous personality. He was almost one of the founder members of the Labs.

Inspected the new large spark modulator for testing large magnetrons made by Michaelis in Macfadyen and Machin's group. A workmanlike job.

CVD meeting all the afternoon. The chief item of interest was Willshaw's talk on 3 cm magnetrons with closed ends to the resonators. They seem most promising but the method of getting the energy out of them is not yet perfected and the output is as yet low on this account.

Heard the news of a tragic crash at TRE when two [some] EMI men (including Blumlein and Brown) were killed.

June 10th

Was delighted to have a letter from Heather—cheery and courageous as always. He has just been allowed to write for the first time.

A CVD policy committee at which two memoranda from Wembley came up: (*a*) valve interchangeability with USA, (*b*) the later stages of valve development. A serious item of policy is materialising involving the putting of entire reliance on USA for supply of certain kinds of valve. The 'lighthouse' valve is the first case. I was able to put in a strong plea for development even though large production does not follow on the ground that each new development is a stepping stone to the next and that the British war radio effort has depended entirely upon having the valve development dovetailed in with it.

Had lunch with Hecht of MAP who spoke frankly with me about disorganisation at RAE radio section and bad staff feeling there—and hinted about MAP. Oddly enough A.C. Bartlett (now at RAE again) called and had 20 minutes talk on the same subject but as usual did not mince his words—he spoke of the difference between all this and the atmosphere at Wembley—it is nice to hear that people value our outlook and our ways, but only if it is to enable us to influence theirs for the better in any way that we can.

June 11th

All day at meeting of Committee on Armaments Development (Dr. Guy Chairman). It is a big job unravelling all the strands and somewhat long winded. There have been already three meetings and I have been unable to attend and I feel somewhat hopeless about the amount of one's time demanded.

Megaw is just back from trials over the W. Coast with centimetre waves on CW. He reports some interesting yet enigmatical results.

June 12th

Marris has had three days at TRE on Oboe and magnetrons and other things. He arrived the day after the accident. Three of the most able of the EMI men were killed and also O'Kane's collaborator in H_2S. O'Kane might easily have been one of the party. Clayton, George and Cherry went with Marris and they were thoroughly initiated and interested. Our stock stands well with TRE and theirs with us and there is a most genuine desire for partnership of the right kind which I for one appreciate enormously.

Marris warned me of the anxiety over 3 cm magnetrons and that Wembley had been doing so little towards supplying practical valves for use at the establishments. We have had the impression that Whiddington has not particularly wanted us in this field and that we are in it at all through our own gate crashing. Our share so far has been research on new designs.

Jolley tells me that Helmore is threatening to link Wembley with him in a 'Cabinet Experimental Group' or some such scheme. I hope he will see me first—but it would be like him not to!

June 15th

GR contract (or ITP) received on basis of £24 000. Have got Espley and Gossling to arrange for Lamont to join Espley's group. Have put pressure on Mr. Rosen (RMA) to see that we have some good men from the pool!

A very encouraging letter from Prain (Diamond Die Control). I was absent from the Advisory Committee but he said that very warm insistence was expressed for the continuance of our technical committee —and even Davidson, whose support one always thought a little grudging, was reported as warmly in favour (Fig. 23).

Telephone: RUGBY 3315-6-7 (3 lines).
Telegrams: "DIAMOND," RUGBY.

MINISTRY OF SUPPLY,
DIAMOND DIE & TOOL CONTROL,
2, HORTON CRESCENT,
RUGBY.

RLP/DIR

13th June, 1942.

Dr. C.C.Paterson, F.R.S.,
General Electric Co. Ltd.,
(Research Laboratories),
East Lane,
North Wembley.

Dear Dr. Paterson:

1. At the Meeting held last Wednesday I was asked to convey to you the congratulations of the Meeting on your broadcast, which appeared to be much enjoyed by those who heard it.

2. At the same Meeting those members of your sub-committee who were present were very emphatic that they would like the sub-committee to continue its work beyond the end of this year and, in fact, indefinitely. I was very pleased to hear this as I am always afraid in committees of a losing of interest. This appears to be quite the reverse with your committee, and the members apparently would like it to continue on the basis of a meeting a month. Davidson, who I should have thought had little time for committee work, is most emphatic that several of them would continue to get together regularly even if the official committee were discontinued.

This is all very satisfactory and I hope that the success of your committee will not be embarrassing to you. The main Advisory Committee were unanimous in endorsing the view that the work should continue and, in fact, the work of your committee is rapidly becoming the main topic of interest to the main committee.

3. Thank you for your permission for Mr. André Cauvin, of the Belgian Government, to visit your Die Room for the purpose of planning a short film of the operations, for the reasons which I explained over the telephone. I enclose a copy of the letter of introduction which I have sent to Mr. Cauvin.

4. With regard to the patent position, I had a talk with

Over/.

Fig.23 Letter, dated 13 June 1942, and enclosure from R.L. Prain, Controller, Diamond Dies, to Clifford Paterson

-2-

Mr. Cornock, to whom you kindly put me through on the phone, and he has undertaken to regularise the position.

Yours sincerely,

R. L. FRAIN

Enclo:

Fig.23 continued.

& TOOL

RLP/DIR

12th June, 1942.

Dr. C.C.Paterson, F.R.S.,
General Electric Co. Ltd.,
(Research Laboratories),
East Lane,
North Wembley.

Dear Sir:

This will serve to introduce the bearer - Monsieur Andre Cauvin - working under the auspices of the Belgian Government, 107, Eaton Square, London, S.W.1. - who is making a film to be shown in the United States showing the application of the products of the Belgian Congo in war.

This film will include sections on Tin, Copper, Rubber and Diamonds, and in connection with the latter I understand that you will permit him to visit and later photograph your Diamond Die manufacturing operations.

Please accept my best thanks, and any courtesy you can extend to Monsieur Cauvin will be much appreciated.

Yours faithfully,

R. L. PRAIN
CONTROLLER

Fig.23 continued.

Preproduction meeting showed the CV 55 (E 1190) as happier and we urged more expansion in view of GR requirements. Gossling had an all day midget valve meeting on Friday and found Exeter demanding midget oscillators.

Went on Saturday with Le Rossignol and Pearce to see the Marconi valve factory at Chelmsford—run by Eisenstein and Young. They are producing about 50 magnetrons a week.

June 16th[288]

Waldram is already off to Anglesey where his 'fan' searchlight beam idea is to be tried out—and I understand he will be flying in order to see what he can see.

'Labour' matters become more and more difficult. The officials seem to want to use every excuse to take women away and when the girls themselves want to leave all sorts of excuses are trumped up. Fortunately the local exchange people are helpful and able to appreciate the difficulties.

Lunched with H.H. who spoke to me about certain IEE matters. He seemed very well and I much enjoyed the talk with him.

June 17th[289-290]

A CVD meeting at which Dr. Moon was back from USA to tell the meeting of progress there. Gossling gave the meeting a full description of Warren's single ended rival to the USA 'lighthouse', [- the] E 1368. It has some better characteristics than the lighthouse.

Returned early to a lunch to honour Castle now resigned from CO of Home Guard and to welcome Chapple—now Major—the new OC.

Winch tells me that the gun flash—'Sali'— device may be wanted 100 off so there is need to see if it can be made in the GEC.

June 18th

Helmore again asking GEC to spend money on 'fish' work without any contract or letter—but merely on his word. Shall have to refer issue to A. Railing.

The 10 cm meeting yesterday (after I left) put strong pressure on supply of 3 cm valves—I understand Birmingham had to take most of blame for non-production of samples and rightly, for Wembley has had to gate crash on 3 cm valves in what it has so far done and Birmingham have been the boys in Whiddington's attitude [eyes].

Armaments Committee all day. Had impressive evidence from Mr. Gibbs director of weapons production—who gave the common sense industrial view point—but gave it very well.

June 19th

A day of Service meetings. (1) Espley's crowd (including Cockcroft, Gray [Grey] and Walmsley. (2) Winch's army visitors on 'Sali'. (3) Megaw's party including Shearing, on the 10 cm transmission (CW) results over sea. So there was a considerable crowd at lunch.

Spent some time making room for Rooksby and his X-ray outfits by making a working space in the gallery for Cates.

June 20th

Winch reported a good meeting on Sali, commonly known as 'Flash-bang' recorder. Our first model and two pilot models are needed—any further requirements being left for the moment—but probably 100 sets. We shall get Claudgen to make the pilot models with us.

June 22nd

The news is not at all bright. The Libyan campaign has not gone well and no one quite sees what will be the result. Ships are being sunk on the USA side. They don't seem to have learnt our techniques. We have still to hear the story—probably depressing—of the Mediterranean convoy which was heavily attacked.

The compact source lamp in higher powers—up to 7 and 10 kw—in spite of splendid work by Francis, Kettlewell and others [-] does not give efficient utilisation in a projector because of electrode obscuration. Pattison from West Drayton was over today and confirmed. He is discussing with Couch and we shall then decide whether to carry on with the work.

MOS want us to do work on temperature co-efficients of insulators and components which cause circuits to get out of adjustment when working under abnormal conditions.

June 23rd

To Head Office with Espley and the rest of the GR group to decide what production and time guarantee can be given to Leedham for the making of the 1500 sets—worth approx. a million. The time schedule is not so much out, if design data and components are to time. We have the impression that DRP (Leedham) does not like having his hand forced in this way and would like to keep GEC Radio [Works] to MAP limited demands. We feel that the Army is most in need of help with its technical gear and we are glad to go all out to help it.

Conference with Goldsmith, Hawkins and Cameron on the slipping back of quality of cinema carbons; we arranged provisional steps.

Suggested to Francis that he should go over to help Espley on GR (with Isaacs and Willoughby) till the functional prototype is complete about September. The suggestion was welcomed. Kettlewell will probably carry on with the compact source work.

June 24th

Visit by Warren, Dr. James and others to STC resulted in very frank and helpful talks on valve design. Warren's E 1368 is very interesting just now in view of its value for local oscillators in our GR, and its equivalence to the USA lighthouse valve. There is no other valve in the running but I expect to meet many expostulations for requiring the E 1368 to be made.

Gossling returned with Donaldson and Jenkins from the new ADRDE at Malvern where they had two days on the proximity fuse. There seems to be a sudden urge there to press forward 500 projectiles for firing trials in July. So the burden of making the valves for them will, I suppose, come on us.

A Radio Committee meeting at HO. I find Robinson very helpful in his backing of broad policies. The meeting went fairly well and Wells gave us a rough statement of Coventry management's view of future organisation for radio and telephones—but I should like to see a more cut and dried plan.

I am going back tonight to witness a run after dark on the 5000 A. Helmore lamp—and which he is attending.

June 25th

An internal meeting with B.S.G., McLeod, B.P.D. and others to see how we can handle the PF valves needed for urgent trials and also the E 1368 (Warren valve) which all emerge from McLeod's shop.

We propose to use Hivac for as much of the PF work as possible.

An OAC meeting at which we opened up the question of tackling all we can with the USA design of fluorescent tube lamp in order to be prepared for after war.

Attended a meeting of the Armaments Committee at which Dr. Rotter was being questioned. An interesting discussion, but I felt I could ill spare the time. However I may be able to be of some help in due course.

June 26th[291]

Gavin is back from the North East Coast where he has had triumphant trials with his 'Mandrel' set. It gave a better result than his most sanguine hope. The remainder of the sets are practically finished—for his group has been working long hours and doing splendid work.

H.G. Jenkins reports arresting results with the UV + fluorescent paint canal navigation trials. We shall have a contract for its development from the Admiralty.

A main CRT meeting all the morning. Whiddington most of the day to see our valve developments—urging 3 cm magnetrons. Was glad to hear from Oatley that he had broached the question with Truscott of the approval of Warren's E 1368 valve for GR so Dr. Black of ADRDE will now formally ask for its adoption by TVC.

June 27th

Gossling reports a helpful talk with Hivac who will help out with various urgent requirements for miniature valves which would otherwise clog up McLeod's shop. This will help us with the E 1368 regarding which I had a discussion with Warren and James today with a view to making the design as much of an engineering job as possible.

We are being pressed to increase our diamond die output to help the Control as well as the extended tungsten production through the new Osram metal factory. We are considering overtime and some sort of second shift arrangement.

June 29th[292]

An output of 1250 valves last week. We are consolidating on two NT 99 groups and Boyland will take on Warren's E 1368—which is very much of a new venture. The little copper concentric valve E 1274 will take a second place for the present.

Had Mr. and Mrs. Miles here (of Phillips & Powis) and showed them much of the Laboratories. They are very great on offensive gear controlled from a distance. A very stimulating man whom I would liked to have talked to at greater length but as they came with Buckell and Neville from HO there was no opportunity for quiet talk.

June 30th

Had Carpenter here from Admiralty to examine our statement of preproduction finance in order to justify the writing off against expenses all the purchases of gear and equipment (approx. £9000). I think he was satisfied. A quiet day with everyone getting on with work. The increase of GEC bonus to £60 for monthly staff is very welcome.

July 1st

Dr. Moon is back from USA sooner than we expected and we had a thorough day of it with him. The lighthouse valve figures in the discussion—which did not however seem to show that the USA's valve had anything over us so far.

Hugo Hirst paid us a visit and spent the afternoon here. It was delightful to see him. He allowed himself to be wheeled all round but was a bit tired by 4 o'clock.

July 2nd[293]

Dr. and Mrs. Wooster here for a consultation on quartz crystals. There seems to be light on one disease of crystal which affects their piezo-electric qualities, and shown up by X-ray, in that elements of the lattices are displaced to a greater or less extent from the dominating orientation.

A valve allocations meeting all afternoon—lively. Pressure for greater production of nearly all types. The E 1271 for AI is the greatest trouble and I expect we must do something about it eventually and break into the NT 100 preproduction. But I fear a serious situation because the dates on the AI Mark VIII are likely to be kept i.e. middle of September to start.

I am preparing a second 'popular industrial research' broadcast—subject luminescence.

July 3rd

Made arrangements to continue—somehow or other—to make the E 1271 valves in the Labs. for the next six weeks.

Gavin is flying with gear to Berwick with all the Army 'Quality' to demonstrate Mandrel and TRE to demonstrate Moonshine. Within three hours of leaving Wembley he is on the site ready for the trials.

Am succeeding with help of Evans (Building Research Station) to cut down the amount of work and sittings of the post-war lighting committee.

We shall try and limit the amount of evidence taken or take several witnesses simultaneously.

July 4th[294]

We have kept up our CVD reports to DSR Admiralty every six weeks for the past two years. Have just been reading the draft of the latest (No. 21)—it is an interesting record and shows very comprehensive field of work—successful in nearly all directions. This series of reports all edited by Gossling will prove a valuable historical record. Am glad to note that our new 'waveguide type' of resonator magnetron is making progress. Megaw and Willshaw are very sanguine. The new big modulator fixed up by Michaelis is proving most useful—and appears to be the largest thing of its kind so far.

A meeting yesterday with J.W. Ryde and the Admiralty infra-red officers showed some surprising lapses. Ryde put the Admiralty right as regards work of ours which they knew all about but seemed to fail to appreciate. They want us to work at vapour lamps rich in infra-red to be modulated. In fact a more elaborate edition of Macfadyen's 'talking down the beam' set.

July 6th

1350 valves preproduced last week.

Admiralty want another infra-red device developed.

MOS are adding to our burden on FM by desiring us to run field tests which they would otherwise do.

On GR, Espley tells me the War Office contingent are now wondering how they want the gear mounted for transport—which may affect our designs.

Tonight I have to attend 'Canal' navigation trials on the Sports Ground—I am told the effects are impressive.

July 7th

The navigation luminescence trials were so far quite promising and striking. They kept us out till 1.30. It is now for the canal pilots to inspect it.

Had talk with Macqueen re. the radio prospects in the GEC—a subject full of uncertainties.

I attended a meeting of the Armaments Development Committee and lunched with Appleton. The issues are not all contentious but some are likely to cause disputing on the committee itself.

TVC meeting this afternoon but without incident. We shall have to meet fortnightly at present in order to deal with the large number of valve applications for new uses.

July 8th

Willshaw and Megaw's modification to the magnetron by closing the resonators is looking very promising with a loop line coupling instead of a slot. The efficiency is at least up to 6% on first trial and the wavelength repetition between the first two samples is very close.

July 9th

Have had a whole day on the Armaments Development Committee and enquiry. Evidence by Dr. Paris leading to much frank speaking and criticism of organisation of design and development. Some sections of the MOS such as ADRDE and SEE are being organised on good lines but armament falls a long way behind. Events in Libya are giving the deliberations of the Committee a good deal of significance and our findings are likely to be important.

July 10th

Megaw reports result of conference at SEE after trials with his 6 cm transmitter with multi-channel transmission and reception fixed up by SEE. He said he did not think SEE were capable of arranging so good a show. If it comes to what is expected it means many E 1231 and E1210 valves. The E 1231 works well as a local oscillator on its harmonics. It appears to be the best valve for harmonics.

A diamond die technical committee—not much of interest reported. We are making about 120 dies per month and Simpson and Davidson are getting good results with our drilling machine. Our design of back piercing machine seems to be meeting with approval. Last financial year I paid off all the expenses of starting up diamond dies—setting them off against profits on other Lab. products. Our diamond die expenses were about £12 000 and our profits on other products (except valves) about £16 000.

Humphreys reported visit by a tough-neck from Ministry of Labour plus a representative of MAP re. our request for a certain number of younger girls. His instructions were evidently to bluster down such requests but Humphreys is not to be taken in by mere tactics.

July 11th

Jesty and Gossling have both had useful visits to TRE; Gossling in discussion with Dr. Lewis on beacon projects for which E 1368 and E 1274 or lighthouse valves will be wanted.

Jesty had a thorough sifting with Bradfield of future CRT type demands with a view to reconciling Dr. Truscott's ideas with reality.

I have next week given up largely to the Armaments Development Committee.

July 13th

A good preproduction week and no really troublesome spots.

Had Rabbin from MAP contracts department to talk over contracts and to look round the place. A very helpful chap. Told him I had reduced overheads for Government contracts from 200% to 180%.

To Cardiff where the Committee on Armaments Development is visiting metallurgical resources at the University.

July 14th to 15th[295]

To Swansea for ditto. I don't feel certain how much practical use these visits are—but the members of the committee are getting clear as to the nature of the recommendations they will make.

The visit to Sir R. Robertson's explosives laboratories at Swansea University was most interesting and informative.

July 16th

Gossling and Le Rossignol entertained a lively delegation from USA (MIT) and the members seemed very interested and impressed.

I hear GEC has got an order for the whole of the GR equipment—some job. There will be a meeting with A.R. tomorrow about it.

A very appreciative letter from Lywood to thank Gavin and his group for the long hours worked and successful result of their 'Mandrel' effort (Fig. 24).

My first GEC annual Board Meeting to review the financial results of the past year.

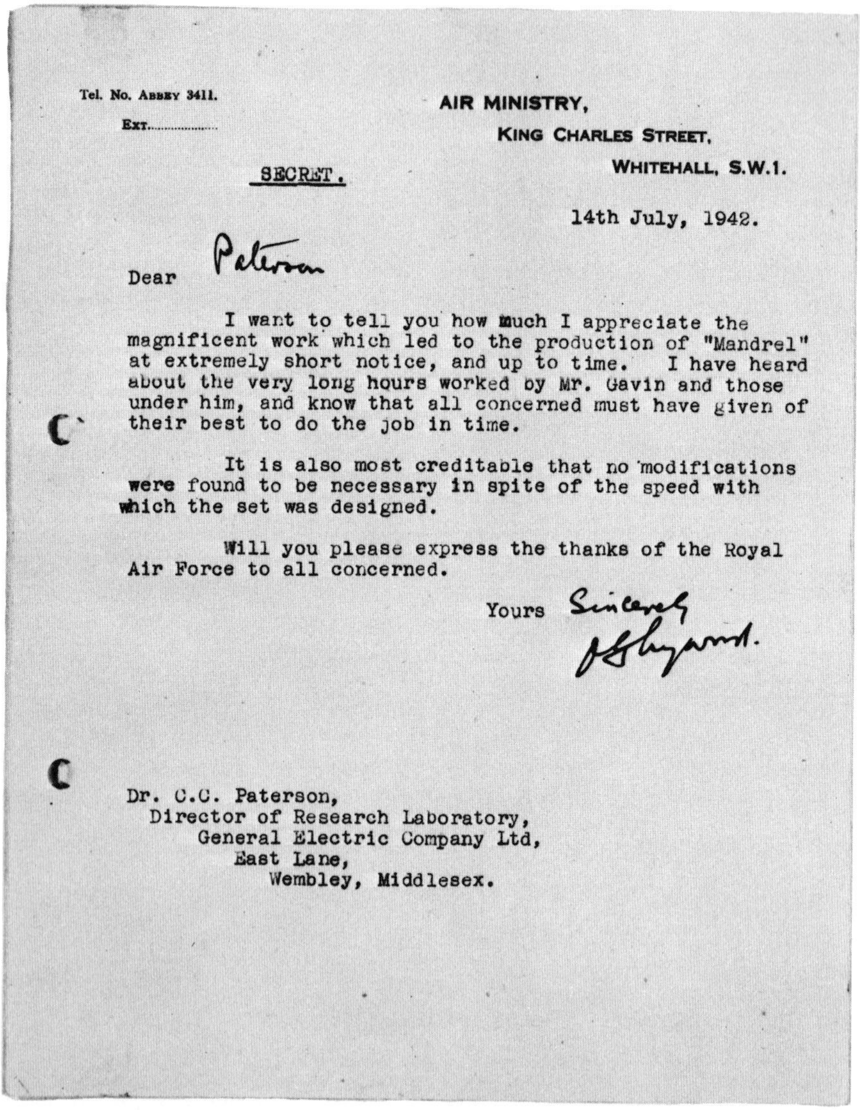

Tel. No. Abbey 3411.

Ext.……………

AIR MINISTRY,

King Charles Street,

Whitehall, S.W.1.

SECRET.

14th July, 1942.

Dear Paterson

I want to tell you how much I appreciate the magnificent work which led to the production of "Mandrel" at extremely short notice, and up to time. I have heard about the very long hours worked by Mr. Gavin and those under him, and know that all concerned must have given of their best to do the job in time.

It is also most creditable that no modifications were found to be necessary in spite of the speed with which the set was designed.

Will you please express the thanks of the Royal Air Force to all concerned.

Yours Sincerely
O G Lywood.

Dr. C.C. Paterson,
Director of Research Laboratory,
General Electric Company Ltd,
East Lane,
Wembley, Middlesex.

Fig.24 Letter dated 14 July 1942, from Air Commodore Lywood to Clifford Paterson on effort put into 'Mandrel'

Lord Hirst told me today of his decision regarding the IEE Benevolent Fund—particulars of which will transpire later. I feel very pleased about it.

July 17th[296]

We have an important joint demonstration with ICI at the Empire Building on our joint paint drying efforts next week. This grows into a major effort.

Had talks with Bligh and Chapple about Heller's altimeter which is being so completely turned down by MAP. Helmore is inclined to take it up and make trials. We find that Heller's device is the same as Salford have been working on successfully for a long time with the support of the authorities. The device is closely allied to Salford's proximity fuse which I am told has produced, on trial, very hopeful results with bombs. Salford have apparently been doing a good pioneer job with this device.

Went to Head Office for a round table meeting of all concerned with the large GR contract to settle final arrangements.

Had talk with S.V. Williams and Chirnside re. Salford iron powder which is getting on well and shows really good qualities.

July 18th

Owing to German break-through on the Southern Russian front and the shipping losses by submarine on Atlantic seaboard (mainly I believe USA responsibility) we are going through a period of grimness. Tails are up for the moment over Libya and Egypt because the Germans there are very far from their base and we seem to be able to hammer them. But it is to be hoped that pressure on the Germans will not be relaxed elsewhere so that they can devote more attention to the Middle East.

July 20th[297]

Holiday breaks are affecting preproduction. I learn from all hands that the manpower position is getting very tight and stringent steps are being contemplated in various directions.

Checked through our preproduction finance with Le R. and Miss Lock. By the end of May we had cleared all our losses and appear at present prices of valves to be making about 8 to 10 thousand [pounds] margin per month. Propose lowering overhead to 180% (from 200%) from April 1st and recommending adjusting prices of valves.

An Advisory Board meeting this afternoon at which interesting points were discussed.

Afterwards saw Lord Hirst with J.S. Highfield. He has decided upon his IEE Benevolent Fund scheme—at which both Highfield and I are delighted. It will be called the Lord Hirst fund with a principal of £20 000.

Stevens has a heavy visitation tonight on canal lighting accompanied by flying—Third Sea Lord threatens attendance!

July 21st[298]

James reported measuring oscillations with the E 1368 valve down to 6.5 cm due to control of the circuit used with it and particularly its sliding contacts.

Attended meeting of [Armaments] R & D Committee at which Air Vice Marshal Baker gave evidence. Very good witness although I could not get him to agree that an inter-Service facility for Research and Development is an anomaly in a three service set-up.

CVD meeting at which two very important valves went a stage further towards notoriety: (*a*) the E 1368 as above, (*b*) the new magnetron design on waveguide principle. Willshaw and Megaw got this to 12% efficiency and a good output and all looks very promising.

The verdict on last night's canal navigation trials is highly flattering to the scheme. We have promised a full trial on the canal over 1½ miles length by middle of August when there is no moon. Z Dept. Head Office is handling.

July 22nd

Things got lively on the Armaments R & D Committee and I shall be in the box tomorrow. Some of the assessors appear to be unduly nervous of giving evidence. I must rush through a memo on our GEC committee system for research and development liaison.

Was able to attend a CVD policy meeting this afternoon. DSR just back from USA looking thin but keen.

July 23rd[299]

An early meeting of [Armaments] R & D Committee at which the assessors gave their evidence. Many signs of hot air and the chairman had some difficulty to get his guidance across. Most people were agreed on 'control' but I was the only one to try to put across a properly established machinery for technical liaison based on our GEC 'regular' meeting plan.

Lunched with the Minister of Production (Oliver Lyttleton) and BSI Executive. It did not lead far but I hope it impressed on him the part which BSI can play.

Was most of afternoon with Engineering Advisory Committee trying to guide the scheme for scientific and engineering advice to the War Cabinet, on A.V. Hill's proposal. Saw our Chairman about it (Mr. Butler, Minister of Education) at 6.30. He is putting our memo to his Cabinet colleagues.

July 24th

Aldous reported a much better outcome of the 'flap' meeting on valves for recognition than we had feared. I am afraid TRE and Ferranti between them had to carry the trouble by making the sets operate better.

Bligh had an 'off the record' talk with Air Commodore Blandy about police radio service for London, which revealed very clearly the need for our sets and which GEC has not production facilities to supply. One chance seems to be to turn it into FM and supply on a development contract! But I doubt whether controls can be got round this way.

August 4th

The writer has just returned from a week's 'break'—the longest time that any senior member of staff manages nowadays.

Most things at the Labs. are going well. Gossling and Bell paid a call on Mr. Hecht and Mr. Dixon at MAP and had a useful talk on many aspects of valve matters. Gavin's jamming activities have been going strong and we are asked to press forward with an urgent new set to counter the enemy short wave service. Humphreys with ICI had a large visitation on paint drying from Government officers. Bligh and his group are occupied on FM field trials. The 10 cm meeting went through last week with much talk and little heat. The impression is that GEC Wembley will have to make most of the early samples of E 1364 [E 1264 or CV 64] (i.e. magnetron for AIS). BTH look as if they would be late with their production.

August 5th

Espley's GR job goes through its teething troubles. The need for the RX trial to establish the 25 cm principle is rather getting in the way as it absorbs components made for the GR functional prototypes.

To C. C. Paterson, Esq.,
The Research Laboratories of the
General Electric Co. Ltd.,
WEMBLEY, Middx.

The General Electric Co. Ltd.
Radio & Television Works,
Spon Street,
Coventry.

Form 105R. Telephone No.: Coventry 2027 (2 lines). SECRET

Your Ref:	Our Ref: W.H.P.	Date 5th August 1942.

Dear Dr. Paterson,

I am enclosing herewith, as a matter of interest, some notes of an informal speech made in our Machine Shop by Sir Watson Watt when he visited us last Saturday. He gave a similar talk to the Engineers and a third one to the Departmental Managers saying much the same things each time.

He indicated that over the last few weeks the number of German night raiders brought down by the previous Mark of equipment made by us was greater than the quantity brought down by all the other defensive devices together. Partly as a result of this the new Mark has been put at the very top of the priority list, not merely in this field but of all equipments required by any of the Services.

The general tenor of Sir Watson Watt's remarks was that everybody concerned was extremely pleased by what had been done and by what was now being done.

Yours truly,

Similar letters to - Dr. Railing,
Mr. Wells, Mr. Macqueen, Mr. Robinson.

Fig.25 Letter dated 5 August 1942, and note of address by Sir Robert Watson-Watt, from W.H. Peters to Clifford Paterson

Enclosure with Peter's letter of Aug 5

RADIO WORKS.
5th August 1942.

Sir Watson Watt started off his talk by saying that the people who worked in the Machine Shop and various other parts of the factory were extremely important. He expected that there had been criticism of the customer in this particular set of jobs on which we were employed because the customer fed them with work for purposes that he didn't explain. He fed the technical staff with information which was sometimes wrong, frequently late and occasionally incorrect and that interfered with the smooth flow of work through the various departments.

The night raider was severely discouraged rather more than a year ago but we obviously had not finished with him then. The German Air Force will unleash on us once more all the aircraft they can spare which will not be so many as last time, fortunately, also the crews will not be so good, but they have a good organisation behind them to do the dirty work.

Sir Watson Watt said he might have mentioned which way the bits and pieces made in this factory are used in building into the system for defeating the night bomber, but there is a good principle which is applied to everybody in this country up to and even including members of the War Cabinet. They are told just enough about the country's secret processes to enable them to do their work effectively because it is better for everyone to be able to say "We know just as much as is necessary to enable us to do our job".

He said that the customer (the R.A.F.) was completely satisfied with the efforts shown - even though some of the people thought the hand-made products rather a waste of skill and a high price to pay. It was worth paying to have this equipment to put into the Air Defences in the coming autumn and winter. It was considered an investment - not a gamble.

He went on to say that the G.E.C., in conjunction with T.R.E. and M.A.P., had turned out equipment not merely from the development stage but from the earliest research stage into the production of a considerable number of hand made models and these models have, in the few months of use, accounted for a very handsome proportion of the planes brought down. The G.E.C. score card is now proportionally very much better than those of the other night defences.

What the workers are doing is to make life just a little easier for the ace night fighters who have to fly in a black sky in the middle of the night looking for black-painted aircraft.

He said that the skilled efforts of the particular jobs which were being done in the Works were the most important single bit in this country's war effort for the autumn and winter of 1942/3. Every ounce of energy put into it will give a better return in dead Germans and live Britons.

At the conclusion Sir Watson Watt said as he had been a Trade Union agitator most of his life, he was used to heckling, so that if anyone had questions to ask he would do his best to answer them.

Fig.25 continued.

A household appliance meeting at HO to prepare for post-war doings and lunch with Macqueen to talk over Head Office, Coventry etc. radio affairs—off the record.

August 6th

Review of preproduction figures with Le Rossignol shows that by June 30th we had covered all our arrears and there was a reasonable margin of profit dating from the start. It is proposed to reduce the prices of certain valves from July 1st to keep the rate of profit about right with 180% overhead. I find the whole effort has come out in a way which the Admiralty ought to be very pleased with.

The problem of heating of wood mouldings by HF current has arisen with MAP and others and we are about to see what is involved in the problem.

A circular request by the Admiralty to increase rate of output on all its programme items by 50%! A somewhat extraordinary demand. We can probably do a good deal with valves—but experimental work is not so easily increased.

August 7th[300]

A letter from Peters recording Watson Watt's talk at their Works showing that Mark VII AIS was bringing down more night bombers than all the other defences put together and acknowledging GEC's research, development and production effort in conjunction with TRE and MAP. AIS Mark VII has now highest priority of all radio devices (Fig. 25).

This means that after all the cold looks we had when we insisted on making tools for preproducing the CV 64 magnetrons, we are going to be relied upon for all the valves needed in the first two or three months for AIS Mark VIII—because as I rather guessed, the BTH are late with their mass production and are not expected to begin delivery till November.

Helmore called this afternoon to put his side of an incident in which Jolley was the innocent cause of letting the Admiralty into Helmore's confidential work on Staines Reservoir.

Had indication that there is doubtful intrigue going on in connection with Leedham's DRP control of the Radio industry—and other, most unsuitable, people are trying to get themselves appointed.

August 8th

Went over our getter manufacturing position with Britten. He is getting on well especially with the nickel coated type. Except for one bottleneck (steel HT pots) he is ready to increase our rate of production some three times—and already we are well up to the MOV requirements.

Discussed diamond die section with Leeds and Dudding. We must now get our financial position flattened out and know where we stand. Works appear highly satisfied with our dies—but we have had to write a very frank letter to Prain about Davidson's unwillingness to institute our techniques and machines but insists on using his own modifications and ideas—which may be all right in their way but we cannot assist him.

Stevens reports that the Admiralty without waiting for the 1 mile Canal navigation trial has decided to go ahead with the whole 35 mile scheme—the trial however to take place next week. This will be a considerable job.

August 10th

Discussed with A.S. Gosling and Marriott re. revised prices for valves to the Admiralty as from July 1st. I think the new proposals should give satisfaction and should leave us with the desired margin.

There is some to-do about the midget valve for the Torpedo Factory which tends to shift characteristics slightly on shelf. It is an impossible proposition to avoid this and TF will have to give their circuit a margin of safety.

August 11th

Bligh's field tests with their FM set in conjunction with SEE, Horsham, came out markedly in favour of FM compared with AM. They were carried out over long distances at wavelengths up to 7 metres. They had the running of the tests.

Had long talk with Levy and West to get some reconciliation of their point of view with that of L.G., J.Y.F. and G.C. I hope I succeeded.

Espley attended a meeting today with MOS and Peters and Southworth to see if 30 GR sets 'of some sort' could be made by January. It seems an impossible idea but I believe the exigencies of the situation have induced Southworth and Peters to say they will 'do what they can'. I think Espley feels much apprehension but we must discuss the matter further tomorrow.

August 12th

A reminder from Admiralty that they are counting on us for CV 55 as they have agreed several new programmes on the strength of our efforts.

A long CRT meeting—but some wasted time. The new double screen techniques which we have evolved with the aim of simplifying production seem to be going well—CRTs are, however, difficult gadgets to inspect and approve to definite limits.

Visited the new MOV controlled valve factories at Alperton with Dr. Truscott. Was glad to see valves for BBC beginning to figure in the extension programme.

Espley's brief report yesterday on GR preproduction seems to be true—but I suspect the predictor will prevent any anticipation of date.

Winch reports good progress towards completion of the prototype of his 'flash bang' equipment for ADRDE. It looks as if it would be ready for trial in a fortnight.

August 13th

We hear that the Ministry of Labour press gang is operating on the site questioning all workers as to the work they are doing—and intending to extract all possible moveable women. We shall be seeing how many girls in this category we have left in the Labs. This will be a headache for O.W.H., B.P.D. and self.

A TVC meeting at which Dr. Truscott showed that under MOV instigation he was persuading everyone to use the 1368 valve in preference to the 1361 (CV 52)—on the grounds that Wembley would be the only producers of the CV 52 MOV showing a strong preference for E 1368.

I have been summoned to a meeting of the Engineering Advisory Committee tomorrow to put forward a list of names from whom scientific and technical advisers to the Ministry of Production can be selected. I wish I knew what they were expected to do.

August 14th[301]

A joint meeting of the Engineering & Scientific Advisory Committees to make recommendation for three scientific and technical people to advise the Lord Privy Seal and the Ministry of Production on the applications of Science etc. to the war. An important meeting presided over by Sir Harry [Henry?] Dale—very ably. I had to leave before the end but the list of names was nearly complete.

A Brundrett and Harley preproduction meeting at which Brundrett urged very strongly the need to pass to Works some of the older stagers such as NT 93, NT 97 and NT 98 (magnetron). We have got four new ones now rapidly approaching: E 1368, E 1274, [the] waveguide cell, and

1210 of various types—an [d] output of E 1231 must be doubled as soon as possible.

There is much intrigue at high quarters in Westminster, I am sorry to say, which is now coming to a head—jockeying Sir Frank Smith, Leedham, F.O. Brown out of their posts. There is proposed a small Radio Board to co-ordinate Service radio, but a crucial point seems to be a dispute over the vice-chairmanship. There seems to have been friction for a long time though why—knowing the people concerned—I cannot understand.

There has been a big meeting all the afternoon on Proximity Fuse valves—Donaldson and Gossling are handling from our end.

August 15th

Marris's group is being asked to help with Mark II Oboe as an emergency measure. There is also a heavy push on H_2S and O'Kane looked in to talk over matters generally. He seems to be having to help the TRE 'preproduction' factory at Bournemouth; there is a lull in the experimental side of his work. H_2S touches us mainly in the supply of magnetrons. By the way there is a scare at the moment, two out of five valves sent to TRE having been advised as non-arrivals there. We put the security people onto it at once.

A somewhat harsh and grim situation at the PF meeting yesterday—a Mr. Rothwell of ADRDE being peevish and unhelpful in the face of difficulties, and not favourably disposed to meeting of all concerned to clear them up. Apparently not a very good type of man for co-operative work.

August 17th

Preproduction about 1200 valves in spite of holidays. We have new high objectives to meet with the magnetrons and CV 52s.

Lunched with Chelioti. He is having serious anxieties over ignorant bureaucratic handling in regard to treatment by the Board of Trade and others of the lamp industry in the hope of finding hidden sources of manpower in such factories as Wembley. The whole personnel engaged on GLS lamps is apparently about 300 in the Osram Works.

August 18th[302]

An appreciative letter from MOS on the conclusion of our work on immunisation of enemy bomb fuses. It is not always [usual] though it is not unusual to receive letters from the Ministries of this kind. It says: 'as you are aware the research and development work under this contract provided us with a most valuable piece of bomb disposal apparatus'.

Lunched with Prain (Diamond Die Controller) at the offices of his Company—Selection Trust—and discussed various matters with him. Incidentally Leeds and B.P.D. had drawn up a provisional price schedule for our diamond dies which we shall ask Lamp Works to agree until December 31st. Prain was not surprised at our prices which are higher than his schedule.

Professor Emeleus came to see us from Belfast to see about a programme of work on gas discharge devices. Ryde looked after him.

August 19th[303]

Had an hectic hour with Humphreys over a voluminous report on heating and ventilating from a committee of the Building Research Station and about to be adopted by a committee of the Ministry of Works and Buildings. As Humphreys described it—the product of incompetency or roguery or a mixture of both. I went to the Ministry to a meeting between the committee and representatives of the electrical industry—at which some very stiff and straight things were said—some I fear by myself. I had to argue that the whole basis of the report, which was financial, was wrong.

Went to see Solomon at Beaconsfield. One ought to be able to drop in once a fortnight and look him up.

The press gang visitor from the Ministry of Labour appeared today to start his enquiries. This means some hard work for O.W.H.

August 20th

A Radio Communications meeting with Coventry and Head Office at which all current and prospective work was reviewed and the policy of forcing the pace with the Ministries was reaffirmed after a somewhat vigorous discussion. The rumoured changes in organisation at MAP on radio are a bit of a cloud, and the industry tends to be incensed that it is not consulted on a matter in which it is so vitally concerned.

August 21st[304]

The trials on the canal navigation job passed off well earlier in the week and Stevens says the navigational part of it was quite accepted. There seem to be certain reservations by the Ministry of War Transport but don't know what they are founded on. A large ship is now to be taken up.

The report of the Armament Research and Development Committee is now out and follows the general line one anticipated from the discussions. It deals to a considerable extent with the higher official posts. The scheme is all right but everything depends on the spirit and if the spirit is all right almost any old scheme will work.

Went to the Exhibition Building to see how it shapes and particularly to see the radiant heat paint drying development. Shall be glad to see component making going ahead there.

Landale was here from Signal School and requires multi-megawatt magnetrons. I expect as usual our good friends who arrange these things will say that GEC is too busy to share in this development!

August 22nd

The 'press gang' is not as formidable as one feared and as usual reasonable and straightforward treatment on our part meets with a similar response. Dudding followed Humphreys in explaining matters and it now remains to go over every man and woman in the place in the light of the work they are doing.

There has been much activity between glass workers and metal craftsmen in the evolution of new types of glass working machine. The machines are evolved as the result of co-operation between craftsmen on both sides and may or may not afterwards be put through the drawing office. It is a capital arrangement.

I gave my second BBC talk this evening on 'Light for Nothing'.

August 24th

At preproduction meeting had to face up to difficulties of increasing output groups in spite of restricted labour supply. We lose as many women as we take on, through illness or other not unreasonable excuses, and we are not on highest priority as yet for labour. I hope we can be if Humphreys's and Dudding's conversations with the press gang man bear fruit.

Warren reports talk at DSR with STC and Telfer (MOS) on CV 90 (E 1368). STC valve (which is double ended) is an amplifier whereas our single ended valve is an oscillator. It looks as if we should be left alone with our type. One CV 90 has reached down to 6.5 cm.

Wood and plywood drying by high frequency heating is to the fore and MAP are in touch with us to see if we can take on anything.

August 25th[305]

Gavin's 'mandrel' trials showed quite definite results and indicated the type of jamming set now needed for specific wavelengths.

Our CV 90 (E 1368) valve appears still to promise all right but we must take steps to eliminate one blind spot in the frequency range over which it will work. At present the blind spot can be pushed out of the way up or down, but to eliminate it requires a different pinch design.

A nice letter from MOS terminating Pye's work on ROF and the supply to them of batteries on MOS contract.

Went over Espley's work and ways and means of avoiding the use of our craftsmen by using outside (Witton) help more. He has got a big job.

A very warm letter from RAE recommending Nicholson for special promotion on account of the standing he has achieved on cooling problems for engines.

Reviewed all workshop priorities with Gardiner. It is indeed difficult to do all we would like—tools, tools, tools for various jobs all the time.

August 26th

Visited with Jolley the Inter-Services Research Bureau premises where they are working with our help on a naval device of considerable interest. They appear to be a red tape crashing place and get going on things with speed. Colonel Dolphin (in charge) seems very capable.

Had Mr. Peat to lunch—joint Parliamentary Secretary to the Ministry of Supply. A thoughful and reassuring person who offered to be of help to us in any unofficial way. Told him of the unfortunate attitude of Board of Trade and other departments to luminescent lamps. Took him all round and was glad of the opportunity.

Mr. Carew from Canada called and gave us a picture of radio things there.

Gossling returned from useful visit to TRE and ADRDE.

August 27th[306-307]

At the Osram Advisory Committee this morning Chelioti gave a sad account of the official attitude to lamp manufacture by the Board of Trade and Ministry of Labour. Injustice and reluctance to enquire as to the facts of situation seems to be their guiding rule of procedure.

Met Sir Harold Hartley, Sir J. Greener and Ricardo at lunch. The former urged my attendance at Oxford in September to a select meeting on education for industry. I must try to go.

A post-war lighting committee on natural lighting of buildings all the afternoon.

August 28th

Two visits by Helmore with heavy Westminster tonnage—Air Marshal Peck, Deputy Chief of Air Staff, and Lord Cherwell, confidant of the Prime Minister. I went back at 8.0 pm to meet the latter. Both were very interested in Helmore's big lamp but its operational advantages have yet to be studied.

I have once again to block the attempted passage of Birmingham academicals (Dr. Wilkins etc.) into our luminescent powder research and manufacture. My frank letter to Wilkins is in Dr. Walmsley's hands for comment.

Truscott and company here for a valve allocations meeting—cheerful; but the load piling up onto us grows continuously. The CV 52 (E 1231) is wanted in numbers up to 500 per week!

Policy meeting on GR at which there was much talk about responsibilities for approval of designs and procedure connected therewith. But the job seems to be proceeding with only the minimum of delay to which we are accustomed these days.

August 29th

Ideas regarding the larger magnetrons are crystallising and Willshaw and Le Rossignol went into them with me today. The 3 cm valves (waveguide type) look very interesting; for with the resonant cavity type there is a large spread of frequency about the maximum instead, as thought, of concentration at the desired wavelength. Hence values of efficiency in the past based on total energy radiated are fallacious as a guide to effectiveness in operational use.

Willshaw's waveguide 3 cm magnetron, although radiating about half as much energy as the older type, has four times more in proportion at the desired wavelength and should therefore be twice as effective in use. This will be tested out in a few days by Signal School.

Gavin has had a message that the 100 sets of his Mandrel are very urgently required and are to be increased to 300. I have no doubt that the loss of 30 planes in the last raid over Germany is the reason for this sudden demand.

There are changes in Signals section at the Air Ministry, who direct our RCM work. Addison, I understand, takes Lywood's place and Tait becomes head of Signals in the RAF.

War savings has made another jump forward on the new cycle just starting. It starts about 30% up on the last one. Mrs. Paterson still handles the work almost single handed—but the magnitude of it now requires a collaborator for tackling this pretty onerous work.

August 31st

A poor preproduction week—partly due to holidays—plus one or two accidents. Much searching of possibilities to see how we can honour the cheques we have issued, on the assumption of increases of production. It looks as if our numbers would have to increase to 1750 eventually and every bit of floor space must still be husbanded.

Some testing will go to the Exhibition [Building] and save encroachment on the production space in the new extension. Some workshop transfer may also be possible.

Ring seal glass bases are still causing trouble in Glass Works although technically they are nearly there. But I hear there is a great prospective shortage and millions a year will probably be needed from us.

September 1st

A CVD (GEC) meeting. There were no rabbits out of hats this time but Willshaw's waveguide 3 cm magnetron was shown to have a very favourable wavelength spectrum and promised to give a better operation efficiency than the orthodox form. Several problems are up to us: (*a*) a high duty cathode for the CV 76 magnetrons, (*b*) the 3 cm waveguide [magnetron], (*c*) the VT 93, 97 cathode collapse.

September 2nd

Discussed our component making with Gardiner for it appears that fly presses and operators for them may be a serious bottle neck for preproduction of valves.

A special meeting with Whiddington, Birmingham, BTH and GEC to consider 5 MW magnetrons; high duty cathodes and life test facilities seem to be the most urgent immediate requirements. The whole problem is a major one but we shall make a start.

Went over Humphreys's list of staff with him in their various categories for the press-gang man—who incidentally appears to be a reasonable and helpful individual. But it is necessary to be scrupulous and careful that descriptions are not language liable to be misunderstood by the kind of people administering the matter at the Ministry of Labour.

September 3rd 1942—4th Year of the War[308-309]

A TVC meeting in the morning. This is the day of National Prayer and the work was suspended from 11 to 11.15 at the Labs. for attendance at the broadcast Service at five centres.

A CVD meeting at DSR Admiralty all afternoon. Some important issues and reconstitutions were settled. But the most important unofficial information is the supersession of Sir Frank Smith and Air Commodore Leedham and their substitution by Sir Robert Renwick and Dr. de Laslo. I know neither personally but reports are not promising. Watson Watt reappears as one of the heads of administrative radio but I am not sure in what capacity. Had a talk with Walmsley to beg him to hang on, but he fears he will suffer because of his partiality to 'industrial firms' and his consequent unpopularity with TRE. A serious shake up altogether. I trust the Admiralty research and development organisation under CVD will stand firm.

September 4th[310]

The canal lighting job has had a sudden hitch. When tried with a 10 000 ton ship and a tug in front, the smoke from the tug seriously interferes with the UV beam. Also in this particular test there did not seem to be time to arrange the direction of the beams. A big ship is a very different thing from the tugs with which the scheme was first tried. The problem has been put back into the laboratory.

A post-war lighting committee at the Ministry of Works and Buildings all afternoon. Evidence from the several electrical associations led by Jones of ELMA [Electric Lamp Manufacturers Association]. A lively afternoon but not a war winner.

Intensive consultations led by Le Rossignol on high duty cathodes for CV 76 magnetrons. Somehow valves have got to materialise pretty soon.

September 5th[311-312]

A very busy Saturday morning. A good meeting under Dudding's chairmanship with Bell as convenor of all concerned with oxide coated cathodes, with a view to progressing the research programmes and appraisement standards and techniques.

Most valves now have oxide cathodes and we must know the fundamentals.

Jolley reports his job with the Inter-Service Research Bureau on its first trials went off with surprising success and the Bureau is to be congratulated. Prince Louis of Battenberg [Lord Mountbatten] tested it personally for upwards of an hour. A special request from Colonel Dolphin to allow Hobson to go to Scapa to help with the next trials.

Marris reports very strong pressure for Oboe.

Letter from Addison (head of RCM) agreeing to everything needed to push the job forward. If it is not available in October the bombing season over Germany will be too far advanced to make high priority for this worth while.

A long talk with Pringle (DEE Admiralty) regarding the canal lighting job. In spite of disappointment over the recent full size trial with a 10 000 ton ship, he feels the job ought to go on and his message to me was to push it from our end.

Had meeting with Stevens, Jenkins and Smyth and found them very contemptuous of Mr. Pringle's attitude—'if he would only recognise the difference between making experiments and giving demonstrations, and not mix them up together, embarrassing situations would not be precipitated such as the present'.

September 7th

A morning with Humphreys on manpower and its use of the Labs. The press gang representative, Mr. Cowleyshaw, is reasonable and helpful and most concerned to represent to his principals what sort of a place this is. Humphreys draft report, prepared over the weekend, certainly gives a telling picture to all who will trouble to study it, and as now finalised I can see no weak spots.

Had Dr. Vick here for one and a half hours to review our different researches for MOS with which he is concerned. Proximity fuse was the chief subject and in connection with which personal relationships between the establishments were not of the best, admittedly due chiefly to one individual at ADRDE! But Cockcroft has decided to hold meetings on proximity fuse at which he will preside—which is excellent. The whole episode shows how well Dr. Vick prepares the background of jobs he handles.

The other jobs were larger batteries—searchlight visibility etc., techniques regarding which he was specially appreciative of Waldram's contribution, and also the magnetic hardness tester (Farren)—and compact source searchlights and luminescent graticules.

I am dining this evening with Sir Philip Joubert.

September 8th[313-314]

Had a very interesting evening with Air Marshal Joubert yesterday. He was very sniffy as to the value of either the Army or the Navy compared with the Air Force. He was exercised on certain questions of wavelength to use for new ASV and I have arranged a small assembly at home on Thursday evening when he is paying me a visit—Gavin, Megaw, Marris, Espley and self. He was rather depressing as to the lethal value of our anti-submarine depth charges although they seem to have a shattering effect on the nerves of the crew. His slogan is bomb, *bomb*, *bomb* German towns hard and all the time.

Much time being spent on the manpower question—the schedule and getting new people too. The authorities of the Ministry of Labour are capable of appalling follies in their present mood.

The CV 58 (E 1273) studied by Professor Willis Jackson, seems to work effectively on 6th harmonic, i.e. at 6 to 7 cm without overloading. So as a local oscillator for the projector transmitter its stock rises.

September 9th

A CVD 10 cm meeting all the morning. Nothing startling. Lunched with Walmsley and was glad to hear that so far he is still in the saddle and also Leedham. O.F. Brown remains DCD—at present.

September 10th

Had the great pleasure of attending a meeting with the other two trustees, Highfield and Ashbridge, in Lord Hirst's office to arrange final passing over of the Trust and the cheque from Lord Hirst for £20 000 for IEE Benevolent purposes.

Pressure from Lamp Works and Jesty for the production speeding up techniques for CRTs which our CRT group has been working on and Works are now keen and contributing their part of this effort. Usual shortage of craftsmen makes it go slowly.

The CV 90 (E 1368) is having to blossom out into varieties led by the USA 'lighthouse' valve—which has a type giving 4 kW output. There seems no expectation of getting any USA 'lighthouses' so we must supply—MOV and probably Cossor taking over eventually.

September 11th

Gossling reports that the last meeting of the decimetre meeting (which now merges into the RDF and CW meetings) ended in an atmosphere of peevishness and he had to speak out to prevent a characteristic brainstorm being worked up out of nothing. The little diode CV 58 and our contact with Cosmos was the storm centre on this occasion.

Mr. Cowleyshaw of the Middlesex press gang made his final visit. We have found him helpful and reasonable. We are handing him a very full report of the Lab. activities and some 15 pages of staff particulars. As we are a unique organisation not fitting any Ministry of Labour schedule it is difficult to get Mr. Cowleyshaw's principals to have a correct picture of the Labs. and their activities.

Air Marshal Sir Philip Joubert came to dinner last night at home. I also asked Marris, Megaw, Espley and Gavin—to talk over Service RDF applications of all sorts. A very useful two hours talk with Joubert from which all indications point to 10 cm waves for ASV. We naturally talked over the state of the war, and had the feeling that it may not be long before the U boat is thoroughly defeated.

September 12th

Had a call from Sir Robert Renwick as a result of our talk with Joubert. He is sending someone to talk the subject over with us. Warren reports adoption of CV 58 diode for GR—and success in getting the CV 90 to oscillate proposals [properly] at longer wavelengths by increasing capacity in the heater circuit.

Mr. Reeves here with Marris and group on Oboe.

September 14th

Preproduction was good because no valves are in trouble—but NT 99 group were on holiday which knocks 300 off the score.

Helmore brought C-in-C Fighter Command (Sholto-Douglas) and General Pile of AA Command to see the large lamp. They seemed very interested and impressed. Brought them afterwards to the hut where NT 100s are being made and Air Marshall Sir Sholto-Douglas talked with the operators and I hope helped them to realise the key nature of their job.

Lunched with Dr. Railing and talked over political and other matters including Sir Robert Renwick who is also CCD [Controller of Communications] Air Ministry as well as [Controller of Communications Equipment] MAP—a good plan if the holder of the office rises to the occasion. Sir Robert is sending his assistant E [Eclair-] Heath tomorrow to talk further about ASV.

September 15th

Mr. Eclair-Heath duly arrived and his visit lasted until after lunch. The ASV problem was talked over and many other matters. I hope this first contact with the MAP 'all highest' radio control will continue happily.

Captain Willett (dear fellow) came to look us up and talk to our groups, particularly in the workshops, and very happily he did it. Afterwards he went all round the place and examined some of our latest valve developments.

Had a visit from a Lieutenant from USA Air Force who wanted advice on an infra-red problem. Rather an interesting contact. I hope we helped him.

September 16th

C-in-C Fighter Command spoke of the Turbinlite as one of his main defensive weapons and is fitting another 150 particularly on Mosquitoes. So far it is encouraging but we all feel bothered as to whether there is real keenness on the part of senior operations officers in backing up the new weapon by personal enthusiasm. Five Germans have been brought down and many missed by faulty operation. But recently enemy activity over this country at night has been very light.

The troublesome high duty cathodes of the CV 76 magnetron are slowly finding their level. Life tests seem to indicate the present technique will stand up to 50% of desired loading.

We are asked to make two more field strength indicators for 3 and 10 cm respectively.

Canal lighting—Stevens reports reasons for disappointment on last full scale trial are becoming clear and can probably be corrected on next trials.

September 17th

A Brundrett and Harley preproduction meeting which covered a wider field than preproduction. There was a strong urge (Truscott defending) to put at low priority the functions of this place as a mere producer of valves and use it for the newer lines of which [such as] 1 to 2 cm and the poly-megawatt magnetrons. We are going badly to feel the shortage of craftsmen in the maintenance of our existing preproduction tools.

Had confidential talk with Brundrett on the folly of over criticism, and vilifying individuals whose only fault is that they may be getting tired in their jobs.

Have formal request from Admiralty to investigate infra-red properties of compact source lamps.

Glad to have Colin Smithells here today for a few hours.

September 18th

Both Brundrett and Harley were very impressed by the look of our Manpower report and asked for copies. I think it is just as well they should have them on the ground that the more publicity there is for the Laboratories' vital war work, the less likely is it that silly mistakes will be made through ignorance in high quarters.

A diamond die technical committee this morning—uneventful. Leeds is mainly short of die polishing machines, but they are on the way. He is concentrating upon speeding up the drilling of large dies and the first stages of fine dies—forcing the pace up to the optimum point where the surface will just not be damaged. The meeting was followed by another to consider prices amongst the die manufacturers. Our costs and prices were put up to be shot at, and I don't think will be found to be far out.

Waldram reports much interest on the theoretical side of searchlight beams (Fig. 26). Mr. Corcoran is being transferred to ADRDE to tackle the theory of searchlight beams and has been here to get such information as we can give him. It is amazing (nay jealousy) that they cannot back us up here and ask us to continue what is our own pioneer field of work.

September 19th[315-316]

Partridge has attended a meeting called by Admiralty (Hughes presiding—MOV and Mullard) to consider steps to improve silica seals for CV 14 valve which is particularly exacting on seals. There are three glasses: (*a*) our own, (*b*) Mullards obtained from Holland but now unavailable, (*c*) USA, probably Corning, glass. There was an effort to show that Wembley glass caused heavier breakage than the others but this was countered by the data. Mullards however were completely unco-operative in disclosing techniques whilst GEC was the reverse. Meanwhile J.H.P. has several lines of attack for producing a stronger glass. The compositions of all three glasses are identical.

A private letter from . . . showing a definite ruling from above that Dr. Wilkins must be satisfied by his attendance at our CRT meetings for his inquisitiveness into luminescent power matters—and there will be no separate study committee on this subject. So that's all to the good in the interests of not wasting time.

Heller and Bligh have for months been urging an aeroplane landing altimeter on the change of aerial impedance principle. We were put off by DCD and told this had been studied and discarded. But we found Salford has been for a long time on the same track and had got an aeroplane at their disposal in Cumberland! So forces were joined and Salford applied further pressure. We hear today that a demonstration is

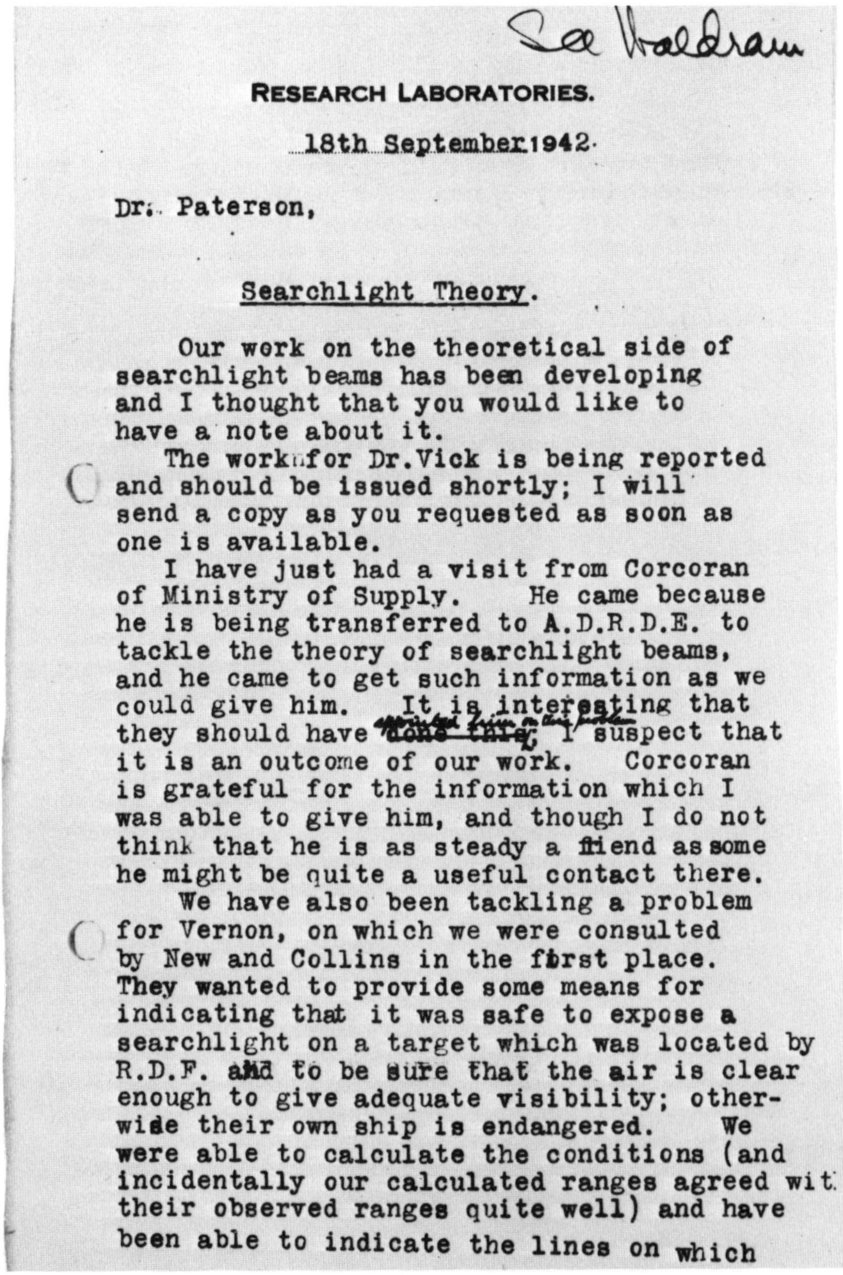
RESEARCH LABORATORIES.

18th September 1942.

Dr. Paterson,

Searchlight Theory.

Our work on the theoretical side of searchlight beams has been developing and I thought that you would like to have a note about it.

The work for Dr.Vick is being reported and should be issued shortly; I will send a copy as you requested as soon as one is available.

I have just had a visit from Corcoran of Ministry of Supply. He came because he is being transferred to A.D.R.D.E. to tackle the theory of searchlight beams, and he came to get such information as we could give him. It is interesting that they should have ~~done this~~; I suspect that it is an outcome of our work. Corcoran is grateful for the information which I was able to give him, and though I do not think that he is as steady a friend as some he might be quite a useful contact there.

We have also been tackling a problem for Vernon, on which we were consulted by New and Collins in the first place. They wanted to provide some means for indicating that it was safe to expose a searchlight on a target which was located by R.D.F. and to be sure that the air is clear enough to give adequate visibility; otherwise their own ship is endangered. We were able to calculate the conditions (and incidentally our calculated ranges agreed with their observed ranges quite well) and have been able to indicate the lines on which

Fig.26 Note from J.M. Waldram to Clifford Paterson dated 18 September 1942, on work of the Laboratories on searchlights

to work. I think that we are well on the way to elucidating what happens with searchlight beams. A second report is written about this.

We are also in touch with A.A.Command over several sides of the searchlight problem, and Col.Bray is a very good friend; he consults us fairly frequently. They are having a further shot at the idea of the "searchlight fan" and I hope to go for some further flights. Col. Bray has just been told officially about the G.E.L. lamp, and as he is the moving spirit in the fan experiments he is sure to connect the two together.

. . . . .

Fig.26 continued.

requested by MAP so Heller and Stenning are up North fitting our gear in the plane and arranging for the trial. Good work. It seems an excellent method for safe landing.

Yesterday we all examined Turner and Smyth's method for accurate centering of CV 55 electrodes by optical projection. I asked Algar to contact the illumination group some months ago and this very promising device is the result.

Our gas burner for producing inert atmosphere for bright annealing is proving very successful and large plants in South Wales are adopting it. I. Jenkins and S.V. Williams are responsible for it. Bright annealing seems to be the only alternative to 'tinning' in the tin plate industry—zinc and tin having to be substituted.

September 21st

This is the long drawn out period when Stalingrad is being so heavily attacked by the Germans and nothing but hand to hand slaughter in the streets and houses is possible.

Winch and Macfadyen have been helping Mr. Aldis to try out a talking down the beam scheme with optical modulation over distances of four miles. The optical arrangements are very perfect but 'shimmer' causes noise in daytime. I am afraid he is up against it and we do not feel able to accompany him into the very extensive investigation to perfect the thing even if this seemed possible. He is proposing a 9 in telescope 5 ft long!

Magnetron life tests are a very live and difficult issue just now. Endless modulators are wanted. Gossling is our conscience on the matter and he rightly points out the ineffectiveness of everyone's life testing.

A Ministry of Works and Planning Lighting Committee. This seems to be progressing as well as one would expect.

Lord Cherwell is interested in some searchlight proposal of Waldram's and he is seeing him this evening.

September 22nd

Visited our 'annex' in the Exhibition [grounds] and saw the anode rolling under Hobday going well. The conditions are very fair and canteen meals are good.

A tragedy alas! Walmsley is going. Had a long talk with him after the TVC meeting today. Sir Robert Renwick thinks it would be best—but he is keeping on his CRT and two or three other committees. It was delightful today to have a real unrestrained talk with him. He advises me to get G.P. Thomson to Wembley as, like others, he thinks all research is done in Service Establishments.

September 23rd[317]

Waldram's talk with Lord Cherwell covered the problem of aircraft flares and their optimum design and operational use.

Rang up G.P. Thomson and fixed to see him on Tuesday for a preliminary talk.

A CRT meeting with Walmsley presiding—very full of subject matter—and a generally good feeling. Walmsley tells me that the meeting has been spoken well of in high quarters.

Bligh reports a good feeling regarding the frequency modulation meeting held at different manufacturers' establishments—and their willingness to co-operate in exchange of information.

An MOV – RL transmitting valve meeting—Smith, Cosgrove and all the well tried men of that group came. What a delight it is to work with them—absolute confidence and mutual backing all round. The best of all our liaisons!

September 24th to 26th[318]

To Salford to spend a day or two in the SEI factories to become familiar with their latest activities. Was much impressed with the many factories and the bold way they had taken on their organisation. Their radio work is but a small part of their effort. Ran over to the Warrington (Veevers) establishment but had only five minutes there.

Waldram reports that Lord Cherwell is putting the flare job to us through Bomber Command. The work has already started and Dr. Dickins and Mr. Stevens have discussed it with us.

September 28th[319 - 320]

Preproduction certainly has its troubles. Tantalum is no longer procurable from USA so that CV [NT?] 93 and 97 have to be redesigned or modified. This will hold them up somewhat.

The NT 99 suffers in a similar sort of way from inability to get oxygen free copper which is the only material permitting of satisfactory machining. Dr. Guy spent this afternoon seeing round the Labs. and was a helpful visitor. He told me much in confidence of changes in high places at MOS—but they hardly concern us.

Called on Dr. Vick to ensure that our liaison with the searchlight section of ADRDE can run as a smooth partnership. Corcoran is going there to work, presumably, under Chivers.

Whiddington spent the afternoon here and he seems relieved at the appointment of the Radio Board attached to the Ministry of Production and from which CVD will presumably obtain its status. He confirmed that he wants us to take on the 1.6 cm valve development.

September 29th

We are somewhat bothered by the activities, or threatened activities, of the Association of Scientific Workers who seem to use the powers given to trade unions to worry and hinder people who are trying to carry the war burden.

Called to see G.P. Thomson, the chairman or deputy chairman of the new Radio Board, and had a one and a half hours talk with him about Wembley and about the whole present field of radio activity. A very useful talk over things and personalities.

Stevens reports the return to port of our 'diffused lighting' ship. It has stood up to all the weather and rough usage of a long trip in service and the installation is in order. So there seems still less a reason for not going on with the scheme. The impression we have is that the authorities concerned want to damn it because it is a lot of trouble but cannot find a technical charge to make against it. But that may be unfair to them. We shall see.

September 30th[321]

Visit from Mr. Allcock, Director General of Communications Equipment, who came to see over Wembley. We are greatly relieved and favourably impressed. The visit was timely and I hope will lead to a consolidation of relationships.

Espley had one of his big days with GR: Cockcroft, Grey, Black, Paris and many more came down to see the 'functional prototype' of the gear. This had been promised, if possible, by the end of September and all who came expressed themselves very pleasantly surprised at the progress. The 'functional prototype' has now to be made to 'function' electrically. It is hoped this will happen in the next week. Meanwhile the changes are being settled for the final prototype which Coventry and Witton will take chief responsibility for, Wembley being consultant. We are also responsible for the CV 55 and CV 90 and CV 52 valves.

October 1st[322]

There is a crescendo of activity centered around Dudding and Jennett regarding quality control and the statistical practice associated with it. They make frequent dashes for evening meetings to other parts of the country and everywhere find practical interest and a desire amongst production engineers to apply the advice given. Frank Gill leads the crusade.

October 2nd

Waldram has gone a step further in his flare experiments and is reproducing the effects to scale in a water tank with glass window. He has written up the subject based on existing atmospheric data.

Commercially our frequency modulation research is coming badly to grief owing to our not being able to exploit GEC developments and salesmanship with the Home Office [and?] police. The business is going elsewhere to a small firm.

Wood drying (3-ply) by HF current came up at my RS meeting today with A.H.R. and L.G. The principle was approved and we shall probably accept a development contract for a prototype installation in conjunction with the Tyne Plywood Co.

October 3rd

These are days when for weeks the main attention in the news has been focused on the attack and defence of Stalingrad.

Had a ring from Walmsley. He has now given up his office duties at MAP and is for the moment a freelance. He has asked to come here for a short time to study cathode ray tubes from the inside. He will be a very welcome visitor.

October 5th

Preproduction this week has been patchy and it looks as if, for one reason and another, we are not working up to our guarantees. There will have to be a 15% uplift in order to do so. Douglas has introduced the idea of bank clerk assistance after hours and I think we shall make a trial of it. [But nevertheless note letter from Harley on CV 64, filed in diary (Fig. 27)].

Lord Cherwell has been here again today with a large party to consider

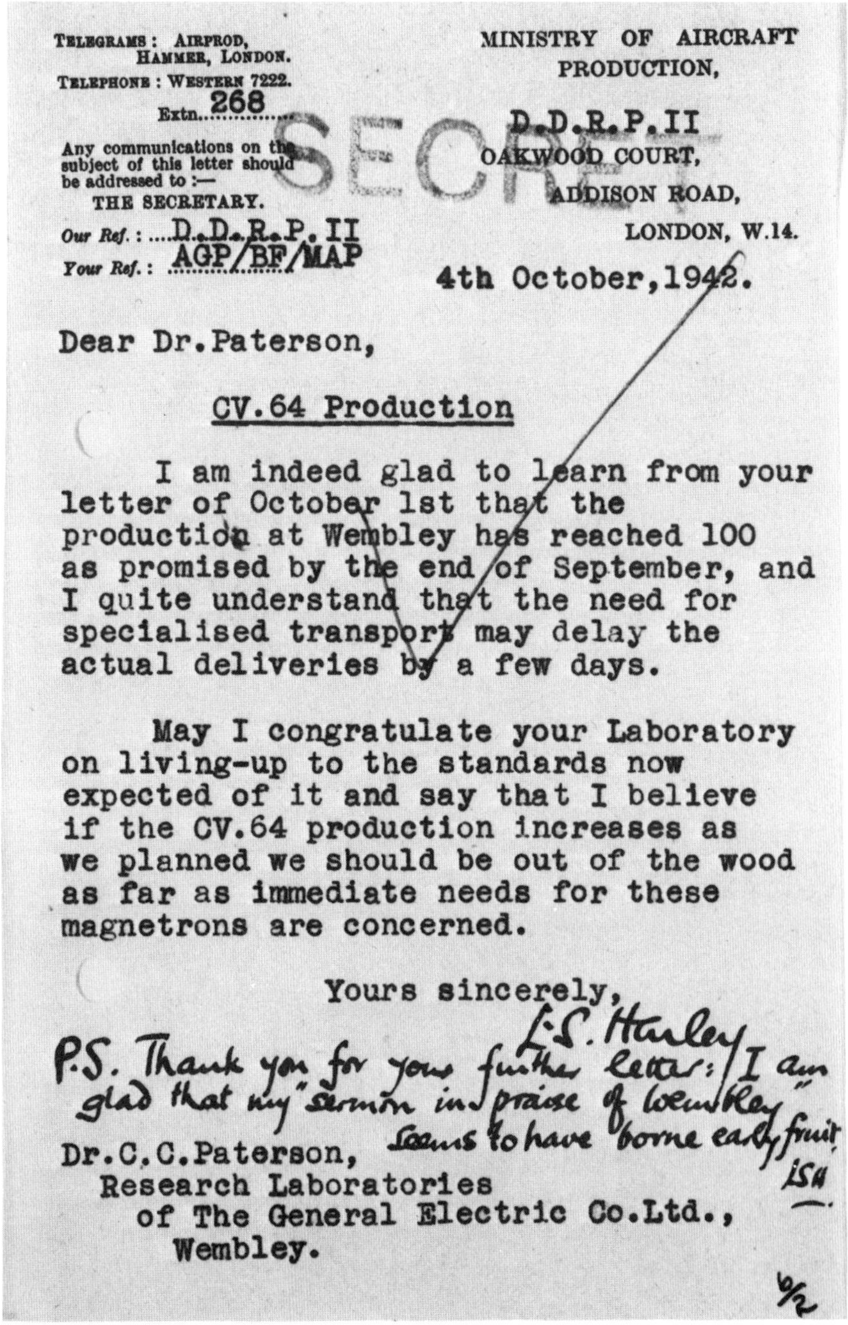

TELEGRAMS: AIRPROD, HAMMER, LONDON.
TELEPHONE: WESTERN 7222.
Extn. 268

Any communications on the subject of this letter should be addressed to:—
THE SECRETARY.

Our Ref.: D.D.R.P.II
Your Ref.: AGP/BF/MAP

SECRET

MINISTRY OF AIRCRAFT PRODUCTION,
D.D.R.P.II
OAKWOOD COURT,
ADDISON ROAD,
LONDON, W.14.

4th October, 1942.

Dear Dr. Paterson,

CV.64 Production

I am indeed glad to learn from your letter of October 1st that the production at Wembley has reached 100 as promised by the end of September, and I quite understand that the need for specialised transport may delay the actual deliveries by a few days.

May I congratulate your Laboratory on living-up to the standards now expected of it and say that I believe if the CV.64 production increases as we planned we should be out of the wood as far as immediate needs for these magnetrons are concerned.

Yours sincerely,
L.S. Harley

P.S. Thank you for your further letter: I am glad that my "sermon in praise of Wembley" seems to have borne early fruit. LSH

Dr. C. C. Paterson,
Research Laboratories
of The General Electric Co. Ltd.,
Wembley.

Fig.27 Letter dated 4 October 1942, from L.S. Harley, DDRP II, MAP, to Clifford Paterson on CV 64 production

flares. I left them with Waldram and his Lordship having a debate on the subject before the above audience.

Have had to try to decide means for extensive modulator life test gear for magnetrons. Herwald, under Machin, may be able to organise the matter, but the business of getting authorisation for purchasing all the gear and placing orders outside will be a job these days.

October 6th

The meeting yesterday on flares concluded satisfactorily and we are to continue with the work probably on our Bomber Command contract.

The only part of the AIS Mark VIII equipment which needed alteration was the TRE acrial, which they had pressed, in favour of our design. TRE now say that Wembley (Clayton and Espley) design is the goods and Dr. Starr requests us, with a very good grace, to incorporate it. TRE and Wembley have ceased to claim the honours of individual components but have come to share credits and blames like good partners.

Helmore turbinlite for single [seater?] plane tactics is wanting smaller radiators—which is an incidental problem for our people and workshop to handle.

October 7th[323]

A searchlight panel meeting and a proposal from Mr. Couch that Morgan Crucible should have a development contract for high strength carbons. Carbon black is imported from USA and a large production is wanted in this country by the rubber industry. We have been showing willing to do the research but there is a question whether there is enough suitable raw materials in the country to enable an appreciable contribution to be made.

Have taken part in a post war planning research committee at the IEE—formulating an Electrical Industry Research Corporation. A preliminary document is nearly complete, but it is going to be strong meat for the crystallised. [those set in their thinking?]

October 8th[324-325]

G.P. Thomson and Jeffcock here on a whole day visit to see the place and to get initiated into CVD, CRT and all the rest of it by B.S.G. I am more than glad of this opportunity to introduce Wembley into the new picture.

I hear Leedham is leaving but have not heard who is coming in his place.

Dr. Railing has been commissioned from the Cabinet through Geoffrey Lloyd to tackle if possible the fog dispersal question on aerodromes—particularly electrical treatments. So it comes to Wembley to consider the whole problem within 24 hours! I rang up Ryde and we shall discuss tomorrow morning.

October 9th[326]

The fog on aerodrome problem has been to the fore. Ryde had spent much of his time, and of the night, studying the literature and putting together a report which, with a small addition, was in excellent shape for typing and sending to Dr. Railing. Air [Vice] Marshal Linnell came, brought by Helmore, and after inspection of the lamp, came to chat over tea. I mentioned to him the fog problem and, with various remarks concerning things which had better not be mentioned, he gave me the background of previous work on the subject. A useful talk and a charming talker.

Mandrel, after being of fluctuating priority, has come to rest at the top of the list and there is likely to be heavy pressure again.

Wilson and Fletcher came to a CB meeting and spent an hour going round the Labs.

October 10th[327]

Bligh reports re. the GEC aerial impedance altimeter (terrain clearance) that the several other devices are meeting with doubtful success. Ours comes into the picture rather late in the day but it will probably be given a run for its money.

Landale (of ASE) has a set-to with Le Rossignol and Megaw re. life of CV 76 magnetron (heavy duty cathode). There seemed to be some difference of opinion, we holding that the life appears less than Landale finds—but anyway it is improving and a revised circuit in our test gear may show a better result at Wembley.

We find we must make our own alumina powder for making experimental sapphire boules. Dauncey proposes to use the incinerator for his furnace. A good idea.

Walmsley, relieved now of administrative responsibility, came here on Tuesday and Friday just to enjoy the study of the scientific and technical side of CRT development. It was delightful to have him here and he seemed to enjoy the work thoroughly.

October 12th[328-330]

Preproduction patchy—CV 58 and other valves show high wastage figures. New opportunities for production to go wrong seem to be constantly discovered.

The report of Dr. Guy's committee on armament research and development is now being implemented. I was an assessor on the committee.

This has been a day on the subject of de-fogging of aerodromes. Ryde, Humphreys and I lunched with Dr. Railing to discuss policy and the memo by Ryde and the others was slightly amended. I got back to Wembley post-haste to have it retyped and was back in town in time to keep an appointment (with A.H.R.) at the Ministry of Petroleum with Mr. Geoffrey Lloyd and Sir Donald Banks—and our old friend Mr. Simon Marks who is helping the Minister. We explained the proposal for de-fogging favoured by us (until it is proved wrong), i.e. filtering the moisture from the air and then warming preparatory to ejecting a stationary layer onto the runway. Sir Donald Banks and the Minister were handed copies of our memo and after reading carefully asked me to compliment Ryde and company on their soundness and resourcefulness. We were instructed to proceed at once with our future calculations and to try to report again in a week. A very plesant one and a half hours conference. I asked for and obtained agreement to a regular 'team conference' as clearing house for the job if it goes forward.

October 13th

More internal discussions on de-fogging are pointing the way. We are now contacting Mr. Lambert of F & C [Fraser & Chalmers] for collaboration on design of ducting. James tells me of success with the CV 90 for operation down to 6 cm on 3rd harmonic and also excellent frequency stability with change of voltage and heater current etc.

A GEC CVD meeting. Rather dull except for James's contribution. It is the last meeting which Shearing will attend as he is leaving the Admiralty. He has been a helpful and loyal colleague and joint chairman of the meeting.

Had a visit from Air Commodore Addison and Wing Commander Rogers to urge the vital importance of having Mandrel and its testing gear ready for the air by December 1st. Gavin faced the situation calmly and with assurance. Subject to clearing one or two remaining uncertain design features in the next fortnight, he said there was every expectation of our coming up to demands and nothing would be allowed to slip. The Germans have their GCI in a good solid block 150 or 200 miles deep—and it needs dealing with.

October 14th

Winch's flash-bang trials go well and several prototypes are to be made. The trials are not complete yet, but there seems no doubt about the result.

A diamond die advisory committee at which Mr. Davidson gave unexpected and most generous tribute to Leeds diamond powders and said they produced polishes on dies that he had never before experienced! I hardly knew which way to look.

Had Mr. Barratt (late of Albright & Wilson) here as an introduction from the new scientific trinity at the Ministry of Production. A most excellent man. He wanted to talk over everything and see if we needed anything to facilitate our work. He has the same sensation as we have that industrial talent is not really brought into the councils of the Ministries, because there is in fact a psychological gulf maintained between civil servants and industrial staffs. His reflections were very interesting.

October 15th

Gossling and Donaldson have returned from ADRDE whence Mr. Thompson has returned from USA and study of proximity fuses for Navy. The latter has been sold the fuse—and further work will be on USA lines and presumably supplies will also come from there.

A useful Coventry Radio meeting which resulted in a meeting with A.H. Railing and L.C. Gamage regarding GEC contacts at high levels in Government to keep the Ministries from forming wrong impressions of GEC capacity for radio production.

October 16th[331]

An allocations meeting in which, as usual, Wembley is asked to push up its output of nearly all groups somewhat further than it can comfortably do. The main extra load falls on Douglas and his groups.

G.P. Thomson rang me in regard to my proposal that Picken might be used in one of the Ministries. I am seeing Admiral Grant about the possibility.

The fog-clearance plans develop gradually but there will not be an interim report by Monday. R.H. Fowler thinks 'this is typical PM action and liable to be a waste of effort'.

October 17th

Mr. Umpleby of DCD (MAP) came to see Bligh and to bring some Americans. He incidentally told us of further replacements and appointments at MAP. They seem to be drawing on Farnborough for one or two Assistant Directors.

October 19th

Pre-production suffers from various diseases—I suppose we can never expect all our groups to have a good passage in one week.

Severe restrictions have come along as to magnetron handling and registration for the sake of security and checking. Pearce nearly threw in his hand! But these things usually look worst at first blush.

I discussed the fog clearance schemes with Bragg today. He is inclined to look on the whole idea as a waste of effort—but keeps an open mind.

I understand the Helmore turbinlite is not so much favoured now on account of getting the attacking fighter in the right position at the right moment. There is also jamming of AI. These flights will however be kept in commission until the new year. The new scheme combining all in one machine is now the chief hope.

Have been arranging for use of the end of the canteen for office work in connection with secret work. Miss Winton will go in there with two others.

October 20th

We are expecting to take a part in Phillips & Powis automatic pilot and Bligh will send Heller to examine its electrical anatomy. The Salford-Wembley altimeter is not getting on very fast with its trials. As usual no plane available and we are waiting several weeks for one.

We are to have a meeting tomorrow on fog dispersal and Sir H. Hartley has asked us to lunch with him in anticipation.

Couch has taken a further step towards collaboration between Morgan Crucible and GEC on carbons.

October 21st

An RDF (formerly centimetre) meeting—Walmsley presiding. Receiving valves morning, transmitting afternoon—45 present! It was simply a bare bones progress prodding meeting.

Professor Rankin and Sir H. Hartley had received Humphreys and Ryde yesterday on de-fogging and supplemented what we knew. I attended a small lunch with Hartley and his Gas Light & Coke colleagues to exchange views before the meeting with the Minister. It was useful and I learned that they were also following the air drying plan using calcium chloride as a cleaner which enables smaller volumes of air to be handled. Sir Donald Banks was very pleased with the meeting and said it was the direct result of our request to have such a meeting and to pool ideas. Dr. Slade of ICI was very helpful.

I listened to a most violent outburst from Rankin about the goings on and intrigue at MAP. He resigned in protest!

October 22nd[332]

Had Walmsley to lunch with me to talk over MAP staff changes. He tells me that Whiddington is leaving, and transferring to help Gough at MOS—so Griffiths will presumably take his job in CVD. There seems to be much lack of confidence in the new management and all are glad to hear of the appointment of Air Marshal (resigned) [retired] Freeman, to be supreme head at MAP of all research and development.

A TVC meeting—not eventful, and afterwards the first meeting of the Lord Hirst Benevolent Fund Trustees.

I hear from Humphreys that all preparations for a report on fog clearance are complete and the writing up will start tomorrow.

October 23rd

Gavin has just returned from three days at TRE—wasted, as far as he can see, by amateur suggestions from TRE staff so that after trials his Mandrel sets emerge in exactly the same form as they went in—but we are three days to the bad. He was distressed by numbers of well qualified fellows wasting their time through lack of proper common equipment and facilities. I gathered they too were somewhat fed up. The place seems to have grown faster than can be assimilated.

We are being asked to collaborate in the training of personnel to use Mandrel. They can probably be centred in the Exhibition Building and also visit Claudgen where assembly is taking place.

A firewatch meeting to consider the invitation to our women to volunteer to joint the firewatch groups.

Aldis came re. his beam telephony scheme and agreed with me that the Admiralty must be approached for a contract if the job is to go on.

A GR progress meeting which I did not attend—all seems going as well as such jobs can.

October 24th

Partridge and MOV have had another encounter with prejudice from the Admiralty on acceptance of silica valves using our glass to quartz seals. Mr. Hughes who has charge seems to feel somewhat awkward about the arbitrary behaviour of his inspector. As usual we seem to be supplying all the work and brains and getting all the abuse.

Megaw is going to operational research HQ MOS for transmission trials at Mr. Starr's invitation.

CVD Technical Committee minutes suggest the elimination of the GEC CVD meeting and the substitution of a sort of conversazione at Wembley. I doubt if this would last any time in the absence of a meeting.

Recent meetings at SRDE (James and Lamont) decided to use the CV 90 on its harmonics for 1200 Army communication receivers to go with our 'projector' sets. This is quite gratifying but I do not know how they are going to get the valves in time—and it is again disappointing that Pye are making the receiving sets rather than Coventry.

October 26th[333]

Preproduction at 1250 again but we have to get up to 1500 before the end of the year. We have been considering the security scheme for magnetrons and do not see what real security it gives even if established. The CV 90 of Warren has been having its ring seal troubles and the promised ten per week has not materialised in October alas!

Bell's services have been begged for to go to an aerodrome to see why Oboe will not work with CV 90s which give out in a way TRE cannot explain.

Espley has been having a job to get his 25 cm GR oscillator to work over the required frequency range. He thinks it bids fair to be all right now but he and Walter have had to face up to some radical changes.

Attended a Regional Production Board talk by Mr. Coulston of Hoover on evils of uncontrolled subcontracting.

October 27th[334]

Saw Boyland re. progress of preproduction group on CV 90 valve. Ring seal cracking put it back to the research group but it is now out of that particular difficulty and Boyland will give it attention again with a view to redeeming our promise to deliver ten per week. Boyland is still very keen on the prospects of the E 1274, which is also being proceeded with but with rather less priority. The two centimetre-wave glass magnetrons CV 79 and 89 are progressing under his superintendence.

Had G.M. Wright and Shearing here, the latter to hand over his affairs to the former. It was very good and sound of Shearing to do this in so thorough-going a way. There was much talk of the 6 cm propagation experiments which all are very pleased with.

Clayton and Marris's aerial policy with TRE is working out well—and having succeeded with the AIS aerial, he has now very hopeful work on the all round aerial for Oboe. Of course it is wanted in a great hurry.

The equipment of a Mosquito with a turbinlite is nearly ready—and the large ground arc is also nearly ready for trials at Heston. The ISRB [Inter-Services Research Bureau] 'vessel' has had its final trials at Heston and has gone off to Scotland.

October 28th

Had to get Ryde and Humphreys to withdraw to a quiet spot at the Exhibition building with Mr. Lambert (of F & C), to finish the draft of our fog clearance scheme. I rang Sir Donald Banks to report the position and he supported me in my policy that the main scheme must come first and any experiments must be subsidiary and not in the nature of demonstrations.

Had Griffiths (CVD), replacing Whiddington, to study the work here on valves. He is a very delightful and knowledgeable man. We hear that Canada has decided to adopt the CV 90 type of valve in place of the USA 'lighthouse'.

Had a very nice letter from Pilkington after Dudding and Jennett's visit there to lecture on Quality Control.

October 29th

Bell has returned from his adventure and done a good job as usual. He found the trouble to be associated with the running of the cooling motor and its effect on the valve attached. So we hope Oboe can get to work again.

A Brundrett and Harley preproduction meeting at which many things were discussed including the magnetron security proposals. I am to try to get an informal meeting with Lord Rothschild. Brundrett urged us to push on with the 1.6 cm magnetron development with Oxford who have pioneered it. Got Megaw along to discuss it with him.

Ryde, Humphreys, Lambert (Fraser & Chalmers), and Cooper have worked hard at the fog dispersal report and got it finished in draft late last night. It is an excellent document. The office rushed through the typing and duplicating today and we sent six copies up by hand at 5 o'clock this evening to Sir Donald Banks.

Brundrett came to today's meeting full of administrative burdens of his own, but gave a clear mind to our troubles. I drove him back to town in order to get an extra half hour's talk with him, and discussed the future of the GEC liaison meeting with him which the CVD technical committee wishes to dissolve. I found him full of sympathy for my scheme of reverting to the original status of the meeting whereby the Admiralty go through our valve work with us, and invite anyone from other establishments to attend who may have something to bring up.

October 30th

Discussion with Colonel Sampson of ICI re. the fog dispersal scheme and particularly the filter part of it. Sir Donald Banks rang up, very congratulatory re. the report and wanting a meeting on Monday.

I looked into the 3 and 1.6 cm magnetrons with Willshaw with a view to giving the latter a better place in the priority list.

The Fire Watch management met with me all the women (a large number) who would come to a meeting to consider volunteering for Fire Watch duty. A happy meeting and businesslike.

November 2nd[335]

Preproduction is down for various causes and one wonders how we are to honour our promises. More stringent specifications are one cause but a surprising one is the devastating effect on a group when it is thrown off the rails through the shortage of some component—it takes weeks to recover, physically and psychologically.

I gave a Pacific Broadcast this morning on Joseph Swan and the invention of electric light.

The effect of the special alumina we make for sapphires at Salford arose in a meeting today with Professor Briscoe who is finding it difficult to interest the authorities in the high killing power of the particular dust for weevils in grain. The thought is to sell the proposal to the cocoa people first. The weevil appears to become dehydrated.

We had Squadron Leader Woolridge to lunch, on fog dissipation. He has taken part in 70 night bomber raids on Germany. Then a conference with the Minister and his advisors at which the GEC had many kind compliments on the report by Ryde, Humphreys, Lambert and Cooper. It seems to me an excellently thought out piece of work—and it stood up to the friendly criticism they offered. We have been given the job of installing the scheme on an aerodrome.

November 3rd

A short wave propagation meeting at Wembley with Mr. Shearing in the chair and many Service representatives. Megaw leads for Wembley and the work done on the west coast and analysed is yielding valuable results for Admiralty and Army. This is Shearing's last appearance as a Civil Servant.

Fog clearance is now known as 'Fido'. Dr. Railing held a GEC meeting with Lonsdale and ourselves and went over the scheme with all concerned. Fraser & Chalmers have accepted responsibility and are not afraid of the job.

Finally a meeting with A.R. on radiant heating, for the gold plated mirrors have given out in our ordnance factory equipments and complete guarantees are required by us that the alternative rhodium plating will be in every way effective. Johnson & Matthey appear to be well on top of this technique.

November 4th

A CRT meeting nearly all day—but went stickily and Walmsley was not very pleased. He has taken on a job with the Air Ministry on aerodrome telephone communications etc.

Warren has news from Dr. Touch (who is returning to Canada) that they are pressing on hard with the CV 90 and want from us 100 [sets of] parts so that they can push on with their assembly.

We have been helping training centre in Ayrshire with fluorescent maps and other fluorescent devices. All the work was done for Canada 18 months ago but muddling on their part sidetracked us. It is consoling to see this coming into its own at last.

I hear that a Fido experiment with coke heating on Staines reservoir site was pretty successful last night.

November 5th

Quite exciting news from Libya over the retreat and demoralisation of the Axis forces—everyone very cheery.

A CVD policy meeting at which discussions took place over the CVD being attached to the Radio Board and having the authority and prestige deriving therefrom. A discussion also on the advisability of carrying on the GEC liaison committee—at which I obtained permission to leave the meeting in order to leave the discussion unembarrassed, having previously expressed my views to Brundrett.

The decision of the committee was good. A regular morning meeting under Brundrett's chairmanship will be held at which senior officers from Establishments will attend. Simultaneously, and also through the afternoon, valve departments at Wembley will be thrown open to visitors from Establishments. I think this will put the GEC meeting onto its proper basis of reviewing the work in its general aspects rather than in its details.

Fido goes ahead energetically, but the engineering of the scheme seems a considerable business. The first thing is to fix the aerodrome, for that decides whether electricty can be used for the main source of power.

November 6th

A meeting with A.H. Railing and L. Gamage to review Service matters generally. Saw Watson Watt who assured me that AIS Mark VIII is still behaving well.

Marris tells me that Oboe No.2 prototype is nearly ready.

GR has functioned as a complete electrical unit but Espley is disturbed I believe about the amount of work in producing it.

November 7th

Increasing interest is focused on the output and performance of 10 cm triodes and their possible performance in relation to klystrons—which are the possible alternatives and capable at present of more power.

Espley had a troubled GR meeting yesterday at which the merits of assembly of complete gear at Coventry or at the predictor makers was canvassed. Also his automatic frequency control which is a great advance in operational simplicity came under criticism as adding to the amount of HF production work.

The group [pre-production?] (internal) meeting today had I understand a stormy session over various things which seem to be holding back the full fruits of our effort. Pearce and Douglas are good prodders.

Partridge showed me interesting developments of silica seal glass drawn off from the furnace through an iridium nozzle—enabling the same furnace to be used several times. The next step is to 'draw' the glass continuously through this die using the caterpillar made by Fortey.

November 9th

The news is electric. Complete victory in Libya and landing of USA army in French N. Africa. Everyone very cheerful. By early train to Reading to visit Phillips & Powis with L.G. They are convinced that aviation requires a large amount of electrical application of the ordinary type and of radio, and they and the GEC might get together in connection with some of them. They have shown much enterprise in very diverse directions—but they are all focused on Mr. Miles himself, the founder of the concern. An automatic pilot is the immediate gadget. I was most interested in their improvised factory for making a cheap torpedo automatic pilot—and in its effectiveness.

November 10th

Willshaw reported apparent explanation at last of the reason for low efficiency of his 3 cm waveguide magnetron. Brazing on the end plates so that they make perfect electrical contact has produced a valve with an efficiency of 20% and magnet of 14 000 gauss as against a previous best of 12% and a much stronger magnet. This is the greatest relief to us all firstly because there had seemed something wrong we could not understand and secondly because we needed the extra power. This points the way still more definitely to an improved 1.6 cm valve although a visit to Oxford by Willshaw and Le Rossignol yesterday has brought back very complimentary remarks re. the set up and the people there.

Preproduction accounts for the first six months of this financial year have shown a surprisingly large margin in spite of reduction of valve prices in July. If we take no profit for the period of preproduction prior to April 1st we show a margin of some £75 000 which will presumably be declared and returned to the Admiralty. The causes are several, one being the larger number of valves called for and the other the better working conditions in the new premises.

November 11th

L.F. Britten reports good progress with his getter production. We have now capacity for supplying outsiders in large quantity but as yet only STC are ordering regularly. Some production is in progress at the Exhibition Building. Our thin walled nickel getter seems to be in large demand.

We had a 'security' meeting here with Major Bramble, Group Captain Archer, Brundrett and Harley—subject magnetrons. I had asked for it on account of an 'instruction' received from the Inter-Services Valve Production Committee without previous consultation with us to see if it was practicable. Everyone saw that the proposals were ridiculous and I was asked to write suitably to the Secretary of IVPC. The affair could be set to comic music if it would not be such a waste of time.

Had long talk with Brundrett re. conduct of GEC (CVD) liaison meeting and its best procedure. It will be based, if possible, upon projects! Brundrett had a difficult passage over this meeting at the CVD policy meeting on November 5th when I left the meeting. We shall have to make a real effort to make it a success under its new management.

Gossling back from Malvern (ADRDE): subject PF. He seems satisfied pending the main meeting tomorrow. Marris also has an important meeting tomorrow re. adjustments in AIS Mark VIII to overcome difficulties with E 1271 valves running away.

November 13th

FIDO proceeds. A visit today to Farnborough by Ryde and Humphreys to examine effect of the air jets on landing bombers. The aerodrome has been practically settled and electric power is adequate.

Had Watson Watt (now Sir Robert) here all morning to examine work in progress on radio. Marris looked after him mainly.

Dudding and Jennett are going strong with their missionary work on quality control and letters of appreciation come in frequently.

Quality control activities now total: 14 papers published, 20 lectures and the like, 14 major industrial contacts to give advice—all by Dudding and Jennett.

Willshaw is not so sure now about the practical value of his 3 cm magnetron advance which he mentioned on the 10th. The efficiency is there but oscillations may be of the wrong brand [band?].

Have finished six-monthly statement of valve preproduction finance. The report explains a margin of £75 000. Most interesting.

November 14th

I end this day in a state of righteous anger at the follies of the Manpower Board which has served us with a list of men, most of whom when taken will cripple vital sections of our work. Our protests will be going in to high quarters at once—but we have mighty little power to exert ourselves before the axe falls. The Board are acting as ignorant officials without knowing the harm they are doing.

Two important meetings yesterday (*a*) on midget valves and PF, (*b*) on the E 1271 modulators and their tendency to run away. Both finally reached some conclusion or appeared to do so.

It looks as if Fido will be very expensive!

A speedy valve allocations meeting registered the position here but landed us with the job of doubling up on our CV 55 programme by next June.

The poignant news came just before the Consultative Board this morning that our dear friend C. [Christopher] Wilson died last night. The man who with Driver invited me to start the great venture of the GEC Research Labs. in 1915 and who has been a member of the [Laboratories' Consultative] Board ever since its start in 1919. We shall miss his personality very much indeed.

November 16th

The news continues exciting. At the moment there is a race to occupy Tunisia—both sides with sea and air communications. The clash in N. Tunisia must be imminent.

Preproduction is still full of difficulties and hold ups. CV 55 is in trouble due to copper machining and difficulty with the small seal. CV 52 which Douglas is looking after has gone to pieces owing to doubts about testing or socket contacts or what not. These things, and others, are such a worry to the workshop that they shrink from honouring cheques we give to desperate Service Depts. wanting CV 90s, 79s and 89s, 1274s and the like. So everyone is worried and patience is at a premium. However, all are patient, but we are in for a pickle if delivery of equipments using the valves are up to time.

Had a talk with whole CVD staff on the new arrangement for CVD 'conversazione' day at Wembley.

Winch has gone down to Larkhill to test our high flash-bang gear. He can detect a 1/25 second flash from a 9 W sidelight lamp at 1500 ft—a distance at which it was only just visible if you knew where to look. This means that a gunflash ought to be visible at 30 000 ft.

Am off to TRE Malvern today.

November 17th

At TRE Malvern. Had a two hours talk with Rowe over all RDF applications on the stocks and in view, and about the prospects vis-a-vis enemy countermeasures. These things must not be set down in writing.

Discussed GEC programme for the future in relation to our contract and its continuance. Continued this with Dee in afternoon and then saw other sections, including SBB [?] in which it is suggested we might be of assistance in the future. I found my visit very attractive and liked the Service in the College Chapel for TRE staff at 8.45 am.

November 18th

At ADRDE Malvern—and started with 45 minutes talk with Cockcroft which was very useful. We touched on many points of policy under CVD and other matters. Rest of morning with Black who took me round the departments. I was specially interested to see the PF work.

November 19th

Espley gives a good account of the designing of GR and his automatic frequency control arrangements. These are now very simple to apply and he has introduced some magic into it.

No. 10 Propagation [i.e. Radio] set for the MOS is in a very different plight. SRDE (and SR 9 at HQ) seem to have made a complete mess of a set which we had devised and helped to get on the rails for them. The situation was so gross that I got Dr. Paris of MOS to see Megaw and Bligh this afternoon and take their evidence on the matter. I have not heard the result.

A TVC meeting this afternoon at which the sad situation respecting insufficient valve production and raw materials was revealed. Only 22 million instead of 44 million valves for 1943. This will restrict radically the fruition of the radio development programmes of all Establishments in the country.

November 20th

Lunched with [Commander] Goodeve who is now Assistant Controller of Research and Development. He wished to discuss the use by the Services of industrial facilities for development and research and methods of assessing the effort on different projects. A very interesting talk.

Visited G.P. Thomson (Radio Board) on status of Inter-Services Technical Valve Committee with a view to being made respectable. There is no doubt that we shall be attached to the Radio Board—which is where we should be.

The visit by Megaw and Bligh to Dr. Paris yesterday did useful work and will enable the best to be made of a bad mess. I fear some four months has been wasted.

November 21st

The first shots of PF with our 10 mm diameter valves have given good percentage result. So far, the usual boost of USA effort has caused unfounded aspersions to be cast on our valves! Winch had some first class flash-bang trials at Larkhill yesterday and obtained the astonishing result of detecting in daylight a gun flash at 10 000 yards. Bragg was there and was duly interested. We have had great difficulty hitherto in impressing the authorities with the sensitivities we have achieved in photoelectric detectors. As a matter of fact the Admiralty Research Laboratory have always come in to queer our pitch. I hope this time the job will come away clean.

I had the great satisfaction today of accepting a framed portrait of myself to be hung in the Laboratories—to mark my election to Fellowship of the Royal Society.

November 23rd[336]

News continues surprisingly good. The Russians have broken the Germans north and south of Stalingrad.

Preproduction topped 1303 valves but we are having a struggle against absences and wasted time of temporary women staff. Had meeting with group leaders to devise steps to remedy situation.

The GR meeting on Saturday demonstrated the electrical gear 'on the air' and our progress surprised the visitors.

Flash-bang meeting to decide next step forward and probably 70 sets will be made. Bragg was very impressed.

Sir Cecil Kisch and Sir Donald Banks come to talk over FIDO and were shown model demonstration with smoke jets. Also certain lighting proposals in fog, which Waldram handled very well with the visitors. They were shown round hastily (Fig. 28).

A message from Admiralty supervisor to tell us not to be over anxious about Manpower Board threat—as our work was well known and its importance outstanding. The latest news that Cripps is to go to MAP and supervise radio affairs will, I hope, secure that radio and valves are given better support than hitherto on the industrial side.

November 24th

Gavin and his group are having a grilling time getting out the Mandrel sets. 2 am at Claud Gen and continuous late work on the monitors. I shall be thankful when they can get away for a break for they cannot stand this for much longer.

Absences through illness and slight indisposition of women are a sore trial to us. They stay out in wholesale numbers, and I sometimes wonder if the doctors are justified in their easily given certificates.

Gossling reports further good results with our 10 cm [10 mm?] PF valves (experimental). Griffiths spent the day here with Megaw on magnetrons and two men from Clarendon, Oxford.

A post war planning (lighting) committee at the MOW and PB [Ministry of Works & Public Buildings?]. A cheery meeting for news is good from Russia although North Africa is going to be a tough campaign.

Diamond die work goes well. Some so-called high grade diamond powders from USA are distinctly inferior to ours on photographic examination. Leeds is very pleased with his new polishing machines.

November 25th[337-338]

The FIDO job is estimated at a million excluding refrigeration. I guess this will kill or delay it whilst various partial ideas are put forward.

A [GEC] co-ordination committee—the first after F. Clark's resignation. Things were talked over very frankly, also proposals by MOS to 'unit' production of radio functional parts.

A No. 10 (beam propagation set) meeting—and a lively raking up of reasons for the mess by which the GEC have been ignored as to the intentions of SRDE in changing their designs. Some straight things were said and it is now best to let bygones be bygones.

An inspection of the 300 channel MAP set worked off one crystal, developed by Bligh's group—Stenning and Heller.

An outstanding piece of work and most ingenious. It will now take a good time to make the next model which must be properly engineered. Much interest in it shown by Service visitors.

November 26th/27th

To Manchester to attend meeting of Diamond Die Technical Committee and to visit the Clemison factory which is just getting into operation. The evening meeting in the hotel continued to 11 pm and was useful. Reports came from USA (through the control). Over there they seem to be aiming at speed before they have mastered quality, but their efforts may teach us something. No evidence of successful '0.0005' dies—and the samples of their best diamond powders are inferior to ours. Clemison factory was well equipped and Mr. Davidson a kindly host. They are not really going yet with trained operators, so it is early to judge.

November 28th[339]

At Bradford Radio Works to see particularly their pioneer work on 'routiner' making and use for systematic testing for faults in radio sets. Visitors are coming frequently to see the developments.

News continues to be exciting—self destruction of the French Fleet at Toulon—Russian offensives and breaks through at Stalingrad and Moscow—Churchill to speak on Sunday evening. The war has definitely entered the phase in which the Allies have the initiative. We have to see what the present generation of Germans is like on the defensive.

Ryde and Waldram had Professor Merton (one of the trinity of advising scientists) to study laboratory models of fog and searchlight. We are making further experiments.

An all day session with Chivers and other ADRDE searchlight workers on Thursday started with their usual aggresive and critical attitude on Waldram's reports to the Home Defence Committee but it ended with greater appreciativeness and less acidity. I wish this group of fellows knew how to work in a brotherly way.

November 30th

Preproduction last week was 1540 valves, beating all records—this was because on the whole none of the types had a really bad week.

Francis reported his attendance at a meeting on compact source lamps at MOS, for the Generals find that 10 kW lamps in 36 inch reflectors save half a man per searchlight. So, many men and girls are to be used in industry producing the lamps for effecting the saving. I think Siemens will be the only lamp maker who is free enough to devote effort to this.

A flash-bang meeting here of some 12 army officers to decide on details to be incorporated in the three prototype equipments. Winch gave an excellent model demonstration to scale of a gun flash under night and day conditions and showed the recorded signals at 15 000 yards.

December 1st

Have been going into the costs of our getter production which is now going well under L.F. Britten. Requests from outside (STC) are coming in and our plant is being increased to cope with all requirements likely to be made on us.

Gavin and company have kept their date—November 30th—for the delivery of the Mandrel sets made by Claud Gen and supervised and tested from here.

December 2nd

An RDF CVD meeting all day. Receiving valves in the morning, transmitting valves in the afternoon. Our main receiving valve

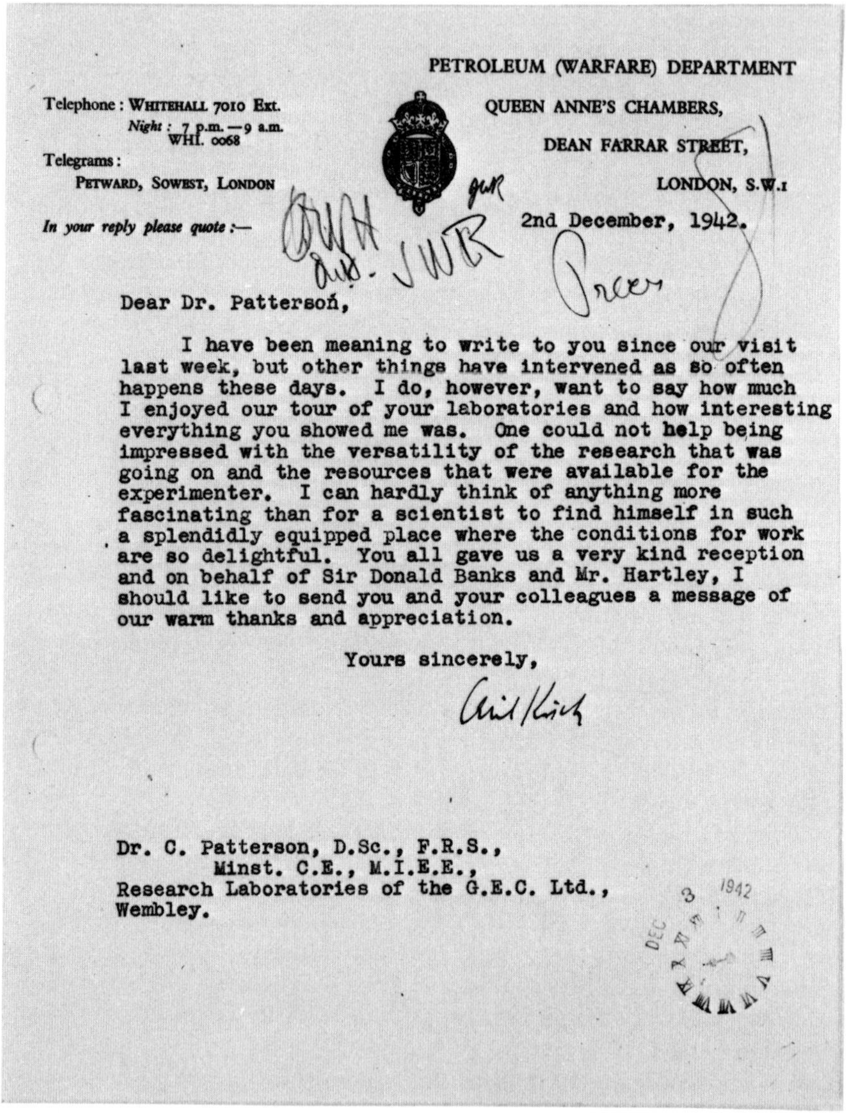

PETROLEUM (WARFARE) DEPARTMENT

Telephone: Whitehall 7010 Ext.
Night: 7 p.m.—9 a.m.
WHI. 0068

Telegrams:
Petward, Sowest, London

QUEEN ANNE'S CHAMBERS,
DEAN FARRAR STREET,
LONDON, S.W.1

In your reply please quote:—

2nd December, 1942.

Dear Dr. Patterson,

I have been meaning to write to you since our visit last week, but other things have intervened as so often happens these days. I do, however, want to say how much I enjoyed our tour of your laboratories and how interesting everything you showed me was. One could not help being impressed with the versatility of the research that was going on and the resources that were available for the experimenter. I can hardly think of anything more fascinating than for a scientist to find himself in such a splendidly equipped place where the conditions for work are so delightful. You all gave us a very kind reception and on behalf of Sir Donald Banks and Mr. Hartley, I should like to send you and your colleagues a message of our warm thanks and appreciation.

Yours sincerely,

Dr. C. Patterson, D.Sc., F.R.S.,
Minst. C.E., M.I.E.E.,
Research Laboratories of the G.E.C. Ltd.,
Wembley.

Fig.28 Letter dated 2 December 1942, from Sir Cecil Kisch to Clifford Paterson following his visit to the Laboratories

contribution was by Dr. James who has surprisingly good results using CV 90 as [on] harmonics. The valve only requires 200 volts and is tunable of course to all wavelengths. On transmitting valves, mainly magnetrons, super-power valves go to high priority and the 3 cm variety run on low magnetic field at 20% efficiency (of which one sample has so far given this result) stands high. The lower wavelength seems likely to be further reduced to agree with USA.

Called with Dr. Railing on the Minister of Fuel & Power (Geoffrey Lloyd) to discuss FIDO. He and his three advisors were suffering from apprehensions when they heard that the scheme might cost a million, and the arguments were the usual ones to be expected in such circles. I think their greatest fear is having to sell the idea to the Air Council. It is left for them to see if they can do so. Our advice was unequivocal: 'Don't waste valuable time in deciding, and if you decide, go the whole hog'.

December 3rd

Espley sent me a message to say that the GR gear (improved functional prototype) travelled to Malvern and went straight 'on the air' on arrival and worked effectively continuously without a hitch—a most unusual and praiseworthy result which brought very appreciative comment.

This was our great CVD day at Wembley—the first under the new conditions—conversazione plus a business meeting. It was voted a real success. Fairly senior people came to the meeting and under Brundrett's chairmanship it tackled the work in order of the projects for which the valves are needed. Perhaps we could have had a few more visitors, but we are likely to have more next time.

December 4th

Had meeting re. Service contracts with Dr. Railing and Mr. Gamage. These I find increasingly useful.

Returned with Brundrett to a preproduction meeting in which we had to probe the 3 cm magnetron position—which was also discussed on Wednesday and which is not nearly as far forward as it should be. It looks as if Wembley would have to come in and carry the burden, for this set is badly needed for the Fleet Air Arm.

Gavin has had an appreciative note from Sir Robert Renwick of thanks for their good work with Claud Gen in completing the Mandrel sets (Fig. 29).

I had some very warm and unsolicited tributes to the Laboratories from Brundrett who said he found no other organisation to touch Wembley for helpfulness and practical objectiveness combined with a grip of their subjects. Very nice.

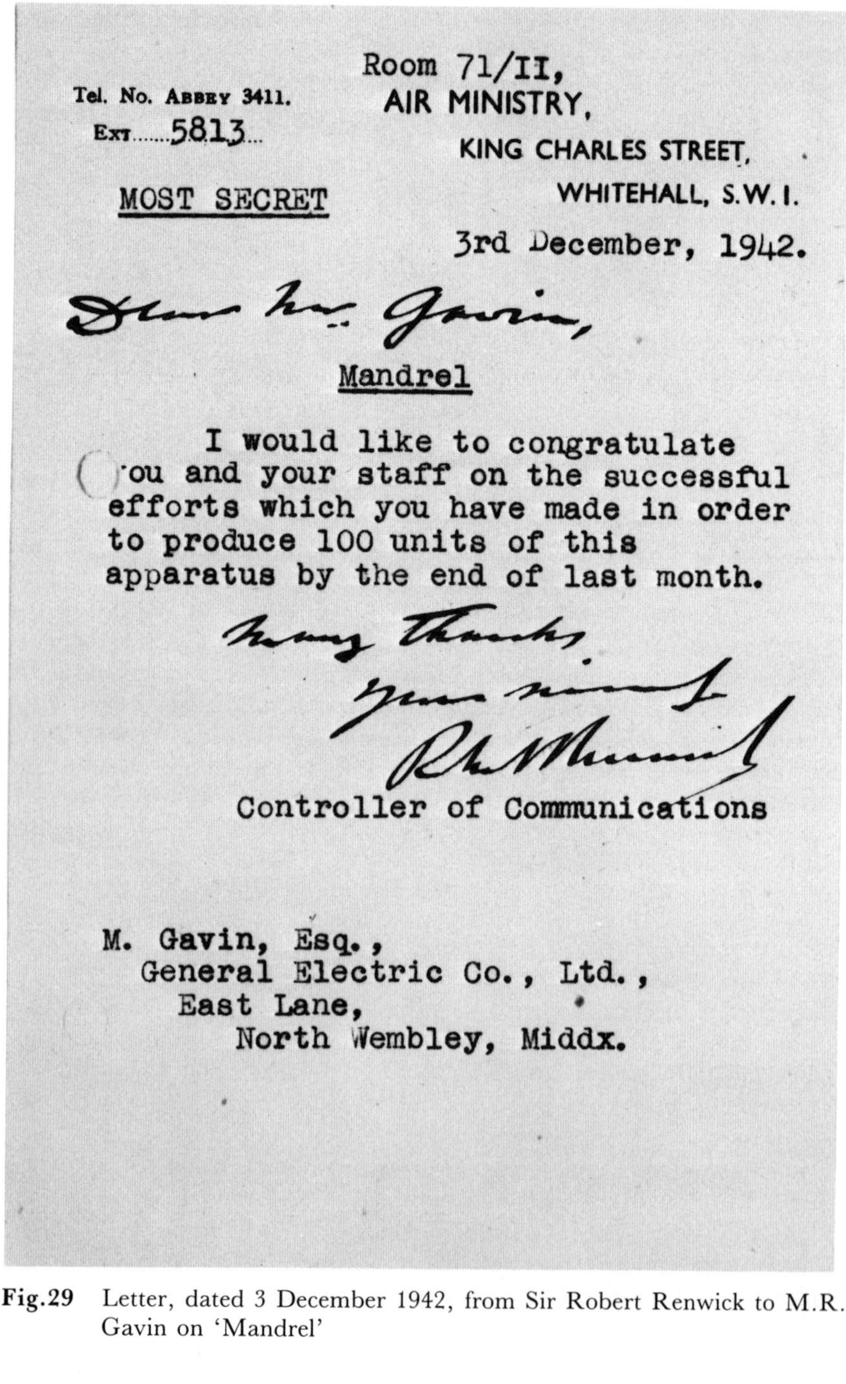

Room 71/II,
AIR MINISTRY,
KING CHARLES STREET,
WHITEHALL, S.W. I.

Tel. No. Abbey 3411.
Ext. 5813

MOST SECRET

3rd December, 1942.

Dear Mr. Gavin,

Mandrel

I would like to congratulate you and your staff on the successful efforts which you have made in order to produce 100 units of this apparatus by the end of last month.

Many thanks,
Yours sincerely,

Controller of Communications

M. Gavin, Esq.,
General Electric Co., Ltd.,
East Lane,
North Wembley, Middx.

Fig.29 Letter, dated 3 December 1942, from Sir Robert Renwick to M.R. Gavin on 'Mandrel'

December 5th

The Inter-Services Research Bureau 'fish' has apparently come through its trials at the Hebrides with flying colours. The pilot, I hear, planted his egg on the battleship all unbeknown! Quite a number of fishes are now on order. [See note of position on Helmore and ISRB work dated December 7th 1942 (Fig. 30).]

Have asked Le Rossignol for an enquiry into how we stand for meeting the various demands for valves using the CV 90 techniques—they seem to be multiplying. Truscott has come out trumps with an offer of 10 000 metres of tantalum wire to help us out with our CV 97 triode problem.

We are saying goodbye to Harley who is leaving his valve post at MAP and going to the Information (Radio) Bureau under the Radio Board. I shall miss him and his friendly courteous disposition.

December 7th

Preproduction is short of last week's total—but no panics.

Dr. Vick called to talk over compact source lamps for Army searchlight services. Siemens with few other research responsibilities have been fostering this project. Personally I think that nationally it ought not to proceed—on account of the load it would put on industry already starved of manpower for much more important projects.

Waldram's main programme of work on properties of the atmosphere is coming to a provisional close—and his contract with MOS will probably continue at a low financial rate for another quarter. Meanwhile he is busy on other things.

December 8th[340]

To the Admiralty to meet representatives of the Morgan Crucible Co. with Mr. Couch who is very anxious to secure the co-operation of Morgans in joint carbon development for searchlights. I was not sanguine but expected to be persuaded into some compromise arrangement. But I was relieved of anxiety by the uncompromising opposition to the scheme by Morgans and the meeting broke up after an hour's argument. It is now up to Couch and Vick.

Dined at the Joubert's with Blackett. A most interesting and revealing evening's talk at which some idols were indicated as filled with straw. Blackett is a man to inspire confidence.

SECRET.

HELMORE AND I.S.R.B. CONTRACTS.

POSITION ON 7.12.42.

Mosquito Turbinlite Aircraft.

The installation has been completed, and the light has been successfully operated in the air during daylight. The machine is fitted with four cannon, and an 80 KW. light, using a 30 mm. H.C.D. positive powered by half the number of batteries used on a Boston.

"Helmore Floodlight" (Big Arc).

The installation at Heston is approaching completion. The disc negative has been installed and run; it requires alteration to the magnetic control system to increase the current to the required value. Most of the radiolocation gear has been installed.

I.S.R.B.

Tests in Scotland have been completed. The vessel has been able to dodge torpedo nets and reach the side of a ship without detection, although all listening devices were installed on the ship. Another vessel with a higher horse power motor is being made.

....

Fig.30 Note of position on Helmore and ISRB contracts on 7 December 1942

December 9th

I ought to have recorded earlier the major share which O'Kane has taken in pushing and developing 'H_2S'. Rowe at TRE told me, and was unstinting in his commendation. I saw the photos when I was at Malvern and I hope the gear will soon be in operation.

Marris reports a rough meeting at E.K. Cole on modifications to Mark VIII AIS to patch up the difficulties caused by the E 1271 hard modulator. Everyone digging in their toes and each wanting to insist on his margin of safety—which renders the design untenable. The meeting appeared to end with honours easy, but I think Marris had to hold the position somewhat alone.

Went round CRT developments with Jesty and his group and was much interested in their researches. A good organisation of an excellent group.

Had an extraordinarily nice letter from Air Commodore Addison (Fig. 31)—thanks for prompt delivery of special Mandrel sets which 'were brought into operation on the first favourable opportunity after the target date we had fixed'. This is a remarkable achievement. Another feather for Gavin and his boys.

December 10th

PF midget 10 cm [mm?] valves are going well now and we are trying to launch the making of them at Hivac. I understnd the CV 61s in torpedoes are functioning successfully, but have no first hand report.

Am feeling somewhat doubtful about the financial stability of our diamond die production and there has got to be a regular searching of records for costing it up.

December 11th[341]

The visit to the Fleet by Stevens and Ackerley takes place next week to advise on its lighting.

The nephew of Langmuir of GE was here all yesterday—a most delightful visitor with a very high coefficient of absorption of information. Nevertheless he will have to come again to see much there was not time for.

PF shoots with our 10 mm valves with a very worn gun gave us a surprise by shaking off their oxide coatings. Otherwise they stood up well.

An SW propagation meeting at Wembley with many officers present from Admiralty and MOS to prepare everything for the Coastal Stations which are being set up for systematic work.

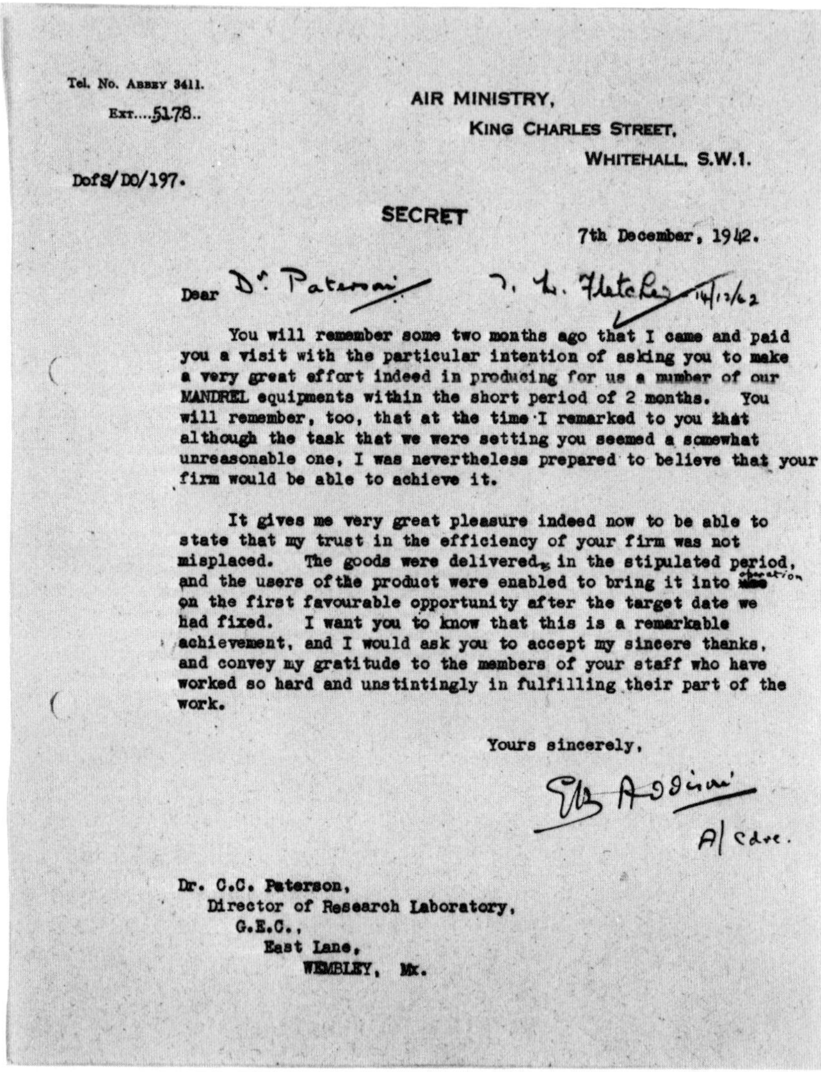

Tel. No. Abbey 3411.
Ext....5178..

AIR MINISTRY,
King Charles Street,
Whitehall, S.W.1.

DofS/DO/197.

SECRET

7th December, 1942.

Dear Dr. Paterson, T. L. Fletcher 14/12/42

You will remember some two months ago that I came and paid you a visit with the particular intention of asking you to make a very great effort indeed in producing for us a number of our MANDREL equipments within the short period of 2 months. You will remember, too, that at the time I remarked to you that although the task that we were setting you seemed a somewhat unreasonable one, I was nevertheless prepared to believe that your firm would be able to achieve it.

It gives me very great pleasure indeed now to be able to state that my trust in the efficiency of your firm was not misplaced. The goods were delivered in the stipulated period, and the users of the product were enabled to bring it into operation on the first favourable opportunity after the target date we had fixed. I want you to know that this is a remarkable achievement, and I would ask you to accept my sincere thanks, and convey my gratitude to the members of your staff who have worked so hard and unstintingly in fulfilling their part of the work.

Yours sincerely,

E B Addison
A/Cdre.

Dr. C.C. Paterson,
Director of Research Laboratory,
G.E.C.,
East Lane,
WEMBLEY, Mx.

Fig.31 Letter, dated 7 December 1942, from Air-Commodore Addison to Clifford Paterson on 'Mandrel'

A Federation of British Industry Research meeting to discuss research effort and mentality of industry generally. Met Griffiths of Mond Nickel and discussed his mission to USA to get collaboration over metallurgical developments.

December 12th

Yet another unsolicited testimonial, this time from Dr. Grey of MOS, chairman of the committee on GR, who gives us top marks on the GR effort so far.

December 14th[342-343]

Preproduction 1450 valves—almost normal. The testing and its complications were only gripped just in time six weeks ago when Dudding pointed out the dangers and we brought Francis into the work.

Had Mr. Graham Cunnington [Cunningham?] (DGMS), Dr. Paris and General Butler here, to talk over liaison and to look over some sections of the work. They were here for 3½ hours and saw a good deal. Cunningham explained that the MOS are third in priority for manpower, so do all the sacrificing by cutting its programmes all round.

Joubert come home to tea on Sunday afternoon and was very insistent on the value of the 'Lee' [Leigh] light and hoped that GEC would maufacture, seeing that the present producer cannot tackle requirements.

The U-Boat menace seems still to be the most serious of all.

December 15th

Marris has got four members of his group down at TRE aerodrome on the Oboe installation which we have every hope is going to prove another success.

In preparation for the CRT meeting tomorrow Jesty has gone over developments. Our work seems up to requirements. We shall probably take on another CR tube in our preproduction group—either Skiatron or a new screen technique tube which the factory has not yet succeeded with.

We must put more push behind our part of FIDO. Humphreys saw 12 000 gallons of petrol burnt to waste to make one experiment yesterday at Staines!

December 16th

A CRT meeting at Wembley with Walmsley presiding. Cathode ray tubes are complicated gadgets and the meeting is correspondingly involved. Now that it is officially attached to the Radio Board I think everyone takes it more seriously. I fancy the official CVD organisation is wanting to exercise some measure of control—but I am not sure yet in what direction.

December 17th

Gossling and Donaldson back from PF meeting at Malvern. They seem to have got the valve position for PFs where it should be. The focus of production is on Hivacs—ours is to try to keep the valve performance and characteristics on the lines.

A very lively CVD meeting at which I had to defend Walmsley as chairman of the [CRT?] committee from having its affairs and status reorganised in his absence. I think I did good and the chairman (Brundrett) backed me to the extent of asking the CRT committee to comment on the proposals of the CVD technical committee of which Black is chairman.

Horton from ASE was present to make the most serious representations as to the priority of megawatt magnetrons over other types. He made a very good case as the result of recent trials with 500 kW operation, presumably using CV 76s.

December 18th

Truscott's valve allocation meeting this morning left me with a rather desperate feeling about who is going to make 3 cm magnetrons for the several equipments maturing. MOV, BTH, Birmingham University all seem weak reeds to lean upon due to shortage of personnel.

A GR progress meeting at HO with Grey in the chair. A cheery meeting and very pleased with itself due to phenomenal progress already noted in this Diary. It looks as if our planned dates of last summer were going to be met.

Peters made handsome offers to get 20 preproduction sets made. Espley has every reason to be proud of the way he and his group have carried the job so far. The focus of responsibility now passes to the production end at Spon Street and Witton.

December 21st

Preproduction nearly 1500 for the week but the prospect of established types passing to the Works does not seem to grow brighter.

Some unpleasant jolts re. call-ups from the Manpower Board which, if carried out, will just land us on the rocks.

Had visit from Prain and some USA diamond experts. The supply of diamonds and boart seems to be likely to be critical on account of USA demands.

December 22nd[344 - 346]

No definite news yet as to the effect of Mandrel on our bombing raids—but the Hun seems to have got his 61 cm AI fairly efficient. Dr. Suits from GE Schenectady who spent Saturday with Gavin was very impressed that wc could produce our Mandrel sets so quickly. He also thought a lot of the E 1274 valve.

Had an internal meeting with Marris, Gossling, Bligh, Megaw, Gavin, (Espley away) on post-war radio applied to aviation. I want a policy for GEC, and Marris, Bligh and Espley are to get out a functional specification to start with.

G.I. Taylor came today re. FIDO to discuss our scheme. Air Marshal Sorley who was with me at home on Sunday was not very enamoured of the idea of fog clearance and preferred to put his money on radio landing devices.

December 23rd

A day for finishing off and clearing up various matters before the holidays. Everyone in good spirits under influence of good news of break through by Russians into Donetz area.

Had return of costs of our diamond die production April—September. There is a considerable loss and we must raise our prices by 50%. We can probably bring down our costs by quicker drilling of the initial 'coning', and other economies—but we cannot make up so heavy a leeway.

December 24th

O'Kane came to discuss various things including H_2S which he is now introducing to Bomber Command and teaching to them. Oboe is being satisfactorily installed at Malvern but has not yet flown. GR is reported at ADRDE to be functioning still with never a hitch—and somewhat better range has been obtained. So our once despised 25 cm system, rejected for AIS, has come into its own.

Partridge reports that his special 'drawn' silica seal glass is giving enhanced strength in valve seals and is up to expectations.

And now for a brief holiday break!

December 28th

We have frequent jolts when testing our new preproduction valves under operational conditions. This time it is CV 56 magnetrons (wave jumps) and NT 99 [(emission)] valves. This means constant losses of finished valves on final test.

I am bothered over the position of our CRT committee and its work, which goes very well but is being assailed by overhead doctrinaire and sometimes prejudiced people. The chief attackers are the 'technical CVD committee' under Dr. Black as chairman who know very little of the efficiency of this CRT committee. The fact that its chairman Walmsley is no longer at MAP is a great weakness. He is coming tomorrow to talk over the position.

December 29th

Walmsley came today and would not take the above too tragically and insists that he will handle the situation as he has had to handle the same kind of intrigue for the past two years. He told me that Leedham is being appointed DCD! Not at all a bad appointment. He will at least not be prejudiced against industry and is an honest man.

I presided at a meeting of the Lighting Committee at Ministry of Works and Planning at which fluorescent lighting as supplementing natural light was studied. The architects and others were much impressed and had apparently never realised the effects which some of us have been working at for so long.

December 30th

Jesty has had contacts at MAP and ADRDE which indicate more reasonable explanation of the CRT manoeuvres and put a cheerier complexion on things.

A GEC radio meeting and a good spirit. Macqueen made various revelations of meeting between RMA [Radio Manufacturers Association] and Government officers. There is a helpful move to make plans ahead for after war television broadcasting. There is general satisfaction that Leedham is back again.

Joined Chelioti and Wells in a call at the Ministry of Production on Mr. Garro Jones and Mr. Chegdwidden to discuss radio production. We all said the usual things and the Ministry of Labour and Manpower demands were blamed for most of the troubles. It was felt that there was a large element of 'false demand' in equipment programmes and that much will be done by 'dehydrating' requests for equipment and components.

December 31st

As 1942 passed out there is a general feeling of confidence and cheerfulness—mainly because the Russians are exploding the myth of German military superiority.

An Osram Advisory Committee meeting at Hammersmith at which manpower was discussed and I urged the consideration of control of the lamp industry of a character similar to that which applies to the valve industry (VPC) [Valve Production Committee].

Had Griffiths (CVD) to lunch and to talk over things in general.

1943

January 1st

Visit by Dr. Railing and Mr. Gamage to the Labs. for a 'Research Services' meeting.

Had visit from Diamond Control to advise that our labour would be reserved—which is some comfort at least!

[S.V.] Williams went over possibilities of powder metallurgy and possibility of collaboration with GEC Works. Our own place in the art is looked upon as quite advanced and important.

January 2nd

It now appears that all three of our candidates for 'honours' are to receive them—Megaw, Breadner and Miss Rowe. They all had to be associated with valves and CVD—and I have no regrets as to choice.

January 4th

Night trial at Heston on the big GSL [Ground Search Light?] Helmore lamp gave satisfaction to Helmore, Jolley and co-workers, but did not appear to throw much credit onto the ELSIE [Radar Searchlight Control?] equipment. There is to be a further trial tonight witnessed by Air V/M Linell.

Heather paid his first visit to Wembley since his illness and spent three hours enquiring into our various radio projects. Gavin has been told by Mr. Hecht that there is some doubt whether he (Mr. Hecht) will remain in control of RCM [Radio Counter Measures] research. There seems to be a good deal of satisfaction over Leedham's appointment.

January 5th

A Glass meeting at Hammersmith. Everything one meets turns on personnel and the difficulties of getting and retaining it with the ignorant administration of the Ministry of Labour. A long talk with Dr. Railing on various matters and carbons in particular which Cameron shows are suffering from worn out tunnel furnaces and ineffective temperature control in mixing and extruding.

January 6th

Schuil called to say that DSR Admiralty wants him to go abroad on camouflage study in other world areas. We both felt he should go. He will probably go to the Mediterranean first and be away three months.

A telegram of congratulation from the Controller of the Navy on the recognition of the Labs. in the New Year Honours (Fig. 32).

Called with Giggins at MOS contracts department to stave off a request for an official enquiry into our research contract costs. I had with me all papers to demonstrate our costing system and calculation of overheads. We left with the assurance from them that they would not ask for a cost enquiry.

Dudding has been holding several internal meetings—emission, CRT and others.

January 7th

A CVD modulator meeting yesterday raised matter of hydrogen filled gas modulator. Ramsay will get something going at our end but BTH are doing the main work. There followed a meeting with Ratsey of ASE on the modulator we are to make for testing the megawatt valve. All is very nebulous and we have decided to get out a specification of what we conceive to be required and let it be shot at.

Have heard from Sir Cecil Kisch that we can charge up our FIDO expenses to Ministry of Works and Buildings.

January 8th

A Mr. Harris of MOS was announced early and Humphreys received him. I went down as soon as I could and found them in a huddle. Mr. Harris informed me that the operations of the Manpower Board were very serious and we were due to lose all our people under 30! He

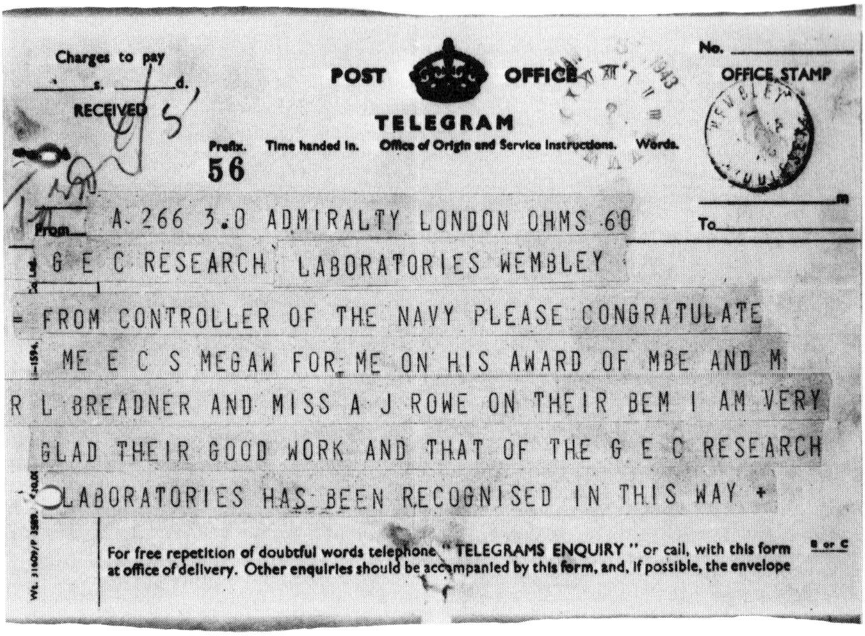

Charges to pay
s. d.
RECEIVED

POST OFFICE
TELEGRAM

Prefix. Time handed in. Office of Origin and Service Instructions. Words.
56

No.
OFFICE STAMP

From A 266 3.0 ADMIRALTY LONDON OHMS 60

To

G E C RESEARCH LABORATORIES WEMBLEY

= FROM CONTROLLER OF THE NAVY PLEASE CONGRATULATE
ME E C S MEGAW FOR ME ON HIS AWARD OF MBE AND M
R L BREADNER AND MISS A J ROWE ON THEIR BEM I AM VERY
GLAD THEIR GOOD WORK AND THAT OF THE G E C RESEARCH
LABORATORIES HAS BEEN RECOGNISED IN THIS WAY +

For free repetition of doubtful words telephone "TELEGRAMS ENQUIRY" or call, with this form at office of delivery. Other enquiries should be accompanied by this form, and, if possible, the envelope

Fig.32 Telegram from Third Sea Lord congratulating members of the Laboratories' staff on awards

informed us that he is the Divisional Officer for MOS—opposite number to Admiral Gurnall whose offices he shares. I gather the three Services are to conspire to save Wembley from being shattered by the press gang, but that it is going to be mighty difficult to do so. We have to afford these officers all the information they require. So here's a how d'ye do.

Ryde attended a meeting of 'eminents' about his 'light chart' for hours of dark and dusk at different times of year and after talking round variants they decided upon his form of diagram which was very gratifying. They are making it a supply to large numbers of officers who have to plan operations.

Heatherington from DSIR here all day to get to know the place.

January 9th

Donaldson was at a PF meeting yesterday at ADRDE and all seems well. Good firing trials with the 10 mm valves—which seem to be rugged enough. Hivac will have an order for 1000 per week which presumably will be of the Wembley pattern and design.

Further serious talk with Humphreys over our policy regarding proposed calling up of some of our clerical staff. The Divisional officers of the supply Ministries will have the main responsibility for fighting for the exceptional cases.

News continues to be good from Russia but whether they will be able to reach Rostov in time to cut off any of the Caucasus army of the Germans remains to be seen. Everyone feels things are dull and slow in N. Africa but it is so easy to overlook the great difficulties and the fact that after all there is no hurry there, provided the Axis is expending its strength there.

January 11th

Preproduction is suffering from many troubles, the chief being low emission from oxide cathodes in several types—the greatest disaster being the NT 99 in which failures are catastrophic. The total for the week is 1238 instead of 1500. I am more than thankful that B.P.D. pressed six months ago for thoroughgoing research into oxide coatings.

Marris learns that Oboe I is 'it' and that 'Gee' is no good. This probably means that Oboe II rises quickly to a high place.

Winch reports great interest by New Zealand in flash-bang and that they will probably want a pilot model. The Admiralty also seem interested after all and are asking for a photometer unit and light amplifier from us.

January 12th

A diamond die specification meeting at which we finished a draft to go to BSI and to the users. A small flap yesterday when Pye complained to Butement at MOS that our CV 90 valves would not work. Butement rang me up personally about it. Warren reports today that the CV 90s returned by Pye work properly in our circuits and Pye have been informed that their circuitry is therefore faulty. We shall send to help them later in the week.

January 13th

A CVD (RDF) meeting all day—the magnetron position is not happy because neither Birmingham nor Wembley have succeeded in designing a satisfactory 3 cm [magnetron]—a most urgent demand. It is true that the main responsibility lies with Birmingham but we are second string in the development and have been doing our best. Every design fails either through wave jumps or through low efficiency and the 'waveguide' design has not yet taken us over the stream.

On the other hand the CV 90 stock is rising fast, not only because it is doing its stuff in the hands of the experimental establishments but because Sutton (just returned from USA) tells woeful tales of the American lighthouse valve and the rejections they are getting there in production. It has also been discarded as a 10 cm valve—and several production establishments in the USA are starting to make the CV 90 instead of the 'lighthouse' because the techniques used are those to which valve makers are accustomed. So that is one up for us and paves the way for the two or three variants of the CV 90 now on the stocks for higher and lower powers.

I wish however we were in the running with klystron tubes and spark modulators—particularly the former.

Donaldson (on proximity fuses) had the satisfaction of hearing that 18 out of 18 fuses using our 16 mm valves on firing trials oscillated satisfactorily. I am wondering what Dr. Black will say now after his formal [former] aspersions upon midget valves as compared with those from USA!

January 14th

Partridge's silica seal rod drawn through an iridium die is showing good results and improved seals.

Espley has returned from GR trials at ADRDE—all successful. There seems no expectation of using anything but the CV 58 diode which no one but Wembley appears ready to make.

A CVD liaison meeting—good in itself but the people coming to the conversazione part were but few.

Had notification that Sir Stafford Cripps is visiting the Labs. on Saturday afternoon—so had to arrange that the whole place keeps in action through Saturday.

January 15th

Went through more lists of young fellows who had received registration notices—we can only release a few. Cornwell is still in great trouble with his LSO personnel through illness and other losses.

A propagation meeting with G.M. Wright in the chair.

Our Junior Bradford party—Stenning, Heller and Bayliss—taking multi-channel set by road has had a crash and all are somewhat knocked about. The gear seems to have come off best.

Attended meeting of the Operations and Technical Radio Committee of the Radio Board to consider terms of reference etc. for TVC and other committees which are now being taken under its wing. The terms of reference for TVC went through all right.

January 16th

Cobden Turner and a Salford deputation came to discuss mainly jewels for instruments. They plan to extend their activities to cover all allied matters and were somewhat covetous of our diamond die and powder plant!

The visit of the great man with his entourage came off and I think went well. Sir Stafford Cripps, Sir Robert Renwick, Mr. Stevenson and Mr. Strauss (two Labour members [MPs])—also Lady Cripps. Sir Stafford played up well and made a very good and sensible speech over the Lab. broadcast which was heard well—thanks to Brittain's attention to the public address amplifier (Fig. 33). I think the visit will have done good internally and I believe it was important and interesting to the visitors. Dr. Railing came and took an important share in the visit, emphasising the great variety and strength of GEC manufacturing enterprises. The visitors went round the whole Labs.—hardly missing a room.

Lord Hirst was unwell and Mr. Gamage had to stay with him.

SPEECH BY SIR STAFFORD CRIPPS TO THE RESEARCH LABORATORIES AT WEMBLEY.
SATURDAY, 16th JANUARY, 1943

I would first of all like to tell you the great pleasure it has been to me to see over these Laboratories this afternoon. I understand that many of you have stayed at your work in order that I might have a better view of your activities, and for that I would like to express my gratitude.

All of you will realise that the war is becoming more and more a technical war, a war of science, and the arsenals of victory today are very much the arsenals of science, research and development, and you, who are working in such an arsenal here, are a vital force behind the pressure of scientific achievement which will undoubtedly be the final decision in the result of the war.

I have witnessed the deftness of your fingers and the ingenuity of your brains, I wish I had the brain and the ingenuity to completely understand the most intricate work you are doing here, and I want to express, on behalf of the Government and the country, our feeling of gratitude to you for the great contribution you have been able to make.

You are the pioneers, of course, in many forms of development. You do not see the manufacture of the article which means days and weeks of research for one productive item, but, nevertheless, many of these things you develop have been used in the various theatres of war, and have actually helped over some difficult problem raised through great scientific changes in the laboratories of Germany or Italy, and we have got to see to it all the time that we keep ahead in this scientific struggle which is continually going on.

There is no doubt that you, who are working in laboratories, doing development and research work, are pitting your brains against theirs, and we can be proud that British development, at least in this scientific warfare, has always been able to hold the lead, and it is your work which has given us this lead, and it is your work which is going to keep us in the lead to the day of victory.

I would just like to say again how grateful I am for the work you are doing, stick to it and I am sure we shall come out on top.

Fig.33 Transcript of address given by Sir Stafford Cripps during his visit to the Laboratories on 16 January 1943

January 18th

The contact Col. Beecher made with me re. help to Col. Dolphin of the Interservice Research Bureau has ended in our seconding Buswell instead of Jolley to him for some 12 months on the grounds, I understand, that Jolley's equatorial proportions might not give him the necessary flexibility in confined spaces. We are glad to be able to help Dolphin.

The Helmore lamps, except the large one, are not catching on very well, as they are not any longer favoured operationally. We had our first raid for some 12 months over London last night and I still think the turbinlite should be useful if 'operations' would try to make it so. The big ground lamp is to be known as the Helmore Floodlight though as far as I can see its technical and scientific features as well as its engineering are mainly due to the GEC.

Preproduction weeks total is 1350. Much trouble still with emission efficiency of cathode coatings. Subject was discussed at Truscott's allocation meeting in morning and Brundrett and Truscott preproduction meeting here in afternoon. Everyone cheery but difficulties in getting suitable personnel are the cause of much inefficiency.

Gavin reports a new RCM jamming job, details of objects of which must not be put in writing.

Order received from MOS for two of Winch's flash-bang apparatus to be made at Bradford.

January 19th

The news continues good. Leningrad is relieved and there appear to be one or two more groups of 30 000 prisoners down the front—along nearly the whole of which the Russians are attacking.

A TVC meeting this morning at which we were duly impressed with our enhanced dignity and new terms of reference.

Lunched with Ackerley and Mr. Guy Campbell over future presidencies of the IES and the possibilities of having sometimes a really eminent man in the chair.

Mr. Thompson of the Diamond Die Control has been here on his return from the USA. It looks as if one of the Companies will get on with fine die making if they attend to putting flats on their diamonds, to sizing their needles and to proper grading of powders.

January 20th

We had the first serious daylight alert today for 12 months. 30 planes over England and 11 down. The bombs were dropped south of the Thames.

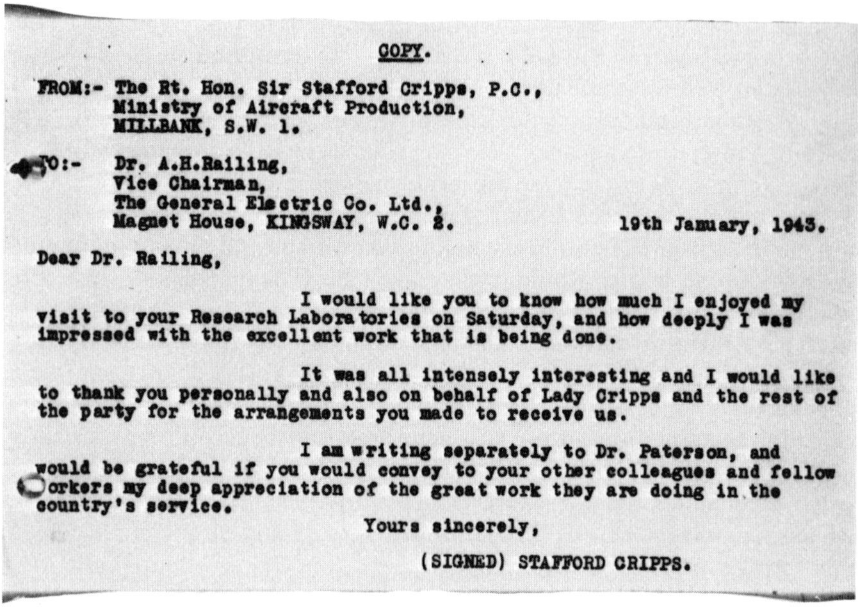

COPY.

FROM:- The Rt. Hon. Sir Stafford Cripps, P.C.,
Ministry of Aircraft Production,
MILLBANK, S.W. 1.

TO:- Dr. A.H.Railing,
Vice Chairman,
The General Electric Co. Ltd.,
Magnet House, KINGSWAY, W.C. 2.

19th January, 1943.

Dear Dr. Railing,

I would like you to know how much I enjoyed my visit to your Research Laboratories on Saturday, and how deeply I was impressed with the excellent work that is being done.

It was all intensely interesting and I would like to thank you personally and also on behalf of Lady Cripps and the rest of the party for the arrangements you made to receive us.

I am writing separately to Dr. Paterson, and would be grateful if you would convey to your other colleagues and fellow workers my deep appreciation of the great work they are doing in the country's service.

Yours sincerely,

(SIGNED) STAFFORD CRIPPS.

Fig.34 Letter of thanks from Sir Stafford Cripps following his visit to the Laboratories

Sir Cecil Kisch came re. FIDO to review progress. He is an acceptable visitor. Ryde went over our doings and the simplification of our original scheme, which no one seems to want to push on with because it is too big an undertaking. [Letter of thanks from Sir Stafford Cripps received (Fig. 34)].

January 21st

A request from Pringle (DEE Admiralty) for help in fog landing on aircraft carriers. We have not had much joy so far working on Morton's suggestions for fog landing, using lights—but we may think up some new idea.

Called at our Exhibition premises which are serving an invaluable purpose. Rycroft and Connell are handling the extensive paint drying developments by radiant heat—and Hobday is in charge of our component making group there—mainly rolling anodes.

January 22nd[347-348]

Two officers here from RAF Training Centre, Reading—on [an] anti-submarine automatically dirigible torpedo device. Farren is helping them with one of the electrical mechanisms for it.

The NT 99 trouble with Met Vick set is now explained and the necessary set alterations decided on.

Commander Carr from Vernon here today. Damant has left and Carr is now exerting his weight and is bluntly expressing opinions as to whether and why not work on carbons is not being pressed on.

January 23rd

Saturday. The very sad and unexpected announcement in the 7 o'clock broadcast of Lord Hirst's death. He had been having one of his usual bouts of illness but he had seemed so well just previously that one did not expect this. We lose the founder of the Research Labs. and a staunch friend and counsellor who by his mere presence as Chairman of the Company caused the balance to be maintained by directing upwards all the forces which might otherwise operate horizontally in opposition to each other.

The Labs. have been closed today (except for skeleton staffs) in view of last Saturday's opening for Cripps's visit. The day off will do no one any harm. There is much illness—more I think than usual.

January 25th[349]

Preproduction nearly 1500 valves last week. Marris had many pieces of inside information from TRE. He is being asked to produce quickly three Oboe ground stations, which we shall probably take on. There is still great keenness about Oboe and accurate bombing seems assured.

But there seems a slump in new RDF projects. No one can think of anything new to go for by way of major projects.

Mr. Balmain was here to get us to make UV lamps for germicide work to supplement chemical treatment of the air.

G/C [Group Captain] St. John from Reading came to ask us to develop their anti-submarine scheme further. They had been very pleased with Farren's and Ackland Snow's preliminary contribution.

Drs. Paris and Vick from MOS came to consult me about a man to help them with certain infra-red applications—considerable projects they are too. I at first thought of Jolley but a whole time man is wanted and I put them on to Frank Gill—who may be able to handle the job.

January 26th[350]

Lord Hirst was cremated today. I attended the very dignified and beautiful service at Golders Green.

Bell and Warren came back from the Admiralty Establishment at Bristol with a more detailed and gloomy picture of the position of the lighthouse valve in USA. I don't know if Sutton's stories of alleged ineptitudes are true.

Had a visit from Dr. Thompson of Bristol to enquire if Wembley would take him on and his little VM valve (local oscillator) since Bristol are stated to be going over to magnetrons entirely. After consultations with Le Rossignol, Bell, Warren, Ryde and Gardiner, we agreed to the principle. Craftsman facilities are the main trouble.

Discussed waveguide cells with Harris and possibility of preproduction of 250 per week. The Exhibition Building seems the only possibility.

We are still in difficulty making boules. We have some USA sapphire in 2 mm rod form.

I seem to be receiving unearned increments of credit! For lending Buswell to Col. Dolphin—which was much to the interest of our work with him—I have received personal letters of thanks both from Lord Hankey and Lord Selbourne [Selborne]—not to mention Col. Dolphin and Col. Beecher! Buswell is already working in his new capacity.

Memorial service for Lord Hirst at St. Martin's-in-the-Fields—a very impressive occasion—and congregation.

January 27th

Had an opportunity of seeing Walmsley before the CRT meeting today and got him to accept a wording of a minute which will give an answer to CVD policy committee on the procedure which focuses departmental responsibility on Mr. Hecht. I hope it will dispose of the innuendos which were thrown out as to 'unsatisfactory' matters [manner?] in the way CCRTD [Committee of CRT Development] had conducted its responsibilities.

Chirnside reports another use discovered by the biologists for his porous alumina—as an absorbant in the process of separating penicillin—the enemy of staphylococci. This contract may show the difference between ours and Sims' alumina for making boules.

January 28th

An OAC meeting with H.H.'s [Hugo Hirst's] place empty. Fletcher spoke very helpfully indeed in paying a tribute to H.H. on behalf of the meeting.

January 29th[351]

A day full of incident. The great meeting of PF took place and after much previous discussion and doubtings the powers came down solid for the Wembley design of valve produced by Donaldson, Jenkins and Gossling. The rival design by Rangaby was naturally supported by Hivac who had been induced by Rangaby to put work into a rival design to ours. How hard I tried to save all concerned from exactly this trouble 12 months ago when Whiddington, Ellis and Telfer had not the courage to tell Hivac that the type was to be settled at Wembley and their function was to turn it into effective production. We shall have much to do now in helping Hivac.

Leeds's diamond powder is in considerable request but his production gets so much more efficient that we can easily cope with the demand.

His special graphite powder is also demanded by all CRT makers now.

CVD policy meeting this afternoon contained some breezes between Walmsley and Black as to the utility of the Technical Committee of CVD. I had to support Walmsley to the extent of saying that the Technical Committee in many technical issues would have more enlightened discussions if there were present people from lower levels who could keep them straight on questions of fact.

They want Mullard to take one of the CV 90 valves in hand and develop for production. I offered collaboration with Mullard. Warren and others question Mullard's ability to co-operate.

Black was announced as going to the USA as a heavyweight on valve and circuit matters—on the whole a good move I think.

C.S. Wright (DSR) pulled me back to his office to say that the Radio Board instructed him to ask me if I could send someone of sufficient tonnage to USA to handle the CV 90 situation there.

One's first reaction was to say that the damage, if Warren were away for two months, would be too serious—but we have talked it over and I await Warren's reaction.

I then got DSR alone and challenged him as to how matters stand with the anti-submarine warfare and whether he really has the situation in hand. If not I suggested to him that he ought to consider whether one of our teams at Wembley could not be of help. DSR took the suggestion very handsomely and I shall probably be hearing from him.

In the evening a full dress show at Heston with the big searchlight. Helmore (now putting up as MP for Watford), Stafford Cripps, C-in-C Fighter Command, Gen. Sir F. Pile [C-in-C Anti-Aircraft Command] and many many others. The show went off well and Helmore was generous in his acknowledgement to our team. I shouldered cost of the dinner afterwards to some 20 people in the RAF canteen! A most sumptuously illegal repast.

January 30th[352]

Dr. Priest from TRE here to talk about valves etc. for beacons and the prospect of getting them made in the USA. He is going there.

February 1st

News is still good, 30 000 Germans eliminated at Stalingrad and the whole of the Caucasus well on the way to being cleared. Real fighting about to begin in Tunisia.

Warren has decided to go to USA re. CV 90 valve. On the whole I am glad of this.

A meeting and lecture by Dunsheath on Industrial Research at Royal Society of Arts.

February 2nd

Discussed with Macqueen policy in Radio Manufacturers Association and industry re. future television arrangements and the like. C.S. Wright

Copy of letter received from the 3rd Sea Lord -
Controller of the Navy.

P.013675/42.

Admiralty,
London, S.W.1.

1st February, 1943.

Dear Dr.Railing,

I have noted with satisfaction the consistently good work carried out by the Research Laboratories of your firm in developing and producing in quantity several types of especially important valves for use in Radio Direction Finding equipment.

I should like to convey to you my appreciation of the valuable service which your Laboratories are rendering to the war effort, and to express my confidence that the high standard of this service will be maintained.

Yours sincerely,

(Signed) C. Wake Walker.

VICE-ADMIRAL.
Third Sea Lord and Controller of the Navy.

Dr.A.H.Railing,
The General Electric Co.Ltd.,
London, W.C.2.

Fig.35 Copy of letter, dated 1 February 1943, from Third Sea Lord to Dr. Railing on the work of the Laboratories

(DSR) told me he is probably asking Dr. Roberts (submarine warfare) to follow up his discussion with me on January 29th and call to see us.

February 3rd

Diamond Die meeting when we heard from Mr. Thompson just returned from USA. Nothing very new. Apparently all our literature sent to America on diamond die had never been circulated there! Buried in offices!

February 4th

The oxide coated cathode research bulks considerably in our outlook. Ramsay and Machin are doing their best with it and Dudding helps the effort in every way possible with meetings and statistical perspective in analysing the results. Nothing outstanding has as yet appeared to enable us to feel that we have the subject under control but there are various clues—including the nature of the nickel alloy core material.

February 5th

Ryde tells me that the FIDO developments are promising a much simpler scheme as a result of interesting experiments showing the effect of one jet 'curtain' on the other and the influence of entrainment of air by the jet.

A meeting with Railing and Gamage at which my contact with DSR over submarine detection was appreciated.

February 6th

A nice letter to Dr. Railing of thanks from the Controller of the Navy for the valuable work we have been doing in the 'development and the making of valves for important Radio Direction Finding applications' (Fig. 35).

Mr. [Dr.] Roberts head of the Anti-Submarine Establishment in Ayrshire [Fairlie] is coming next Thursday and as a first step wants us to take on an audiometer.

Telephone message from Coales (ASE) asking if we would take on another gun ranging job for shorter wavelengths and suitable for combined operation[s]. We must see how the Roberts interview goes before committing ourselves.

February 8th[353-354]

News from Russia, particularly, is most heartening. The Germans are virtually out of the Caucasus and are giving way over most of the front south of Orel. Churchill has just returned from Casablanca meeting with Roosevelt. Everyone expecting something to happen soon.

In preproduction Algar's two valves CV 55 and CV 8 are under the weather—and the CV 90 is not going very well in its teething troubles.

A meeting with Home Guard authorities to see how Research [Laboratories] and Union Works can contribute another 150 men. There seems no reason why they should not do so.

A visit from Air Marshal Babington and others re. the demonstration which Farren and Ackland Snow and the Air Force officers have worked up to demonstrate torpedo guiding by phase difference. As a laboratory demonstration it was excellent but I think there is a long path to travel between this and a supersonic [ultrasonic] system under water. I pressed Air Marshal Babington to contact Coastal Command and DSR Admiralty and get the proposition vetted by the existing experts.

Mr. [Dr.] Roberts and Mr. Carrol [Professor Carroll] from DSR are coming on Thursday morning.

Lunched with Dr. Railing and discussed 'staff' trade unions and particularly the Association of Scientific Workers with whom the Engineering Employers Federation are discussing recognition.

February 9th

A serious talk with Le Rossignol, Warren and Gardiner about the slow progress on CV 90 in preproduction. It was agreed that things would go better if Craig went over to this group for some weeks.

At the HO Battery Meeting (Heather present again) Sanderson exhibited our layer battery design using his 'tray' construction. It received general commendation. GEC has an order for 2000—to keep them in the picture.

A visit from Wing Commander Hiscox and Group Captain Evans from Fighter Command—interception experts. A very interesting conversation on their results and tactics. Hiscox could not see how the big searchlight was to be used operationally. He feared the same ill considered decision as had been taken in connection with Helmore's turbinlite where there was much wasted effort. I fear it is Helmore's salesmanship which is causing the trouble.

February 10th

Post-war lighting planning and Diamond Die Specification Committees all day. The latter is fortunately now ready to pass to the BSI.

February 11th[355]

Visit from Coales of Admiralty to ask us to take on (for Admiralty and MOS jointly) a sort of small GR [became CRF— Compact Range Finder]. We can probably do this but must make sure there is not conflict between this and any new submarine warfare work.

Dr. Roberts and Mr. Carrol came and we had our preliminary talk over certain detailed and fairly easy problems first and later probed him with questions which would help us to grasp the problem. It was a good plan of his coming here with fairly definite tasks within our compass. Two or three of us will now go down to Fairlie near Largs, on a return visit. Those present today were Marris, Farren, Espley, Le Rossignol, Cornock, Francis and C.C.P.

GEC Board meeting—the first after Lord Hirst's passing—and very impressive in tributes and resolutions for carrying on the tradition. A.H. Railing and L. Gamage appointed Chairman and Vice Chairman of the Board.

Welcome news from Chirnside and Dudding that Dauncey has at last started to make good sapphire boules like the Swiss. It is apparently a question of making our own alumina powder from the beginning using 99.99% pure aluminium.

February 12th[356]

A Consultative Board Meeting held downstairs to save Heather from having to make two journeys upstairs. We revived the original terms of reference of the Board and decided to ask the new Managing Directors to confirm them. It is very interesting to go back over the old records and realise from them Lord Hirst's sound sense.

Have been trying to find with Chirnside a place where his special boule alumina can be made. Probably it will be in a corner of one of the battery rooms. We have decided to accept both the submarine work and the extra GR job.

February 13th[357]

A meeting with Bell, Bligh, Williams, Humphreys and Gardiner re. induction heating at all frequencies. Gardiner had got out an excellent graphical chart showing the different fields of induction heating, associated with the scale of frequencies. This will probably be an active after-war development.

Chirnside had a difficult time at MOS yesterday when he went up at short notice to answer for Salford's stewardship on making boules. Fortunately he and Dauncey had just turned out their first successful boules the day before—and the immediate situation was saved.

The news from Russia is still of triumphs. The Germans are moving back rapidly to some line and all the Caucasus and much of the Donetz is now reclaimed. Stalingrad was a Sedan for the Germans and their prestige has suffered badly.

February 15th

Preproduction 1406 this week. Miss Lock has got out costs for period October, November, December 1942 on the new prices we quoted. They show we are all square within £4000 in a turnover of £90 000 which is pretty good. There are losses on several valves but the period is too short to split hairs.

Gossling has had two days at Malvern—also Warren—and talked usefully over various matters with many people.

We still get occasional complete collapses in one or another preproduction valves. This time it is the CV 8 which has been running 12 months. Recently it was the NT 99. After last weeks 'prodding' and adding Craig to Boyland for a time the CV 90 promises to have no major difficulties. Small assembly jigs and the like are what have been needed.

February 16th[358]

Visit from Prof. Carroll to tell me more about anti-submarine matters—as he did not feel that Dr. Roberts had been very expansive. He looked over Wing Commander St. John's arrangement which his assistant (Griffith) is working hard over. But we feel they are working without a bit realising the real difficulties.

An allocations, and preproduction meeting to follow—Brundrett, Truscott and Griffiths. Very cheery occasion with delightful flashes of candour. Brundrett referred to indiscretions of the Manpower Board and the likelihood of being able to protect skilled experimental glass workers. He accentuated the determination to call up everyone under 30.

February 17th[359]

Willett turned up first thing for help in 'RG' described as the best kept secret of the war, but I have known about it for some time. It is associated with seeing in the dark. Willett seems to be responsible for getting the vital part manufactured in a tearing hurry and EMI who have developed it can't tackle all the assembly because of shortage of workers. Several went to EMI to visualise matters. They have an immediate bottle neck in glass working which we have offered to relieve. But disastrous disturbance of our work would follow the tackling of assembly on a large scale. We may be able to do some of it.

A TVC meeting and then lunch with Blackett who is very outspoken and helpful in his war and its equipment commentaries.He raised with me after-war problems concerning the handling of all the scientific staffs swollen by war developments and the ability of industry to absorb them.

February 18th

A GEC radio co-ordination meeting with Heather again in the chair. A very frank and useful discussion. FM is active in many directions and there is a strong urge to take on some more old fashioned radio, i.e. low and medium power broadcast for the Dutch Government.

We are asked to give all assistance to Mullard to enable them to manufacture the CV 90 valve. It would be smoother going with MOV if Mullard had not in the past insisted upon one way traffic in their favour.

Serious trouble with AIS flashover phenomena in service. The voltages generated seem to be greater than expected and sets in operation are having many valve failures. There was a meeting here for four hours today of all concerned to chew over the difficulties—I am not sure that there is a royal road yet out of a very sticky position. It is too bad—but we hope for the best.

February 19th[360]

R.H. Fowler came today and we discussed counter-submarine measures and tactics. Fowler was very glad that C.S. Wright had adopted the policy of bringing Wembley into things. Farren is at Fairlie today. Marris, Espley and I are following in about a fortnight.

Col. Phillips (Managing Director of Research Enterprises Ltd of Canada) called today and was an interesting visitor, and capable.

Jennett is making many industrial contacts on quality control—the last being the GEC Magslip factory at Leicester. Each contact seems to induce several new ones.

February 20th

The need for using only home grown wheat has given a good opportunity to those who want to kill weevils with our special alumina. So Prof. Briscoe and Mr. Cobden Turner are now together in making plans for producing the special alumina powder.

February 22nd[361 - 362]

The news in Russia still keeps favourable. In Tunisia the enemy seems to be driving the 1st Army northward somewhat, whilst the 8th Army is also working northwards in their rear.

We have an order for several Oboe ground sets which will be made here with the help of Claud Gen.

Air Commodore Leedham accepted an invitation to lunch with me and I heard the whole story from the beginning of the amazing doings starting with the Du Parc report—a travesty of British impartiality and apparently a nasty piece of work—Leedham's dismissal by Sir Robert Renwick and his return as DCD, after some MAP officials must have realised the real character of their actions. Leedham tells me he is engaged in a detailed scrutiny of the TRE projects and programmes.

Farren has returned from Fairlie after studying the requirements for their audiometer, which he is now proceeding with without delay.

February 23rd

A meeting at MAP on high frequency heating of plywood seems to have regularised the position and to have started the necessary liaisons between the electrical and the aircraft firms. I hope we may have our contacts with de Havillands and with Phillips & Powis.

Valve preproduction finance has been put a little out of joint by our having included in our current receipts £24 000 for valves in stock which should have been regarded as in suspense. We are having to go carefully over matters to see how we stand in consequence. We shall probably set off some of our valve profit for the year against the item.

We are still in the doldrums over the CV 8 valve. The emission has gone completely to pieces and there is at the moment no clue.

February 24th

An RDF/CVD meeting all day. A good exchange of technical information particularly in the afternoon when Megaw gave some interesting accounts of magnetron work on behalf of himself and Willshaw. Lunched with Hecht.

Bell tells me that trials with the 50 AIS modulator valves CV57 under 'equivalent to operating' conditions show results very much better than the reports lead one to expect from the MAP testing centre. If what he reports is true much of our troubles with AIS Mark VIII would appear to be explained.

Chirnside and Cornock have their heads together over the patent situations on alumina for boules. I have encouraged Chirnside in the idea of making here sufficient alumina for the process of penicillin making. It appears that this might be within the compass of the Labs. It has to be very pure and perhaps 100 lb a week.

February 25th

Dr. Black is here from ADRDE for two days. He is going to USA on behalf of CVD.

Messrs. Grey and Butement were here yesterday to review some of Bligh's work—and to press the urgency of a set Bligh's group has in hand for fixing the desired frequency without messing about with successive approximations.

February 26th

Went to see Benjamin reflector etc. Works at Edmonton. Meetings at Labs. on radiation heating. We have strengthened up our group appreciably for tackling what is likely to be a large technical effort.

February 27th[363-364]

The BM 313 magnetron for ASVx does not appear to be a very firm rock on which to build—and although one or two samples of Willshaw's alternative appeared somewhat better, we have more work to do on it.

The CV 57 (Bell's hard modulator) for AIS is still much under fire. MOV are hopeful that a promising experiment of substituting ceramic for mica will turn out well. Meanwhile hot airs are blowing and lurid tales of short lives which we don't repeat in the laboratory.

PF development seems to be nearing the melting pot again—but a meeting on Tuesday will clear the air and show what, if anything, must be changed.

March 1st[365-366]

Preproduction (valve) numbers are not rising as we hoped—partly because existing types CV 8 and CV 55 are down through epidemics.

Had Sir Alex Ramsay and Mr. Ben Smith (MAP Chairman of Production Board) to visit—with Dr. Railing. An interesting occasion with clashes between the visitors as to the merits of private enterprise and state management. Mr. Ben Smith seemed well up in our various Government activities as he had met them in his departmental duties.

March 2nd

Marris returning from TRE has much to report as to various projects of 10 cm standardisation of T & R boxes and the like—and we are disappointed that GEC engineers are out of it and that EMI only seem to be associated with the projects. Oboe and Pathfinder projects are meeting with operational success and these seem to be our main forward interests. We have a bomb timing problem for another MAP Dept. It looks as if Edwards, O'Kane and George would remain on field work with TRE and not return here. I very much wonder how much such men are being spoiled for systematic research and development work in the Labs.

Lunched with Mr. Prain who consulted me about several things including Leeds—more diamonds per month from Wembley—trial of Cleminson 'ball' machine—quartz crystals etc. I joined the directors of the 'Selection Trust' afterwards at their delightful directors' lunch.

March 3rd

One of the PF meetings yesterday, when the air was full of alarms about radical changes of programmes and scrapping the valves just developed. But it was all traced down to a Mr. Lennox whose ideas were put where they ought to be.

Jenkins tells me that luminescent 'bats' for the Admiralty which we developed, are much favoured and 400 are ordered from us. They are used on aircraft carriers for hand signalling of landing aircraft.

A GEC liaison CVD meeting. Good attendance including Lewis of TRE and Black. There were no kicks but some useful points ventilated including the CV 57 for AIS, in connection with which Sutton of Bristol is being somewhat headstrong.

Had to go to an RCM meeting at Air Ministry with Lord Cherwell, Sir R. Renwick, Watson Watt and others. It was largely operational and reactions of jamming. We are to press on with as high a powered scheme as possible. Renwick dilated with satisfaction on the merits of Oboe and H_2S.

March 4th

Had visit from Col. Chapman of the new Tank Study Centre at Virginia Water to establish contacts and exchange experience. Also his assistant Capt. Brand. There are probably many ways in which we can mutually help.

The mysteries of the CV 8 seem to be taking refuge at the moment in some property of the pumps. Dudding is taking a hand, with others.

March 5th[367]

Horton, Coales, Shire and others came to discuss the new combined operation gun laying project. It looks as if it would come to us to make up as we judge best but TRE rather want to force the design into standardised component moulds.

GR is apparently now entirely on its legs technically. A change in the circuit has increased range and given a good performance over the whole wavelength range. So Espley has rid himself of quite some anxiety.

March 6th[368]

Bell gives an astonishing account of Mr. Sutton's rather pathetic and unwise attack on MOV at Hammersmith yesterday over the CV 57 modulator which is really nothing to do with him. There was nothing about the meeting which could be described as helpful either practically or psychologically.

Machin tells me that he and Ramsay have their measuring techniques for oxide cathodes thoroughly established and they are now about to start on their main programme.

I have asked Cornwell to convene a Laboratory Stewards general meeting for Monday morning. We must start these again now that the Labs. have taken up their war time form and have to work it to a maximum efficiency. We might well have done this six months ago.

Pressure for Oboe is strong.

March 8th

At the moment war news is rather stationary. Eighth and First Armies are prodding and being prodded in Africa and the Russians due to

SECRET 8 Mar. 43

DR. PATERSON, A.L.F.S.

This is just a note to let you know the position to date on A.L.F.S.

The week before last you will remember that I took the experimental apparatus down to Lark Hill xhsrexwx and officially handed it over to the School of Artillery with the approval of A.D.R.D.E.

While there, I instructed the Flash Spotting Group in the use of the apparatus, after which we took part in a tactical trial, allowing the Military personnel to operate the apparatus; we only stood by to make sure that they handled it correctly. We carried out day and night trials in this manner and we were pleasantly surprised at the aptitude shown by the newly instructed personnel. The apparatus worked quite well and Col. Hemming and Maj. Jessop were very pleased to take xyxxxxhxx it over as they are anxious to commence a series of tactical trials so that by the time the Pilot Models are ready they will have a thoxxxgx team thoroughly trained and able to complete the "User" trials quickly and efficiently.

I understand in strict confidence that Maj. Jessopx is hoping to be allowed to take two of the Pilot Models out to the Middle East to try out their possibilities under battle conditions. The Pilot Models will probably be finished about the middle of April.

. . . .

G.Phillips of A.D.R.D.E. who has until recently been associated with the A.L.F.S. job has now been transferred to Malvern on some other work. In this connection he wishes to measure some atmospheric transmissions over a mile range in connection with an urgent job which must be finished by 15th. March. Ixhxxxxxhxxxxhxx He has been here and talked to Waldram and myself and decided that the most suitable appa ratus is a photoelectric one which I have just had made up experimentally. In view of the urgency, I have agreed to take it to Malvern on Wednesday, show him how to use it and do some night test measurements with him after which I shall probably leave it with him for a week.

I hope that you see no objection to this proposal.

I shall be glad of some experience with this apparatus as I am evolving a design ultimately for use in the FIDO jobs, as I understand from Ryde that there is likely to be a demand for such apparatus.

Fig.36 Note from G.T. Winch to Clifford Paterson dated 8 March 1943, on 'ALFS'

weather are having to give the enemy time to establish themselves.

A Japanese convoy has just been completely wiped out.

Preproduction results today were up to over 1600 valves—a record. We resumed Laboratory Stewards meetings this morning and shall hold them monthly for the present. There is much to talk over. Lunched with Dr. Railing and discussed matters of administrative policy and the like.

The AIS modulator is still the major 'flap', but the ceramic insulator trials look promising—nine out of nine valves being better than all previous production valves.

Dr. Guy rang up and wants me to help him with an enquiry at Farnborough into the value of their electrical engineering programme and development. Sir Stafford Cripps has asked him to undertake the business.

Winch has delivered his two engineered flash-bang pilot models to Larkhill and saw them into proper use there. His report to me is attached. It all seems quite promising (Fig. 36).

'Wings for Victory' week started on Saturday. The Lab. target figure is £2500 and money is coming in fairly well. Mrs. Paterson and Mr. Ogden are handling the doings under Cornwell.

March 9th

Wings for Victory total £1500 already after three days. Good going.

Said goodbye to Schuil who leaves today for camouflage work in Mediterranean and W. Africa.

March 10th

A lively CRT meeting at which Mr. Bradfield seemed to get across most people, but did not get the support of the chair. The machinery for getting things done seems interminable at MAP mainly because TRE and RAE insist at each turn on every job passing through every bottleneck.

March 11th

The valve preproduction accounts for the period since October 1st have gone through various vicissitudes but I hope Miss Lock and Le Rossignol have straightened it all out sufficiently to show that the costs and payments break about square. This looks as if it would be the case taking the whole period from April 1st.

Meeting with Mr. Sheppard and Mr. Balmain at HO re. aerial

disinfection against germs. The medical people seem keen on this for school and industrial and transport vehicle application. The chemical treatment looks more important than the ultra-violet.

March 12th[369]

A Consultative Board Meeting at which many matters of Radio and Valve policy were talked over and the relation of the Laboratories' valve research to post-war orientation of the industry.

Had a two and a half hours talk in the evening with G.M. Wells of the extreme Left, and some of his colleagues with Le Rossignol and Gavin, on the laboratory structure and ideals and the impact on their trade unionism and particularly the AS[c]W [Association of Scientific Workers]. It was a very friendly talk designed to clear the air and establish an understanding of our respective outlooks.

March 13th[370]

Gossling reports a useful visit to Admiralty (ASE) at Witley where he went over the whole organisation.

Gavin is off to Dunwich on his jamming programme. He spoke warmly of our meeting last evening and its helpfulness.

Our Wings for Victory week at the Labs. has realised £5150—so doubling our target. Everyone has shown great enthusiasm and the result is a thrill all round.

Bell has just returned from an aerodrome in NE [SE?] England where he and Edwards and Aldous were examining conditions of use of CV 57 modulator valve in AIS Mark VIII—where over-voltages have been causing so much loss of valves. It is true the valves have a stiff testing in the set on the ground, and this can be somewhat ameliorated. The party came in for some real excitement for the squadron was up at night after raiders and had 17 contacts and brought down five.

March 15th

Went with Marris and Espley to Fairlie (Ayrshire) to Dr. Roberts Anti-Submarine Establishment to take our collaboration in this work a stage further. We were there by 10.0 am and spent most of the day with Dr. Roberts and various of his staff discussing the submarine seeking torpedo and the pros and cons of certain possibilities. It is agreed that we carry out with them some fundamental experiments to test the chance of dodging reverberations, which is the chief bugbear. We are also doing some minor jobs re. [e.g.] Farren on an audiometer. The new work will mostly come Farren's way. It will probably mean another development contract.

March 16th

Humphreys and Rycroft had success at Egham yesterday in putting forward a quick acting thermostat as a detector of flame for locating at vulnerable points in tanks. The authorities seemed delighted with it.

Preproduction last week was 1700 valves—a record.

March 17th

Bell tells me that the larger clearance CV 55 which was supposed to be a degraded version appears to be equal in performance to the fine clearance valve. If so it will mean a new lease of life for the CV 55.

Jesty is not finding the direction of things on CRTs at TRE and elsewhere particularly helpful. It seems to be guided more by prejudice and intrigue than objective fact which is all very trying—but it is not the first time we have had to try to carry on with incompetent leadership.

A CVD policy meeting at which Oliphant in his naive way said he thought everyone was getting stale in their jobs and ought to have a general post and that anyway the Birmingham people didn't like having to do the dull work of finishing off their researches, which I had to point out was much more difficult than launching the dash for glory idea.

March 18th

A very full TVC meeting with certain contentious issues which made me feel as chairman that I had some justification for being there. The functions of this committee become more and more responsible.

Lunched with Good—always inspiring—and discussed present and post-war issues—both IEE and industrial.

Air Chief Marshal Sir Philip Joubert came to see the Reading party's anti-submarine model. I took the opportunity to get him to ring up DSR Admiralty and invite him here to see the device. Joubert told me many things about the war and Marris came in to discuss AIS Mark VIII after three hours discussion at a big meeting to consider the modulator and its valve. I gather the meeting went over the ground covered by Bell recently at Fighter Stations and the improvements expected in the valve through the use of the ceramic insulators we are rushing through.

March 19th

As a result of Joubert's visit yesterday DSR is coming down on Tuesday to see the phase angle device of St. John's. Meanwhile we have confirmation from Fairlie of the fundamental phase angle experiment we arranged with them.

Bell reports a weighty meeting yesterday on AIS modulator. Dee was there and appeared to be impressed by the fact that we really had been doing everything conceivable to increase the value [valve] factor of safety.

Contract came in from MAP for Gavin's latest ground jamming set.

March 20th

The news generally is not so good. Everyone speaks as though the Allies are on the offensive and the enemy on the defensive but the Germans have done a powerful 'come back' in Russia, taking Kharkov and driving the Russians back a lot.

The Savings Cycle just starting has gone up from £115 to £163 per week. This has been helped by an increase in GEC War allowance just announced. But apart from that we had increased 30%. The collectors are a most enthusiastic group.

Megaw reports that he has had an invitation to serve on Appleton's committee on radio propagation—so absorbing much of the operational activity of the informal committee.

March 22nd[371-372]

Over 1700 valves preproduced this week but those we most want to succeed are not yet in the clear—CV 55, CV 79, CV 89, CV 58. This emission variability still dogs us.

We have had more Russians here today—from the trade delegation, including Mr. Borosenko—and showed them non-radio activities and some valves.

Gavin is negotiating his new air jamming equipment contract—probably £3000. It has been arranged after discussion with the Clarendon Labs. at Oxford.

March 23rd[373]

The PF programme has been undergoing modification once more on the basis of large rather than restricted use—but the valves themselves seem better than the circuit employing them. It means that Donaldson and those working with him have their work cut out to fill the programme.

A searchlight carbon committee at which, at Couch's request, we arranged for a comparison of the relative qualities of GEC and Morgan carbons.

DSR Admiralty, C.S. Wright, came to review anti-submarine possibilities with us and the MAP [RAF?] (St. John) party. A satisfactory conference. It was nice to see Wright here again. He looks a hard worked and an anxious man but keeps cheery and patient. Told me of his 'midnight follies' experience when he pulled to pieces certain treasured features of the organisation—but said he put in some very warm praise of our valve doings at Wembley.

Had long conference in afternoon with Marris, Espley, Macqueen, Peters and Jesty on post-war television and the alternative courses to follow.

Churchill has made a long speech on after-war intentions and prospects and gives 1944 (sometime) as a possible date for ending the European part of the war.

March 24th[374]

I looked over the ground jammer and receiver developed by Gavin, Houldin and Heathcote and about to be sent for trial to the East Coast—they seem sanguine about its effectiveness. On all their searching receiver techniques this Group has been receiving visitors from competitors to help them in our construction methods, i.e. Marconi and one or two others.

Dr. Bowden has been here from TRE on beacons and visit to USA. He is leading the expedition and is a very good type.

Oliphant has emitted a memo. on research which shows how we are poles apart in our views as to the place of research men in the war effort.

Industrial Heating meeting at which we reviewed the GEC metallurgical and furnace activities. It showed what steady and successful work has been going on under S.V. Williams and that Ivor Jenkins has been doing specially well.

March 25th

The general registration of engineers who have served apprenticeship is required, but the instructions are not capable of interpretation. So we are told to register 'everybody'.

The requirement that MOS should cut down its production programmes by 20% means I fear that 'flash bang' and similar developments are likely to be in jeopardy.

I think we ought to have a record of the devoted band of group leaders and assistant leaders who are handling our valve preproduction work amidst great administrative and technical difficulties. These are appended herewith (Fig. 37). In six months time they may have recrystallised into new groupings.

March 26th

Discussed possible Workshop personnel re-arrangements with Gardiner and Warren to meet present war pressure on Gardiner's own office.

Lunched with Watson Watt who had a mission with me on disclosures of CRT information in parallel with EMI who have a contract from MAP. I could not quite gather the inwardness of his request but he concentrated on our disclosure of luminescent powder techniques, to which I had to counter with a request for information as to why such disclosures are desired. It ended by his promising to spend a full half day here studying luminescent powders.

A diamond die meeting this afternoon at which we considered the means of speeding up general production of sizes from 0.0015 to 0.004. There appears to be a prospect of a serious shortage.

March 27th

Full details are now being worked out by Marris with TRE of our extended Oboe programme.

The temporary workshop changes seem to be working out—with Foxon joining Gardiner's more administrative staff and McLeod taking Foxon's job for the time being and Organ being charge-hand of receiving valve workshop in place of McLeod.

March 29th

Good news from Tunisia—Rommel has had to give up the Mareth positions.

Preproduction 1820 valves last week—a good ending to the financial year and record for any one week. Chirnside's oxide coating seems to be producting good valves where the factory oxide could not, but we have failed to discriminate any difference in composition between them.

31.3.1943.

Valve Pre-Production.

Supervising Committee meeting. Dr. C.C. Paterson., every Monday morning at 8.45 a.m. Dr. Dudding, Messrs. Holland, Francis, Douglas, Pearce, Le Rossignol, Gardiner.

Valve Group	Leader	Assistants	Supervisors	No.of Oper. F=Full H=Half	Equiv. full time no.
Triode Group N.T.97,99.	Mr. Wright	D.H. Read (Tech.)	Miss Shaylor " Harris	23F 18H	32
Med.Triode CV55 Diode C.V.8	Mr. Algar	A.Cutting C. Chant	Mrs. V.Turner	12F 19H	21½
Rectifier Group. N.G.T.8 C.V.12	Mr. Douglas	G.Hubbard(Student) K.Parrish(") Mr. Walsh(Tech.)		22F 18H	31
Small Triode C.V.52.	Mr. Douglas	Mr.Southon Mr. Lane		13F 22H	24
Receiving Triode CV90& derivative.	Mr. Boyland	Mr. Armstrong. (Part time)		6F(Inc 2 boys) 3H	7½
Glass Magnetron	Mr. Boyland	Mr. Stannard. (Temp.)		2F	2
Magnetrons.CV38, 56,64,69,76.	Mr.A.G.Pearce	Mr. Brewer Mr.Hartnell	Miss Worth	17F 18H	26
C.R.T. NC12 VCR87, VCR514	Mr. Jesty Mr. Harvey	Mr. Nixon	Mrs. Freeman Miss Gornall (Clerical & Dispatch)	19F 8H	23
CV114 & 115	Mr. Harris Mr. Shaylor	Mr. Hills	Mr. Knott.	1F 6H	4

Fig.37 Note dated 31 March 1943 of valve preproduction group leaders and staff

Valve Group.	Leader.	Assistants.	Supervisors.	No. of Oper. F=Full. H=Half.	Equiv. full time no.
Hand Modulator. NT100	Mr.Solomon.	Mr.Armstrong. (Part time)	Mrs.Smale.	19F 7H	22½
CV58	Mr.Peachey.		Mr.Brooks.	10F 17H	18½
Crystals. Red, Orange, Green Star	Mr.Ryde. Mr.F.H.Brittain Dr.Ransley.	Mr.D.E.Jones.	Mrs. EYre.	4F. 17H	12½
Testing.	Mr. Francis. Mr. Holland.	Mr.Cheeseman. Mr.Willmot. Mr.Hammett. Mr.Fricker. (Mr.Falconer) (Mr.Isaacs.) (Mr.Willoughby)	Miss Fildew. (Clerical work re contracts etc) Technical Supervision.	20F 16H	28
Packing.	Mr.Holland.		Miss Parslow.	5F 5H	7½
Press Work.	Mr.Gardiner.	Mr.V.Williams.		6F 7H	9½
Anode Rolling.	Mr.Gardiner.	Mr.Hobday.		2F 15H	9½
Inspection & Gauging.	Mr.Heywood-Hunt.	Mr.A.D.Allen.		11F 12H	17
Glass Workers on Production.	Mr.Breadner.	Mr.Sims	Miss Butler.	21F 10H	26

Fig.37 continued.

The CV 55 optical projection device for centering electrodes is just about ready to be used by the group and looks as if it would be a success. Haines and Herring took over the design from Turner in the Illumination Group.

Chelioti came for the whole afternoon to see round things of interest—the latest in diamond dies and cathode ray tubes.

March 31st

Two post-war planning meetings on research at IEE and FBI. It is difficult to meet the claims for time of both war and peace at the same time, but a certain amount of peace-time work has to be done.

A valve preproduction meeting with Truscott and Brundrett—at which the needs of CV 90, 89 and 99 were pressed—for we are somewhat behind our promises. The attached schedule forms a record of the devoted band of workers on valve preproduction. Some 30 members of the scientific staff and as many assistants as supervisors and about 300 equivalent operators many of whom count as two half timers.

April 1st

The changes in the workshop seem to be generally approved by all concerned and I hope will introduce the fresh outlook which always comes with changes. The die for the ceramic insulator for the CV 57 (AIS modulator) is complicated but I hope will be in use today so that the new and better CV 57s can appear without delay.

Had Mr. Coales from ASE to see me with Megaw. Coales is wanting us to make some propagation experiments to see what conditions exist near the surface of the sea for low angle RDF ranging. It looks like a case of tackling it quickly now, before the approaching propagation work begins under Mr. Cobbald's committee.

April 2nd[375]

DSR rang me to say that the way was clear for our collaboration with Fairlie and a formal contract for general work would be negotiated at once.

Jeffcock called today for initiation into RCM work with Gavin and backed our proposal for a resumption of the Wembley RCM meeting/visits.

Had an 'RS' [Research Services] meeting at HO with Dr. Railing,

Gamage and Heather and went over the list of the more interesting items of the past month. Railing and I discussed *modus operandi* for collaboration on plastics with Mr. Barron.

Breadner reports entire satisfaction by the Gramophone Co. on the use of W1 glass for our 'RG' seals. It has apparently already reduced their breakage.

April 5th

There is an air of suspense for we are all expecting a W. European fighting front to start up at any time and no one knows where! So the atmosphere is one of lull before the storm. The RAF are hammering hard.

Another record preproduction week: 1960 valves. This will be difficult to beat and has some fortuitous features in it.

Oboe is being pressed hard but the T^2R box is being passed to EMI to develop on 'standard' lines. Winch reports moderate progress with his flash-bang and three pilot models. They are not showing by *daylight* the range promised by earlier experiments. He is going to make a new attack with improved photocells.

A TVC meeting this afternoon. These become more important and more interesting. The CV 55 and the CV 58 were in the limelight. It looks as if we should have to make the small glass magnetron in the Labs. (CV 79, 89) as it looks as if the factory would be a long time getting under way with their design.

April 6th[376]

The rush insulator for putting the AIS modulator valve (CV 57) on its feet has come out to plan and MOV are now making up valves with them. Good work by our craftsmen, Foxon, Williams and others.

We hear a ruling has been made which virtually protects all our student assistants. Miss Winton also has indefinite reservation—so we are getting on.

Dr. Constant of Farnborough came today. He is expert on aero engines and consults with us on Erith problems.

April 7th

A day at CVD (RDF) Committee—receiving valves in the morning and transmitting in afternoon. Pleasant meetings but not very eventful.

CV 55 in its larger clearance improved form was to the fore. It appears that for most purposes the larger clearance valve is quite as good as the fine clearance one. This is probably Warren's last appearance before leaving for USA.

All his reports on the CV 90 and its legitimate and illegitimate children were favourable. In preproduction this valve is steadily working through its growing troubles and has still some way to go.

April 8th[377]

A radio co-ordination meeting at HO at which our post-war programme on radio transmission and television came under review. Maskell has prepared a review memo. of all possible activities and this will be studied piecemeal at future meetings.

Winstanley discussed with me electrical cooking equipment for workers' houses and some new and good ideas which he has.

We are in the midst of elections to the Leading Staff. The number of available men is great and I am not sure how the appraisement is going to come out but I feel thankful for our system of election, as against appointment. It is a good one.

April 9th

These days go by relatively quietly. Plenty of visitors from the Services today about our many jobs. A new one from Carrol of DSR on anti-submarine and also Captain Cooper of a similar department to acquaint himself regarding our programme on this subject.

Coales and Shire here on the modified GR combined operations unit.

A Consultative Board meeting at which I described our internal laboratory structure and its impact on senior staff prospects through their having no 'titles' to their posts etc. It is quite interesting talking over the form of our organisation which has been the result of natural growth and evolution and regarding which I am having awfully appreciative letters from members of the Leading Staff.

April 10th

Ramsay reports that the gas triode is coming to the fore again on account of success in the USA with the hydrogen filled triode. He finds that his E 1290 triode appears to work well when filled with hydrogen and I have encouraged him to follow this up energetically for a few weeks and see how the matter develops.

Attended a radio educational conference under Lord Hankey's chairmanship. Sir Stafford Cripps spoke—not very inspired I thought. Lunch after at which I sat next to Air Commodore Addison. He tells me that a very favourable report has been received on the operational value of Mandrel, to the disconcerting of Mr. Jones [Dr. R.V. Jones?] (Lord Cherwell's protegé) who somewhat prematurely told him that Mandrel was no use and ought not to have been used! It is very nice to hear of the success of a weapon upon which so much crash effort was expended here and at Claudgen.

April 12th[378]

I thought our last week's valve preproduction figure was an artificial record but this week we reached 2150!

The 'CRF' job is the follower up of GR and we are searching for members of the groups to handle this and also the anti-submarine jobs which are rising in a crescendo of work. A request that we would tackle aspects of the phantom submarine problem came out of a visit of Mr. Lemon last Friday.

Warren did not leave after all on Saturday for USA.

April 13th

I have spent all day with Dr. Guy at the RAE enquiring into the use of manpower in their Electrical Dept. (Dr.Bairstow). I hope I was able to help. My main criticism was the non-use of junior girls of the WAAF (matriculated) type. The most interesting item I saw was the use of high frequency for engine ignition. Apparently the experiment on an engine using very small currents which Campbell and I published from the NPL in the last war used—according to RAE—a very high frequency and hence an inadvertent result.

April 14th

Discussed with Le R, Marris and Francis how we could collect a group to work on the new AS job. Our inclination is to get Proctor, Dauncy and Sloane with Brittain's consultative help to handle the matter attached to Espley and Marris.

The Oboe job had another heavy prodding yesterday at a large meeting in town and the question of how it was to be made was canvassed. There is every indication that MAP people look upon Coventry as just a sundry model shop job establishment—which makes everyone somewhat anxious.

A GEC CVD liaison meeting with Brundrett presiding. A very happy occasion with plenty of discussion but few kicks. Had heart to heart talk with Dr. Lewis afterwards on matters involving certain personalities on our two sides—including how to progress CRT work in the light of the present people who have fingers in the pie.

April 15th

Hopkinson and King Brewster have been using the Helmore turbinlite with an infra-red screen for taking dark photos. It appears only suitable for static use as exposures required are too long for airborne use.

Our getter production is now being required for supplying Cossor with a large quantity and Britten tells me that special measures will have to be taken to increase production. We moved part of it to the Exhibition Building—but the labour difficulty is always a handicap to expansion there.

Oliphant has been here all day and brought with him Commdr. Whitaker, RN, who had much to say about maintenance of radio equipment in the Eastern Fleet. Simplicity is the golden rule.

April 16th

We had our meeting on AS phantom submarine job today and agreed Dauncy, Proctor and Sloane with all the others including Ryde and Brittain to advise. It is linked up to Fairlie through Marris and Espley. Ryde and Marris threw a good deal of light on the essentials of the problem.

Sir R. Watson Watt came to lunch and to spend the afternoon on luminescent powders and their application to CRTs. He is most conscientious in looking into everything he is shown and I cannot but believe that he really enjoys going into things. He was told some of our difficulties—mainly political and I hope good will come of our talks. The whole story from the powder to the final appraisement of screen quality is very impressive to me and much credit is due to all concerned in both groups. Jesty and Ryde and Jenkins made good showmen.

Contract cover with DSR Admiralty for AS is likely to be about £12 000 per annum.

April 17th

Ramsay and Machin are making good progress with their fundamental research on oxide coated emitters. They do not appear to follow the usual laws of other emitters and a new basis for design and performance would appear to need working out.

Espley has auto frequency control for GR now working satisfactorily. It is a mechanical automatic adjustment.

April 19th

Preproduction still goes well. The CV 90 is about 40% efficient [40% yield?] but most people are sanguine about its prospects. Even the CV 79 and 89 are looking up somewhat. They appear to depend on a proper insulator to reduce losses.

A Glass Advisory Meeting at which glass breakage was the main topic—rough and careless handling—a psychological problem. Partridge showed a courageously conceived molybdenum spider and extruder die for quartz tubing. It was promising enough on its first trial to make people sanguine that the scheme will work.

Discussed with Chelioti the management of the site committee and the status of the committee—whether advisory or executive. Both Phelps and Warren are now retiring, which leaves a more junior set of managers to represent the establishments.

April 20th

The first of a new series of RCM meetings with Service people at Wembley. I have not yet heard from Gavin how it went.

Wing Commander Cooper and Mr. Holt Smith from DCD department had a good look round the Labs. today. Mr. Picken was also here for the day with Gossling.

The major event for me was a long estate meeting—the last, I regret to say, at which Mr. Phelps and Mr. Warren will be present. I am chairman this year and expect to have plenty of situations to help to handle.

Tomorrow I go for two days to ASE Haslemere and then for a few days break and rest over Easter.

April 27th

News from Tunisia indicates the closing phase—but it is rash to prophesy. The Germans are fighting to the last but are losing ground.

Donaldson reports good shooting trials on PF at a target using the 15 mm valves we made 12 months ago and which were so abused by

Dr. Black at the CVD Technical Committee! The pressure on us for 10 mm [diameter] valves is very strong pending production at Hivac.

The CRT meeting which I missed last week went I understand very well, mainly because Dr. Lewis of TRE was present. Lewis is undoubtedly one of the level heads and honest collaborators. The great thing is to make sure that he has all the facts right.

We have an enquiry with Levy and West, at request of Watson Watt, to look into luminescent powder stocks for all users and Mr. West is drawing up a preliminary memo.

Farren's anti-submarine work is gathering strength both with the homing torpedo (with Group Capt. St. John) and with the fundamental conditions in conjunction with Fairlie.

April 28th[379]

Tragic news that A.E. Schuil has been killed whilst on his special Admiralty mission to the African theatre of war. It was only about a month ago he came to say goodbye in his new naval lieutenant's uniform. He had earned golden opinions from the Admiralty for his camouflage researches on ships and he was out East on the same work.

We had an internal anti-submarine meeting this morning to clear our ideas on the phantom sub. problem.

It seems clear that our first job is to make a laboratory apparatus to enable tests to be made of reflection characteristics without having to do them at sea.

A CVD policy meeting, very quiet and relatively short—but we learned that G.P. Thomson's OPTEC committee of the Radio Board is dying and will no longer function. Some say this is due to intrigue and others to a habit which the chairman had of dragging out proceedings and tiring members of his committee. It is sad, for this committee had important functions.

April 29th

Ramsay reports further promise in connection with his hydrogen filled E 1209. EMI are very interested in it and also the Admiralty. The high repetition rate of which it is capable gives it a considerable advantage. The life is at present of the order of 250 hours.

Magnetron research with us has been having very bad luck compared with that at Birmingham, who have been successful in bringing off many striking successes. Their latest is the copy of the dwarf high efficiency USA type which seems to promise useful advances. Their very high

Church House,
Dean's Yard,
S.W.1.

Ref. CCE. 3/7 &
4S/31.

29th April, 1943.

My dear Paterson

I feel that you ought to have more than the official acknowledgment of Report No. 8193 on the Valve Type CV.55 (E.1190) which you sent us. I should be glad if you would say to your research staff how grateful we are to them and their co-workers for the very concentrated work they have put in to produce this Report in such a remarkably short time.

I do not pretend that I am competent to judge the content of the Report, but I have read it with great interest.

Thank you for your continued help in this as in so many other things.

Yours ever
Robert Watson-Watt

Dr. C.C. Paterson, O.B.E., F.R.S.,
G.E.C. Research Laboratories,
Wembley,
Mdx.

MAY 1 1943

Fig.38 Letter dated 29 April 1943 from R.A. Watson-Watt to Clifford Paterson on Laboratories' Report on Valve Type CV 55

power magnetron has also beaten us in time, although ours apparently promises a better performance.

April 30th

Admiralty are wanting urgently a copy of the special telephotometer which Schuil devised and made for his camouflage researches. It is going to be difficult to manage although Waldram and Morse have handled Schuil's instrument. It is said that Schuil was making another telephotometer and had some unfinished parts at his house.

I was glad to take a share on the IEE Council yesterday in getting Megaw nominated for the next Council.

A meeting here today of Kodak and MAP and others to consider use of the now grounded Helmore turbinlite for infra-red photography.

An amazing piece of effrontery by the Office of Works and Planning in demanding, without pre-consultation or any preliminary, possession of a large piece of the Laboratories' frontage for a huge water container, placing it right in front of the architectural elevation of the place. We are handling the matter diplomatically and trying to suppress our abhorrence of such Nazi methods.

May 1st[380]

It is impossible to keep up to our full output of diamond dies without one or two reserve personnel to substitute for those away ill. There is always someone away. Efficiency is improving with the quick rough drilling machines and more effective polishing machines. Apart from other dies for copper and large tungsten we turn out some 25 dies per month between 0.0004 and 0.001 [inches] which is equivalent to three times that number of larger dies. Nevertheless the control are pressing us hard for more output.

Cornwell has met the Ministry of Works man who wanted to commandeer our lawns in front of the Labs. for a 100 000 gallon water tank. He has demonstrated the folly of putting such a mass of water as this close to our basement and the search is now for a suitable position on the undeveloped frontage of the site.

Mrs. Harvey of the Factory Department was here this morning to see me about the relation of our preproduction localities to the Factory Act. I suggested a half and half arrangement with them.

The 'election – selection' for men to be invited to join the Leading Staff has resulted in a decision to issue eight invitations.

The following were clearly indicated: Cornwell, Chirnside, Cameron,

Gavin, Jesty, Stevens, Rooksby and James and I shall be seeing them next week about the matter.

A very pleasant note from Watson Watt in appreciation of our reports—attached herewith (Fig. 38).

May 3rd[381]

Over 8000 valves despatched from here in April—a record.

Winch tells me that he is being pressed to develop his 'flash-bang' device particularly for indirect [flashes in] daylight since the direct feature can be met to a large extent by hand working. So he has further research in sensitivity in order to get his gun flash to register in brighter sunlight.

Mr. Grey from MOS (Radio Development) called on me to impress on me the importance of the 'anti netting' research he has in hand for tank sets.

It is part of an Army series of developments for the proper engineering of their whole signals equipment. It is apparently the stickiest of all the problems and is up against [to?] the GEC.

A TVC meeting of a more difficult nature than usual; late submissions and crash programmes plus acute shortage of valve production capacity are the ingredients of frayed susceptibilities. We finished with a scrutiny of British and USA equivalent types with a view to a comprehensive schedule.

May 4th

A household appliances meeting at HO to review postwar projects—and showing our latest properly designed radiator for 180° equal heat in the horizontal plane—mainly Rycroft's work.

A post war Lighting Committee (of the Ministry of Works & Planning) at which we passed the artificial lighting sub-committee's report on domestic lighting for small houses.

May 5th

A very warm [copy] letter [received] from Prof. Swift (Engineering at Sheffield) appreciating Dudding's series of quality control lectures at the University and praising the GEC attitude in facilitating such activities (Fig. 39).

Mr. Gamage came to lunch and for the afternoon. He spent his time with B.P.D. on quality control, Chirnside and Dauncey on alumicide and

COPY/

Newton, Chambers & Co., Ltd.,
Thorncliffe,
Nr. Sheffield.

9/LAB/P

1st May 1943.

Dear Professor Swift,

"Statistical Methods in Industry".

I attended the last of these lectures in the University last night and I feel that I cannot let to-day pass without sending you my best thanks and those of the staff of Newton, Chambers & Co. for the privilege you gave us of attending.

The University is doing a most useful work in laying as it were the foundation stone for the avoidance of waste in industry. For my part it is the exact thinking which the use of Statistics necessitates which is so valuable.

I should have much liked to have thanked Dr. Dudding personally from the audience, on the occasion of the vote of thanks, for his personality had so very much to do with the success of the Lectures.

My knowledge of the subject is a theoretical one and, accordingly, I was particularly interested by the slick way in which the Lecturer handled the masses of results.

I talked to the Coke-Oven Managers Association in March last on the "Use of Simple Statistical Methods" in the Coke-ovens Industry. If the "Gas Journal" should publish anything of this I will let you have a copy to send on to Dr. Dudding. The remarks I made were of no theoretical interest whatever, but they do indicate that there is no end to the application of the methods so admirably described by your Lecturer.

Yours sincerely,

(Sgd) A.H. Dodd.

Chief Chemist.

Fig.39 Copy of letter dated 1 May 1943 from A.H. Dodd, Chief Chemist, Newton Chambers & Co., to Professor Swift, Sheffield University, on B.P. Dudding's lectures on 'Statistical methods in industry'

boules—Cooper on Fido and other details. A very appreciative visitor.

Our new members of the Leading Staff have now accepted.

May 6th[382]

GEC RL allocations this year look as if they would come out much as last year—but the great question has not yet come out of the bag what are our overheads and how do they agree with the 180% applied throughout this year 'on spec' to our large Government transactions.

The large number of young fellows we have taking exams is causing inevitable trouble through shortage of staff on highly urgent jobs. PF is one of the cases and on Donaldson's suggestion I am writing Prof. Cockcroft to see if we can have here one of their men to join our group.

Wing Commander St. John on the AS weapon is not allowing difficulties to stand in his way and wants to send two mobile workshops and two mobile drawing offices here to work with us on the job. Good work.

May 7th

Flash-bang contract is being extended. Winch was at Larkhill and showed in trials the possibilities of telescopic reception for increasing sensitivity and making reception of indirect flashes possible even in very bright daylight.

He has also promising results on his IR 5 project i.e. all round infra-red signalling from the masthead. Over two miles is apparently practical. Altogether a good episode.

Visited, with Leeds, Messrs Temco at Lydbrook near Ross-on-Wye with the rest of the members of the diamond die technical committee. We had a meeting in the evening at Royal Hotel, Ross-on-Wye and [which] seemed to be voted by all as one of the best we have had for the exchange of experience and general helpfulness. I made a suggestion which to my surprise was cordially supported. Each maker of diamond dies is to give returns which will enable his efficiency in die 'units' per worker per 100 hours to be determined. Each size of die will be assessed as x units the sizes between 0.004 and 0.006 (about) being unity. The scale of units which Leeds and Simpson are proposing is to be circulated for comment—and once agreed we hope to have a yard stick for each factory to assess itself as against the other. The letter from one of Leeds diamond die customers—Messrs. Connolly—has given him encouragement (Fig. 40).

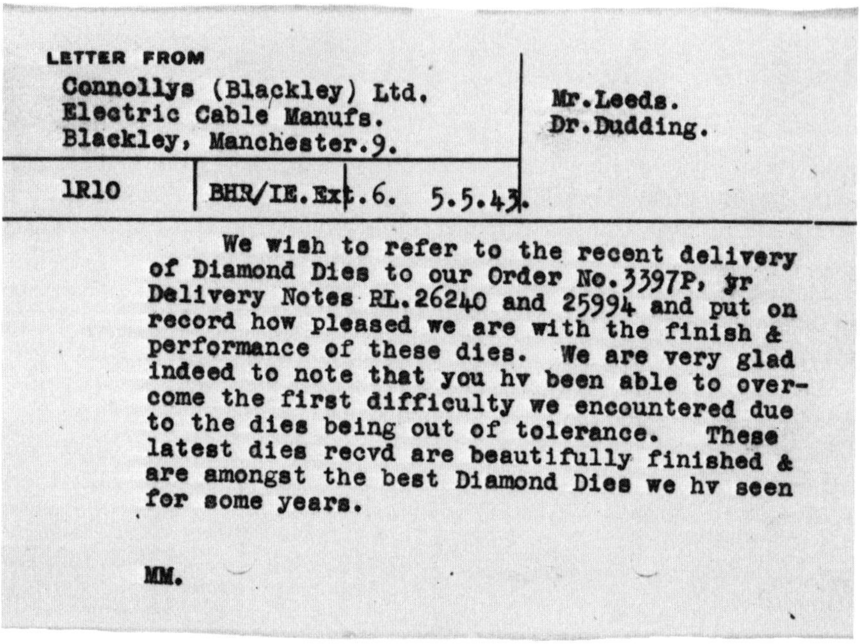

LETTER FROM

Connollys (Blackley) Ltd.
Electric Cable Manufs.
Blackley, Manchester.9.

Mr.Leeds.
Dr.Dudding.

1R10 | BHR/IE.Ext.6. 5.5.43.

We wish to refer to the recent delivery of Diamond Dies to our Order No.3397P, yr Delivery Notes RL.26240 and 25994 and put on record how pleased we are with the finish & performance of these dies. We are very glad indeed to note that you hv been able to overcome the first difficulty we encountered due to the dies being out of tolerance. These latest dies recvd are beautifully finished & are amongst the best Diamond Dies we hv seen for some years.

MM.

Fig.40 Precis of letter dated 5 May 1943 from Connollys (Blackley) Ltd. to R.E. Leeds on quality of diamond dies produced in the Laboratories

May 10th[383]

The news from Tunis is marvellous—complete collapse of the enemy armies. A good opening to the summer campaign.

1850 valves last week. All types going reasonably well according to their various stages of development. The investigations on cathode coatings show definite differences between batches of Chirnside's powders—which seems to give hope of a clue for further advances. But the general level is greatly improved already.

Layer batteries are being made at Wembley and Witton, and Nattrass is at the factory helping to get going there and is doing well.

Oboe crash contract is now taking shape and RL will take responsibility for technical direction whilst Reliance Telephones will do the assembly.

May 11th

Discussed with Bligh his resources for tackling the MOS tank set contract to avoid 'netting' as desired by Mr. Grey—there is not much we want from the Ministry and we know our programme.

Kersey has his 1942 – 43 preliminary accounts statement. It looks as if we should come out reasonably right but I can't tell 'til our valve preproduction statement by Miss Lock is complete. It is intriguing these days when the annual accounts sift themselves out for we have no profit to take up + or - balances.

Our 28th CVD report is just out and shows as much activity and progress as usual. B.S. Gossling is editor-in-chief and has to collate everyone's contributions.

May 12th

Ryde reports a marked improvement in crystals by using a 'C' process alloy which has put up the effectiveness by some 2 decibels. The work indicates that there may be still further progress, but as we are not far from the theoretical best possible it will be achieved by cutting down the downward spread of distribution.

Allocations and preproduction meetings with Brundrett and Truscott. A happy occasion partly because we are well on top of our guarantees.

Excitement just as I left this evening with Dudding by Rooksby's and Chirnside's reports of an oxide coating powder for cathodes higher than anything so far experienced by making up the powder in a different physical form. But we do not build our expectations on these sort of 'first night' results.

May 13th

Sir Robert Renwick in touch with us to hasten Oboe and get it into production.

Saw Mrs. Schuil today who has come back to live in their own home. How universally the death of her husband is deplored.

Lunched with Dr. Levy and Chelioti to talk over his USA visit. He seems to have had a rough reception from the IGE [International General Electric?] over there. We agreed about preparing a recommendation re. stocks of luminescent powders for this country and also on transmission of information on powders to Australia.

The fight in Tunisia is over and there is no longer an enemy there. I fancy the quick collapse is going to prove embarrassing to our military time table!

May 14th[384]

Oboe to the fore again—MAP do not like the idea of Reliance Telephone and RL doing the job and want it to go to Coventry, so there is another twist around the loss of time!

Bell reports favourably of the CV 57 modulator position—the lower insulator when degreased behaves according to plan and I hope the valve is now OK for its job.

Lunched with Good to talk over diamond die specification and taking over by the BSI of its finalisation.

Visited Air Vice Marshal Tait at Air Ministry with Jesty to try to get a move on with our study of operational requirements of CRTs. Tait gave us a helpful reception and got Hecht to use him to cut any impeding red tape.

May 15th

Aldous has been working with Bell on the CV 57 (hard modulator) and thinks the valve is now equal to saving the situation—the length of life is so far uncertain, but the actual required length of life may not be more than 50 hours flying. MOV have had instructions to put any other product on one side however urgent if it will help to increase production of the CV 57—verily a crash situation.

Oboe No. 2 [Mark 2] which Marris is handling is also No. 1 priority of all MAP projects. The urgency comes from the likelihood of Oboe No. 1 being jammed. Espley gave me an interesting account of his solution of the blind landing problem—it appears to me very simple, and worth while following up. We agreed that if put up to MAP or its establishments pride and prejudice may tend to deflect their opinion from the purely objective path. So we shall probably try to work it out here experimentally using the sports field in place of an aerodrome and a car in lieu of a plane.

If it works up to expectations it can be combined with Heller's altimeter to make a very simple combined instrument.

May 17th

Preproduction over 2000 valves—but there is still a mystery hanging around our costs for the last three months, which has to be clarified. We have an apparent loss to account for.

Advisory Board meeting this afternoon at which valves were the principal item—the urge being to get in on the small button seal miniature type such as are designed for pack sets.

May 18th[385]

The Illumination Group has been doing good work on flares dropped from aircraft—an investigation fostered by Lord Cherwell. Waldram has photos of gear at Knockholt for getting polar curves of light distribution from the new shielded flare. Beggs and others have been engaged in this.

Attended a conversazione by the Americans on Radar (RDF) and was glad to meet all the radio people there (Fig. 41). I hope I was able to put one of Gavin's troubles over RCM right by speaking to Air Commodore Addison asking him to continue to treat Wembley as one of the family and not as a 'mere contractor'. I think the trouble arose because of a difference of opinion between the Cabinet influence plus the Clarendon versus Air Ministry Signals plus TRE.

May 19th[386]

A main RDF meeting at MAP—but not very eventful. I constantly feel regret that GEC is 'out' of the velocity modulator field which is completely absorbed by EMI and we are now virtually out of any magnetron work below the 10 cm region.

Discussion with Sir Stanley Angwin and Coventry people at Post Office on how to get telegraphic picture transmission sets made—everyone trying to pass the burden of engineer[ing design] designing onto the other. Meeting fixed at Dollis Hill for Monday. Our Illumination Engineering

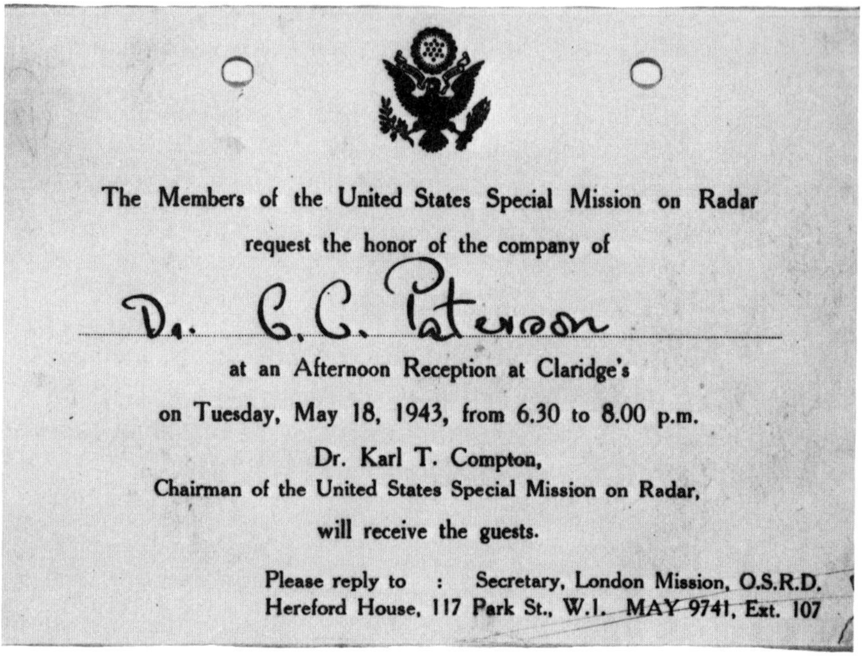

The Members of the United States Special Mission on Radar
request the honor of the company of

Dr. C. C. Paterson

at an Afternoon Reception at Claridge's
on Tuesday, May 18, 1943, from 6.30 to 8.00 p.m.

Dr. Karl T. Compton,
Chairman of the United States Special Mission on Radar,
will receive the guests.

Please reply to : Secretary, London Mission, O.S.R.D.
Hereford House, 117 Park St., W.1. MAY 9741, Ext. 107

Fig.41 Invitation from US Special Mission on Radar to Clifford Paterson to attend reception on 18 May 1943

Dept. may be able to help. Although the job comes with impressive priority sanctions it seems to me to be mainly needed for Press propaganda and so it is one of those many cases in which the decision of what relative priority is in fact given to a development project is left to the industrial establishment to decide.

May 20th

A TVC meeting with many projects to consider and some knotty points. But the meeting is enthusiastic and gets through plenty of work.

Went to Royal Society to hear Professor K.T. Compton on American organisation of science for the war. A full attendance but I was so sleepy due to disturbed nights with raid alerts plus the effect of lunch that I took very little of it in.

May 21st[387]

Air Commodore Addison visited us yesterday to see Bligh's FM multichannel set and also to see Gavin. The latter had a happy time with him and all clouds and misunderstandings were removed arising out of an incident at Drifford [Defford?] aerodrome when some rough and unwarranted things were said to two of our people, Drs. Houldin and Lamont.

Mr. Sims was here with S.V. Williams today. The special tungsten alloy and Heavy Alloy progress satisfactorily so far as production is concerned. But investigations of new alloys needed for more forward development go very slowly. Mr. Sims hardly understands how impossible it is these days to undertake investigations which are not connected with some immediate war effort.

Edwards looked in from TRE where he is still active in helping to understand and adjust our AIS and Oboe equipments under operational conditions. He is doing a very good job there and is a valued ambassador of Wembley at the experimental and operational establishments.

A searchlight carbon panel today at which Cameron was able to show, and ADRDE and Vernon confirmed, that GEC carbons for searchlights were not inferior in burning rate, light output and strength to those of the Morgan Crucible Co.

May 22nd

A letter from Warren from USA giving plenty of information about receiving valve matters over there. The Navy Dept. seems to think that it is all right for USA lighthouse valves and will not want the CV 90. But plenty of people over there are very interested in the CV 90.

Ryde showed me a brief report from Bell Labs. on attenuation at 1 cm in which their practical results agreed extremely well with Ryde's calculations from his theory. Truly a triumph for him.

Discussions have been taking place re. the Helmore large searchlight for which 6 metre mirrors are being talked about! However the arrangements for the six sets are going according to plan which pleases everyone concerned but we all feel that this is a heavyweight programme but with chancy success and in any case is a defence weapon as against offensive devices, which we prefer to be working at.

Megaw went over the CV 79 and 89 (glass magnetron) valves with me. MOV do not after all seem so anxious to throw over the project entirely but the rate of production may be reduced. Our own valves made at RL at present show considerable wastage owing to low output—at present a mystery.

After the Tunisia collapse of the Axis there is at the moment an hiatus with violent activity of our Air Force against the Axis aerodromes and air resources.

May 24th

Professor A.V. Hill here all day. An interesting visitor whom I was glad to give an opportunity to learn something of the place.

Preproduction was about 1850. CV 90 is going on well and getting over its troubles. CV 79 is slower in its progress.

A workshop craftsman's lecture by Dudding and Jennett on quality control.

May 25th[388]

To the Exhibition Building which is proving useful for many purposes. The CV 114 is going there for preproduction at the rate of 100 per week.

Had considerable difficulty in trying to meet Espley's fears of interference from the new valve life test huts. His small temporary hut which he uses for receiving work tends to dominate the locality! We shall find some solution, but with a lighter heart if there had not been so many scares and unnecessary expenditure of money and effort on this kind of thing in the past.

May 26th[389-390]

A day of important visitors to the CVD (GEC) meeting. The meeting itself was good. The various service radio projects involving our valves were run over to see what difficulties or complaints there might be. Discussions were snappy and to the point. I had good opportunities to speak to Hecht, Griffiths and Oatley separately about Wembley position in regard to CRT research and development work. Whereas the CRT group at Wembley had from the beginning acted as the inter-Services expert research team detached from commercial considerations and at the service of all the Experimental Establishments, there were recently distinct signs that Wembley was being regarded only as an adjunct to a GEC Cathode Ray Tube Factory and on a level with concerns which have no similar research facilities to ours. EMI is the only equivalent concern. I believe I did good.

Walmsley has resigned from the CRT meeting chairmanship and I believe Oatley will take it on.

Griffiths and Telfer called to ask if Wembley would accept a commission to develop the button seal 10 mm valve, bringing in MOV as far as they were able to enter. I found F. Smith open minded but have not yet sized up what this would mean here.

May 27th

Have had a message from DSR that favourable developments in the phantom submarine problem have transpired and we can probably ease off the work. A meeting takes place with them next week to review the matter.

If we have to take on the button seal job it looks as if we can 'man' it at Wembley.

May 28th[391-393]

Gavin was at Dunwich this week on the night of the mass bomber raid on Dusseldorf—and tried out his 'Grocer' which is the operational name for one of the RCM projects. He was given a good deal of inside information which cannot of course be written down. He was told that Mandrel has been and is still of the greatest use.

Farren (Superintendent of RAE) and Perrin [Perring] came for the day to see the place and particularly Fido. They are the most charming of visitors and we had a very frank talk on the subject of RAE and its outlook before the war.

Breadner is in some difficulty over the glass discs he is making for RG and which we are supplying to EMI. There are minute cracks which we must find the reason for.

May 29th[394]

The Oboe crash programme is getting into full swing here and indeed it is some undertaking. Knox at the centre of the constructional side is invaluable. We have TRE workers here and a roof station for co-operating signals with Malvern. Marris took me over and is having a job of it keeping all the parts of it moving, including contacts with Westminster and TRE. The Oboe operational scheme is very promising—and its stock is rising rapidly.

Had discussion with Megaw, Le Rossignol and others. CV 79 and 89 are in an unhappy state at the moment. We shall have to have meetings once a week for the present between the research and the preproduction groups.

May 31st

There is a hiatus in the war—everyone expecting the European invasion but preparations not yet quite complete.

Preproduction 2000 last week and all types (except CV 79) looking up well. Even CV 58 rose to 85% [production] efficiency due to better control of sealing in.

Jesty has spent most of the week on operational trials and subsequent meetings, and the old anti-jamming tube no longer appears to need the extreme AI qualities. So the tubes we thought we should have to make are not now needed, and our VCR 87 can go ahead on present performance. We still press in varius directions for 'operational' tests. Every time they have come off important facts have been learned.

Espley tells me that the performance of GR brings constant praises and satisfaction. It seems a shame that the predictor is now to be 12 months late—although the respite is probably welcome to Witton and Spon Street.

Mr. Eclair Smith [Heath] called from Sir Robert Renwick to be satisfied on the prospect of Oboe and to see what he can do to help. There is a possibility of one of the sets flown over Germany having been lost and if recovered by the enemy the appearance of Mark 2 becomes the more urgent.

June 2nd[395]

A cathode ray tube meeting (RDF) all the morning. Walmsley has resigned the chair and Oatley—his probable successor—presided. We got on well and have decided to cut out from the meeting all contract and acceptance questions which are departmental and keep the meeting to research and development questions. Oatley is putting up to CVD policy committee a proposal to split the meeting, having a general meeting in the afternoon and a limited one with GEC, EMI and Cossor in the morning.

An FBI research meeting in the afternoon. Very lively. The directors of Research Associations attended and Dr. Hadfield conducted vigorous 'intruder' tactics.

June 3rd

Farren and Ackland Snow have just returned from Fairlie where preliminary phase difference underwater trials have been made by them and by the Air Ministry group. They seem to find the proposition promising and are going ahead with the next step in the experiments.

Prain to lunch and discussed diamond die affairs which seem to be progressing smoothly so far as the control is concerned. He will send me the the prices now ruling for the other production groups. We seem to be getting a monopoly in the supply of diamond powders and Leeds is catering for large stocks.

Members of the Research Lab. Consultative Board waited on the Managing Directors to obtain their confirmation of the functions and status of the Board. This was obtained and it was agreed to ask Chelioti to join the Consultative Board.

Jenkins reports interest by ADRDE in discharge lamps as infra-red sources for night photography and we are likely to have a development contract for an investigation of it.

June 4th

A Research Services meeting at Head Office, I had a list of all the main Government contracts for research and development. The amount received from all these during the year 1942 – 43 was about £180 000. We are spending about the same sum per annum on these researches as we spent before the war on GEC research. Attached is a list which shows the proportion of activity on each project (Fig. 42).

We had a large meeting—Horton presiding on the Compact Range Finder project at which people were horrified at the delivery date in 18 months time—largely owing to delay in supply of components.

This was followed by a heavyweight meeting—Leedham presiding—on Oboe at which every phase of it was gone through in detail. Leedham explained the redoubled urgency owing to the fact that the enemy is

RESEARCH LABORATORIES
of the
GENERAL ELECTRIC
COMPANY LIMITED.

CB.655.

LIST OF RECEIPTS ON VARIOUS GOVERNMENT RESEARCH CONTRACTS

FINANCIAL YEAR 1942-43

Contract No.			£
351/C21c	Anti Creep	Metallurgy	203
294/2	Atmosphere	Illumination	4131
Br/4D/83792	A.C.S.	Lamps	1846
S.B.24085	A.B.Contracts	Radio (Gavin)	8464
294/8085	B.D.	Bombs	616
4/59027	C.V.D.	Valves	43804
S.B.3022	C.R.T.	Cathode Ray Tubes	14858
G.37837	D.L.	Illumination	1103
294/10626	F.M. Contracts	Radio (Bligh)	12072
	F.I.D.O.	Fog	1353
S.B.31837	G.E.L.	Helmore	27900
S.B.20061	H.E.L.	"	10668
S.B.24828	G.R.	Radio (Espley)	17099
294/8872	25.G.	" "	1642
S.B.3110	25.I.	" (Marris)	18655
20356	I.S.R.B.	Secret (Jolley)	2740
S.B.22976	I.R.	Flash Bang (Winch)	3930
294/12254	M.H.T.	Elect. Eng. (Farren)	975
4B/39156	U.S.W.	Radio (Megaw)	6261
294/7170	V.B.	Batteries	1461
294/2	P.F.S.	Photo Flash (Winch)	1438
59646	D.S/Can	Illumination	2064
Sundry Contracts			1902
			£185191

SUMMARY

	£
Illumination	9150
Radio	64193
Valves C.R.T.	58662
Helmore	38568
Sundry	14618
	£185191

Fig.42 List of receipts on various Government research contracts for financial year 1942–43

starting to jam the Mark I Oboe. This is a striking example of a low priority development project coming into the forefront of urgency in a way which could not be foreseen and illustrates how important it is to let developments have a run for their money and not to choke them off in the early stages because of possible production difficulties.

June 5th[396]

The CV 79 and 89 report is only slightly more reassuring. There seems to be some major unknown factor which is hiding itself away.

We still do a certain amount of the sort of work we handled in peace time. D. Jones is on tungsten wire and working with the factory on alumina doped wire. The results look promising at the moment. Dudding is putting work in on this and had a big report under way—with his usual energy.

June 7th[397]

Preproduction of valves, 1662—magnetrons are in temporary difficulty with leaks and the total is 100 down. We have, after all, to keep up our output of NT 99s which have become very urgent but we shall drop NT 97s and CV 8 will be transferred from Algar's group to Wright's. This will enable Algar to increase CV 55s. We have now made 10 000 NT 100 since the start. These are giving place to CV 90s which are to be established in the hut.

Went with R.O. Jenkins to MOV to see what is to be done about executing the GEC contract for making battery type button valves. MOV means to make them whatever exhortations may be received from Dr. Truscott about not absorbing their facilities in this direction. Research Labs. are ready to act as camouflage if necessary.

Mr. Hackbush, senior management of Research Enterprises Ltd. of Canada, visited us and spent a very full afternoon seeing as much as possible of the valve work.

A Workshop lecture in the evening by Rayner on fine measurement and the like. Very interesting.

June 8th[398]

The multi-channel set (Bligh, Stenning, Heller) is finding growing favour at MAP and MOS. We are making a second model for Bradford who will then make the final prototype.

A small and select luncheon party at Claridges to meet Dean Mackenzie head of the Research Bureau of Canada.

Mr. Lemon from DSR Admiralty to explain that the phantom submarine project is so promising due to recent experimental success in producing the desired effect that Wembley can let up on this part of their development work.

June 9th

Called on Brundrett to raise a question of the tendency of the CVD Technical Committee to become a Star Chamber deciding matters off-hand which the other committees should handle, under conditions in which the people concerned in industry are allowed no hearing.

A TVC meeting—very lively. The committee is in a constant difficulty because it is the only body which can warn establishments developing gear that the production position respecting given valves is so difficult as to make it dangerous to adopt them. We cannot however insist on alternative types being used and Dr. Truscott presses hard for us to exceed our terms of reference in this direction.

A CVD policy meeting in the afternoon at which the new arrangements for the CRT meeting were approved and we shall have a morning research meeting and a general meeting for all in the afternoon.

I had to register disagreement with a policy to hand over certain valve research fields to USA and leave British industry dependent on the USA. I think the point was well taken.

June 10th[399]

Dr. DuBridge (head of Radar Dept. of Massachusetts Institute of Technology) here all day to see over the place with Dr. White. They were kept pretty busy and I feel sure they were interested.

We turned out 35 triple naught diamond dies last month—a record for us and I should think for this country. Some of them were 0.0004 in.

Had interesting examination of work by McKeag, Jenkins and Ryde on a new series of luminescent powders they have been working at for several years past. The new series reproduces the results of the existing zinc beryllium silicate powders with an apparent improvement in falling off of CP [candle power?], with cheapness and simplicity added. I am glad we, and not an outsider found them out. The good work continues.

June 11th

A CB meeting—the first with Chelioti as Member of the Board—with detailed review of the past year's accounts.

We have got out our six months' valve preproduction financial statement up to March 31st for the Admiralty. It shows an apparent loss of £20 000 in about £200 000.

And now a brief break for Whitsun.

June 15th[400]

A well needed Whitsun break has improved everyone's grip on their responsibilities and tempers. The calm before the storm still continues and everyone is still guessing where our blow is to fall. The Italian islands in the Sicilian Straights have just fallen.

The first Board meeting of Lumifax Ltd. of which Chelioti and I are GEC representatives and Gamage chairman—but intends to attend as seldom as possible.

Hastened back to meet Sir Cecil Kisch and Mr. Crawford to show them our progress on the fog clearance scheme. The main demonstration was in the Exhibition building on a 1/12th scale model. It showed entire confirmation of what had been found out on the very small model. The whole time of investigation is most successful and brings credit to Ryde, Humphreys, Cooper and Lait.

The data observations must now be rounded off but it appears ready now to be followed up with a revised scheme to be got out by Fraser & Chalmers. We have an impression that there are one or two people at the Ministry who would like to see the proposal abandoned so leaving the petrol burning scheme to hold the field.

Warren is advised today as ready to return from the States after revisiting Rogers Tube Co.

Great pressure for preproduction of CV 76 magnetrons. Rang Willett who, after enquiry, says that the equipment in which it is being used is a winner—and as it is going to remote localities a more liberal policy of spares is needed.

June 16th

Very heavy pressure by MOS (Mr. Grey) for our prototype Series III Vehicle Set (anti netting). Bligh is in session at MOS all the week on specifications and schedules. Our Oboe is also being pressed for increased equipments (100 instead of 20), so our Works, as well as ourselves, are feeling it!

June 17th

Settling with Pearce and Le Rossignol that we can meet an enhanced programme of CV 76 magnetrons and oddly enough I ran into Captain Willett in the Carlton Grill for lunch and was able to tell him.

Our diamond powders are increasing in popularity and are recognised as the best. We are establishing the grading on a sound and firm basis.

Our valve preproduction financial statement for six months ending March 30th comes out about all square without profit. So really we are down by about £20 000. The total involved in this six months is £237 000—half a million per annum! It is a credit to all concerned that an amateur group of scientists and financiers at a Research lab. can run such a show relatively smoothly. I wish I could mention specially so many who are working conscientiously and quietly at all branches of their job—producing the results, and coming up to guarantees and all with a clear statement of its finances.

June 18th

A long Radio Co-ordination Meeting at HO attended by Dr. Railing on (*a*) present and prospective war radio equipments (*b*) post-war radio and formulation of GEC policy.

Rankin and others came to inspect FIDO from the Ministry. The figures of petrol consumption by the rival petrol burning scheme are apparently astronomic (Fig. 43).

June 19th

Willshaw gave me an account of his efforts to remove wavejumps from 3 cm magnetrons. He seems to be largely successful using a 12-segment block—but there have been many disappointments before and we are counting on nothing until improvements are well established.

June 21st

Had message that Warren had arrived back in England from the USA—welcome news.

Leeds has had a weekend with Simpson on diamond dies and has settled with him a recommendation for 'unit' values or dies from 0.006 in down to 0.0004 in. Unity is at 0.004 and 0.0004 has a value of '8'.

Gavin tells me that RCM 'Carpet' is not looking promising at TRE

PETROLEUM WARFARE DEPARTMENT

Telephone: WHITEHALL 7010. Ext.
Telegrams:
PETWARD, SOWEST, LONDON

QUEEN ANNE'S CHAMBERS,
DEAN FARRAR STREET,
LONDON, S.W.1

In your reply please quote :—

19 JUN 1943

SECRET.

F.I.D.O. INVESTIGATION.

Dear Dr. Paterson,

The Director General has asked me to inform you that the Rt. Hon. Sir Archibald Sinclair, Secretary of State for Air, has expressed his appreciation of the excellent work which has been done in connection with the Fog clearance investigation, described in the Progress Report. He is pleased that the project has been pursued so forcefully and that the experiments are continuing.

The Director General would like to take this opportunity of thanking you once again for the expert assistance which you have given the Department in the course of the investigation.

Yours respectfully,

Stanley G. Crawford

Fig.43 Letter from Petroleum Warfare Department on FIDO investigation

which puts higher importance on his 'Air Grocer'. It appears that he has not previously had much TRE support for Air Grocer so this is a welcome change of attitude.

Espley is back from GR trials using auto frequency control. They appear to have been quite successful and the gear is left with ADRDE for a decision.

Jesty is also back from TRE where he had a peaceful passage and left for operational trials at Plymouth on the higher brightness daylight tube—also successful.

A very moving lecture at the Labs. this evening by one, Flight Captain [?] Warne, on his experiences after torpedoing and being eleven days in a small boat.

June 22nd[401]

Warren came in. His experiences and activities in USA though useful are on the whole disappointing not through any fault of his but because the USA mess up over their lighthouse valve has poisoned the ground for any valve of that type and has caused the USA developers of short wave radio to seek their salvation in a new magnetron they have developed. The trouble over there is that they have never made the NT 99 nor the CV 55—which means there is a bad gap in their resources.

Ryde's night illumination chart has been taken up and published by the Admiralty—but no acknowledgement of any kind made in it to Ryde!

June 23rd

Mr. Thompson of Diamond Die Control here to settle up various things including the USA automatic drilling and needle polishing machine. We shall probably try to make one. He said that tests of our powders in the USA showed them to be better than any others for polishing.

Messrs. Dixon and Shaw of MAP were here to talk over with Warren particularly the USA position on beaconry. The position is anything but happy. Progress is very slow and security plus secrecy are choking the developments. It looks as if it will be impossible to resist the urge to use [the] 'lighthouse' [valve] instead of our CV 90 valves simply because the former is getting into production and no one wants it over there. There seems no doubt that the CV 90 is the better proposition technically.

June 24th

A meeting at the Royal Society on a post-war abstracting service. The Chemical Abstract organisation is anxious for a given solution on the lines of their publication, but I hope the proposition will be put up in terms of the needs of the professional users of all kinds and an informal meeting be called to discuss these needs with the Science Abstracts Committee.

June 25th

Jervis-Smith is leaving us in due course to become Assistant Secretary to the IEE but I have made an interim arrangement by which he is not released from his war job here but continues his reponsibilities in full although he may be able to get to the IEE two afternoons a week.

We have decided that the USA diamond die needle polishing machine must be made outside much to the regret of the workshop who would like a 'proper' job like this.

June 26th

Bell showed me the first double pentode needed for Bligh's tank sets. We shall have to apply to CVD to have it made a recognised development. He also has samples of Aldous's vacuum condenser suitable for the AIS Mark VIII feedback. The evidence is being confirmed that if a heavier push had been put behind the feedback matter, the trouble and hot air over the Mark VIII*a* could have been avoided. MOV are going a further stage towards giving the CV 57 a larger factor of safety by gold plating the grids—which suppresses grid emission and the consequent liability to large anode currents which in some sets gives short lives.

June 28th[402]

Valve preproduction 1830 but we are having much difficulty over women glass workers and output, and generally it is not easy to keep up enthusiasm amongst routine workers. The NT 100 has sunk to virtually nothing in favour of CV 90, which has not yet mounted to a preproduction stage in the hut, though Boyland is carrying on in his lab. CV 76 which is being crashed in front of everything has responded through much late work and troubles with the test gear.

Mr. Ashley Cooper and Mr. Gamage came to lunch and to see over the place afterwards. He was very appreciative and is making another date with Dudding over quality control.

Jenkins's luminous bats for aircraft carriers are always impressive to visitors.

A GEC sales conference on infra-red radiant heating today and tomorrow seems to be going very well. Dr. Settle, Humphreys, Heather and myself were present for part time.

Small FIDO lab. model demonstrated by Cooper is now very clear and good and quite an interesting thing to show more important visitors.

June 29th[403]

Willshaw has at length got a 3 cm magnetron design apparently free of wave jumps. Efficiency approaches 20%. If it had been available four or five months ago it would probably have been adopted for TRE equipment but it comes late and the CV 108 with its imperfections has been adopted. But ASE (Admiralty) are interested. The USA 'Millimag' is also an important rival with its 50% efficiency but unknown production difficulties.

An FBI research meeting and interesting exchange of views on first draft report.

June 30th

A CVD–RDF meeting all day. The discussions were useful though nothing outstanding. Willshaw's 3 cm valve was ventilated and it was agreed as probably better than the CV 108 which is getting into production—so the new valve is probably too late for TRE applications. But ASE (Coales) are very keen on it because they are very dissatisfied with the CV 108 and very sceptical about the USA 'Millimag' valve until it has been established as a thorough going concern.

July 1st

We are still awaiting the great Allied offensive and the secret seems to be so well kept that some are asking whether in fact it is going to materialise.

Bligh's anti-netting set (Series 3) is going through vigorous pre-discussion with a view to a specification of requirements which are actual possibilities. Mullard are in the discussion and are competing with GEC for the development work. There is a serious question of getting USA to play, because the tanks shipped to Europe are delivered direct complete with radio—but the new set is a marked improvement on the present American one.

Flights are taking place with our preliminary model of the multi-channel set for MAP and our people have been out over the Labs. in a bomber, testing communications with the lab. under flight conditions. That is the way to get work pushed on!

Diamond die spec. (BSI) now under way. Settled the spec. today in full meeting—good work.

July 2nd

Research Services Meeting with Dr. Railing and Mr. Gamage at which we agreed that a policy in regard to alumicide must be settled at once particularly in regard to its manufacture in places like Australia.

A diamond die technical committee in the afternoon at which Leeds's & Simpson's unit value curve for dies of all sizes was approved with slight modification. We are to express the efficiency of our different die making groups in terms of units per worker per hour. It was appreciated by some that this would have a direct bearing on correct assessment of die costs.

July 3rd

There seems to be considerable difficulty in our Home Guard setting a watch each night on the Wembley site. Chapple, the CO, is losing men fast from call up and everyone except those properly exempted appear to be in the GEC company. So we are at an impasse! A Colonel from Home Guard is coming to see me on Monday.

Owing largely to Humphreys's sympathetic handling of matters with the labour and manpower authorities we have so far lost vitually no one from any grade in the Laboratories who is important. We took on nearly 200 juniors of about 17 and naturally as they attain military age many of them get called up but all who are working for degrees and have prospect of getting them are left with us.

It is true we have been in constant anxiety but amongst Government and industrial circles there has been a strong enough body of opinion to counteract the more ignorant circles handling manpower matters.

July 5th

Preproduction 1750. If we can keep our weekly output above 1700 we shall, as far as we can see, remain solvent. We are again in emission difficulties with oxide cathodes; NT 99 and CV 8 are both suffering. The trials using CV 79 and 89 in a transmission link indicate that many rejected valves 'in the cupboard' may be acceptable.

Bell did not have too happy a meeting on CV 57s at MAP on Friday;

TRE seemed to be again weakening on the fitting of feedback in AIS hoping that other remedies in the valve will assist sufficiently.

Harris saw Ekco re. their taking over from us the CV 114 valve. No one seems keen to do it, we are having difficulty in getting together a group of workers to make this valve at the Exhibition Building.

July 6th

Bligh had preliminary trials for an inspection by Lywood and Walmsley who are coming tomorrow to inspect communication sets for their aerodrome and general RAF ground gear. Bligh tells me that expectations are more sanguine as to the Series III tank set and we think of allotting space for the work at the Exhibition Building.

Gavin's RCM jobs are impressive. His 'Grocer' is now doubled in strength and 'Air Grocer' is making good progress. His use of a receiver crystal as a noise generator is very pretty.

Had a meeting with Goldsmith and the Carbon group on cinema carbon quality which has been indifferent. The main troubles lie in plant and supervision (control) defects, but much forward research is needed on a number of aspects.

July 7th[404-405]

A very busy day on CVD (GEC) and preproduction meetings. Mr Hecht looked in to tell me that his sphere of activity had been broadened to cover communications and RCM. We are very glad to have him back. Brundrett took the chair as usual and there seemed very few kicks and much activity going on smoothly. The same appeared to be the case in the preproduction meetings.

Sir Geoffrey Burton and General Butler called to look over our valve preproduction activities and left after two hours inspection under Bell's guidance.

Partridge is off to Lemington to look into various jobs and get a number of things done. He is a tremendous worker whom difficulties cannot beat!

July 8th

Further 'prodding' to hasten over Oboe! Long talk with Wells over the service given by the scientific worker and its reward—basal rates and war allowance and all those factors which the fault finders and discontented love to make trouble over.

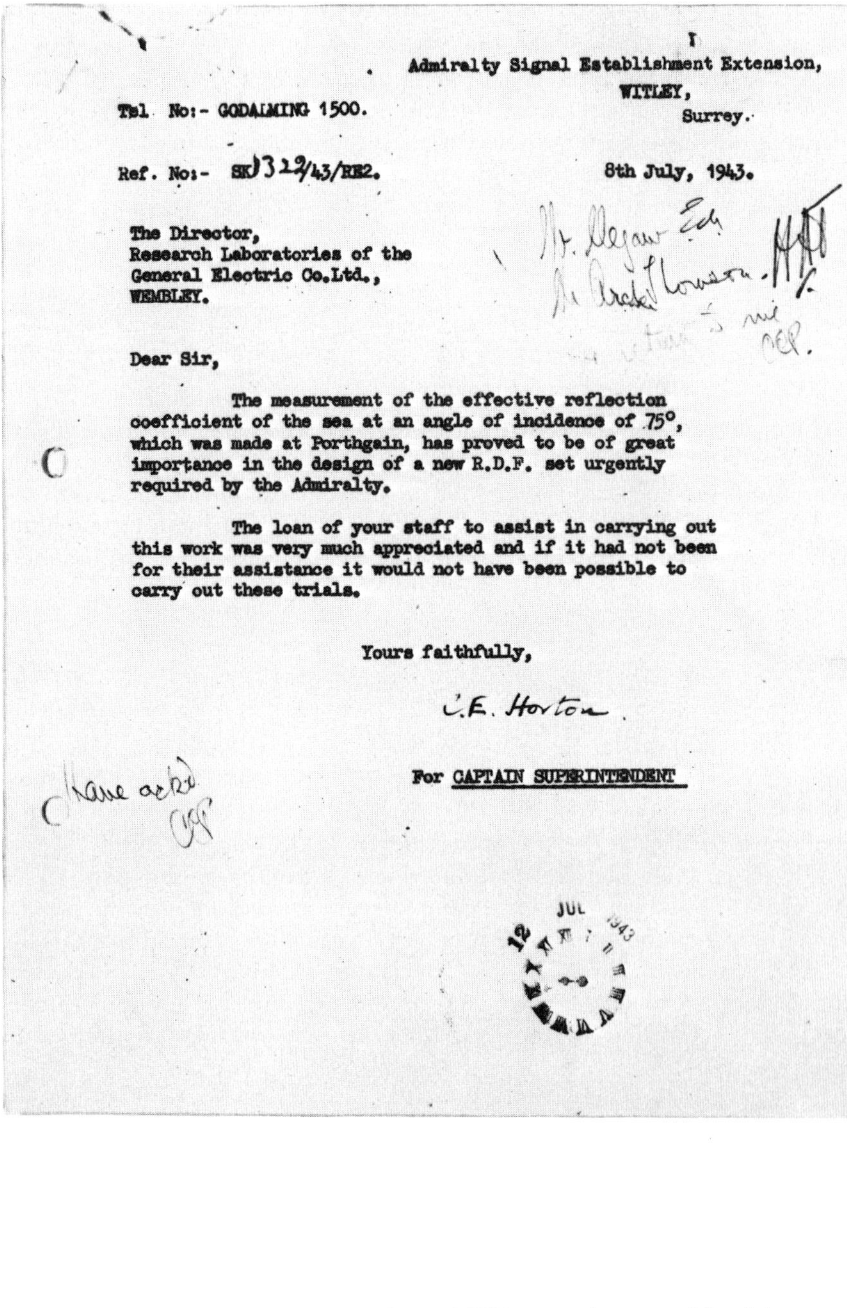

Admiralty Signal Establishment Extension,
WITLEY,
Surrey.

Tel. No:- GODALMING 1500.

Ref. No:- SK/329/43/RE2.

8th July, 1943.

The Director,
Research Laboratories of the
General Electric Co.Ltd.,
WEMBLEY.

Dear Sir,

The measurement of the effective reflection coefficient of the sea at an angle of incidence of 75°, which was made at Porthgain, has proved to be of great importance in the design of a new R.D.F. set urgently required by the Admiralty.

The loan of your staff to assist in carrying out this work was very much appreciated and if it had not been for their assistance it would not have been possible to carry out these trials.

Yours faithfully,

C.E. Horton

For CAPTAIN SUPERINTENDENT

12 JUL 1943

Fig.44 Letter dated 8 July 1943 from ASEE on assistance with trials

July 9th

A CB meeting at which we faced the present figures and deadweight laboratory expenditure. If the war stopped tomorrow it would be £400 000 and if we reduced numbers to pre-war it would be £250 000.

Attended at ERA Council to talk over the IEE research report for which I was to a considerable extent responsible. I think we secured a fairly favourable reception for it although it goes hard with the crystallised. [Letter received from ASEE, Witley, thanking us for assistance with trials (Fig. 44)].

July 12th[406]

The attack on Europe (Sicily) has begun not inauspiciously.

I went round the Oboe gear and effort with Marris and have the feeling that everything possible is being done to expedite it and win the date; Knox and his boys are doing well in the construction of the prototype.

Lunched with Dr. Railing and discussed with him Erith research and GEC deadweight expenditure as compared with Research Labs. financial obligation.

July 13th

We still make satisfactory alumina for boules and Salford appear to be making good boules from it. The Ministry of Supply seem inclined to set their own standards which do not apparently interest Salford who are able without difficulty to use the boules which MOS want to reject.

There is a similar rather impertinent attitude on the part of the Ministry of Food who ask for a complete specification of how the powder is made before they will have any tests made with it on killing weevils.

Had visit from Air Marshal Sir Richard Peck with Mr. Gamage to lunch and see over the place. He is an attractive visitor and seemed genuinely interested in most of what he saw—particularly the radio equipment.

July 14th

Ramsay's hydrogen modulator and rectifier seem to promise well and to have likely uses. He is having to pay less attention for the present to the oxide coatings research for emission characteristics. That is still a matter in which fundamental knowledge is sadly lacking.

Dr. Paterson. July. 1943

Mr. Good's Job.

~~I am sorry - I neglected to tell you about this job, and I should have done. I regarded it as at least connected with the Sub-Committee work, on which I report to you periodically, but it is more important and I ought to have let you know.~~

The job concerns the prevention of attacks on dams similar to ours on the Möhne dam, by the use of smoke, or dazzle, or both in conjunction, using searchlights to illuminate the smoke. Stiles and I have been advising on the dazzle side and on the possible illumination of smoke; the Porton people on the production of the smoke. The matter is at a high level and comes from Dr. Merton; I had a visitor today from A.A. Command who tells me that a trial is being arranged at very high priority. I have undertaken to help with some preliminary lab. experiments if necessary. I am putting it under the sub-committee work so far as contracts are concerned; though the matter is very important, it does not involve much work here at the moment. If it grows later we may have to take it as a separate matter.

Bomb Firing Tank. The investigation of the lighting for the bomb firing tank is going nicely. Turner and I have visited the tank, with Lawrence of Glasgow, and the experiments on the lighting have bee fruitful. The G.E.C. has been asked to supply the actual equipment and we had a long meeting today with Head Office (G & R, Osram Tech. Z Dept and Messrs Pritchett & Gold) and the experimental people and the M.A.P. to discuss preliminary details. It involves a battery equipment like Jolley's big one; he and Hawkins were also present and have helped very much. The tank itself is a collossal affair, after Jolley's own heart; I should like to tell you about it some time. As soon as their small experimental tank is ready, there will be a programme of experiments in Scotland on which I propose to send Turner and Morse, which will take two or three weeks probably.

Meteorological Contacts. We have had recently a further contact with the meteorological people, who have a programme of work to connect visibility from the air with general meteorology, and have been asked to contact us. I have had a preliminary discussion with Sir G. Simpson and the officer concerned, one A.W.Brewer, who has as his assistant Bastell's son, now in the R.A.F. as meteorological air observer. They are coming here for further discussion on Thursday; they may borrow one of our nephelometers and make further high-altitud observations with it. I hope that this may be the beginning of an expansion of our experiments on the lines which we had hoped it might follow.

Vernon - Haze Meter. You will remember that I told you of the project which Vernon have been engaged on in collaboration with us for some time, to provide a haze meter for gauging whether the atmosphere was suitable for exposing searchlights at sea. They have had one of our nephelometers for some time now, and have established their point and completed a report on it which is about to appear. I have brought them in to touch with the meteorological group mentioned above, as some of the problems have very much in common. Our relations with Chesterman of Vernon have been particularly cordial and are a great improvement on those with Harcombe. I have been meaning for some time to show you the enclosed personal note from Chesterman in reply to one of mine which I wrote to help him to clear up a point in his report. Their apparatus, by the way, has the code name "loofah", for no reason at all.

Fighter Command - Homing Conditions. While with Sir G. Simpson I met a meteorologist from Fighter Cd. whom I know, who brought the problem of helping the fighter aerodromes to decide when the weather is operational in the sense that returning aircraft will be able to pick up the approach lights. We had already had a very similar problem, and were able to make some useful suggestions which we hope to try out when the nights are a little darker. The problem

Fig.45 Summary by J.M. Waldram of Illuminating Engineering work in the Laboratories in July 1943

involves detecting and assessing haze which may be above ground level at altitudes up to 500 ft, and I think that it can be done with the aid of a projector like a cloud base projector but inclined at a suitable angle; the brightness of the beam seen from the other side of the aerodrome is explored with a very simple photometer.

Flares. We have reported on the polar photometry of flares and have had a very nice letter in reply to which I drew your attention. When the film is ready (it has been a little delayed through an unsuccessful sub-title) I should like you to see it. We are going to arrange a film-show for the benefit of those concerned as soon as we can. Beggs is working on a more general report.

Schuil Photometer This is well in hand; the instrument itself should be finished so far as construction is concerned by the end of the week. I am in close touch with Morton Williams at Admiralty about it, and I hope that it will be all ready somewhere close to the time I estimated.

Low-Brightness Photometers. A meeting took place at N.P.L. a few weeks ago, at which we were represented, on the design and construction of photometers for low-brightness work. Stevens went in my place as I was away. The idea was to discuss between various research institutions who were interested, to avoid the dangers of everyone making arbitrary photometers and getting results which could not be coordinated or relied upon.

Sub-Cttee G. The big summary, which I mentioned to you, is passed and is in production; I will let you have a copy to see as soon as one is available. I think it should be a very valuable document.

This is a review of the principal matters which I am concerned with and doesn't include Stevens' work. I have been in the habit of reporting to you at intervals our progress; there are a number of small enquiries and consultations which come in fairly frequently which we have usually settled out of hand and which I haven't always bothered you with. We have been able to advise and to put people in touch with information in other establishments which would otherwise be unknown to them

Two others which I had overlocked -

Submersible boats A rush enquiry from Admiralty. We were able to work out for them the visibility of some special and very secret submersible boats, give them the proper camouflage colours, and show them the effects in the lab. I should like to describe the device to you some time.

Fighter Command and A.A. Command - Searchlights. Another rush enquiry, settled over one lunch hour. Some S/L officers and an adviser from Fighter Command O.R.G. wanted to report quickly on a better disposition of searchlights to meet the F.W.190 raiders being used against us. Almost all the information was already in our big searchlight report.

Jmbs.
10. July '43

Fig.45 continued.

Last evening I gave another talk to twelve of the girls (on radio). This time it was the glass workers and they seemed to find it a revelation to be having a share in so amazing a matter.

A CRT meeting today—the first under Oatley's chairmanship and the new regime, i.e. research questions in the morning—GEC, EMI, Cossor—and other questions in the afternoon with Cinema Television, Ferranti, Cosmos and Mullard present. The plan is good and should work out well. The industrial representatives are united and co-operative but there are rifts in the lute where some of the service representatives are concerned. Today's meetings were spent in haggling over matters in which some of the Government people were plodding along paths long left behind by the industrial groups.

July 15th

Sanderson has at last got his new layer battery separators and can now make up batteries for the GEC contract from MOS. We met at Head Office to try to settle policy for larger scale manufacture away from the 'scarlet' areas of Birmingham and NW London.

Met Levy and West and Chelioti at the Ritz for lunch and discussion of Lumifax matters.

GEC Board meeting for review of year's accounts. A record year with sales up from 26 million to 31 over last year. Dividends the same of course. The first meeting attended by Garrard and Chelioti as Directors.

Met C.S. Wright at Royal Society who said Canadians are successful and keen regarding 'diffused lighting' and rather sore that our authorities were so half hearted, after our successful trials.

Farren has just returned from Fairlie where the second series of trials on the 'phase' A – S method for the underwater weapon look well worth following up and we are recommending this to DSR.

July 16th

Coales came to tell me that the Admiralty Staff are not now backing the Compact Range Finder equipment which Espley is working at—as an operational item. Very disappointing but the decision was carefully weighed. The development is probably to proceed so far as the range finding part is concerned but not at high priority.

He wants, for the 3 cm band work by EMI, sample magnetrons of the Willshaw pattern and special waveguide cells which Harris can make. We shall probably rush samples through which EMI can try out within six weeks.

July 17th

The Admiralty contract for collaboration on anti-submarine with Fairlie is in, and is put at £12 000 per annum. Wright told me a few days ago that he was most anxious to foster this liaison to the point of ensuring that Fairlie had the latest resources and knowledge and experience at their disposal of an industrial research establishment.

Attached is Waldram's summary of the position in connection with various war projects with which our IE group are concerned (Fig. 45). The dam job and the bomb firing tank are very interesting and major undertakings.

July 19th

Holidays are with us and people are taking a fortnight this year. Consequently preproduction was only 1550 last week.

Espley's blind landing experiments (a GEC development) seem to make him sanguine—and we propose to make demonstration equipment if it promises well. The easing of pressure on C.R.F. may give him a chance to get on with it.

An Advisory Board meeting with full attendance and very informative. Lunched with Chelioti and discussed build up of Glass Works technical organisation.

July 20th

Gavin's group is getting on well with its jamming gear. 'Air Grocer' is ready to take the air as soon as TRE are ready for it.

July 21st

A full meeting at Head Office with Heather presiding on television after the war to formulate a GEC point of view. Marris, Espley and Jesty had written an excellent review of the facts and issues involved and Heather kept the meeting to the point so that all agreed it to be a statesmanlike discussion. Certain conclusions, leading to an 800-line system, were come to and the meeting adjourned for a month.

A CVD policy meeting this afternoon—lively and interesting. I approached Brundrett on post-war liaison between Admiralty and GEC on valves and radio and we are lunching together in a week to talk over this and other matters. He gave me the strongest assurance that if he and the present DSR are in office the arrangements for liaison which were started before the war will be continued, naturally on a much reduced scale from the present. He reminded me however that he and C.S. Wright were not popular in some quarters and may have disappeared by that time from the management of these things at the Admiralty.

July 22nd[407]

Osram Advisory Committee meeting and much indignation at the building up of demands from the Government for lamps and no corresponding assistance to industry to get labour.

To the RAF training centre at Reading at Air Vice Marshal Baker's invitation with Wing Commander St. John and the AS group to meet informally over the next steps in the anti-submarine development. Air Vice Marshal Baker reminded me that Sorley would have the final say in whether the scheme should or should not go forward.

Subsequent meeting of Science Abstracts Committee to consider co-ordination of abstracting services. Not a very sympathetic committee to the main ideas.

July 23rd

Air Marshal Sorley came for the morning—and particularly studied FIDO, AS, Helmore and Illumination, in fact our researches for MAP other than radio. It was well that he came for he has considerable influence in MAP projects.

We have various undeserved setbacks in our efforts to carry on! Douglas had a Home Guard accident which has laid him up indefinitely and Brewer, who is Pearce's right hand man on magnetrons, must take a sedentary job. This sort of thing is very hard when we are trying to expand rather than contract.

July 26th

Mussolini has resigned—the first major crack in the Axis machine and a big one. The Sicilian invasion goes well.

Gavin's 'Air Grocer' seems to be destined for multiplication before it has been tried. He had a message that 50 off will be asked for at the earliest moment—so here is another crash programme in the making on RCM.

Deputation here from Admiralty and Marconi for a multiplied by four

NT 99 glass – metal triode—very like the E 1149 of two years ago. We can certainly design it and supply samples but finding a factory to produce will be the catch.

Have been taking initial steps to try to get an electron microscope from RCA but fear we shall find a long waiting list.

July 27th

A Glass Advisory Committee at which Chelioti initiated a new regime for the technical advisory and liaison machinery for the Company's glass activities. A very good move which I hope will bear fruit. Partridge will be the main RL representative and Dudding and I will be alternative as second representatives.

Gavin now tells me that another 600 Mandrel sets are being ordered which is an eloquent comment on those who initially condemned the scheme as operationally unsound and useless!

A Ministry of Works and Planning Committee at which we passed the draft of the National Lighting Sub-Committee's report. The age old question was prominent of whether the lighting code is to be mandatory in bye-laws or a code of good practice. I am afraid the compulsory element was in the majority.

July 28th

Have had no little trouble through some outside agitator (Transport Workers) misleading some of our diamond die workers. The majority of them are genuinely disturbed and sorry at malicious reports which are spread about the group. The one or two who seem to be misguided are far from pleased at the loyalty of the remainder. I don't suppose we are at the end of the incident yet.

An MOV transmitting valve meeting at the new Perivale factory. We had some four or five new triode valves to report—when we thought that triodes had reached the end of their development!

July 29th[408]

The Navy Torpedo Factory has come back to us about the CV 61 valve used in their latest equipment. It does not seem to be uniform or really reliable and seems to me to bear out the apprehensions we had when first our help and advice was but grudgingly received by them. Donaldson therefore has a further job on his plate.

Had Dr. Barron here for first talk over plastics and the way we shall organise the machinery for contacting all and sundry in the GEC.

July 30th

G.M. Wright came yesterday to a propagation meeting and disturbed me (as he was disturbed himself) at the things he hears, about handing over research and development projects to the United States and accepting a defeatist position which will vitally affect British Industry after the war if not earlier. These matters appear to me to be handled here at high levels by people who have no experience and little knowledge of practical things.

A TVC meeting. Great concern at the large number of magnetron types due to efforts to use different wavelengths for every project. It points again very strongly to the need for a tunable magnetron and I had a discussion with Willshaw and Gardiner which I hope will result in starting up something seriously on the subject.

Willshaw tells me that the Birmingham 3 cm valve CV 108 is very short lived which adds one more to its bad marks. It seems to me that TRE will have to adopt Willshaw's 3 cm valve which seems to be proving out just at the right moment.

Marris has more reports of American invasion into the RDF field. American equipments are being pushed in front of our crash Oboe units and are likely to be put in the air first. The authorities do not seem to realise the effect of this on teams of men they are driving to extremes of effort to achieve early dates for these emergency British equipments.

We are starting the August Bank holiday weekend in a heatwave and I am thankful everyone is going to get a break. There is much suspense over Italy's policy. Will she surrender and [or] will she let her half-hearted war efforts continue.

August 3rd

Preproduction in a short holiday week—1650. We have again serious pressure on our workshop facilities and competition between craftsmen for CVD experimental work and craftsmen for valve preproduction. One of the troubles is that the tools used by valve preproduction require continuous reconditioning which absorbs a large slice of our workshop facilities. Any new work connected with preproduction seems therefore to suck facilities away from our all too limited experimental valve work.

A meeting at Head Office (A.R. presiding) with Salford on 'predictors' and shall GEC cut out a new development line of their own. The sentiment was favourable.

August 4th

To Witton with Humphreys to see over Kaula's insulation factory and discuss HF preheating of pellets.

August 5th

Welcome news that we have at last received a (very high) priority for labour—so we may succeed in getting a few younger women and we shall try for a skilled craftsman or two to help out with our experimental valve work which seems to be pressing on us.

Willshaw reports as we expected that TRE (Dr. Skinner) is very interested in his E 1487 magnetron (X band) which TRE thinks will meet the bill in place of CV 108 (Birmingham valve). ASE Witley are also wanting it for an additional set—so we are likely to be busy on it.

To Royal Society where we discussed universal science abstracting.

August 6th[409]

A research-services meeting at Head Office at which A.R. and L.G. examined figures of laboratory staff etc. expenditure when Government development contracts and the like cease—over £300 000.

Brundrett lunched with me at the club and paid me the compliment of discussing with me his departmental and staff difficulties and relationships. He is one in a thousand and I do not know what we should do without him. His confidences are not for record.

Coales attended a CRF equipment [meeting]. CRF is to go forward to a definite production date in 1945 instead of being cut off at the development stage as hinted a fortnight ago.

We have 12 or more B.Sc. passes from our student assistants and two or three 2nd Hons. in the London General exams.

August 7th

Saturday. News of advances in Russia (Orel etc.) and Sicily is good and there is a feeling that the Germans are being stretched.

The CRT programme requirements are stretching both staff and space and Jesty presses for a decision either to cut down programme or plan extensions. I am consulting Hecht of MAP before deciding.

James reports the E 1440 (Small CV 90) now in order and to be recommended. E.K. Cole want the CV 114 modified for more suitable manufacture—and we rather agree.

August 9th[410-412]

The Italians are proving slow in deciding on an unconditional surrender and on Saturday night the Pirelli Works at Milan are reported completely destroyed by bombs. The Russians are doing well at Kharkoff and seem near to a break through.

The Birmingham valve group is breaking up. Oliphant has gone onto new work and takes most of the brighter men—and Randall goes to Cambridge to teach at the Cavendish.

Megaw is editing a *magnum opus* of a magnet report in conjunction with Birmingham, Oliver and Sucksmith. The typing and main work is being done here.

The CV 79 and 89 valves are still in the doldrums and in fact the whole of our valve preproduction is there. In the middle of summer we have more illness than in winter and 'nervous debility' shows either the quality of health [of] the women we have now—or their state of morale—or our own inability to gauge their limitations. Only 1225 valves last week.

The security man who has been threatened for six months made his attempt on us today and succeeded in walking straight through the front entrance without challenge. We are all most depressed.

August 10th

There seems to be a fair prospect of fitting in the various new valve projects in our workshop programme by some reshuffling and possibly by moving some more component making (press work) to the Exhibition [Building].

Meeting on primary batteries at HO and particularly layer batteries. A minor flap over the method of making contact with the plus terminal but we agreed to send in our sample to MOS using our tried method and have a claim to alter the process for something still better after the first 2000.

August 11th

A visit to BTH with others from GEC re. unification of luminescent tubular lamps between GEC and BTH. Good progress was made because everyone wanted the auxiliaries as well as the tubes themselves to be a joint product. This drawing together of GEC and BTH is very welcome.

A message came through from Megaw that Willshaw's visit to TRE and ASE had very much raised the stock of Willshaw's 3 cm valve and there is a general demand for more samples.

An RDF CM [centimetre] valve meeting was held today at MAP but I could not be there. I shall hear about it tomorrow.

August 12th

The news re. Willshaw's 3 cm magnetron E 1487 is even more interesting than yesterday. The efficiency is of the order of 30% so that in addition to its immediate purpose as a substitute for the CV 108 it is almost qualifying for the high power magnetron required by ASE. Apparently TRE get a considerably increased range with the equipment and the valve will work over the whole range of its power without wave modes.

To HO to discuss Alumicide and policy about it with A.R., L.G., C.T. and Chirnside.

A useful meeting staged mainly by Humphreys in consultation with Dudding with the heads of all valve groups on the question of labour, its priorities and its treatment for greater efficiency of output! Our labour loss is at the rate of 20% per annum but the Ministry of Labour have so messed their calculations as to have made no allowance for this—so that they are compelled to require that new labour is only to be used for new requirements and not for making up wastage!! Our sickness waste is 4% working hours and absences for other causes about 5—6%. We have now 510 women on production and preproduction at the Labs., about half being half timers.

August 13th

A Consultative Board meeting. We have interesting discussions these days since all members are 'live' and have strong mutual interests. Afterwards Humphreys and I discussed with Heather the arrangement of Committees and sub-committees on Electrical Heating and on Radio. We are starting in a gentle way with post-war preparations within the ambit of the Advisory Board machinery.

A further CRT 'operation' visit to study 'smoothing' of tubes. Hopkinson reports a 50% success.

The demands of Coventry for crystal valves and our need to supply them with the silicon crystals therefore has necessitated an increase in our chemical capacity for making them. Chirnside is fixing this up together with barium azide making—thus we manage to surmount one difficulty after another. They are all minor ones but everyone is most anxious to make a helpful contribution.

Megaw went over his big magnet report with me, now ready for duplication—a monumental effort.

August 14th

Marris seems satisfied with the progress of Oboe 2 with its various ground stations and airborne equipment. The whole project is difficult for anyone who is not soaked in the subject. It certainly requires a great and sustained effort. The main difficulty has been with the aerials and their ability to function inside a Mosquito. They will have to provide a blister on the plane to ensure their proper functioning.

Gavin announces a very simple circuit for generating jamming energy in place of all the troublesome and second best devices which have so far been the only ones to be thought of. He seems very pleased with its possibilities.

Saw B.S. Cooper to ask him to take on the electron microscope—if we get it from the USA. He is, as one thought, very pleased at the prospect.

War news is of good military progress in both Sicily and Russia—but no cracking up of the enemy yet. Even Italy seems to want to continue the fight although she has officially discarded Fascism.

August 16th

Preproduction is back at 1800 valves for the week and a more cheerful and sanguine atmosphere exists. We still have not received confirmation of the prices asked for our valves as from April 1st—something like £160 000 is being held back by sluggishness of Admiralty, Bath. In spite of what Bath says, I know definitely that the papers have only just been sent to DSR for approval.

Dr. Skinner of TRE writes to tell me how impressed they are with Willshaw's E 1487 magnetron.

Gossling, Warren and James have been to TRE, and Gossling to ADRDE also. Jesty has also just come back. These visits do much good in keeping up technical contact and maintaining personal relationships.

It looks as if we must now make the best we can of our present scientific staff and spread them over our jobs as efficiently as may be. There seems little chance of getting any more.

Now that things are going badly for the Germans and satellite states they are wondering where they stand, and now that there is more activity against the Continent, over the Channel, we have to be on guard for enemy parachute attacks and all the rest of his devilry.

August 17th

An internal meeting to see what sort of crash programme might be hoped for, for the E 1487 and E 1494 magnetrons if such were desired. We should hope to begin preproducing valves in October and be making ten per week by the end of the month. Testing might present difficulties.

Breadner and Simms have some interesting new glass techniques. One is the cleaning and rubbing of tungsten wires with special hard glass (in order to prevent oxide formation) prior to normal sheathing. The oxide film appears to be the key to several troubles including the non-wetting of the tungsten by the glass. They have also devised neat and cheap little low power ovens for annealing components as they come off the machines. 600 watts instead of 6 kW.

We had the security men Captain Bennett and Captain Lewis here to look us up. They seemed tremendously interested in all they saw. Tomorrow is our CVD and allocations meetings and Brundrett comes to me tonight to talk over staff and other matters before the meeting.

August 18th

A day of CVD valves and preproduction—very busy. Brundrett had two hours with me talking over research organisation questions and staff conditions—an interesting comparison. At the CVD GEC meeting there were several very live questions to discuss including our E 1487—94 valves (X band). There was a very lively recognition of its merits and, I think, of Willshaw's perseverance in sticking to his efforts to get a reliable design in spite of months of disappointment and of the prior responsibility of Birmingham in the magnetrons for X band. Coales came up to me afterwards to congratulate us on a victory!

The preproduction meetings in the afternoon were also lively and exhibited renewed pressure for valves from Wembley—particularly magnetrons. The programmes of the professionals MOV and BTH have been delayed and the position is serious. So our crash programme for the E 1487 and E 1494s must go forward and we shall probably ask Wright's group to take them on.

August 19th

Peters (Coventry) came to talk over the state of the various development jobs which are going on or half going on, i.e. AIS, Oboe, GR, Series III for MOS, CRF, blind landing (GEC), television and other minor activities. 1 and 2 are going on hard, 3 is going on but is delayed, 4 is problematical, 5 is going to prototype. 6 and 7 we are all excited about, being something for after the war.

Warren learns that the USA lighthouse valve is again under some cloud, and there is pressure again for the CV 90 to be made over there. We shall see.

The glass workers' room is crowded out with an influx of new workers and something must be done to ease the pressure; I have been looking round to see what could be possible in the way of extra space. To start with we must move the RG work somewhere else.

August 20th

The daily administration of the details of the place takes up much attention of those of us who previously were able to devote almost exclusive attention to the scientific and technical side of things. One of the great troubles is that people are not free to leave if they want to and we are not free to get rid of people when we want to. It is a detestable system which takes away people's freedom in these matters.

Mr. Lonsdale of Erith came to see us and look round and we had a useful talk on post-war research facilities at Erith. We found our ideas very similar on all points.

Mr. Butement of MOS radio called to discuss procedure for the Series III set and wants the whole focus of the development moved to Christchurch together with a zoo of other people under a member of the SRDE staff. The ground for this is stated as the necessity for rendering the set hermetically sealed for tropical service. I must see Bligh and Dr. Railing about it.

August 21st[413]

Foxon showed me his very ingenious sewing machine attachment which he had devised for the use of a home group of seamstresses he has organised. These are mounting luminescent fabric on landing bats—also Foxon's design. The experiments have been in [the] charge of H.G. Jenkins and seem to give the Navy much satisfaction. There are one or two hundred to make.

Marriott came to discuss possible scheme for the post-war organisation and control of the British radio industry. I am very glad this is being attempted.

August 23rd

Preproduction 1750 valves—discussed crowding in some of our groups, i.e. Douglas, Peachey and glassworkers. The RG job will go into Vacuum Physics 1 and so ease the position. That technique is going well but they now want four instead of two wires through the glass.

Peters and Macqueen came to discuss the position created by Mr. Butement's visit last Friday. We all felt it was once again a 'plant' by which an incoherent group of people are assembled to do a coherent job without any experience of its engineering. I don't know our attitude yet.

Am off to TRE with Rowe this afternoon by road.

August 24th

Had a very interesting three and a half hours drive with Rowe, with long talk over all the projects—RDF and RCM. Bombing with accuracy and with small loss of planes is the objective of most of them. The whole scheme of projects seems to tie the enemy up completely. I am glad to find that Wembley is on the more important ones—Oboe, RCM, H_2S—the latter mainly through O'Kane who, for a time, was its main focus.

I visited Defford Aerodrome (TRE) where Edwards and Cherry are at present engaged on Oboe Mark II. I tried to get the idea home that we were beginning to need our marching orders for the jobs for TRE at Wembley to follow Oboe. Rowe told me that they had many more projects on the list than they could tackle at TRE. Saw Skinner on the subject of Willshaw's valve E 1494 and told him of the priority programme for preproduction we had started. His problem is how to release the improved E 1494 when the CV 108s are still the preponderating supply.

August 25th

Dr. Black spent the night at home and had spent the day here exchanging information regarding advances at Wembley and in the USA.

A CRT meeting with Americans present—no special comment except that the meeting seems to be serving its purpose.

We have been reviewing rates of pay of student assistants over the past four years to see what advances have actually been made at the Labs. The result shows rapid advances of basal rates apart from war allowances.

August 26th

Dr. Black here again having been all round the valve work here before returning to the USA.

A large GEC meeting at HO on television at which our ideas on policy, technical and otherwise, were further crystallised. It is clear that some development work will be needed before we are sure that our technical ideas are well founded.

Had discussion with Heather and the others on the Butement mission. The more we think of it the more incensed we feel at the suggestions and the way they were put up to us. I don't think the GEC can offer to co-operate on these lines. They cannot take the necessary responsibility to ensure success in their share of the work and nothing but a mess is in prospect.

Gavin tells me that his jamming circuit was welcomed yesterday by Hecht and his assistants but we have heard nothing yet from TRE.

August 27th[414]

Have taken preliminary steps to sort out new arrangements for handling the Helmore big searchlight job—MOS are asking us to accept a development contract from them to do much the same development as we are doing under Helmore for MAP. MOS of course want the development to be focused under Paris and Cockcroft in their own organisation. Helmore appeared to recognise this yesterday and our plan is for Jolley to discuss with Cockcroft the approval by him of a small committee of ADRDE and Wembley organisations, and if possible Helmore to guide and foster the development. Jolley is seeing Helmore with the letters this morning.

Had Wing Commander Cropper and Farren to discuss political diffifculties, i.e. their A – S work, since Cropper is afraid the RAF people working on the job will be recalled to their teaching work because Fairlie are taking a conservative view of the experiments in progress. I have asked Cropper to try to get their *status quo* accepted for a few weeks longer.

A thorough talk with Mr. Hecht over our CRT contract and programme of work. Apparently MAP would like a new contract in line with those they are giving EMI and Cossor but Mr. Schoenberg is making certain conditions which would make difficulties. (Our contract is likely to be put at £20 000 for 1943 – 44.)

Cropper's anxieties appear to be a mare's nest.

Waldram brought me in Schuil's nephelometer which has been built by him out of Schuil's half-made parts and is being sent out East to be used by his colleague who is carrying on with Schuil's work. It is a most compact and excellent device with an enormous range of brightnesses.

August 28th

Winch has just returned from trials with our first Bradford made prototype of his gun ranging set, at Larkhill. We[He] had a happy time there and demonstrated full intensity signals at 15 000 yards direct flash by day and night and indirect by night. He cannot overcome the condition of the indirect flash in bright daylight, although it works on an overcast day.

Ramsay has developed to a satisfactory stage his hydrogen filled diode for large powers and TRE are keen on its use in the AI Mark 9 and probably for other purposes. It needs no temperature control and is compact and efficient. We have to consider at our preproduction meeting on Monday whether we can possibly make it.

August 30th

Preproduction 1900 valves—discussed how we might preproduce some of Ramsay's hydrogen rectifiers. Unless we can be relieved of NC 12 mercury diodes we can only produce a trickle of about six valves per week. We must raise this question again at the preproduction meeting.

Was in the chair at a meeting to decide how we can formulate a 'code of practice' to keep back the use of discharge lamp lighting. Pretty ironical that I should have to be the one to do it! I hope it will put this type of lighting into such a desirable position that every one will want to use it after the war.

Dudding has difficult consultations on in regard to Quality Control pamphlets. There appears to be rather a squalid rush on the part of certain Ministry officials at various levels to cash in on this and gain credit for what they have had no hand in building up.

August 31st

Jolley reports that there must be operational contacts regarding RDF use with searchlights in order to ensure proper operation of the large Helmore lights. The six sets which are to be 'copies' of the present set require, according to him, better remote control arangements than the normal Army types.

Stevens reports collapse (at a Naval staff meeting) of diffused lighting as a project for use by the British Admiralty. It looks as if the Canadian Navy will go further with it.

Lunched with Lord Rothschild under solus conditions in his flat. He wants us to devise an anti-sabotage bomb device for detecting workmen—particularly foreign ones—carrying such bombs into factories. I have asked Humphreys with Rycroft to look into it and handle whilst I am away next week. There are probably one or two methods which can be used.

September 1st

A meeting of Research Committee of FBI [Federation of British Industries] to finalise our report. These are very vigorous and instructive meetings. We practically finished the report and it is not bad except it is too full of Research Associations expounded in wordy felicitous language!

A CVD policy meeting at which visits to America by Jesty and (perhaps) Le Rossignol were pressed. Our CRT contract was recommended for renewal, probably £20 000.

September 2nd[415]

Jolley called me to see a film of the operation of ELSIE made by using a telescopic lens on a ciné camera attached to the radiolocation director on the searchlight. The targeting was not good and showed the desirability of using visual following as soon as the RDF succeeds in picking up the object.

Mr. Grey of MOS called in relation to correspondence he has had with me to loan him someone to work with Professor Willis Jackson at SRDE Christchurch on electrical measurements for six to eight months. Proctor is prepared to go and he will spend one or two days here a fortnight coming home for week ends. Proctor will be missed from here but it will probably benefit everyone all round.

The second matter (Series III tank set) to discuss with him was unpleasant and unsatisfactory. Macqueen and Peters came to join in and it appears (what Butement did not say) that RPC has ruled that GEC has not got manufacturing capacity for this gear and therefore development cannot take place at GEC. There was no prior consultation with GEC and we were somewhat unco-operative, not out of pique but because we cannot contemplate success in a difficult job carried out in a disassociated atmosphere such as is proposed at SRDE. When I arrived back at my office I found that Peters and Macqueen had suggested that the functional prototype should be completed at Wembley and that we should work with Plessey as the manufacturers. This obvious arrangement had not occurred to Grey as he said he did not realise that Wembley would be prepared to work with any other manufacturer than the GEC.

September 3rd 1943—Fifth Year of the War

Invasion of Italy announced this morning. And so we begin the fifth year of the war. There was a special National Service and [of] prayer this morning which I attended at HO. I am very glad that all work was stopped throughout the country for this quarter-hour service from 11.0 to 11.15.

The Russians seem to be doing great things—almost a German collapse at some points. Truly things are going well—far better than our materialistic and individually grasping tendencies deserve. Trade Unions lead the way and everyone else follows the Gospel of 'Get all and give as little as you can'. But there are thousands who have not bowed the knee to these Gods and they are the people who are saving the situation.

September 20th

Am back after a fortnight's break at Witley, Surrey, close to the ASE (RDF) establishment. I visited them and also Liss (life test) and ASE at Eastleigh where I was shown the latest radar gunnery equipments and the arrangements for trial installation on board ship. Very interesting and impressive.

Bell has left me the report on the group of new valve design projects which he is mainly responsible for (Fig. 46).

News came through to Le Rossignol ten days ago that his son had been killed flying (fighter pilot) in the near East. A sad blow which we all feel keenly. This may affect his projected visit to USA. I understand from Picken that Jesty's departure to USA is likely to take place very soon.

A record number of preproduced valves last week 2277. All the jobs seem to be going as expected. Marris says that Oboe is gradually finishing off and that TRE must be approached for our next job for them under our AI contract.

Bligh is back after an alarming illness—but has no news as to how the Series III tank job is going to be handled by Mr. Grey and SRDE.

September 21st[416]

A magnetron meeting with the BTH at which they had no use for Willshaw's CV 108 substitute and declared that their version of the small USA valve is just as easy to manufacture in quantity. So there is likely to be some sparring at the RDF meeting tomorrow.

I have been asked for a Friday evening discourse at the RI. Leeds will help me with a talk on diamond dies. Our 'unit' value curve for dies, agreed at the last diamond die meeting, has just caused a considerable reaction on the part of the other makers. All with one accord begin to make excuses and do not want to produce their figures! I think patience must be the word at present. Our own output figure appears to be of the order of 0.1 unit per worker per hour.

Dr. Paterson, Sept 18 43

As my going on leave coincides with your return I am leaving these notes on the state of the various items on my programme.

J.B.

E1496 Megawatt triode for A.S.E. Witley.

The first valves are on the pump & are looking promising. Reeves & Coop (A.S.E.) came up to discuss the programme on the 16th. They urgently require three valves by the end of Sept. We ought to be able to meet this requirement. It is likely that a further 24 valves will be wanted by the end of this year and I think the workshop arrangements already made will cope with this.

Aldous is handling while I am away.

E1495 High power C.W. valve.

The tools for making the electrodes have been made & some electrodes – cathodes & grids & parts of the anode – have been turned out & look quite sound. The first samples are going to be held up for the glassing jig which has been designed and drawn but is not yet started in the workshop. It is going to take a minimum of one month to make this jig & I have asked Gardiner to get it started as soon as he can.

Dr Griffiths came out to see how this valve was getting on just before he left for U.S. & Mr Picken has been out to see the electrodes etc. since. There is a good deal of interest all round in it & I think it will be a pity if the arrival of the first samples is too long delayed.

P.T.O.

Fig.46 Summary by J. Bell of state of items in his programme

<u>E1457.</u> Large CV90 Type.

L. Tremlett, with assistance from the workshop, has been struggling with the problems associated with the cathode seal & the flexible diaphragm. We have changed from the flat disc to a feather-edged cone for the cathode seal attaching the diaphragm by means of a gold seal. Dummies have been made up successfully & Tremlett is now tidying up some loose ends before having a shot at a complete valve again.

<u>E1489.</u> Double pentode for Bligh.

Going according to plan.

<u>CV57 & CV174.</u> Mk <u>VIII</u> A.

Aldous & I in company with Smith & Windsor from HOV visited RAF Coltishall taking with us 25 CV174's having the new Titanium di-oxide coated anodes (instead of the usual blackened nickel). The tests on these valves both on the ground & in the air showed a marked improvement & we left behind enough valves to equip four aircraft. We were promised a detailed report in a fortnight & if this is favourable Hammersmith will change over all their CV174 production to the new anode. Hammersmith are anticipating the result & are making all the preparations to <u>change over now</u>.

I can come back for the CVD Liaison Meeting on the 29th if necessary

Fig.46 continued.

September 22nd

Disappointing in some respects. Mr. Grey (MOS) writes that they intend to adhere to their plan of centering the Series III set in SRDE Christchurch and suggest our carrying on in parallel at Wembley, but nevertheless lending them Nelson. We shall have to have a conference on our resulting policy.

A CVD – RDF meeting all day. The BTH have pulled up their socks and got on with their 'millimag' X band magnetron to the stage at which they can show most of their troubles surmounted. So preproduction is to begin on both our E 1487 (Willshaw) and the BTH valve so that reliable valves may be available at the earliest possible moment for operational use.

September 23rd

Waldram read a paper yesterday at the APLE [Association of Public Lighting Engineers?] on street lighting and was like old times. It was well received by all.

FIDO has now reached the report stage and the scheme will be formally reported to the Ministry next month.

Ryde tells me that Spon Street have now orders for crystals on their books amounting to some £200 000 i.e. 200 000 crystals at £1 a piece. It seems an amazing number.

Winch is following up another sound ranging scheme using a cathode ray tube in a scheme for detecting rifle or mortar fire.

September 24th

Espley reports that the production prototype of GR is out in the field ready for its trials at Coventry.

A TVC meeting, very full of important issues and many submissions for use of new valves. It is most trying to have to tell Establishments who have been developing equipment that production position does not permit of their counting upon certain valves which six months ago they were recommended to adopt.

Had conference with Drs. Paris and Cockcroft at (Jolley present) MOS to discuss ways of carrying on the Helmore searchlight development under an MOS contract. We want to bring Helmore fully into it but not to make him the go-between between ADRDE and ourselves. We strongly urged a monthly meeting presided over by MOS and ADRDE which Helmore will attend and which will be a clearing house for all phases of the development.

September 25th

Considerable advance in our powder metallurgical efforts through the Ordnance Research organisation (MOS) who wants us to experiment with spongy iron shell bands in place of copper. Also iron for piston rings is considerably to the fore. We welcome this contract, which can be handled as a legitimate war job.

Pressure has suddenly come on our CRT filter making which B.S. Cooper has kept going steadily for the past two years. We increased capacity 50% at the beginning of the year but this has now been absorbed and new demands are virtually 50% up.

Commander Bolt (Admiralty) rang up to say that supply of CV 108 (the Birmingham BTH X band magnetron) has practically failed and he was told that the only chance was to get the Wembley crash programme pressed harder. It is difficult to apply more of a spur than we have been applying to ourselves already. It is a matter for our Monday meeting.

September 27th[417]

Preproduction still over 2000—reported Commander Bolt's message about absence of CV 108s and 'it all depends on Wembley'. The fitting of a 6 volt instead of a 3 volt heater is a nuisance and may be a real catch.

Lunched with the Secretaries of the Royal Society over *Science Abstracts* and the pressure being put on us including the spreading of prejudice and worse to make us relinquish one half of Section A and virtually pass it over to *Physiological* [?] and *Chemical Abstracts*. Quite unacceptable. Egerton was helpful and responsive to the offer that Section A should be expanded to include some 1200 more abstracts annually. I fancy I am in for difficulty in quelling some evil passions and prejudices!

Meeting at Head Office with Dr. R., L.G. and all the GEC valve interests to discuss a new British Valve Association. A statesmanlike meeting and helpful. Things in the GEC are shaping not so badly for the future in these respects.

Attended meeting at Wembley Town Hall with Mrs. Paterson, Mrs. Turner and Mrs. Schuil to commemorate Wings for Victory achievements. Viscount Bennett spoke excellently on British Empire.

September 28th

Helmore wants GEC Research to take on his 'fish' and its internal design and installation. Bligh may be able to do so in his group but we must have a meeting to appraise present position of the job before judging what is possible.

Proctor finds that the job he is wanted for at SRDE is mainly low frequency testing equipment—and a certain amount of advice to other groups on centimetre wave techniques.

The CV 114 makes slow progress towards a more foolproof design. The tools and jigs are being made outside—and meanwhile we are not likely to get into the clear till our design is changed. Increased production is urgent.

Bligh brings good news that his multi-channel set developed by Stenning and Heller has—after long discussions—been adopted by MAP as the communication set for all airborne uses including, we understand, the USA. If so it is a major triumph.

September 29th

Marris and Farren have returned from a conference at Fairlie and the Torpedo Factory on the RAF homing torpedo. Wing Commander St. John is extremely anxious to plunge at once for a weapon notwithstanding the tests staged in a fortnight's time to establish whether the system is worth going on with from the reverberation point of view. It was stated by the Admiralty that a sensitivity of only 200 yards would justify proceeding, and 400 yards would make it really important.

A CVD GEC meeting (Brundrett in the chair) with preproduction and allocations meetings to follow in the afternoon has made a full day. The critical urgency of the E 1487 (X band substitute for CV 108) was emphasised and we shall have an order for some 6000. The hydrogen diode (Ramsay) met and withstood Truscott's challenges and we can now let up on our gas triode CV 12 preproduction to the extent of one half in order to enable us to incorporate the gas triode in Douglas's efforts. The battle over the double pentode for Bligh's multichannel VHF set was more difficult, but Truscott submitted himself to a demonstration and to a cross-examination of Bligh which I think convinced him that there was nothing else which would do the job. There was also strong pressure put on the CV 114 (waveguide cell). These are wanted for use with the X band magnetrons.

Humphreys has gone off on sick leave and will be sadly missed. The strain of his work has been too much for his nervous system and his doctor has warned him.

September 30th

F. Smith (MOV) reports a conference with Sir Robert Renwick at which he is urging an increase of magnetron production from 100 000 per annum to 400 000! It is easy to think of a number!

An Osram Advisory Committee. These meetings improve.

All preparations in hand for diamond die technical committee here tomorrow. An American machine (needle grinding) is here and will be running and also one from Smith's. We had a private meeting with Simpson, Davidson and Walters to consider the first returns of the three fine diamond making groups involving the 'efficiency figure' for each. These figures range between 0.06, 0.09, 0.12 die units per worker-hour which is interesting. Altogether the scheme is as promising as one hoped, but it will at present be confined to the three of us.

October 1st

Everything in the war seems to be favourable—Naples practically taken—airfields in Italy for bombing Germany and Austria from—Russians all along the Dneiper and Japan getting the worst of all actions. But Germany and Japan are not seriously cracked yet and hope to 'defend' till we knock ourselves to pieces and give over due to exhaustion.

A Research Services meeting (Dr. Railing, Leslie Gamage and Heather) this morning. Gamage said that Sir Robert Renwick was inclined to suppress a whole lot of projects such as miniaturisation, GR and anything which might interfere with facilities for magnetrons and the equipment they are used in.

Diamond Die Technical Committee meeting at Wembley—a pleasant occasion. The single head USA auto needle grinding coning machine just arrived in time to be shown operating. The efficiency figure of merit unit value die scheme was accepted as we hoped; the actual returns from Simpson and Davidson being confidential amongst the three of us.

Had Megaw, Willshaw and Bligh to decide how we could reinforce the magnetron study group. If possible we shall get McBride, when he returns from sick leave, to join up with Stainsby on the examination of experimental magnetrons.

October 4th

Another good week of preproduction; 2174 valves out of 2015 guaranteed. The CV 114 is the main anxiety for valves appear to get worse and more difficult to make every week. The receipts for the first five months are £160 000.

I hear from him that Air Commodore Leedham is coming on October 20th at my invitation to familiarise himself with our work. He has not been here for a couple of years.

I attended a Ministry of W & P [Works & Planning?] meeting of

Lighting Study Committee. Our report is nearly finished and I think we shall all be thankful when it is done.

October 5th[418]

Battery and Carbon meetings at HO. It is good to keep these GEC liaison meetings going.

Marris and Bligh attended a discussion with Group Captain Helmore and afterwards at RAE and Staines to consider whether we could take up the radio side of Helmore's fish. We have definite views as to how it ought to be carried out and shall put them up later in the week.

Mr. Griffiths, our new acquisition for helping our student and technical assistants and supplementing Dudding's supervision, has got Dr. Wolfenden of Uppingham to come and talk to our young men in November on the general question of youth and vocations.

October 6th

A GEC radio co-ordination meeting attended by A.R. and L.G. Very useful and a comprehensive discussion. The matching up of development and production facilities involves some difficult problems, since the Ministries are adverse to having development done where production cannot be guaranteed. This was one of the troubles with Series III set.

Am off to TRE this afternoon. Have left a CRT meeting in session at Wembley.

October 7th

A most interesting and helpful day at TRE with eight opposite numbers to myself from STC, BTH and all the others. Rowe, Lewis and Ratcliffe looked after us all day and showed us operational films and RDF presentations. A very interesting inspection of their 'operational results' room with charts and data about the bombing, the submarine and other branches of the war. It was a graceful gesture from TRE to industry.

October 8th

Representatives of Ekco and Ferranti have been here to take manufacturing particulars of CV 114 waveguide cell which Harris and Shaylor are getting into preproduction. It is being redesigned for easier making but this will take some time to finalise.

The CRT meeting on Wednesday apparently went off well and without special incident.

Chirnside reports that Dauncey has made some crude sapphire rod following a lead from USA. We very much want to perfect this technique, in place of boule.

Partridge's high temperature glass furnace has been improved by increasing the input of electrical conduction through the glass and improving the furnace top. Much larger and more successful melts are secured.

October 9th

Had Mr. Lang Brown here from DSR Admiralty. Brundrett told me recently what a hole they are in with regard to naval camouflage since Schuil was killed, and I volunteered to see what we could do to help out. We suggested instead of seconding a man as before, that we undertake the camouflage duties, which Schuil did, from here. The proposal has been welcomed by DSR and it is now virtually fixed that we organise to do this. Waldram's work for Ministry of Home Defence on visibility is now practically complete, so that another job of this character comes in very well. Harper will probably take as much part in the plan as Waldram; Stevens will also keep close touch. I feel very pleased about this new outlet for our activities for it sets a new precedent and I feel confident our fellows can deliver the goods.

October 11th

A TVC meeting all the morning. The really contentious items get pushed out of the agenda so we are going to have a special meeting to take them. The Admiralty and MAP representatives do not seem to find it easy to reconcile their views on nomenclature.

Lunched with Dr. Railing—discussion on short term and long term post-war research in the GEC.

October 12th

A Glass Advisory Committee at Hammersmith in its new guise and scheduled terms of reference. It was recommended that Osram Glass Works in future will make their own tank blocks. Various interesting matters and techniques were discussed.

The afternoon was spent over SRDE matters. Professor Willis Jackson first called in order to tell me how heartily he disagrees with the policies of SRDE and of the MOS direction. They appear determined to build up SRDE into an establishment which is independent of such places as [the Research Laboratories of] GEC, EMI, BTH etc. He was followed shortly after by Mr. Grey and Mr. Butement who came solely on a courtesy visit to make sure that I was aware of their appreciation of our attitude in subscribing both Proctor and Nelson (on Series III). Grey wanted it to be understood that when he arranged for Series III to be transferred to SRDE it was under a misapprehension regarding the readiness of Wembley to pass on designs and developments to competitive radio manufacturers. We had some friendly but plain speaking.

October 13th[419]

N.R. Campbell sees an application of some of James's experimental results to his valve noise theory and presses the need for following up the clue by a mathematical investigation. He was retiring from the Labs. anyway next April and wants to cast off much of his patent work now. He had indeed stuck to a distasteful job out of loyalty for many years and some rearrangement of work must be made, but, not being a reserved occupation, it is most difficult to carry this work on.

A Diamond Die Advisory Committee at which the future of the industry vis-a-vis control was discussed. A trade association with a technical committee and a Government representative was favoured. Discussed the USA 'coning out' machine and the group efficiency determination scheme.

A CVD policy meeting—friendly and effective.

October 14th

A somewhat difficult special TVC meeting on the subject of bringing CRTs into the CV nomenclature system. It sounds simple but was another illustration of how nomenclature seems to rouse all the most evil passions in the human breast. The meeting ran from 10.30 to 1.30. There was a spectre of a compromise hovering about towards the end.

The security officer Lieut. Commander Tomlinson called on an inspection and found us immaculate!

October 15th

A day at Wembley with the BTH and GEC of all three estates from both concerns. We met in the library where Jenkins, McKeag and Bowtell had staged an excellent exposé of our latest luminescent powder research, culminating in the calcium phosphates and chlorosphosphates with antimony activators. It was a great day for the powder group and very complimentary remarks from Mr. Sporberg and Mr. Warren of BTH; Ryde must have felt gratification.

Foxon showed me his complete luminescent 'bat' equipments for HM Ships. He has made an excellent and tidy job of these and of the organisation of their production under 'home' conditions.

Dudding has been awarded a medal by the Institute of Production Engineers for best paper of session. Subject—Quality Control.

October 16th

Had good talk with Griffith over the care of our student and technical assistants. He is, I think, doing very well indeed and can be of the greatest help, with Dudding, to our 200 youths and to their part in the Research Labs.

October 18th

Preproduction 2270 equal to the record. All types (nearly) are up to a good figure. The CV 90 is turning into a well behaved production type.

Ramsay reports having got his hydrogen triodes and diodes into control and is sanguine as to their future. We must put them up to the equipment designers as a rival to the trigatrons.

We have had the flash-bangers here, to report a mistake by ADRDE who instructed us to proceed with an experiment on the mortar detector without MOS permission—and therefore MOS do not 'feel about to implement the instruction'. It amazes me how one department will let down the Superintendent of another.

Colonel Boyd of our local Home Guard Company came to discuss their taking over the place in the event of a German sabotage invasion. As Research Laboratories of GEC is the only really A priority vulnerable point in the NW of London we are receiving a good deal of attention and shall be getting swelled heads.

The new War Saving cycle (the seventh and still being run by Mrs. Paterson) is up 32% in subscriptions and 9% in subscribers. The cycle is £220 per week for 30 weeks and there are extras to come onto this.

October 19th

Visit yesterday by Gossling and Bell to Marconi (Chelmsford) appeared to be a pleasure to both sides. The high power short wave CW valve

WHITEHALL 1234.
EXT.270.

PRIVY COUNCIL OFFICE,

WHITEHALL, S.W.1.

19th October, 1943.

Dear Dr. Paterson,

RADIO TEACHERS COURSE:
ST. ANDREWS.

I want to tell you how much the Wireless Personnel Committee and I myself appreciate your kindness in arranging for Mr. M.R. Gavin to come to St. Andrews and give such an interesting lecture on "H.F. Valves."

Mr. Gavin's talk was most instructive and was greatly appreciated by all who heard it; and his help in the laboratory work of the Course was also of the utmost value.

We are deeply indebted to him, and to you in consenting that he should come.

Yours sincerely,

Hankey

Dr. C.C. Paterson, OBE.,
Research Laboratories of the General
Electric Co., WEMBLEY, Middlesex.

Fig.47 Letter dated 19 October 1943 from Lord Hankey to Clifford Paterson on lecture given by M.R. Gavin

which Bell is developing looks promising and all are sanguine.

An estate meeting at which Fire Guard accommodation and Home Guard recruitment figured; also our inability to open Hirst Hall for food outside working hours.

Jesty has safely arrived by air in the USA; Le Rossignol leaves tomorrow by sea.

October 20th

Air Commodore Leedham (DCD) has been here at our special invitation from 10 to 3. I felt that he ought to see in greater detail the scope of the work we are covering particularly for MAP. We went over Gavin, Marris and Bligh's work first on equipments. Then to valve preproduction where Holland and Pearce had made a capital exhibit of all our current preproduction types. Then to experimental valves and CRTs. He is a very considerate and nice man to show over the place and I think he really enjoyed his visit. As regards a new job to be given us in connection with our open contract for MAP (TRE) he says that TRE staff are themselves almost entirely on applications rather than research which is the result of a deliberate policy of nine months ago when a large number of new projects were blue pencilled. He took the opportunity of discussing the new Helmore job (Helmore's fish) which is now to be called Helmover. As usual the project has been born in an entirely irregular way and is now in process of being made respectable.

A letter from Lord Hankey regarding a lecture by Gavin to a teacher's conference at St. Andrews (Fig 47).

I have heard both from Hecht and Hughes that the phantom of a compromise on the nomenclature issue has materialised into something more substantial and I hope that our next TVC meeting on Monday will open in an atmosphere less charged with tension than on the last occasion.

October 21st[420]

A propagation meeting all day with the usual large group of important technical officers present, with G.M. Wright presiding. Everyone seems to think that very useful work is being done.

Inspected the CRF functional prototype. Espley is away looking after the first production model of GR. Both these jobs have had excellent careers so far—largely because of the determination and vigour in overcoming each difficulty by Espley and Walter, as it arose. Cockcroft comes tomorrow to discuss these and other projects.

Gavin told me of an unsolicited testimonial to Wembley from

Cockburn of TRE: 'Wembley is the only industrial research place which does not need spoon feeding and which is not more trouble than its employment is worth'.

By the way yesterday Leedham had some very nice and encouraging things to say about the 300 multi-channel set originated and developed by Bligh's group. A full conference is to be held shortly at Wembley to finalise requirements and settle what may yet remain to be done on the prototype. Leedham even hinted at the possibility of the set being an international standard for both war and peace use!

We (Rooksby and L.A. Thomas) are taking on the clearing up of the Salford selenium rectifiers—involving some experimental work and then securing adequate control over manufacturing processes.

October 22nd[421]

The winter war conditions are now getting established. The enemy must be feeling somewhat anxious for the Russians have badly broken his line below the bend of the Dneiper and are outflanking him to the South.

Cockcroft's visit today was short and to the point. He inspected CRF and the relation of its separate components to Army needs. He told me that PF type devices for other projectiles and bombs were likely to be studied. (I thought they had been already.) He is a silent and enigmatical man.

Have been thankful this afternoon to have secured a measure of peace over the *Science Abstracts* trouble at the Physical Society Council. Other members of the Council played up well and I hope the field is set for a constructive policy in regard to the use of the Universal Decimal Classification.

October 23rd[422]

Have a fire control project for DEE Admiralty through Dr. Railing to look into. Am sending Dunham to examine it at West Drayton. If it comes off it is likely to be a joint effort with Witton.

Willshaw reports, from visit to ASE, that the E 1494 magnets job (which was moving over to BTH 'because we cannot do the preplanning') has now come back to us.

Harris has shown me the first satisfactory new design of CV 114 valve using gold seals.

Winch reports the new mortar detector to be more promising than was at first thought likely. Have written to tell Cockcroft.

October 24th

Sunday—and a major sabotage airborne invasion practice with attack expected against the Labs. as the most vital target in this part of London. The call up was at 4.15 this morning and all Home Guards and ARP personnel are at their posts to protect the place. The attack (10 am) has not yet developed.

The spirit of the whole Fire Guard, ARP and Home Guard at the Laboratories seems to me very happy and ideal—and the leadership is one of example and service rather than of precept. Neither force has yet had to come into serious action, but I fancy our people will acquit themselves well if and when they do.

October 25th

The invasion practice yesterday led to no excitement—to everyone's disappointment! But I am told it did a lot of good.

The preproduction groups think that a 'pep talk' is due from a gold braided officer. I have asked Group Captain Helmore who has consented and who will, I know, do them very well.

Dunham tells me that the fire control project from DEE Admiralty is one we have studied before and will not be a difficult one to handle here.

A TVC meeting at which Hecht and Hughes lay down with each other like lion and lamb and the compromise is agreed on the application of CV nomenclature to CRTs. Our careful planning of nomenclature in the past is helping to solve a good many difficulties as British and American valve store articles have to be brought into some sort of system.

October 26th

Our CVD bill to the Admiralty for the past six months is £24 500. It is an anxious matter when one compares this with the MOV situation in which this amount of investigatory work is clearly needed but in which Dr. Schoenberg (with whom I lunched on Saturday) takes a very individualist view of industrial research, with no response to any plea for an industry uniting to constitute a national effort. I must discuss the matter at my next Research Services meeting.

Discussion on plastics here with Salford and Dr. Barron. Quite useful.

October 27th[423]

Air Commodore Leedham rang me to confirm that he has agreed with Group Captain Helmore for a contract for our work on his fish

(Helmover) and we shall now finalise details.

A demonstration of double pentode sealing in (9-pin base) scheme to MOV seemed to convince them that there is no difficulty in a technique for this. Having the process now available at Wembley means that we can readily provide prototypes to establishments in reasonable numbers.

An MOV receiving valve meeting. There appears to be real light on the oxide cathode emission enigma—one more stage towards the perfection of this process. Miniaturisation of rectifiers discussed for the first time.

October 28th

We have virtually finished our post-war lighting committee's report covering dwellings, offices and schools. A most friendly and happy task in which members worked hard and constructively. The last meeting, for the present, met today.

October 29th

Winch reports good tests at Larkhill with the mortar ranger with an accuracy of less than 3° at 6000 yd. and detecting both mortars and machine guns. The officers there appeared to me genuinely enthusiastic.

Leeds is very sanguine over his latest drilling of fine dies using springs as buffers on his rams. Speed and quality are both up—so far. The American who came yesterday with Mr. Prain was very complimentary. He said that the British had taught the USA industry how to see the work on diamonds by use of microscopes and that had been worth everything to them.

The Infrasizer for powder separation which we got from Canada and installed as a laboratory tool is proving invaluable for powder analysis of iron dusts and the like.

Bligh had three Australians here to take careful particulars of the multi-channel set which they may want to adopt over there.

October 30th

The Russians are at present setting the pace and have broken the German line north of the Crimea. Many reports of reaction inside Germany to their parlous position. Report today of good agreement in Moscow between Britain, USA and Russia. So things appear to be shaping well for the eventual collapse of Germany.

There is at the moment a distinct feeling of falling away in the demand for urgent experimental work.

CRF is now not to be proceeded with, which leaves Espley free to think over the best way to keep the group usefully occupied. Marris and I have had much discussion upon the inability of TRE and MAP to indicate any major job to follow Oboe and we cannot yet see far into the coming months. Donaldson on PF finds his main problem pretty well solved and we have to decide what he shall do if the Services do not cough up some problem for us to tackle.

This by no means applies to valve and CRT activities but is chiefly concerned with radio equipment projects.

Leedham told me last week that the Services would soon be considering a new forward programme on radio.

November 1st[424]

A good beginning for the month on valve preproduction—a record of 2468 valves. We shall not easily come up to this again yet awile as we have cashed in one or two reserves. All groups are going well technically and the only main trouble is disorganisation due to illness of workers. The E 1487 (CV 108 substitute) is getting going under A.G. Pearce and we had hoped to have ten good valves last week but out of fourteen attempted only three emerged good due to every kind of teething trouble. Marconi people were over today to get particulars for their manufacture.

Attended a meeting of the group Machin, Herwald, Michaelis and others who have been organising the new magnetron life test huts which are being put down at the extreme NE corner of the Lab. site, at Admiralty expense. The effective screening of the two Nissen huts against emission of interference has been an elaborate affair and will give us much useful experience. We ought to have our first test unit working there by March. It is being well engineered.

Visited the Exhibition [Building]. That too is developing into an excellent place. CV 114 is well established. The anode rolling section has moved upstairs and a section of CRT and other 'overflows' show promise of great utility. Everyone speaks well of the canteen service and other conditions there.

November 2nd

The extension to Jervis Smith valve life test is now practically complete and is a very workmanlike place. It extends facilities to new transmitting and pulse valves and such short wave valves as CV 90.

Mr. Wolfenden (headmaster of Uppingham and other things) came today to give a talk this evening to our student and technical assistants. Everyone voted it an excellent occasion and discussion afterwards was lively.

November 3rd

An RDF (CVD) meeting all day. In the morning our main contribution was from James who gave performance of CV 90 valves with feedback in the circuit instead of in the valve—increasing the output considerably. There was again considerable liveliness over the agreement to hand over to USA the high power X band valve for development and production—because this had become an urgent requirement at ASE and one of Willshaw's valves virtually meets the requirements. There was obviously considerable feeling amongst Service Establishments and when I suggested that it was extremely difficult to be loyal to the letter of agreements made by people ignorant of the facts there was quite a murmur of assent.

November 4th

Had a useful internal Radio Group Leaders meeting to discuss broad outlook towards peace-time programme and recrystallisation of our work—and particularly the priorities which the GEC should put on the various aspects of radio.

There was a good deal of heat generated at a Radio Propagation meeting (sub-committee) charged with seeking that the stations round the country are in proper working order. Some people were very dissatisfied with alleged muddled management of this part of the work by MOS. I hope the little 'flap' will bring its own remedies and that the trouble will not have repercussions at higher levels.

November 5th

Fairlie anti-submarine tests are starting up again and the party goes to Scotland this weekend to make the final preparation trials on whether it is worth while to carry our scheme further with both RAF and Admiralty.

November 6th

Our 'minor products' do not diminish in activity. Getters (L.F. Britten) are at a high level and we are supplying Cossor and STC as well as MOV and Shaw. I understand we are the only people in the country making barium. Glass for silica seals keeps up in spite of threats of disappearance of silica valves in favour of the glass-metal type. Filters (B.S. Cooper) are keeping at a high level and run efficiently.

The senior officers on the Station where the Jenkins–Foxon luminescent ‘bats’ are being introduced seem to be all over them. Foxon tells me that Churchill, on a visit there, has been examining them very appreciatively.

Ramsay and Willmott have just organised a visit complete with apparatus to 60 Wing, to demonstrate that a change from square wave modulation to trapezoidal form makes no difference to the reception efficiency, whilst making a great difference to the life of the gas triode modulators. The demonstration appeared to be most convincing and we must pass the word round to other establishments.

He also left with me a sample of the final form of the hydrogen diode which is simple to produce and better in appearance and smaller.

War news continues good. Word has just come in that the Russians have retaken Kiev. The Germans are hinting at which [what] they will do if they have to capitulate! Our bombing of Germany after going slow for a month on account of weather has now started up again. I do not see how Germany can stand it much longer.

November 8th

Valve preproduction again 2400.

I am busy (in odd moments) with Leeds preparing a Friday evening lecture at the Royal Institution on diamond dies and their making. We have the October returns from Simpson and Davidson— 0.25 and 0.04 die units per worker hour; ours is 0.13. Davidson is in trouble due to absences of his skilled workers—don’t we know what that is like!

Campbell is going to concentrate (before he leaves in March) on signal–noise problems in valves in order to apply his theories further. We are therefore having to share out much of his work. The indexing of secret reports may be shared between Clayton and Chapple, who now does the Most Secret USA reports. The patent work will have to be farmed out, except for that originating inside the Labs.

November 9th

I have a note from Captain Willett head of ASE to say that he is leaving ASE on December 15th. This is a great shock. Willett has always been a helpful and very sympathetic friend to us whom we could approach and be sure of support. He asks to come here to say goodbye on November 30th. [Reply to letter, from Paterson (Fig. 48).]

Visited Fraser & Chalmers, Erith, to talk over post-war laboratory accommodation and certain staff questions. Found Mr. Lonsdale's son just returned as escaped prisoner from Italy with an exciting story of his adventures.

November 10th

Have had all the Service people here for GEC – CVD preproduction and allocations meeting. A lively and interesting day. Apart from magnetron activity on which the pressure is very heavy the main panic situation lies with the CV 114 waveguide cell for which the minimum demands are twice our present objective output. We are going to stretch ourselves in order to see what we can possibly do to meet the situation.

A major fight is going on between TRE and IVPC over our hydrogen diode which TRE badly want. IVPC are refusing to find production facilities for it after the preproduction stage.

Hecht has talked over with us the 'Helmover' job and decided that we shall leave RAE to finish the long wave version themselves and we come into the shorter wave pulse solution.

Griffiths told Gossling that he found one of our valve development charts hanging up and being used in one of the USA administration offices!

November 11th

I hear that work on British Oboe production is to be suspended in favour of USA equipment—this will be a disappointment to those who have worked so hard to get it going. No further details as yet.

A meeting with Dr. Railing on fire control for the DEE Admiralty at which it was agreed in principle to go ahead. It will give us experience in the dead heat [beat?] control of heavy moving masses.

Helmore came to talk over his various projects and also to be briefed for a visit to all the valve groups to increase interest for the preproduction groups.

November 12th

Alumicide, i.e. special alumina, is more to the fore again. The USA are making experiments without cavil. The Ministry of Food are actually

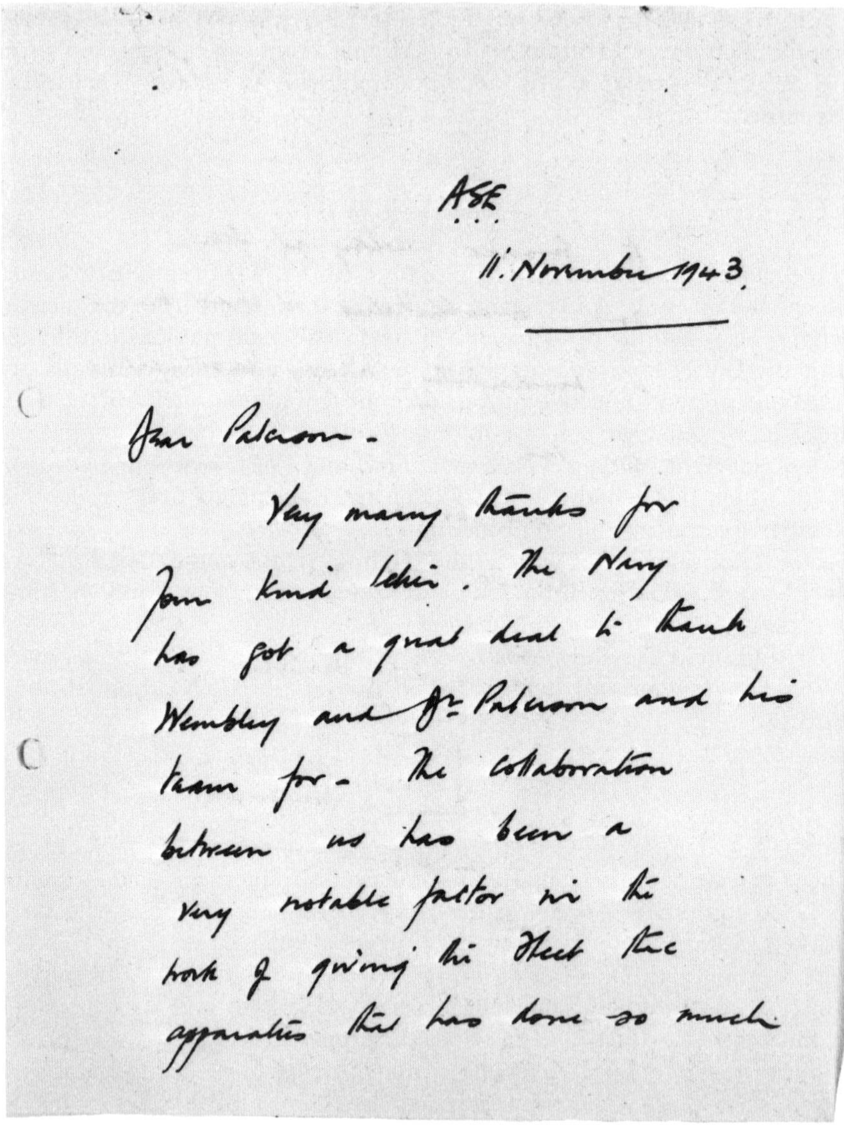

ASE

11. November 1943.

Dear Paterson.

Very many thanks for your kind letter. The Navy has got a great deal to thank Wembley and Dr Paterson and his team for. The collaboration between us has been a very notable factor in the work of giving the Fleet the apparatus that has done so much

Fig.48 Letter dated 11 November 1943, to Clifford Paterson, from Captain Willett, RN, on his leaving ASE

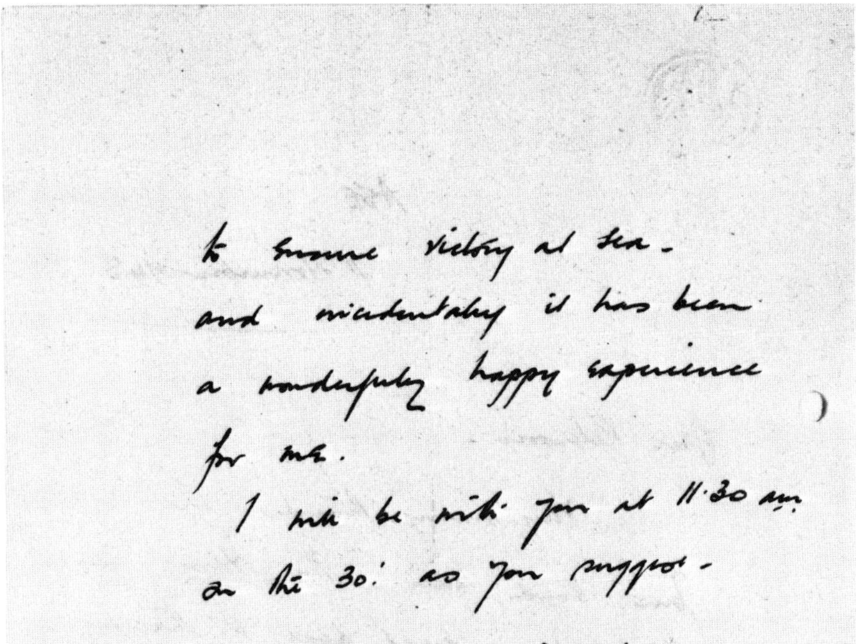

to ensure victory at sea.
and incidentally it has been
a wonderfully happy experience
for me.
I will be with you at 11.30 a.m.
on the 30th as you suggest.

Yours very sincerely
[illegible]

Fig.48 continued

asking for a 30 lb sample to try on 5 tons of wheat. Their attitude to beings like us, as a lower order of life, is amazing—not to say insulting.

November 13th

Leedham has called a major meeting to discuss the general adoption of our multi-channel set—next Thursday. Am trying to get Coventry to take a broad view of minor projects which may lead to more. The latest is another of Grey's from MOS. Peters writes very helpfully.

Insurance money for Schuil's death at last received from Underwriters. Shall set up a small trust in names of Stevens, Waldram and Brewer to hold the capital sum.

November 15th

All well with valve preproduction. Pearce is in difficulties with supervision for magnetrons now that Brewer has to go over to valve life test. The future of magnetrons is clearly going to be a very active one.

At HO discussed with Winstanley and Humphreys his efforts to get certain competitors to join in standardising cooker components for interchangeability. There seems a chance of making a beginning.

Chelioti to lunch, to discuss staff and craftsmen consciousness vis-a-vis trade unionism. I think he felt that the Laboratories had started a 'better way' 20 years ago and were proving it out.

An Advisory Board meeting, very full of problems. Meetings of this kind, I feel, cut more ice than they did.

November 16th[425]

Internal discussion with B.P.D., O.W.H., J.C.C. and H.W.B.G. on reconciliation of rates of pay to technical assistants (youths as they approach 20 and 21 years of age) with craftsmen apprentices. All our groups of people are adjusted on a scale of 'ranges' rather than on fixed figures. Propose to bring technical assistants up to the 'labourer to craftsmen' range at 21 and thereafter treat them more as craftsmen than as student or scientific assistants, with special status for exceptional men, and merit pay for National Certificate; student assistants remain as at present. There are hints that many may now be de-reserved.

There seems no inclination on part of MAP to find us work to fill our AI group properly. Oboe is now discarded except for the receiver, the American makeshift having been adopted.

GR is going strong and a big meeting on the multi-channel set is to be held at MAP on Thursday. Even CRF still has considerable interest attaching to it.

Brundrett tells me that Dr. Sayers is being taken off magnetrons at Birmingham, and a question of someone to take charge of work there will arise.

Another TVC meeting which I hoped would be short and noncontentious. But the old devil of nomenclature reared its head again because other Services want to adopt the CV nomenclature and MAP are too jealous of their own nomenclature to agree a second CV number for their own valves. I left the Chair at the meeting to Mr. Harley (at 4.30) who looked as if he had two hours of it yet.

November 17th

CRT services meeting all day under the 'two meetings' arrangement. A good series of discussions and good feeling on the whole. Sharp, Jacob and Hopkinson are jointly substituting Jesty (in USA) and are managing well. GEC seems to have a very large share in all the experimental work and to be on top of their job.

November 18th

Winch reports that a mechanical instead of an electrical mortar detector is looking the more promising and he and Hill are hard at it.

Gavin whispers a rumour that the Germans, having captured an ASV (old type) set which used the VT 90, i.e. the first of our micropup series, are busy making micropups. They seem to be two years out of date!

Megaw is back from one of his propagation tours in Wales visiting stations on high points, but that on Snowdon is not yet installed. In the absence of telephone connections they improvised their 6 cm channels over 50 mile links and got first class speech.

Chirnside reports having (with Dauncey) made 3 cm sapphire rod which has been sent to Salford for trial.

Cameron reports final discarding of the 230 A. searchlight project by the Army but the Navy are still keen.

Air Commodore (newly appointed) Helmore came today to go round as many of our valve preproduction groups as he could manage in one and a half hours and to talk to the group members. It caused considerable excitement and I wish he could have spared one more hour to complete the tour. He apparently did the job well.

November 19th

Bligh and his group had a field day at MAP yesterday when they had a service operational meeting under Leedham's chairmanship to settle finally to adopt our multi-channel set. They would like finally to have twice the number of channels and double the wave band tuning—but we had to urge that that was a gamble and could not be guaranteed.

November 22nd

Another preproduction record—2681. The main pressure is for the high altitude square block magnetron for Dr. Fraser at TRE. The CV 108 substitute X band magnetron is getting made but more slowly than we hoped. Eight were made good last week.

A PF meeting last week gave Donaldson some more rugged valve development—for the batteries to be used with the valves have just been defined! Truly the Admiralty may complain at the pedestrian progress of the War Office cum ADRDE methods.

The telephotometer, Schuil pattern, has arrived at Colombo where the experiments are in progress and seems acceptable. Salford will probably now put through a batch of 20 or 30.

Breadner and Simms brought me their first specimen of RG discs with four pins and robust pin seating. These are going to the Gramophone Co. for comment. Hitherto they have not been too willing to adopt improvements.

I attended at the Board of Trade today on behalf of FBI [Federation of British Industries] with Larke and Slade to enlarge on our recent report on research in industry. A lively discussion in which I seemed to be providing much of the provocation.

November 23rd[426]

We (RL) had an invitation to Union Works to witness Watsons Electric [Electro-] Medical exhibit by Mr. Caldwell of X-ray equipment for tuberculosis testing of the population. Everyone was most impressed by the originality and effectiveness of the equipment exhibited.

I succeeded in presiding at a good special meeting at the IEE of the Science Abstracts Management Committee. I had been apprehensive of harm from prejudice and intrigue.

November 24th

A CVD policy meeting but nothing outstanding except a discussion on how to get centimetre frequency bands suitably allocated to the many projects and groups of Service users.

Have been summarising to the Admiralty the first six months valve preproduction activities which have [been] analysed by Miss Lock. There is of course a large margin for the moment because many invoices due to be received [in] next six months are not yet in for payment. It looks however as if, for the next six months, we could reduce prices by 5 or 10%. But I have to look into the matter more closely yet.

November 25th

Discussed at length with Marris the work for his group in view of the absence of any proposal from MAP for a follow-on job to replace AI and Oboe. There are several jobs including television looked at as a gap filler.

G.M. Wright put forward yesterday a proposal for a proper research into the effectiveness of differently modulated transmissions judged on signal/noise results.

Donaldson is being recommended for sending to USA on PF valves.

A photographic CRT meeting was held today with Kodak and Ilford representatives and these firms readily agreed to co-operate and apparently to adapt their emulsions to the kind of powders we desire to use.

November 26th

Espley reports that GR is going through its blast bump and crash trials with flying colours—so far. It has still a few more ordeals to survive—but it is earning more good marks. There seems to be just one enigma which has turned up once or twice which Espley will have to resolve.

The Helmore contract for £13 500 has come along. It will probably be handled more by Marris than by Bligh who has more than he can tackle at the moment.

Goodeve lunched with me to talk over things in general and gave me an account of his recommendations for post-war research as it affects the Admiralty. I raised with him the question discussed with Brundrett recently as to the continuance of the Admiralty CVD research and development contract with GEC Research Labs. Goodeve expressed himself warmly favourable.

November 29th

NT 99 is really leaving the place at last. It begins its last week of preproduction here after two years during which 26 467 valves have been made at a price of £4 10s. each.

Farren is up from Fairlie and says the A – S experiment still shapes well. There are other ideas also for similar weapons there! It is a case of one concern going ahead and stimulating others to have brain waves.

Had talk with Laycock re. possibility of work on new photocells in order that GEC may not get still further behind. It appears that he can do something about the problem and I hope we shall be able to pull up some leeway.

Jesty is back from USA this morning and had a great welcome from his group. He speaks most enthusiastically of his reception over there and of the general goodwill towards the GEC and its Research Labs.; Dr. Black was a good friend to him. Many disputed matters were talked over and contacts turned from impersonal to personal ones.

Report on our rod sapphire from Salford which has been cut up and polished is quite favourable and we must now take the idea further. It probably means an entirely new furnace and technique (Chirnside, Dauncey and B.P.D.).

The anti-creep alloy which Metallurgical Group have been trying (using thoria) is quite in the running and MAP representatives (Mr. Waterhouse and Dr. Sutton) have been here to discuss ways of speeding progress.

November 30th

Willshaw and Megaw feel that we have here unique centimetre wave measuring gear and technique which Salford should set to work to establish as a GEC product and we shall draw up a memo. on the subject for Salford.

The large magnetron for ASE requirements is not going very well either here or at Birmingham. We are going to scale up from X to S band our E 1487, E 1494 (CV 208) to see how far on the small type of design we can approach towards the high output desired.

Willett came today to say goodbye on leaving ASE and joining up with the Fifth Sea Lord on Fleet Air Arm Radio. He told us that C in C Home Fleet had said that our radio [radar] had revolutionised sea warfare in a comparable way to the change from sail to steam.

On the surrender of the Italian fleet the Italian Admiral told one of Willett's officers that the British radio was so good that the Italian fleet had been forbidden to meet the British Navy either by night or in bad visibility—very encouraging. We are sorry Willett is going from ASE and so is he.

We had a demonstration from Megaw of the 6 cm propagation work in progress by the Inter-Service Scientific Committee on which he sits. I fancy that 50% of the work is falling on us.

Ramsay brings from the CVD modulation committee a determined attitude by TRE that our hydrogen diode must go forward. They report it as entirely satisfactory.

Bell's high power CW heavily tooled triode (first specimen) is on the pumps and looks a good job. He has three valves in hand reaching up to greater heights:

E 1457 pulsed transmitter for centimetre wave (CV 90 development)
E 1496 high power CW triode of the NT 99 pattern (Gt. Dane)
E 1495 Big CW heavily tooled valve

December 1st

Heavy bombing of Germany, dogged progress in Italy, and rapid progress in Russia are the features of the war at the moment.

Winch, Macfadyen and Michaelis have a field day at Larkhill on their mortar detector when the brass hats are to be present.

An MOV meeting on transmitting valves. There seemed to be much progress with new types and an enterprising atmosphere.

Much discussion on use of staff after the war and numbers required and very diverse views as to whether we shall be up against a gross excess or a serious shortage of people and jobs.

Evening rehearsal of my Royal Institution discourse on diamond dies.

A Fire Control project meeting at West Drayton attended by Dunham and Marris put greater importance on the job we have taken up with Witton.

December 2nd

I was met at 8.10 this morning with the unwelcome news that a security 'crash' man was in the building and asking to see me. He had walked straight past our three gatekeepers. A bad comedown, which leaves me despairing.

MAP Valve and Components mandarins are getting bitten with the dispersal complex again and Mr. Harley of IVPC etc. is putting up to me the need for securing the very vital activities which (to them) are centred in valve preproduction at Wembley. So here is a tenth-hour threatened iconoclasm to be handled!

December 3rd

A Research-Services meeting with Dr. Railing and Mr. Gamage. These are useful occasions. Spent remainder of day at Royal Institution with

Leeds, Bowtell, Miss Anthony and Miss Munro, preparing for 5 o'clock lecture on Diamond Dies. A good and friendly audience paid many tributes to our outstanding work on dies—and as usual the team organised the lecture splendidly.

December 6th[427-428]

Valve preproduction is suffering severely from epidemics of flu. Some groups have 20% away.

Professor Truman and Mr. Stringer (his assistant) called to talk over Breadner's 4-pin RG construction which appears obviously so much better from a strength and production point of view that Dr. McGee's objections to it appear as being (very) special pleading. Prof. Truman had a difficult task in striking a balance between Hayes and Wembley but we felt it necessary to be stiff on the point for McGee seems to have no technique for co-operation but 'do what I tell you'. We would not mind this if we had respect for his techniques and judgment.

Mr. Pearce and Mr. Minns of Watsons Electro-Medical came to talk over post-war developments. They are very much on the map through their tuberculosis mass examination sets and there is a healthy desire to extend the markets along similar lines post-war. A regular liaison meeting was agreed to with us which Mr. Minns will convene and Gossling will act as the focus here.

December 7th

A propagation meeting with G.M. Wright in the chair. I have not heard its doings yet. Lunched with C.S. Wright (DSR) who came to champion the cause of Dr. McGee of Gramophone Co. but stayed to say that we should carry on with our device. I showed Wright Breadner's 4-pin disc construction and he was obviously impressed with its mass production features. Several samples stood plunging alternately in boiling water and liquid air.

December 8th[429]

A heavy morning's argument with Mr. Harley and Dr. Truscott helping the former to inspect the laboratories with a view to his instructions to see about arrangements for removing the most vital valve work out of the London area. It was certainly an eye opener to him to see the place, and at the end of a morning's inspection he declared himself more impressed than ever with the vital work we are doing and equally baffled at the impossibility of moving us. The matter was left that Truscott and Brundrett should meet to discuss the subject. I think I know that Brundrett's influence will never be given to an act of folly.

Discussion with Dr. Railing on Fido—Humphreys and Ryde present to review the costs and advantages of our scheme in comparison with the petrol burner solution of fog clearance now in favour. Our scheme seems to have an overwhelming case financially—and a strong one technically. We are to get Mr. Geoffrey Lloyd and Sir Donald Banks along to see the demonstration.

Jesty gave me a talk on his visit to the USA on CRTs. I think he showed the flag well and did credit to his Lab. and country. His reception there showed clearly what warm friends we have in the USA. The next thing will be a meeting with the Services here to review all information.

Macfadyen speaks very warmly of Michaelis's share in Mortimer II alias Muriel alias Mortar detection. He has been handling the electrical solution whilst Winch and Hill have been perfecting the optical solution.

Megaw has had a party from USA to see and discuss the centimetre wave propagation work.

December 9th

Had telephone message at 8.15 am from Farren at Fairlie reporting 'highly successful' trials on his A–S under water investigations. The staff at Fairlie are equally interested so there is likely to be some activity in the A–S field for a time. I shall be delighted if we strike success in this project.

A GEC Board meeting at which Dr. Railing discussed post-war plans for the Company—the first time I remember a discussion at the Board on the future as distinct from the past and the present.

I have agreed with Mr. Marriott my financial report to the Admiralty on valve preproduction. This time it is a non-committal statement for the first six months of the financial year.

December 10th

Le Rossignol rang up this morning to say he is safely back. He will be in on Monday.

Had a heavy morning with Mr. Freeman of DSR Depot [Dept.?] Admiralty who watches the RG job at Gramophone Co. Everything in the job seems to hang round the sacrosanct opinions of Dr. McGee. There was some very straight talking with but a partially successful ending. Freeman is completely sold upon the immovable nature of Dr. McGee's outlook and this ends every argument.

Jolley was shown by Helmore the minutes of the first Service Searchlight Committee at which apparently the Helmore large ground searchlight was accorded but low priority and spoken of somewhat slightingly. I understand Helmore does not like it—and is going to take steps.

December 11th

The war is going as fast as can be expected I suppose. Ding-dong fighting in Russia and Italy. Excellent results against U boats and the bombing of Germany (day and night) when the weather conditions permit. The Germans keep 'game' in their fighting, which is good. They must be well beaten and this won't happen unless they keep fighting to the end and realise they are smashed.

December 13th

Le Rossignol back from USA and Canada where his study was magnetrons—with Rushforth of BTH. He had the warmest reception everywhere, particularly from the Schenectady folk under Coolidge. He was impressed by the receptiveness of the Canadian staffs and the reverse of the USA staffs. Tuneable magnetrons were one of the chief novelties. Wastage in production and manpower per valve made are substantially higher than with us at Wembley.

We have fairly severe flu epidemic, 15 to 20% absence. Valve preproduction suffers but not proportionally—2323 for the week.

Farren now back from Fairlie. Their model appears to do much better than was expected—possible range 700 yd with only 4 watts in the transmitter when 20 watts is available. Squadron Leader Cropper has put up an excellent effort with Wing Commander St. John and others and 30% of the result is due to them—30% to Fairlie staff and 30% to Wembley! All are delighted and very generous congratulations from Dr. Roberts and his team at Fairlie.

A TVC meeting all afternoon (three hours) with nomenclature again the chief bone of contention between Hecht and Hughes.

An English Speaking Union meeting in evening where one meets a number of overseas and USA scientific representatives.

December 14th

Espley reports that the gun and all auxiliary gear connected with GR is designed and executed up to the highest standard. I hope it will be in time for this war but it seems to me heavy for a portable piece.

The first of a new series of meetings with Watsons started yesterday Gossling looking after matters from our end. The object will be to consolidate and extend gains made recently by the Company in military X-ray equipment.

Wing Commander St. John and his colleagues spent the morning here discussing results and prospects of the successful A – S trials at Fairlie last week. They had plenty of adventures including the smashing of their gear under water. But all conditions for the trials were ideal and this contributed to the success. Talks are taking place behind the scenes and a proper conference will be held when the punctillios are settled between the Admiralty and MAP. We shall probably be asked to make the 'brain' of the first six prototypes. The Government seemed to *have passed over* development to the USA of all automatic torpedo projects—a rotten idea.

December 15th[430]

A CVD (RDF) meeting all day. The chief interest in the receiving meeting was triodes which would reach down to 8 cm and below—if possible to 6. James showed our latest (smaller) model of the CV 90 which can reach to 8—but on harmonics lower wavelength may be possible.

The transmitting valve meeting in the afternoon was enlivened by a number of incidents. Le Rossignol gave some of his USA experiences. ASE (Mr. Ratsey) was insistent on his valve deliveries of CV 69 in spite of our pleas of influenza epidemic in preproduction groups. He also came back on us over the CV 76 life which is nearer 400 than 1000 hours which he asks. ASE were warned all through that they were straining at a result which was beyond present techniques. They don't like it when they have to share the responsibility for the gamble!

December 16th[431]

A fiery meeting at the Gramophone Co. at Hayes over the windows for Project 1967 (RG). It was virtually decided to run through a trial of the Wembley design but several months was claimed for this. We agreed meanwhile to show the same measure of good faith in trying to repeat the Dr. McGee design as they show with their trial of our design. It will be interesting to observe the result.

I came back to Wembley to find the canteen crowded out with visitors from all quarters—a not infrequent experience these days. Dr. Fraser of TRE button-holed me about the hydrogen diode and the square block CV 64. Both urgently needed, and depending exclusively upon Wembley.

A meeting with Dr. Kahn and others of arc welding at Wembley at which a programme was fixed up.

December 17th

A meeting at HO to review progress towards a GEC television policy. T.W.H. presides and A.B. and L.G. present. Altogether a satisfactory situation in that the course was made clear and broad lines of work were defined. It was agreed that we must tackle the transmission side plus iconoscopes actively and as soon as war conditions permit; Marris, Bligh and Espley from Wembley—Jesty unfortunately away on other duties.

A conference in the afternoon at S. Kempton [Kensington?] on the utility and objects of students' vacation courses.

December 18th

C.S. Wright rang me re. A – S. He appeared very delighted at the report from Fairlie of our experiments and tells me the matter is certainly going forward. He said he would have to let me down over the Project 1967 (Gramophone Co.) issue since he was quite sure they were not going to play fair with us and he could not insist on their doing so. He refrained from saying on the the telephone what he thought of them. I promised Wright that we should not be childish over the issue.

Dr. Railing has now heard from Mr. Geoffrey Lloyd (Ministry of Petroleum) re. a FIDO demonstration here. This will take place early in the New Year and the Minister will bring Sir Harold Hartley, Mr. Hartley, Prof. G.I. Taylor and Sir Donald Banks.

A liaison is just fixed up by HO with the Gramophone Co. for the application of compact sources to cinema projection—a promising development plan.

Ramsay is just back from TRE re. his hydrogen diode which is required for a crash programme on H_2S. We can probably meet the requirements for Dec. Jan. and Feb. off Ramsay's shelf and larger quantities after that from preproduction but rub will come when they want quantities from a factory. As far as I can see we shall hope for the best for what follows our preproduction.

The Ministry of Food are getting busy on Alumicide and ask if we can give particulars of the amount of raw material we should want for large scale supplies.

December 20th

Preproduction keeps up but some of the magnetrons are difficult to get into their strides because of the cussedness of the things. There are now three new ones competing for priority, CV 208, CV 69, CV 99, and these and others will have to come up at the allocation meeting on Wednesday.

A Lab. Stewards meeting. Very useful for ventilating all sorts of matters connected with the running and administration of the place. Flu epidemic still pretty bad.

The first Lab. Musical Society's performance took place this evening under Macfadyen's conductorship. An excellent and intimate meeting.

December 21st

A rather disturbing letter from Brundrett saying that, in spite of all said, he must support the move to require us to prepare to move our magnetron preproduction in certain eventualities.

Squadron Leader Cropper tells me that the Admiralty are making big proposals for manufacture of our A–S device. A major meeting I understand will take place on Friday.

Jesty reports first meeting of Hecht's 'Screen Panel (CRT)' yesterday. It apparently went off very well and Hecht was complimentary regarding the work done on the subject at Wembley.

December 22nd

A busy day. Simultaneous meetings here of CVD (GEC) and also a special CRT audience to hear Jesty's American report on his study of USA practice in CRTs.

Had talk with Brundrett and Truscott on the dispersal issue. We are only required to concentrate on the carrying on of CV 76 and getters at Shaw, should this be necessary. Brundrett assures me that it would only be required if London was to be put out of action, these two items being key products.

There was much talk of the hydrogen diode and the objections of the production group—also of the double pentode. The productionists (IVPC) are taking the attitude that no valve required in large numbers can be tolerated if it requires bench pumps and other than machine sealing in. This is because it will interfere with equipment for magnetron making which has also to be handled on bench pumps and has been trebled in demand.

December 23rd[432]

Carroll came for a conference on A – S from DSR Admiralty, to arrange machinery for carrying the research as a major piece of development. There are a number of links to maintain—DTM, Air Ministry, MAP, Fairlie and one or two other departments I do not as yet know. We requested the usual progressing committee to meet regularly to facilitate linkages. Squadron Leader Cropper and the Halton training centre at Wendover will probably take a share—the former on the quartz side and the latter on the design of the body. The 'brain' which is our main function is relatively straight forward to us but an apprehension to all the rest.

It may be decided to use USA bodies and regard the Halton design as a second string. Farren had some pretty oscillograph photographs of the signals taken during the last Fairlie trials. This work will keep Knox and his party busy and will help to fill the vacuum resulting from the cessation of work on any major project, such as AIS and Oboe for TRE.

December 24th

Everyone very glad to be breaking off at mid-day today for the Christmas break—though no travelling and limited food are making this a more unusual Christmas than ever.

Freeman from DSR came today to tell me that Gramophone Co. and EMI are not going to show good will if instructed to put through tests on Breadner's mass produced pins.

December 28th

Back after the holidays and everyone cheerful at continuance of mildly good war news. I think people have been prepared for some Nazi terror during Christmas and nothing has matured except the sinking of the Scharnhorst and another of the many break throughs by the Russians!

Lunched with Good just back from USA on screw thread standardisation and inter-allied machinery for standardisation generally.

December 29th

A CRT meeting all day at which Jesty repeated some of his information from America. A generally lively proceeding but interesting. Some of Dr. Bradfield's attention to minutiae made one ask where the winning of the war comes in!

A telephone message from Mr. Pollard (Ministry of Supply) to mind our step over Winch's work as between the Department of Survey at Larkhill and ADRDE who alone must handle our contacts. The difficulty is that Colonel Hemmings of the Dept. of Survey is efficient and able and ADRDE Dept. there is heedless of time and orderliness and cannot run the trials needed. Colonel Hemming's people appear to prefer to deal direct with us.

December 30th

R.F. Proctor up from SRDE. He seems to have quite a major job of it at Christchurch in charge of their measurements section involving organisation as well as technical guidance.

There seems to be much 'atmosphere' about our A – S job, so much so that USA refuse permission for anyone not in uniform to see the American weapon at Alexandria (Scotland) to which we shall probably have to fit our device.

Lunched with Gamage and discussed (*a*) industrial research (*b*) education in the GEC.

December 31st

Alumicide is making progress towards bulk trials and there is marked interest in it in several quarters. The problem of how to manufacture and also that of lowering the price are not so easy.

Gavin reports 80 'Air Grocer' sets started at the Labs. and probably 400 to follow, the contract to go to Claud Gen by request provided Wembley will test and inspect.

Macqueen called to discuss request from new Television Enquiry Committee for evidence from RMA on post-war policy. The political factors are not to be easily reconciled.

The last day of a year during which Providence has greatly favoured our cause. It has been one of almost unrelieved successes and dogged progress and wearing down of the enemy. It looks as if Russia had almost got the Germans on the run in some areas of the front. In the West our great task has yet to start.

1944

January 3rd

In spite of short week, preproduction was 1800 valves. In the past 12 months we have made 88 000 valves against 52 000 for the previous 12 months.

Bell has gone down with diabetes—bad luck for him. Aldous will carry most of his load whilst he is away—but he is a very key man; all the Government Experimental Establishments and CVD will miss his constant and uniformly efficient help.

Had Lt. Commander Jones, Professor Macrae and Lt. Abrahm here representing Admiralty Operational Research on A–S problems. We discussed (Wing Commander St. John, Cropper and Griffith present) the problem at length in order to make sure that the lobes, ranges and circles etc. gave optimum conditions. Interesting—and it looks hopeful provided the 'head' of the transmitter–receiver can be streamlined.

January 4th[433]

It seems very difficult to get sorted out the Admiralty–MAP machinery for handling the A–S project—so many departments and people with a finger in the pie. We have not yet got our committee appointed for progressing the project.

A TVC meeting this afternoon with no contentious matter leading to asperity. I hope it is an omen for the New Year. Our double pentode valve was christened CV 221 [222] and the CV 215 (hydrogen diode) was even tolerated with a caution. Ramsay has some ideas for a simpler alternative.

January 5th[434]

McLeod our popular foreman of the metal shop has a British [Empire] Medal award in the New Year's Honours List. I am sure it will be generally approved.

Went to Exhibition [Building] with Ryde to see striking FIDO demonstration using point source of light projecting a beam through the hot air streams onto a screen. A most effective show. Was given a very nice little luncheon party there. 'XL' is a happy ship and enthusiastic.

A CVD policy meeting. Lewis was there back from USA. It has been recommended that high power magnetron development shall come back to England but they do not add that it was a pity it was ever exported! All's well that ends well. The Beaconry and IFF [Interrogation Friend or Foe] is in the middle of very troubled waters at high levels. The E 1440 valve will probably be required and the CV 55 holds the field for transmission. The CV 90 always had a poor chance on account of the politics promoting the lighthouse valve—but its chances were further jeopardised by the National Union Company (who had the contract to make it) being in very low water technically.

January 6th[435]

Our high power magnetron work which Willshaw is mainly handling seems to be going fairly well both in the S and the X bands. Coales at ASE expresses himself interested in our three times CV 76 power—a proposal which Ratsey told us a fortnight ago was of no interest. He insisted on ten times the power. This is not the first time that Establishments have wanted to take steps of 'ten times' which are usually too great.

I spoke to C.S. Wright yesterday about control of the A – S job which he agreed was having a sticky time and I told him that we should have to refuse to play if the Admiralty refused us a full share in knowledge of the work in which we are engaged and full access to those who have to deliberate and come to decisions upon it.

January 9th[436 - 438]

Have spent the weekend among the red socialists at the Nuffield Kremlin at Oxford where I led off the first of the three discussions emphasising the need for scientists inside industry as distinct from industry being helped by external research organisations. All the people whose opinion one values approved and supported, but the Oxford foreign economists vilified industry generally and tried to base their vilifications upon their misrepresentation of my remarks. I was very glad they had been typed and circulated in advance.

January 10th[439]

Still a lot of flu but the groups on preproduction do well nevertheless. Mrs. Stirling, chairman of the Middlesex manpower board, came to see our work. Middlesex is one of the most difficult districts of all for labour and no priorities seem capable of getting people. Circular from MAP saying that the priorities which have so far protected a number of our younger men no longer run and we may have difficulties with some of them. So far we have had the double priority of 'valves' and 'MAP'. One of these is now going.

Ivor Jenkins has a job at Whiteheads and we shall miss him very badly from our strength on furnace design and operation for metallurgical work.

January 11th

One more visit from TRE (Dr. Warren) on CV 215 (hydrogen diode) in view of IVPC objection to getting it made. Once again we point out that this is no more 'difficult' to produce than a relatively simple transmitting valve and that most of the RDF valves of the past three years are more 'difficult' to make than the CV 215.

On the A – S issue the winds blow from all quarters. The latest is that its surface vessel uses are much more important than its U-boat uses. It is stated that Blumlein worked out, and reported on a system like this before the war!

To FBI industrial committee where we decided on the establishment of a bureau of FBI to foster research in industry and all that that implies.

January 12th

Cameron and Thomas showed me the latter's Meccano gadget for chopping light from a source of which the polar curve is required and putting it through a photocell and after image [amplification displaying it on a] cathode ray tube so that the slowly circumnavigating mirror round the lamp gives a polar diagram trace on the CRT.

A meeting of GEC Radio Liaison Committee to thrash out post-war radio programmes. We covered half the ground and meet again in a month.

January 13th[440]

Another visitation from DSR Admiralty to ask us to undertake developing a high supersonic transmitter and receiver for a special purpose not to be mentioned. It looks interesting. 'K band' pressure by both Megaw and Willshaw. It is a promising line for bombing Germany in view of success of H_2S in recent bombing expeditions. We discussed K band gear with all concerned and must see how we can take this up more energetically.

In the H_2S gear we hear that our E 1487 is more acceptable than the BTH MX 57 the latter being powerful enough to cause breakdown.

January 14th

Ward came from TRE to urge importance of production on E 1516 (the improved waveguide cell). It appcars that 1500 per month are wanted as soon as may be, i.e. four times our present rate. It is going to be impossible unless labour is sent to us—but we shall see at once about getting the gear and hope that the urgency will produce the people.

Had discussion with B.P.D., O.W.H., J.C.C., H.W.B.G. re. a better organisation of our personnel arrangements at the Labs. The education issue, as well as the calling up and ultimate demobilisation and also the income tax handling, all require very systematic records and handling.

Had meeting at short notice with Carroll, Captains Gladstone and Richards and another from RN torpedo factory about the management of our A – S job. The objective is to be surface vessels and we are to have the small working committee I asked for, with Commander Clark from RNTF as general secretary.

The BCURA [British Coal Utilisation Research Association] came to get ideas as to building design for the new laboratories at Leatherhead.

January 15th

James is pleased as a result of his talk with Dr. Lewis (TRE) on the E 1440 disc seal valve. The report from USA is very favourable and the valve will be made there. Exchange of visits is favoured; meanwhile 2000 are likely to be wanted from here.

January 17th

Our CV 208 (X band magnetron) is showing no more than 50% wastage which is good considering its early stages. It is not easy to trace down cause for low efficiencies which is our major trouble. It is not centering any longer. It may be lack of symmetry of holes and slots.

We reviewed the exhaust pump service for maintenance in good condition. On the average one pump per day goes wrong or is broken.

Conference with Megaw and others re. K band experiments with their centimetre wave transmissions. It is very desirable to include them but we are at our wits' end to see immediate means of developing the CW sets and techniques and of making the K band magnetrons.

January 18th

Hecht was here yesterday over Helmover and the weight of work for DCD on sets. He said that the advance version of the multi-channel set was being approved and he presumed we should be required to handle the development.

A message this evening that Heavy Alloy for projectiles is up in some form for discussion tomorrow.

January 19th

Meeting at Royal Society to select Superintendent of Physics Department at NPL. An interim appointment of Higgins decided on.

Heavy Alloy is being proposed apparently for the nose of projectiles to hold the surface while the hard tungsten carbide pierces the armour. Mr. Gamage had seen the Controller of Materials who wants to know if we can supply. I rang Admiral Pridham, the President, Ordnance Board, for inside information. He promised that his Colonel Temple would come and give us a talk.

Dr. Black is over again from USA and is visiting us tomorrow and Friday.

January 20th

Royal Society Fellowship selections.

I finished a letter to Humphreys explaining the views of myself and others on the introduction of a Trade Union regime into the Laboratories in place of the methods we have followed so far of regulating staff relationships (Fig. 49). The ASW are pressing as a Union for establishment of salary scales and the like.

Sir George Thomson (now Advisor to the Air Ministry) came to see Fido and was much impressed both by the visual demonstration and by the result achieved. He favoured the scheme only as post-war effort and said the flying personnel appeared to like the greater height (than 100 ft) afforded by the open burner scheme now being used on a number of bomber aerodromes.

CCP/NKB.

January 21st, 1944.

Dear Humphreys,

I feel there is a danger that my attitude to the proposal to introduce Trade Union methods and traditions into Wembley may be misunderstood; that it may be felt to be based on a die-hard and antiquated opposition to Trade Unions, whereas in fact the reverse is true.

To get my point of view aright we need first to remember that Trade Unions were born (during the first half of the last century) of the loss of jobs and depression of wages produced by the Industrial Revolution, through the introduction of labour saving machinery and consequent over-production. Workers in different trades banded themselves together in protest against the self-seeking and unsympathetic employers; riots were widespread and the strike was the only weapon available even to the more orderly. During the hundred years which have since elapsed Trade Unions and Employers have developed a much more satisfactory modus vivendi. Agreed negotiating procedures have been substituted for the law of force, and a spirit of compromise has largely taken the place of the bitterness and hostility of the early days. In the meantime, the lot of the worker has been improved beyond measure.

But, although I have fully realised all this, it has seemed to me that the Trade Unions labour, perforce, under two serious handicaps. The first, that they were born of antagonism between two sections of the community and that however much that antagonism had been pushed into the background it remains, inevitably, the essential basis of the relationship between Trade Unions and Employers Federations as we know them to-day, and as a result those in industry whom Trade Unions and Employers Federations purport to represent are supposed to suffer from the same antagonisms. The second handicap is the unavoidable concentration of Trade Unions on the mere material aspects of the lives of the men they represent. I say unavoidable, because the essentially legalistic basis of the relations between the organised representatives of labour and of management necessitates that discussions and agreements are confined to those things which can be defined so precisely in words that they can form the basis of arbitration or legal procedure.

Now let it be clearly understood that I by no means belittle the importance and value of the material things of

life/

Fig. 49 Copy of letter dated 21 January 1944 from Clifford Paterson to O.W. Humphreys on staff relationships

life, but I know that I shall be on common ground with the great majority of my colleagues at Wembley in contending that once one has enough of the material things to banish want, and to put one in a position comparable with that of the average citizen of equivalent ability and responsibility, thus enabling one to achieve for one's family a standard of living which permits them to take a proper share in the life of their social circle - once this stage has been reached I say, things other than the material begin increasingly to predominate and control one's outlook on life. These things are difficult to define, but they include such things as professional standing and influence, loyalty, - loyalty to our own ideals, to our friends and colleagues and to the establishments in which we work - happiness in our daily associations and the satisfaction of knowing that we can unreservedly put all we know into our jobs and make our contributions to the well being of the community and to the advancement of mankind at large. Let us call these things the imponderables.

Now when I founded the G.E.C.Research Laboratories 25 years ago in company with able collaborators, I determined to see if it were not possible to establish new conditions and relationships which should not involve the two handicaps of the Trade Union Regime, to which I have just referred, and without sacrificing any of the beneficial features of Trade Unions. My vision of the Laboratories was that of a community of workers united by the community of interest in the application of Science in industry led by a Director who regarded it a primary responsibility to safeguard the material interests of his colleagues, so that relieved of inhibitions on this account, they could, unitedly, concentrate on the more imponderable values of life - on the building up of the sort of place in which they would like to live and work - a national asset of whose achievements and standing they could be proud as their own creation - in which discipline would arise from the team spirit imposed by the community itself, from a reluctance to "let the side down" rather than from a fear of unpleasant consequences of breaking rules and orders made by an autocratic management.

It was clear that in an establishment of the kind which I envisaged, a man would measure his contribution to the output of the team in terms of what he was and what he did, rather than of the hours during which he worked, and that similar considerations would guide the Director - the leader of the team - in determining (in consultation with colleagues on the staff) the material reward which was each members' due.

In a productive concern there may be a fairly clearly defined relation between the number of hours worked or the number of articles made and the value of a worker's services. But with a Research Laboratory Staff where no such relation exists this is anyway impossible. I, in common with others, was convinced that "hours" and "output" as the factors determining a man's value and reward should be replaced by what a man is and what he does, what he contributes to the pool materially, intellectually and spiritually! The imponderable contributions of an individual being

more/

Fig.49 continued.

-3-

more weighty than the ponderable ones.

Such was my vision and that of many of my colleagues and it was soon apparent that things at Wembley were working out in the way that one had hoped. Our conception of the "better way" for industry - however unorthodox or impracticable some people may have thought it was - was proving itself in practice. As the team increased in size each new member tended to absorb the principles and traditions which the teams itself had created and it was clear that a lamp had been lighted which was going to shine throughout our country and beyond. In fact I know that the influence of Wembley has been felt very far afield, and 21 years plus 4 years of war is a long enough period to establish my contention.

Shortly after I started on my venture there was a more widespread movement amongst Scientific workers which had as its object the development and spreading of the Scientific outlook and the broadening of its influence on the life of our people. This movement - the foundation of the Association of Scientific Workers - I warmly welcomed and I was one of its earliest members. Its ideals appeared to be those of the team at Wembley and it looked as though, with its much wider membership, it would help to speed the adoption of the doctrines which I believed we had in common.

But gradually the A.Sc.W.changed its ground. It moved further and further from the cultivation of the imponderable interests of its members and of industry, towards attention to the material ones.

And now, 25 years after I founded these Laboratories, it is suggested, in effect, by a certain number of my colleagues that I throw aside all that we have worked for during a quarter of a century, that I admit our successful venture to be a failure, and substitute for our "better way" the traditional Trade Union regime.

I am sure that no-one having heard this explanation will fail to realise my sincerity or to understand a sense of concern that some newcomers who have joined us under the stress of war should have found it difficult to appreciate the assessment of values which has governed our affairs as a scientific institution. This sense of disappointment may have been contributed to by me on account of the lack of opportunity to make these ideals clear to newcomers. Even at this late hour I can still find it in me to hope that wiser counsels may prevail and that Wembley may yet be spared the divided loyalties, the inevitable frictions and loss of internal harmony in which the introduction of the old Trade Union methods into staff relations must result. I would ask my Trade Union colleagues to take thought for a moment and to ask themselves if they are really justified, either in their own interests, or in those of the much larger number of our staff who are outside their ranks, in throwing away what we have cultivated. in exchange, at the most, for some small

material/

Fig.49 continued.

material adjustment to certain people, of a kind which, if justified, has so far been obtained by other means. For let this be clear - the staff conditions, and relations between staff and management which we know at Wembley must soon wither and die in such an atmosphere, and, I fear that those now most strenuously pressing for the change would soon find that they had sacrificed the substance for the shadow.

Yours sincerely,

Fig.49 continued.

A contract for Joker (A – S) has not yet come but we have a note from Captain Gladstone (DTM) [Director of Torpedoes & Mines] confirming the arrangement of a proper committee and who shall attend it.

January 21st

Visit by Sir Harry Railing to inspect FIDO demonstration in view of the prospective visit of Mr. Geoffrey Lloyd next Friday. Edwards, who has been for nearly two years on our jobs at TRE, came to discuss his position for he feels he can now be spared and can return to Wembley. He has been fostering the Oboe receiver side. The whole equipment has undergone many changes—probably sound but demoralising to those who are carrying the load.

Sir Harry Railing told me of serious difficulties in securing an emergency labour supply in Birmingham to [for] Dunlops—and most upsetting withdrawals are being suggested. Tests by Dix on CV 79 and 89 on No. 10 set link took place this week but as usual SRDE and Butement (as far as I gather) produced a mess up and after hanging about for three hours our people were given one hour to test all 50 valves they brought for examination. Only ten were tried. The upshot—as far as can be judged—is to show a margin of safety in performance on limit valves.

January 22nd

The hydrogen diode again on the MAP carpet yesterday—also the E 1516, and their possible production to aid a crash programme for H_2X. We made the same statements as before as to our expected preproduction. The same parrot statements were repeated by Dr. Truscott's representative as to the 'difficulty' of the hydrogen diode. Shaylor is confident his E 1516 waveguide deliveries will materialise if the labour is available.

January 24th

We had bombs on Thursday night between XL and the Labs.—the liveliest night for 12 months.

Preproduction last week over 2400 valves and CV 208 is up to our guarantees this month already. Altogether a good week.

Marris and Farren are back from a meeting on 'Joker'—application of our A – S work to surface ships. A. [Admiralty] apparently voted it a successful meeting. The other A – S application 'Dealer' is apparently also going forward.

MAP have a signal that CV 92 (NT 99 as was) is not going well in the USA—short life and bad vacuum. They proposed to hold a transatlantic conference by some new system and D.H. [A?] Wright and Davis of MOV will take part. The E 1440 will also come under consideration.

January 25th

Visit from Colonel McNair from Ordnance Board to explain to us their work in using Heavy Alloy for projectiles. An interesting talk, but the whole project is entirely experimental at present and it seems premature to discuss production of Heavy Alloy yet for this purpose.

January 26th

An RDF meeting all day—not very eventful and a bit troublesome.

Had meeting with Mr. Wheeler at HO from Ministry of Supply re. Heavy Alloy. Mr. Isaacs and Mr. Trapp were with him. They agreed with what McNair had said but we tentatively agreed that we would collaborate if desired with Brown–Firth. We are pushing on with experimental projectiles weighing some 40 lb.

January 27th[441]

An 'Atlantic conference' took place last evening over E 1440 and NT 99. Le Rossignol says it was a slow and dull affair over cable.

Rooksby and Thomas tell me that the work on Salford selenium rectifiers is promising and it looks as if we should get through the wood.

The Confidential History of the Labs. is taking up a good deal of my time and I fear it is going to be too long for reading the whole. The coming of age celebrations are waxing more important and will be a great occasion.

An A–S (Joker) meeting—the first of the regular ones—went off smoothly and well, Marris taking charge.

January 28th[442]

A Fido day and visitation from A.R., Mr. Geoffrey Lloyd, Prof. Taylor, Lord Suirdale, Sirs Donald Banks and Harold Hartley and Mr. Hartley. Ryde gave an excellent demonstration and all were thrilled at the shadow graph projection demonstrations of ours and of the petrol burner schemes. It was accepted that the matter is mainly for post-war use—but very desirable to get something going.

Captain Richards from RNTF [Royal Naval Torpedo Factory] sent, along with others, from Captain Gladstone to propose yet another project for GEC to handle—'Bowler'. It looked as if Coventry and we between us could handle it. Meeting at Coventry fixed for Monday.

A meeting of RP 60 gear (the fire control) job also being looked after by Marris. That also seems a good going concern. Dunham is the focus with Southworth.

January 29th

Here is the end of January. Great expectations and optimism regarding the Second Front. The Germans are toughly resisting everywhere and have their fighting forces in hand—constantly retreating from the Russians without undue loss, considering the enormous scale of the operation over there. Our air raids must be crippling the German industrial effort but they are not apparently breaking morale which is now maintained by fear of consequences. The relatively light losses—say 3%—are a tribute to all our counter measures.

January 31st

At valve preproduction meeting today, it was clear that the inability now to get us any women workers is not only stopping expansion on certain valve types like magnetrons and waveguide cells, but inevitable losses are not replaced and output will begin to go down. We have had exceptional requests in regard to both CV 215 (hydrogen diode) and the CV 114 class of valve. It is hopeless to consider them as things are. The RAF and other Services ought to be made to release thousands of girls in uniform whose time is largely being wasted.

The preparations for the Laboratories 21st birthday celebrations (of the opening of the building) go on well and I am hoping they will have a good effect in stirring up activity and general keenness. The Social Club is certainly handling things well and in the best traditions of the place.

February 1st

Marris was at conference at Spon Street yesterday to see if they will join in, and produce, 'Bowler'. All fixed up.

The second Lumifax Board meeting at their new works on the

Dr. Paterson

ASSOCIATION OF SCIENTIFIC WORKERS.

WEMBLEY BRANCH.

REPORT OF MEETING AT G.E.C. RESEARCH LABORATORIES 31/1/44.

Present. Mr. C.W. Humphreys and Mr. C. Vaughan Williams for the Director.
Miss. L.E. Grey, Messrs. K. Jackson. M.E. Ceal. H.V. Sacks. S. Levenson. G.B. Townsend. M.R. Gavin. W.J.C. Chaple. G. Ellinor. L.H.B. Knox. and Dr. E.P. George for the Branch.

Mr. Humphreys opened by reading a letter from the Director which gave his policy and method of building up the Laboratories on a basis of mutual co-operation, rather than upon the formal relations between Trade Unions and the Employers Federations. Dr. Paterson felt that much more could be achieved by this method than by those owing their origin to the Industrial Revolution and consequently still retaining some of the conflicts between the employer and the employee. He felt that much more had been and could be achieved by the atmosphere of close collaboration between himself and his colleagues than by the intervention of outside bodies.

Mr. Humphreys then went on to clear up some misconceptions on both sides during which all the members present took part in the discussion. It became apparent during this frank and extremely interesting exchange of viewpoints, that practically all of the anomalies which were the origin of our claim had already been or would be rectified retrospectively. Some difference of opinion still exists as to the relative emoluments of the juniors who live with their parents and those who have to be entirely self-supporting. There are also conflicting views upon whether technical and student assistants should be paid on scales comparable with those of the A.E.U. for their apprentices or with those paid by similar firms in this area to their staffs. During this part of the discussion Mr. Humphreys stated that the minimum gross salaries paid to our junior unqualified members would never be less than the minimum gross rates laid down for the corresponding A.E.U. age groups. The existing salary ranges were also examined and although the figures suggested by us are above the mean of these it should be possible for our members to obtain salaries equal to and in some cases higher than those recommended. Several concessions regarding discussions on any matters affecting our members were made. We may, at any time, discuss with Mr. Humphreys, as a Branch, anything which we may feel necessary knowing that it will receive his fullest and sympathetic consideration. Further, if any member feels that his salary is not commensurate with his responsibilities (in the widest sense) he should preferably see Mr. Humphreys himself, or should he feel that disinterested assistance advisable he may ask a Branch official to discuss the matter for him. It should be emphasised here that the discussions now proposed will only take place on the strict understanding that all official intervention (on the part of our Association and the Federation) or any threat of same is in direct opposition to the spirit of this agreement. Mr. Humphreys then stated that the guiding principle of those responsible for the direction of the Laboratories had been and would continue to be maintenance of the best interests of all concerned.

In conclusion Mr. Humphreys stated that he was extremely grateful to the members who had taken part and that he would take away from the meeting much useful information and wished the Association every success in building up its membership in these Laboratories, emphasising, that although the Branch is strong and representative the fullest possible use of the proposed machinery would be assisted by the greatest possible representation.

Although the direct reason for the initiation of these talks no longer exists we, for our part, feel that we have gained concessions of considerable value. These now include permission to use the main notice board for announcements without having to do so anonymously.

Fig.50 A. Sc. W. Branch notes of meeting with management of the Laboratories

Exhibition grounds. They are getting it well equipped and production is under way.

Dr. Rankin and his Fido colleagues from Earls Court called for a field day with Ryde's projection scheme. Rankin is very refreshing, and is immensely impressed.

I asked Humphreys to meet some members of the Association of Scientific Workers on my behalf last evening. They are wanting a salary scale for juniors and approached us as a Trade Union. I have been very apprehensive for months, of the effort (made I believe mainly in ignorance) to introduce a Trade Union spirit into our regime here and so to degrade our tradition by introducing the idea of an essential antagonism between 'management' and 'workers'. I have written Humphreys a letter on the subject which he read out last evening (Fig. 49). Vaughan Williams accompanied him, and the two of them had four hours talk with twelve representatives of the Association. I am delighted to learn that the greatest good sense and helpfulness was shown on both sides and all present showed the desire to put intolerance aside and work for staff loyalties for which this place has always stood (Fig. 50).

February 2nd

I was unable to attend Diamond Die Advisory Committee, so Leeds went and gave a film show using the film I had for my Royal Institution lecture. He also discussed the American machine.

GEC – CVD meeting this morning and allocations this afternoon. We are likely to be relieved of the CV 8 and the CV 52 before long. This will help us in tackling the hydrogen diode (CV 215).

A visit from the Minister [Ministry] of Transport under Sir F. Cook (our old friend of the street lighting days) Perrin and Taylor and many others were there—street signs were, I believe, the subject, I must see the demonstration.

February 3rd

Luminescent powder discussions with Jenkins and Jesty to ease the way for more research by moving routine suspension making to XL or the Works. If we don't hasten this we shall lose pre-eminence in the art.

February 4th

A brochure designed to found a Radio Research Institute has come in for comment. There are ample facilities for research and experiment in the industry already.

A TVC meeting entirely taken up with submissions for use of valves and with Dr. Truscott (IVPC) objections to many of them. The whole situation is so dominated by the inability to produce essential valves that hardly any other consideration counts.

February 5th

Sanderson reports good progress along all the lines of battery development—i.e. layer, tropical and arctic types.

The demonstration of luminous street signs so as to prescribe optimum brightness was very prettily done and exhibited by the usual projection technique. It is a relief to see even one investigation which is not associated with killing. Winch does a little towards colour photometric standards on a background of mortar fire detection which apparently goes well.

An ARP dance and social tonight at Hirst Hall. Usually a jolly occasion.

February 7th[443]

Bell is back part time, after a sharp spell of diabetes. Everyone very glad to see him. Sir George Gee here today having taken a post at MOS, supervising radio matters at high levels. He is approachable and helpful. Saw Cherry who is still working at TRE on Oboe projects. He will probably stay there anyway for another month or so.

The Russians are pummelling the Germans pretty badly—first one part of their line and then another. I wish we could do as well in Italy where defence has the advantage at present.

February 8th

Had to hold internal meeting on CV 192 magnetron in which our preproduction team feel that they are being 'messed about' by having to make speculative changes in the above valve whilst we are in ignorance of the circuit variables with which it is to be used. After getting our experts together on the subject we shall approach Dr. Fraser of TRE who is senior on this subject.

The Lighting report of my Study Committee (Ministry of Works) in its final form is not meeting with entire acceptance—too diffuse and wordy in many places—and we may have to reconsider some parts of it!

February 9th[444]

The Cathode Ray Tube meeting today was lively and interesting—it mainly concerned itself in getting a sense of proportion on powders for screens and the need or otherwise for the considerable variety which defaces the records of TVC and its standardisation committee!

Fixed up with Murdock to join us from MOV to help Pearce with magnetrons. This will be a great help.

Gavin has been asked by Peters to tackle 50 more advance delivery 'Air Grocer' sets.

I handed over today to Humphreys all my cards and records of non-scientific staff as a step in devolution. He will handle them as part of the personnel duties which he is taking over. I shall still hold the scientific staff records and handling.

February 10th

Sanderson's inert HT battery for Air Ministry seems to have met with acceptance. It has a multiple filling device.

Alumicide is being pressed (in anticipation) by the directorate of raw materials at MOS who want to know how it is made and all about it. This kind of proprietary attitude tends to be the fashion on the part of the Ministries these days but is not being encouraged by us. It sometimes arises from a request for information by USA.

February 11th

A large meeting with DTM [Director of Torpedo and Mines] group on our A–S work and its new branches of investigation. Marris is carrying this liaison with Farren associated.

Espley has an eleventh hour spot of bother with GR over the ability of the final prototype to pass one of its tests. He will have to spend a week or two on it with Walter helping him.

Prain speaks unfavourably of the USA progress on fine diamond drilling and has induced one of the group to come over here to study the art in Britain. Our own record of group efficiency lies between those of Bournemouth and Salford and is more jumpy from month to month. The comparison is proving valuable.

A GEC industrial heating meeting at which high frequency heating was discussed. We are stopped badly by inability to get one or two machines from Witton.

February 12th

The five Audiometers made here for Fairlie (for testing ABs or supersonic listening) are ready for delivery and look a good job. They are the work of Midgley and Farren.

A photo was taken today for the coming 21st anniversary celebration. It was of all the present members of the staff who were at the opening of the building in February 1923. I have been busy with a promised write up of the 'Confidential History of the Laboratories'. It has been a very interesting task.

A largely attended meeting was held here yesterday with MOV, Glass Works and RL to get the miniature valve project launched. Each detail was reviewed and allotted to one establishment to perfect.

February 14th

Preproduction remains round about 2000 per week with wastage decreasing in CV 90 but rather serious with CV 192 and CV 208. Also the CV 114 being made at XL has considerable electrical wastage—but there are hopes of better things.

Lunched with Dr. Railing. He is interested to continue the diamond die making here as a Lab. product supplied to the market.

February 15th

Our research contracts get more numerous and it is difficult to note them all in this Diary. The supersonic [ultrasonic] series tends to grow and Edwards is now back to look after one or two of them.

Humphreys has been up re. manpower decisions to withdraw staff from a number of activities over the country, but I don't know yet how we will be affected here. Our efforts to get permission for a small coach to take our workers back and forth to XL is met by all the red tape MAP can muster—and is a queer contrast to their urge to get more production out of our XL group!

February 16th[445]

Breadner showed me a remarkably good piece of mechanisation of the pins for the RG Project 1967. He has surmounted the difficulties in an ingenious way and incidentally found a technique for glazing pins which are to be sealed through glass which constitutes a very promising advance in the general art of seal making.

A CVD policy meeting—quiet and interesting. Main subjects—reconciliation of USA and GB characteristics and long 'memory' devices.

February 17th

A busy day at HO on post-war radio and radio research. The war has started between Bureaucracy and the Radio Industry over whether radio research shall or shall not be centred in Industry. The RMA are having to give evidence to the Hankey Committee and we have been helping Macqueen with the GEC point of view. My memo. on the subject of the proposed Radio Research Institute has been accepted as the GEC view point but there seems considerable doubt if it will be quite acceptable to RMA.

February 18th

I presented the Lighting Committee's report to the Building Research Board today and had a lively discussion on the load which its recommendations would add to the cost of small dwellings—based on any mandatory recommendations which might be pressed for.

Dudding is keeping up vigorous activity on quality control and statistical handling of research data and it is bearing fruit inside and outside of the Laboratories.

Professor Menzies (Southampton) turned up last evening on a photocell problem. He is the focus of operational research at Air Ministry and a useful man to contact. He is coming again.

Humphreys had an interview today with the Man Power Board and MAP at high levels—an astonishing volte face! Valve categories, for some reason unexplained and idiotic, are no longer sacrosanct but radio equipment is! We therefore change the categories of all our people and start again until they find their error when we shall presumably change back again.

February 19th

The sharpest raid for two years last night over London. Warren's home was damaged with incendiaries and everyone seemed short of a night's rest on arrival. Two or three HE bombs were within a quarter of a mile of the Laboratories and several houses demolished.

February 21st

Another weekend London raid. The enemy is using 'Window'. This may account for the relatively few of his planes brought down. Various incidents and casualties especially associated with Lamp Works. Benjamin's family has been hit again—two brothers and his mother have been killed.

CV 208 still gives a large wastage. Valves will not age up properly and solution not yet clear.

Edwards is now entirely back on our work and is handling 'MFU' and underwater 'Commando' jobs of sorts given to us by DSR Admiralty.

An Advisory Board meeting and informative discussion on many issues, i.e. radio transmission, industrial heating—particularly metallurgical.

February 22nd

Gossling reports a good liaison meeting with Watson's Electro-Medical yesterday when deep therapy sets for cancer treatment (post-war) were under discussion. It means a combined 'built in' HT X-ray tube and transformer.

A BSI meeting on the 'Mark' [Kite Mark?]. Position very backward and needing some strong and influential action to bring it into prominence. We shall have to tackle that inept Ministry, the BOT [Board of Trade].

A TVC meeting all afternoon—a peaceful passage. Hecht, Hughes and Truscott all sent substitutes; probably that was the reason.

February 23rd[446-447]

Two incidents today, E 1496 valve with ASE and CV 192 with TRE, where we have to point out that we are merely being treated as a concern whose business is to make experimental valves as we are told without using our own intelligent co-operation. Had a talk with Fraser of TRE and Coales of ASE and hope this will restore our special partnership status with these establishments under CVD.

An MOV receiving valve meeting—a good feeling amongst all concerned.

Raids of appreciable severity are just now a nightly occurrence and various factories if not hit are a bit handicapped due to effect on staff. Mr. Doxey, Lamp Works chemist, has been badly injured and his wife killed. Two of Mark Benjamin's brothers have been killed.

February 24th

Espley has returned from an effort to straighten out GR prototype tests so that they can be approved by ADRDE. The trouble is that they have not a sufficiently capable or informed team available to do the inspection so that their results have any meaning from the point of view of acceptance or rejection.

March 4th[448]

This (Feb. 27th to March 4th) has been—from a domestic point of view—an outstanding week in the life of the Laboratories. The 21st anniversary of the Opening on February 27th 1923 of the present building—so far intact after four years of war and blitz. The idea, and carrying out of the celebrations, was from the Social Club Committee and independent entirely of the Director, Leading Staff or any of those senior elements which might be associated with 'management'. The programme consisted of evening events each day of the week and started with a Religious Service of thanksgiving on Sunday afternoon and a formal recognition and presentation of and to the 34 'original' members of the staff who were present at the opening in 1923 (Fig. 51). The dance on Monday, a 'Confidential History' by C.C. Paterson on Tuesday, and a Social on Saturday were the principal events. The effect of this upon the general *esprit de corps* of the place can hardly be exaggerated. With 1000 wartime people the original staff with its traditions had become diluted—although it was standing up well to the strain. But psychologically one feels we have had a very great reinforcement.

From the technical point of view the week has been uneventful.

Silica valves have now largely given place to the glass-to-metal variety which we forecast to Dr. Dale Bussell in the first year of war. So our silica glass making has been cut off with a knife.

Winch's 'Mortimers 3 and 4' are now ready for trial. The mechanical version (4) due considerably to F.H. Brittain is a very elegant device from the presentation point of view—it has been executed by Norman Hill.

March 6th

Enemy raids have ceased of late but not so ours over Germany. Otherwise the war is quiet except for an occasional 'break through' by the Russians on 100 mile front which on only two occasions has resulted in a substantial 'cut-off' of German troops. The Germans nearly always extricate themselves to the rear.

Fig.51 Photograph of the members of staff who were present at the formal opening of the Laboratories in 1923

Valve preproduction has been steady at about 2100. Bell ([with] L. Tremlett) is having more luck with his experimental large 10 centimetre RDF valve and has changed the design to one which brings them down to 12 centimetres with good power.

A meeting on Alumicide at HO at which we decided probable price (5 to 6 shillings per lb) and conditions for production.

March 7th

Our CV 221 waveguide cell (Ryde, Harris, Shaylor) is progressing. They have evolved a window technique which stands up to considerable powers. The new gold seal design is also coming along successfully and promises well though it has not yet completed its tests.

A belated meeting of my Ministry of Works Lighting Study Committee to meet the wishes of some members who desire to improve the presentation of the report—although it was already approved in October. On the whole it is to remain as it was.

March 8th

The camouflage research arrangements with Leamington seem to be going well. The illumination group has devised an excellent model for studying ships at sea which will be transferred to Leamington in due course.

An RDF (CVD) meeting all day. Transmitting valves were the liveliest, particularly high power magnetrons. The meeting was held in Church House rather than MAP—not a good room though large. Our live objectives are clearly (*a*) higher power magnetrons, (*b*) higher power waveguide cells, (*c*) K band magnetrons.

March 9th[449]

There seems to be doubt as to the strength of the CV 208 design of magnetron to stand up to some of the severer vibration conditions in the Fleet Air Arm and we are taking the matter up immediately offering to change design.

[Air?] Commodore Erlanger (civil flying) came yesterday to discuss blind landing but also saw Fido. The talk on the former will be helpful. The issue is between a system which helps the pilot to land and one which cuts him out except as an automator. The latter does not appeal to me. On Fido he was discouraging on the grounds that a width of cleared space of 100 yards is now asked for, which increases difficulties and cost very much. An aerodrome under these conditions might cost a million pounds to install.

March 10th

Breadner and Co. have completely overcome the problem of the pins for project 1967. All the pins required for the week can be made in two days on the roundabout machine and the cracking off of the glass on the underside of the discs is almost equally automatic.

It looks doubtful whether more than the 30 000 discs now required of us will be wanted—but they will take three months to complete.

A conundrum re. enemy purchase of caesium has been put up to us—somewhat mysterious.

An all afternoon meeting of the progressing committee on Trumper and Bowler. Captain Gladstone presiding, and very well he does it. Everything is directed to the trials in about a week's time on Loch Long.

March 11th

The E 1496 (Gt. Dane valve) has been accepted by ASE at its highest possible loading as an interim measure and on the understanding that we design a valve of twice the power. This had been stated by us to be a long and troublesome job—but Bell has designed an unexpectedly straightforward valve which is a break away in external construction from the micropup series—and we must now see whether ASE will encourage the new design sufficiently to allow the E 1496 to become a second string.

Sanderson has a story of battery changes and advances in several directions—which however must be an embarrassment to the factory. M of S [Ministry of Supply] urge turning the main production of the factory on to our new multi-inert unit HT block—and also talk of adoption of the 'arctic' battery just developed. These will not interfere with the 'layer' type battery which is being made in a new centre in Wales.

Freeman came from DSR to see Breadner's project 1967 pins.

March 13th

Dr. Whittaker from USA (diamond dies) here with Mr. Thompson to study our practice, he will be here for a day or two. Our efficiency value still keeps a little below 0.1 whereas Parkstone is 0.125. 30% of our dies are now made for outside users.

March 14th

A Diamond Die Advisory Committee meeting at which it was agreed to go ahead with the recommendations to the Minister re. a post-war association.

A Fido meeting in Holborn around their new model aerodrome showing all details in lighting and heating of atmosphere. There seems to be a concentration upon providing a wider path between the 'curtains' of hot air both to allow for larger planes but also to give longer for the hot drifting air to be in contact with the moisture particles and so to evaporate both the large and small ones.

March 15th

A meeting at ASE yesterday—Bell and Ratsey showed consensus of opinion about proceeding with 'Ursa Major' the larger valve than the E 1495 (micropup design). ASE seemed very satisfied.

Mr. Trapp and Mr. Isaacs called re. Heavy Alloy projectiles to see what sort of place and people we are.

CVD and Allocations meeting all day. It went off smoothly. There is much to settle and to disclose at these meetings.

Jerry started his Blitz again last night. A parcel of incendiaries beyond North Wembley station destroyed three factories and five houses. It's the sort of thing that may happen to anybody. The incendiaries were widespread over London but I do not hear of any other large fires.

March 16th

Had bad news from Holland that he has to go into hospital and may have a long recovery. This is a serious blow to our valve preproduction organisation for Holland is the business manager of it.

Have been comparing our valve costs with MOV on NT 99, VT 90. The MOV figures are Government costed so cannot be inflated. Research Laboratories are inflated because £62 000 was returned to the Admiralty and we also are allowed to include cost of much of our preproduction plant. Nevertheless the sales price of the MOV product is only about 30% less than ours on these two valves. So preproduction which depends on peoples' brain does not seem more expensive than production which depends on modern principles of management.

Had discussion with AID in order to get this place to fit into the framework of AID inspection system. Seems late in the day.

March 17th

Settled with Holland and all concerned who should carry on in his place whilst he is away. Cheeseman will take responsibility and Richards will transfer from Warren's group. Miss Fildew will run the charging side, as she has, but taking fuller responsibility.

A talk with Espley shows that he will take on consulting work in connection with a GEC manufacturing effort on scanners: Witton as principals.

His [Bligh's?] and Espley's blind landing effort is not so promising in the light of the official proposals for post-war scheme with X band beams.

A meeting at Glass Works to initiate actual work on pressed glass. It is difficult in war time to do anything of this kind on account of both materials and staff but as some valves need pressed bases etc. we may find facilities.

March 18th

The IE Camouflage Gear for the Navy seems to have been much appreciated and the set is now off to Leamington for installation there.

Waldram and Miss Anthony's film of night flares showing Waldram's experiments with them is a good effort and will be shown to Service people. A difficult piece of photometry to carry out.

Sanderson reports that 10 000 per week of his HT inert batteries are being asked for from Battery Works.

March 20th

Valve preproduction keeps up to about 2100 per week.

James is going to try to organise the proper examination of our mixer crystals made by Ryde's group. So far we have been too dependent upon 'reports from establishments'.

March 21st

Gavin reports possible urgent interest in our large E 1495 valve for RCM crash project. If this comes on it will be a grief to ASE.

I attended my first BSI executive committee. It is the hardest job to get Government Departments to keep hands off standardisation and at the present time the BOT has the bit in its teeth and seems inclined to set up a standards department there.

Had Prain to lunch. He welcomes my enquiry into a basis for diamond pricing.

In a CRT conference at Lamp Works Jesty reports happy and congratulatory atmosphere over the auto pump development. Sharp and Hill at our end have contributed.

March 22nd

MacQueen tells me that in discussion at MAP on the subject of the deputation to USA on the multi-channel set MAP volunteered the remarks that this set is the best piece of radio work turned out by this country during the war! Somewhat over high praise but MAP is most anxious to sell this set to America in place of the 10-channel set they are working on over there.

The fire a quarter of a mile away from here the other night destroyed quite a store of valves many of them our CV 90 and CV 79 and 89. We may have difficulty in resisting a desire for us to continue making the latter. We have just shut down the group with a sigh of relief.

March 23rd

A long TVC meeting involving increasingly important issues of secrecy and allocation.

We find that it ought to be possible to recover the above 'fire' damage on CV 79 and 89 if MOS will be sensible about their specification and use some of the 1400 valves, alleged rejects, which we have in the cupboard.

March 24th

TRE tells Bell that the samples of E 1496 transmitter have pulsed at 50 cm and they are very pleased with the valve. We have not yet got them to oscillate at 1 metre on CW but must find out how to do so.

Chirnside has been elected to the Council of the Institute of Chemistry—real merit.

Jesty has had a talk with Mr. Hecht on Orthotrons and 'memory tubes' since we feel that we ought to be in on these developments—which border upon iconoscopes.

A BSI Mark Committee. The old spectre of the proving house is rearing its head again from the Supply Industry side. So we shall have all this fight to go through again, as we did after the last war.

Harris and others have been to TRE to discuss T & R cells (CV 114 and 221). The present cells are much improved and no breakdown has yet been achieved by TRE. The future work is extensive particularly in the K band and our group feels they have a full plate.

March 27th

Saw Ashbridge for information re. warning devices for marine civil use. He had no first hand information.

Watson Electromedical meeting to consider further the quarter to a half million volt X-ray units. Mr. Caldwell had produced new designs which were all very interesting.

It appears that Rabone's AC sorting vibration table serves an excellent purpose in sorting diamonds for shape and we shall have one made.

We have suddenly an embarrassing supply of new labour and we have had a special preproduction discussion to settle how to keep it employed till needed. Holland is unfortunately going off for an operation and may have a prolonged absence. Cheeseman has now taken his place.

March 28th[450-452]

Professor Trewman [Truman?] here with four people from Messrs. Lines to study Breadner's technique for the project 1967 and pin discs which is going excellently.

Air Vice Marshal Tait, Air Commodore Leedham and Air Commodore Cardell came to lunch to examine the 300 (multi-) channel set (Bligh, Stenning and Heller) and then a quick run round the Laboratories. They were very appreciative visitors. The prototype being finished at Bradford is requiring a certain amount of nursing. A Fortress bomber is waiting to take it ready installed to the USA for demonstration there. Heller and Fairbairn go with it. Stenning has done good work in perfecting this design and initiating many of its best features.

March 29th

L. Jacobs's trial of a barium pellet in the gun of a CRT seems to make it possible to rejuvenate cathodes. 'Memory' tubes are to the fore and we have suggested that we should be included in the research. This was confirmed by the CVD policy committee this afternoon and the work will come onto the CVD contract.

Professor Nottingham and Dr. Leverentz are here from the USA on CRT matters. This is their first place of call. Jesty from all reports made an excellent impression on his recent USA visit.

A CVD policy meeting today at which Dr. Lewis insisted on the RCM priority of Bell's E 1495 valve which is still proving out very well at TRE as pulse amplifier. There was also a long discussion on the pedestrian nature of 'break away' research on [in?] the magnetron field. Blame was laid upon the diversion of the Birmingham group to other things and I thought less than justice done to Oxford and Wembley. I was able to insist that bright break away ideas are more common than the craftsmen to carry them out and this causes the more dead certainty projects to have the craftsmen's help.

March 30th[453]

Discussed with Enticknap the incidence of company taxation upon the inclination of firms to undertake development projects. Also went through Humphreys's document showing comparison between pre-war and post-war laboratory costs. Roughly the cost of the laboratories has gone up from £200 000 to £325 000. He has already deduced that the cost of all establishments was up 50% compared with pre-war.

Had meeting at Royal Institution. F.E. Smith, Cockcroft, Rankine [Rankin?], Oliver, Whitehead and Levy, to consider desirability of a new Physics in Industry Journal. A favourable decision—also support for a Physical Society Technical Physics Journal.

March 31st

A somewhat lively Radio Co-ordination meeting at which Macqueen tried his best to debunk the idea of an increased post-war demand for radio and at which I had to challenge the Works as to the greater suitability of some other type of organisation to handle the smaller and highly technical projects in the first instance. A depressing meeting.

April 1st[454]

.... and the start of a new financial year. The war is not going specially well. [In] Italy we are held up. Bombing of Germany very expensive—coal strikes—and people behaving as if the war is over.

We are pressed to more and more work here. Two new triodes of Bell's, E 1495 and E 1457 (large disc seal for 10 cm), are finding favour for RCM by Gavin and TRE and we have had a meeting of all concerned to see how they could be progressed. It is craftsman shortage as usual which is the trouble.

I have had to promise a lecturette at an ELMA luncheon to architects in May. I do not at all welcome this amidst so many urgent matters. Had a conference with Ryde, Stevens and Jenkins about the experiments. Lord Woolton is to speak.

We are told to cut down our gas consumption by 20% on account of coal shortage! We may be able to do something towards it.

Our 'Salute the Soldier' week with a target of £5000 has realised £6381. Mrs. Paterson assisted by Mr. Downing (poster producer) and Mr. A. Ogden are again the responsible party for this excellent result.

April 3rd[455]

Valve preproduction 2377—a good beginning for the new financial year—and an anxious discussion on how to make CV 240 (E 1495) and E 1457. Bell and Gavin go to TRE tomorrow to discuss their use for RCM and other things.

A visit from Mr. Merren (should have been with Mr. G.M. Wright—prevented through sickness) to review our development contracts and to discuss possibilities of making arrangements for similar contracts in peace time. I welcomed the latter proposal—which I made a proposal to Brundrett about some months ago.

Attended at Dr. Snow's office with Bragg to be talked to unofficially about the engagement of a number of specially bright men in industry. According to Snow industry is such a place that no scientist wishes to go there unless compelled! It was a relief to return to one's cheery and buoyant colleagues and the inspiring atmosphere at Wembley. I gave the best of my advice and am asking some people here who they think are the brightest of the Service people we have contacted.

April 4th[456]

Gas consumption edict from Ministry of Production is that all industrial gas must be cut 25%. Taken literally such a requirement would shut the place down for at least one day per week. As usual I expect valves will be exempted from the arrangement.

The Helmover gear for controlling '?' made by Foot, Angel [Angell] and Campbell (Bligh's group) has now its functional prototype ready for trial. It looks a good job with its special modulator box working with some of the Oboe gear.

Ryde is not entirely despondent about FIDO although for so wide a span our advantage over line business [?] will be substantially reduced.

April 5th

A full Lab. Stewards meeting to consider the fuel economy situation; a number of useful suggestions emerged. Foxon has agreed to be chief fuel economy officer.

Examined our new Pay-As-You-Earn wages system organised by Humphreys who has now the wages team downstairs near his office. It is proving a pretty heavy task to get launched.

Marris is finding his various Admiralty A – S jobs a big plateful. Each one covers such a wide range of techniques and problems, from the stresses due to dropping to the quartz crystals and radio gear.

April 6th

Bell and Gavin have spent a day at TRE with incidentally a very happy and genial reception from everyone. The virtues and uses of the E 1495 and E 1457 hardly needed salesmanship so that a situation requiring urgent preproduction of these is obviously developing. Gardiner is entrusted with the task of getting GEC concern[s] to help; Union Works, Chamberlain and Hookham or Gallizia at Rectifier Works may be approached. If MOV could take over CV 155 Algar could handle the E 1495 in his group and possibly Boyland the E 1457.

An RS [Research Services] meeting with A.H.R., L.G., and T.W.H. at which I ventilated the subject discussed with Snow and Bragg on April 3rd.

April 11th

We start again after a very nice Saturday and Monday Easter break. The Russians are the only ones making military progress having smashed up all the Southern German Armies and taken Odessa and entered Rumania and Czechoslovakia. The Navy has fairly bombed the *Tirpitz* and lunching today with Captain Willett, the Fifth Sea Lord gave us a graphic account of the action. We used no bombs powerful enough to pierce her main deck armour but short of that we did all there was to do.

In preproduction the only real anxiety is the CV 114 (T & R box [cell]). Mallett promises sufficient components to rectify the position in large measure.

A TVC meeting all afternoon with all the contenders there—so it was a long affair. It is all to do with Hecht's long fight to prevent CRTs being included as valves in the CV series!

April 12th

Have examined valve cost records with Le Rossignol. There is no doubt that our costs have been decreasing, due to higher outputs by the same numbers in the groups and we shall have a considerable margin to return to the Admiralty for 1943–44. I shall make provision in this year's accounts for £100 000 to pay back as we shall not know the actual figure till May or June.

Have British Overseas Airways people here to talk over the radio future. All is uncertain until Watson Watt's co-ordinating radio committee reports on the basal requirements and conditions.

Winch and company are down to Larkhill again for further field trials on improved forms of Mortimer 2 and 3, i.e. the electronic and the mechanico-acoustic versions.

Colonel Boyd and his adjutant came to say that as invasion of the Continent is imminent he has instructions to take all scheduled precautions regarding vulnerable points, of which this and EMI are the only ones in NW London. He has to draft fifty men in at night as protection against parachutists—so we must find conditions for sleeping and food. I feel sceptical about the usefulness of this but them's the orders.

I am giving this evening my talk on the History of the Laboratories to the Workshop staff and others—in the library.

April 13th[457]

A visit from Mr. Chapple of the Production Executive to enquire into the valve production position and what handicaps it is having to face. This as an independent overhead enquiry. Humphreys had all his facts and figures to hand and I gathered that our position compared with others is a satisfactory one. A tour round the laboratories completed his admiration for what we are doing at RL.

Lunched with ELMA Chairman and BTH and GEC colleagues to arrange particulars of procedure and demonstrations on the great ELMA luncheon to all the quality on May 3rd—Lord Woolton being chief speaker. I think I got them to agree my plan for my talk on post-war lighting.

After great trouble and prolonged planning and jockeying, Cornwell has staved off the Ministry of Fuel silly plot to make us change our oil heating of the laboratories for creosote fuel. We know this would eventually have to happen but the officials insisted on pressing for changes which would only have been justified in the days when submarines made oil economy the most vital of all the objectives.

April 14th

A Consultative Board meeting at which the financial analysis of Lab. expenditure was before the Board which Humphreys has been very ably preparing. It shows a present expenditure exclusive of valve preproduction people of £330 000. If we went back to 1939 strength it would cost about £250 000. With reasonable allowance for growth it would cost now £303 000. These compare with £195 000 before the war.

A debate—the first of its kind—at 5.30 on the pros and cons of part time education based mainly on our own experience. Very few of our Student Assistants turned up and I am afraid the talk was contaminated with a certain amount of the current cant about the cramping effect of scientific training in part time education and other catch words.

April 15th[458-460]

A whole Saturday spent with other research leaders at the Dorchester thrashing out the final draft of the Nuffield College report on Scientific and Industrial Research. It was an interesting discussion and G.D.H. Cole in the Chair did admirably. Lunched with Larke, Slade and Clark (of Unilever). Industry seemed to be on top for once. Chief contributors besides myself were Sir Charles Bartlett, Captain Bennett, Slade and a few others.

April 17th[461]

Bligh tells me that the Fortress plane mounted with the multi-channel prototype set just finished at Bradford has now set off and will be on its way to the USA either today or in the next few days. The set is reported to be doing all it should.

Visit from Professor (Colonel) Marsden from New Zealand—very full of interesting and practical facts and proposals on radio and other things.

Also Professor Polanyi to talk over patent reforms which he is keen on.

April 18th

A Glass Advisory Committee—much better in its arrangement and discussions than it used to be. The home made tank blocks are behaving well on life and Partridge is to be congratulated.

Dr. Whittaker of USA and all the Diamond Die Technical Committee met here to talk over things in general and Dr. Whittaker's experiences in particular. The discussion was very full and interesting and dealt mostly with experiences re. speeds of removing the diamond from cone and channel. Everyone had something to contribute and the meeting seemed to be voted one of the best—probably the best since we met at Ross-on-Wye. Ward was here from TRE. He said that X band magnetrons for operational use still mainly depended upon Wembley preproduced output. Unfortunately out of 100 blocks only thirty go out as good valves—as yet.

April 19th[462-463]

Warren reports that the Miniature CVD meeting yesterday was something of a riot when the Fleet Air Arm policy was disclosed of using the USA voltage of 6.3. Sir George Gee whom I met this afternoon says that the Army has been allowing for nine spares for every valve in every infantry set! He is now cutting them down!

Bligh reports that the Helmover 'directing' set (Forte) [Foot?] works well on the water and that trials have been quite successful.

We had a CVD (RDF) meeting this morning (Griffiths presiding). Nothing special to report. After lunch Jolley drove me to Hayes in Kent where trials were to be run on the first of the six [?] which Witton are making, Sir Geoffrey Burton, Butler, Lee, Sproule and others were present. Trials went off splendidly.

April 20th

Dr. Fraser and O'Keefe of TRE called to chew over the unhappy position of the CV 192 which still shows heavy losses in preproduction and is not married up to its circuit in a way which gives any confidence for future factory production. They said they could and would not consider any modifications of the equipment and the valve must somehow be made to fit. We got little further than to agree cordially that the position is very sticky and that TRE and we must hang together through whatever attacks may be launched against us.

Lunched with Gamage—very useful and helpful talk.

April 21st

A GEC television committee at which we discussed the evidence on transmission to be given by Heather and myself to the Hankey Committee.

Lunched with Settle who wanted to talk about a successor to himself in his department at HO at such time as he may wish to hand over. He suggested a certain member of RL staff. I do not think this is likely but I promised to enquire.

April 22nd

Waldram and Stevens have been going over with me the recent work of the Illumination Section. The group has been working under Stevens's administration, Waldram backing him and concentrating himself on certain special problems. The Schuil telephotometer has achieved no little success and many replicas are in demand. Turner's work at the great bombing study tank at Hellingsboro near Glasgow has been a major job and one of special difficulty on account of the lack of local facilities. It has required considerable resourcefulness. The flare work is being rounded off. The camouflage work which Waldram and Harper have been handling has quickly yielded a crop of results. The work on visibility has now eased up because the principal parts of the problem are now sufficiently solved, except for the visibility of submerged objects for which we are using the Wembley baths. Smyth has had charge of the work on road signs for MOT (Fig. 52).

There is much post-war thinking and much to be done if only we could put experimental weight onto it.

April 24th

Everyone expecting Second Front to begin any day by invasion of the Continent and all Home Guards are bracing themselves up. Otherwise there is a lull in the war.

Preproduction 2280 valves—all groups pretty good except for high wastage on one or two including CV 208 X band magnetron. There is a mystery in the ageing procedure.

A Lab. Stewards meeting at which Foxon gave a good report on fuel economy. If everyone keeps up the good beginning made we shall achieve our 25% gas and 10% electricity [economy].

Had Dr. Thompson here from ASE Bristol to see about triodes for use in frequency stabilisation of which Admiralty are making a great point. Our improved version of CV 90 goes down to 8 cm and we are likely to have demands for it.

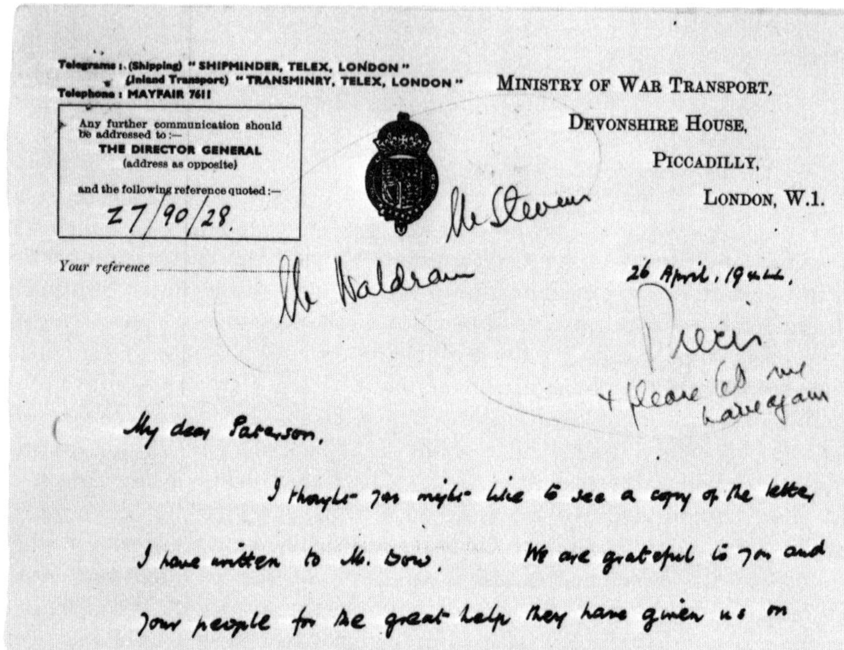

Telegrams: (Shipping) "SHIPMINDER, TELEX, LONDON"
(Inland Transport) "TRANSMINRY, TELEX, LONDON"
Telephone: MAYFAIR 7611

Any further communication should be addressed to:—
THE DIRECTOR GENERAL
(address as opposite)
and the following reference quoted:—
27/90/28.

MINISTRY OF WAR TRANSPORT,
DEVONSHIRE HOUSE,
PICCADILLY,
LONDON, W.1.

Mr Stevens

Your reference .... Mr Waldram

26 April, 1944.

My dear Paterson,

I thought you might like to see a copy of the letter I have written to Mr. Dow. We are grateful to you and your people for the great help they have given us on the illumination of traffic signs, and I trust that in due course you and they will be satisfied with the results of their labours!

Kindest regards,

Sincerely yours,

F.C. Cook

27 APR 1944

Fig.52 Letter dated 26 April 1944, and enclosure, from Sir Frederick Cook to Clifford Paterson on work on illumination of traffic signs

26 April, 1944.

Dear Mr. Dow,

<u>Traffic Signs Committee.</u>

Further to my letter of the 3rd April, I write to tell you that at their last meeting the Traffic Signs Committee discussed the two Reports submitted by your Society and agreed to make recommendations in our Report based on your findings.

The work which you have done for us has been of great value not only in pointing the way to improvement in existing signs but also in demonstrating, to the personal satisfaction of the Committee, that many of the existing signs are quite adequate to their purpose.

The Committee resolved that their appreciation of your Society's helpful contribution to our deliberations be recorded in the minutes and that their thanks be conveyed to the Society. I am also to express the Committee's appreciation of the work done and the facilities made available by the G.E.C.

/It

Fig.52 continued.

2.5.44.

Mr. Stevens.
Mr. Waldram.

3R5.
Hv. replied.CCP.

It gives me great pleasure to do this on behalf of our Committee and to add my personal thanks to you for the trouble which you took on our behalf.

I am sending a copy of this letter to Dr. Paterson.

Yours sincerely,

J.C.C.

J. S. Dow, Esq.,
Hon. Secretary,
The Illuminating Engineering Society,
32 Victoria Street,
S.W.1.

Fig.52 continued.

April 25th

Have been considering GEC radio with Marris and Bligh and one's feeling that on radio transmission engineering we are not making a bid for any place for the GEC nor using the very considerable sales potential which we have built up during the war.

Modifications of the small CV 90 types of valve E 1440 and its varieties are being asked for. It looks as if the largest and the smallest valves would be our objectives in preproduction.

Had visit from Sir Wilfred Garrett, Chief Inspector of Factories and Dr. Merriman who saw round the place.

Message from Westminster to warn us that we were to be asked to increase preproduction of CV 208 magnetrons and to be prepared to discuss matter tomorrow. We shall hear, I expect, what has failed that such an emergency has been precipitated.

April 26th

A very busy day on CVD at Wembley. The morning meeting on CVD proper (Brundrett in the chair) is about one and a half hours. The preproduction meeting (C.C.P. [Paterson] presiding) 1.30 to 3.00. Allocation meeting (Truscott in chair) 3.00 to 5.00. The extra demand of CV 208 is owing to the success of the equipments. The E 1495 and E 1457 as well as two CV 90 offsprings will soon have to be introduced to preproduction. Altogether our showing was good. I wish we would get on better with the largest magnetrons.

The party went to inspect the new magnetron life test huts erected at Admiralty expense and which we are running. They are a good job and the screening is of particular interest.

James reports meeting at MOV re. post-war valves which shows a real urge to get together again on these. I also hear there has been an MOV meeting between Agate of EMI and Peters where it was agreed that to get the best valves there must be circuit co-operation—also a very satisfactory sign.

Rehearsal this evening for my talk at the great ELMA lunch next Wednesday. They appear to have promise of an influential gathering.

April 27th

H.G. Jenkins and group have been working at a high power instantaneous discharge flash lamp for photography and similar purposes. MOS (Dr. Vick) want us to take a contract from them for developing and supplying 2 sets.

Had an approach from Oliver to collaborate technically with BSA in metallurgical developments. This will undoubtedly be useful now and in the future.

IEE Council nominated Sir Harry Railing today for the Presidency.

April 28th

My discourse for Wednesday next is taking shape. The boys have got excellent demonstrations.

We have as yet no news of Heller in his flight to USA but we are told the machine went by Southern route and we may not hear for several days more.

We start our guard of fifty men in the building at night tomorrow—so the invasion cannot be far off. Bombings of Europe from this country are continuous day and night.

May 1st

Willshaw has visited BTH and reports considerable plans and facilities for CVD experiments and experimental production.

Lunched at Ryde's invitation with Wembley Rotarians and gave a talk on 25 years of industrial research. I do not find these occasions at all easy.

A meeting at Watsons Electro-Medical. Half the Leading Staff seemed to be assembled at Gossling's instigation. Mr. Minns is back from USA where he has done useful work with their associates over there. We discussed in detail with Mr. Caldwell his designs for a British equipment.

May 2nd

There seems to be a growing possibility of the hydrogen thyratron being wanted by ASE particularly in a size about twice the dimensions of the E 1209 (which ADRDE want). Ramsay is confident of the result.

Our fuel economy goes well. We are down to our full economy requirement.

Gavin goes to a special RCM meeting this week at TRE on Hecht's invitation to discuss possibilities and programmes in RCM. After talking with Espley he has good proposals for wide band circuitry.

A Lumifax board meeting and a rehearsal at the Savoy for my ELMA lunch talk and demonstrations tomorrow.

May 3rd

A CRT meeting in which the anti-'window' device figures considerably. It is touch and go at the moment whether in its present form it is of sufficient help to go on with.

The great ELMA lunch went according to plan. The demonstrations of luminescent lamps and comparison with the carbon lamp of the early days of the century were very well done by our Wembley team—as usual; and I think everyone was pleased. Dixon (of Crompton), Lord Woolton and Sir Robert Renwick spoke.

May 4th

Visit of USA Naval officers to see all they can of Radar etc.

Captain Brooking called and had a good look round with G.M. Wright. I raised with him briefly post-war collaboration and he expressed the view that 'CVD must continue after the war'. He seemed apprehensive whether it would be focused in ASE so much as in the past and mentioned ARL [Admiralty Research Labs.]. We here have little admiration for ARL.

May 5th

A Research Services meeting at HO. No special excitements.

A TVC meeting all afternoon. Hecht was absent and Hughes, being the only one of the two present, we had a quick and relatively peaceful meeting.

Found Captain Brooking (Superintendent ASE) at the Labs. with G.M. Wright. Megaw had given G.M. Wright a good look round and had had some useful talks on the USW propagation work which everyone feels is going rather slow. Probably Landale will take chairmanship of the committee in place of Wright who is too occupied to attend regularly. This will be a good arrangement.

The pressure for more manpower help and no one available is very strong. Brewer (valve life tests) and Megaw (USW) both need men.

Bell was at Marconi yesterday with ASE people to thrash out E 1496 (Great Dane). We found indirectly that Marconi had been put on the job of developing a valve for the same job and had copied our design with the exception of using a copper gold plated grid to reduce secondary emission. We have not yet received our platinum coated molybdenum which will do the same thing, but it was agreed that there was no justification for adopting the Marconi version with all its consequential changes. It is rather irritating to find a situation like this and I do not see that we have merited this lack of candour.

Gavin was at a special meeting at TRE with Hecht on RCM. The meeting was apparently useful for fostering the existing programme, but longer term work was hanging fire somewhat pending the indications for the future which will become apparent when we see how the invasion of Europe goes.

The whole world still awaits the start of the invasion and it is amazing to be carrying on normal existence and work on the brink of such events.

Hecht opened up to Gavin on the general subject of development contracts—and he gathered that, except for Wembley, the view of DCD is that firms have not made their operation easy—some thinking mainly of their profits—some being overbearing and wanting to dictate to the Services—some lacking the necessary flexibility—and many having too much interference from overhead industrial executives instead of leaving the technical staffs a freer hand. I shall be very sorry if the development contract idea comes into disrepute. I have been preaching it extensively as a means by which Government can use and encourage science in industry.

May 8th

Still no invasion.

Preproduction valves were a record at 2442. All groups doing well, even the CV 114 is doing better.

Went through our 1938–39 records with Humphreys to decide how much work of the Labs. was research which the Works would not have done even though the Laboratories did not exist—compared with the work which the Works would have to have done themselves anyway.

Humphreys and his personnel office (Pay As You Earn taxation included) are doing excellently as far as I can judge and is relieving me of a considerable burden—not to mention Kersey also.

Discussed with Heather and Marris the GEC evidence on television to be given in a fortnight's time to the Hankey Television Committee. A long document.

May 9th[464]

Gossling reports a useful proximity PF [proximity fuse] meeting yesterday when members were out to help. Major Prince (CIRES Dept.) was in the chair. There seems to be still some catch in the arrangements for himself and Donaldson to go to USA—some difficulty on the other side.

Attended a Radio Board meeting on the nomenclature question and the chair ruled there could be no compulsion of DCD who, with Hecht, was present. So it decided to recommend for consideration of TVC that MAP CRTs be excluded from the CV series but otherwise the CV series would apply to all valves and non-MAP CRTs. I expect there will be excuse for further bickering now, probably from Army stores end.

May 10th[465-466]

The new multi-head diamond drilling machine (USA design) has arrived and is working. It appears to be coming up to expectations and Leeds is so far pleased. Our efficiency figure from November to March had been falling badly but in April it has jumped up due to (*a*) a talk which Dudding and Leeds had with the girls, (*b*) we have a uniform type of die to make for the factory new molybdenum production. There is a general feeling of things looking up in the diamond die section.

Mr. S. Cautauld arranged to see me at the club. He asked my advice about Rowe as a Director of Research, Rowe having referred him to me! I had no reservations.

A CVD policy meeting and afterwards to tea with C.S. Wright to talk about the outlook and the future of CVD. The main subject of discussion was USA and British miniature valves and the unwillingness of USA to accept the features of British developed valves. An inevitable situation when we are the suppliants and USA the suppliers.

I gather from the CRT group that their 'colour discrimination' scheme for 'window' frustration is gaining favour.

May 11th

The past year's finances are beginning to show their form. We have an excess of about £175 000 for the 12 months on valve preproduction which is too high. I hoped for not more than half this, as a safety margin. There are several contributory causes, but the chief is that several types have suddenly run into smooth waters and the yield for the same labour has gone up suddenly—in some cases probably 50%. There is also much discussion about disposal of our profit (7.7%) and administration (5.8%) margins. We here are not interested in them but a touch of finance makes the whole world squabble!

Bligh reports that the Helmover control on the Staines Reservoir worked very satisfactorily in a demonstration, air-craft to launch. All are to be congratulated—range 3 to 10 miles.

Had talk with Espley on work he is doing on circuitry for wide band transmission and generally for applying new principles to the art. It seems to be real pioneer work.

May 12th

A Consultative Board meeting—discussion on new industrial research efforts—(British Aluminium, BSA, Courtaulds etc.)—relations between GEC and EMI vis-a-vis MOV—the year's finances.

Heather lunched in my office to discuss many things but particularly GEC radio, regarding which—and the last co-ordination meeting (the debunking meeting)—I had some strong things to say.

At Rabbin's request I called at MAP to hear him explain that after holding off the dogs for four years he felt that he could no longer refrain from having some sort of examination of the overheads applied to our research and preproduction contracts. This is a blow because it takes up time and attention.

Man came from MOS (seeing Bligh and Espley) re. non-metallic mine detection and wanting our collaboration to help Cinema Television on the centimetre wave side of their problem. We agreed to help at least in a consultative way.

May 13th

Winch reports that our two 'Mortimer' solutions (electrical and mechanical) behave about equally and give 1° precision with 1500 yd sensitivity which is very good. MOS appears likely to prefer their more orthodox microphone sound ranging method.

Clayton is just back from TRE where he has been helping on an aerial problem and, as far as one can judge, doing well. There are various jobs of the kind which seem likely to come our way from TRE.

May 15th[467]

Still no invasion.

Preproduction normal. CV 114 is not yet in the clear and CV 215 is held up for pressed bases. CV 221 (the gold seal CV 114) demands high grade machining and gauging and seems to me some way off giving all round satisfaction.

Had a very eulogistic letter from USA re. Heller. He seems to be making a great impression there for clear exposition and helpful handling of discussions.

Lunched with Chelioti and Advisory Board meeting to follow. Post-war issues press hard in a number of fields and in several of them we seem to be struggling to our feet.

The annual finance is at present absorbing a great deal of my attention —particularly in view of impending scrutiny by Government auditors.

May 16th[468-469]

The valve 'fragility' meeting yesterday attended by Bell indicated CRTs as the chief offenders. NT 57 and NT 59 (the old originals) came next. There were no Wembley made valves involved. A solution by MOV using better packing has been awaiting trials by RAE for several months.

Bell reports that the E 1495 (wide band valve) has fully come up to design expectations and all looks promising. Ratsey from ASE called to discuss priorities and particularly to urge development of a large hydrogen thyratron. This is on the whole welcome.

Couch (from Bath) called to discuss our work on large compact source lamps. It is an imposing record in which Nelson (under Francis) has figured, together with Willoughby and [?]. He confirmed that Admiralty are dropping searchlights for fighting purposes, and will content themselves with small lights for seamanship purposes only.

May 17th[470-472]

Megaw had a thorough survey at ASE of our USW programme for the near and the distant future. Two new long term items were added including millimetre waves.

Spent morning with Kersey and Humphreys looking for snags in our overheads computations.

Coales of ASE is bothered at our failure to produce a large S band magnetron for their EMI project. Stainsby was at Hayes yesterday on the subject. I have not heard if BTH have had better luck than we.

Had Lonsdale to lunch at the club to talk over things in general— among them Mr. Pochobradsky's attitude and lack of co-operative leadership.

Dudding's Quality Control pamphlet is being printed as a GEC document by Head Office. His visits with Jennett to different centres are almost weekly for giving addresses to factory staffs. GEC Coventry is one of the enterprising ones.

'Butterfly Circuits' make good progress and Warren tells me that James's variant is the best yet and with the E 1440 valve gets down to 20 cm.

Ivor Jenkins says goodbye this week and starts with Whiteheads in South Wales. His work and influence have been invaluable.

May 18th

Annual accounts show that our overhead must come down this year from 180% to 140% owing to the carrying of our fixed charges over a much greater turnover. This means our Government contracts have been overcharged and I must put into a suspense account a sum equal to the difference—to be paid back to the Supply Ministries.

Mr. Savage, LCC education, came to speak to the Staff and Student Assistants on effects of the new Education Act. A lively hour and very interesting.

May 19th[473-474]

In the chair at BSI Electrical Industry Committee where the hard worn subject of having a second (flatpin) British standard plug was formally raised by IMEA. It was turned down flat and a resolution of mine reaffirming the single standard for each rating was passed.

The GEC education committee met under Heather's chairmanship and appears to have had a good progressive discussion.

May 20th

Spent all morning on the annual accounts with Humphreys—they are by no means in good shape and I have a difficult time ahead of me with them, owing to an error in allocation which affects the overhead figure.

May 22nd[475]

A whole morning on accounts with A.S. Gosling and Humphreys to find out where is the money corresponding to the £40 000 to be reimbursed to the Services (on 1942–43 accounts) on account of change of overhead from 180% to about 135%. I felt sure that the money must still be somewhere in our accounts for no payments away have been made. It was eventually found in hidden reserves in the form of 'stocks', valve work in progress, unallocated overhead on salaries expended and not charged in claims—and such like. Am writing to the Contract Departments to advise them of repayments due.

Lunched with Peters for heart to heart talk on outlook on GEC radio engineering which I felt was too restricted to match up with the position we intend to win for the GEC. He reassured me convincingly and we agreed that he would convene the first of a regular set of meetings between Works engineers and Wembley research staff.

Smithells came for the night. He is getting his bearing in his new job as head of British Aluminium Co. research.

May 23rd

Heather, Marris and I gave evidence before Hankey Committee on GEC recommendations on Television. The Committee was most appreciative of the report and the thought put into it.

Lunched at special BSI meeting to bid farewell to Le Maistre and on to newly constituted RDF (CVD) meeting on oscillators and amplifiers etc. (Brundrett in the chair.)

May 24th[476]

CVD magnetron meeting, Dr. Lewis presiding. The rearrangement of the CVD meeting appears to be an improvement. Technical issues are properly ventilated.

The accounts are still causing us bother as one detail after another has to be ferreted out. In the allocation of burdens to the various classes of work and to the Departments which pay us there is only very rough justice. But having decided on the roughness of the justice accountancy requires that the calculations shall be meticulous. It seems to me to be equivalent to making measurements to an accuracy of plus or minus 10% and giving the results to 0.1%.

Willshaw reports joint work with Manchester (Prof. Hartree's dept.) which at last seems to throw light upon 'mode jumps' in magnetrons and the cause thereof and points the way to their avoidance. It concerns the conditions for 'build up' of oscillations as distinct from their maintenance.

May 25th[477]

The Hellensboro tank for study of the dropping of projectiles into water is now illuminated up to a point and Turner (who has had a very trying time on the job) reports preliminary trials.

Marris and I saw Dr. Carrol [Professor Carroll] at DSR Admiralty about getting a better priority onto our 'Bowler' work. Hawes is wanted both here and at Greenock and has no effective understudy at present. We shall probably put Kaye onto the job.

Household Appliances meeting in afternoon, the first for a year! Hot plates and simmering devices were the chief items.

Had an hour with Enticknap and Humphreys to see how we should now show valve stocks and work in progress stocks in the Lab. balance sheet. He confirmed desirability of seeing Mr. Touche tomorrow to get his guidance. There was much searching to find out where a quarter of a million pounds appeared to be secreted—and which is to be paid back to Government. Most of it was found in the Research Laboratories account at HO. Humphreys has studied with Kersey where the hidden reserves are. They do not in effect need to be touched in order to pay all we owe.

May 26th

MAP inform us unofficially that USA has devised a substitute for our E 1440 valve and the 'Beaconry' party over there are deciding not to use it—having decided six months ago to do so. This naturally enhances resentment against the prefabricated attitude of the USA on all these matters.

Mr. Touche came and gave the final touch to our financial presentation.

Reeman and Thomson and Jolley came over to discuss the new arrangement by which Power Jets Research and Development Co. is to take over the Farnborough staff (including Reeman) working on this subject. Reeman may continue to be loaned to them from here. This is a very interesting experiment.

Dr. Vick here to see what we are doing in general physical research.

May 30th

Tuesday after Whitsun. Very hot indeed. No Second Front yet—but progress towards Rome. Valve preproduction 2300. Had Mr. Whitehouse and Mr. McLennan (MAP audit and contracts dept.) to examine our overheads and what we include in them. Rabbin's assurances to me were quite sincere and they contented themselves with ensuring that our system was a proper one and was being properly maintained. I think they were surprised at its thoroughness. They also did not desire to investigate past years. They left at the end of the afternoon satisfied with our arrangements and gave me the impression that they

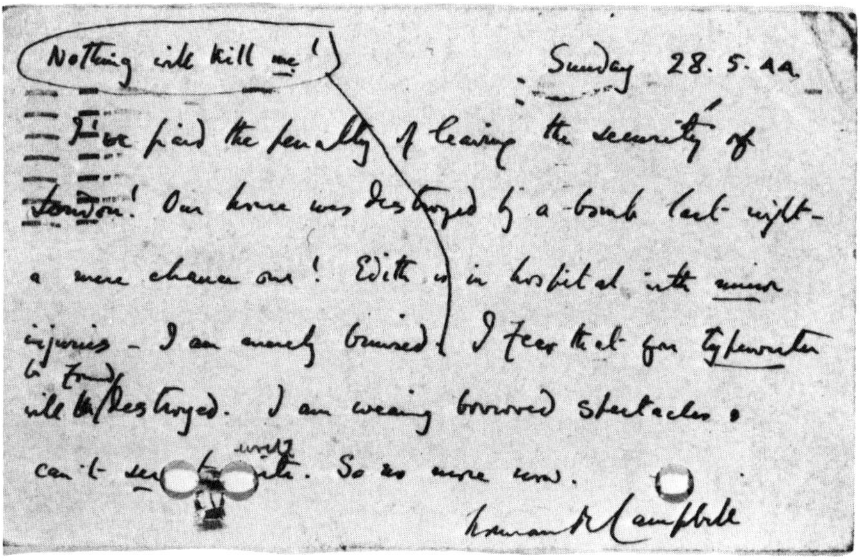
Nothing will kill me!

Sunday 28.5.44

I've paid the penalty of leaving the security of London! Our house was destroyed by a bomb last night — a mere chance one! Edith is in hospital with minor injuries — I am merely bruised. I fear that your typewriter will be destroyed. I am wearing borrowed spectacles. can't see to write. So no more now.

Norman R Campbell

Fig.53 Postcard dated 28 May 1944, from N.R. Campbell, recently retired from the staff of the Laboratories, to Clifford Paterson

were not going to harry a place like ours which put frankness (and incidentally integrity) in a high place. In future we have agreed that research and development contracts shall have progress claims, and at the end of the year we shall have a settlement which will be on the basis of the overhead as determined on the year's accounts.

Had tragic postcard from Campbell from Dorchester that his newly acquired cottage home in Dorset to which he had just moved was destroyed on Sunday night. Mrs. Campbell has minor injuries and he is bruised only (Fig. 53).

May 31st

All day at BTH Rugby on luminescent lamp meeting at which Siemens joined for the first time. There was a good measure of agreement on most points.

Went over with Warren and Davies [Davis?] their new set up for magnetron preproduction and experiment. It will be a good unit when it is properly going.

June 1st

The disastrous drought seems to have broken—partially anyway. The suspense concerning the Second Front however still continues.

I have been seeing Donaldson regarding a possible post as understudy to Caldwell at Watsons Electro-Medical. Donaldson will now see Caldwell.

There seems to be a further request for 600 sets of 'signalling bats' which Foxon handled for aircraft carrier landings. If so, this will keep Foxon busy when added to his fuel economy duties. There is a tendency for use of electricity to climb again. Good resolutions tend to lose their edge.

June 2nd

Campbell writes from a Dorchester address. His house was demolished by a very large bomb, and one gathers that there is virtually nothing left. He was more or less buried, except for his face, in clay which was a protection when the roof came down.

An RS meeting—mainly on the past year's finances, and general relief that a satisfactory Government enquiry has taken place. Sir Harry impressed his warning to make sure that the assessment of overheads from time to time is such as to ensure that there is no deficit.

A TVC meeting at which the nomenclature question was once more fought over—but good temperedly. There is to be another main meeting with the stores people—so hopes may yet be dashed.

June 3rd

Price has finished the trial Heavy Alloy projectiles each with a tungsten carbide core in one of three different positions. These are 1¼ inches diameter. He has made ten billets for the large projectiles (3 in) but there seems to be some hitch at the high levels—probably forgetfulness.

The order for 600 more luminescent 'bats'—if we can take them on—is not entirely unexpected and there may be a change in the method of 'serving'.

Leeds tells me that we turned out 178 diamond dies last month. They were aiming at 200. The new start is partially psychological. Leeds is segregating the very fine dies from those from 0.0015 and upwards.

June 5th

Visit by Joseph Lucas management to talk over their proposals for extended research facilities. ERA Research committee in afternoon to consider new research director of ERA in place of Wedmore, retiring. We made I think a good arrangement in inviting Dr. Whitehead to be assistant director and to act on Wedmore's retirement until a new Director is appointed.

June 6th

The invasion has come at last—a cool blustery W to NW wind. One would not have thought it possible in such weather.

June 7th

Actual facts are meagre but we are assured that progress is satisfactory.

Partridge reports that Glass Works have agreed to fit electrodes into their L_1 glass tube furnace so adapting to practical production the technique we have found so satisfactory on a small scale.

This has been a CVD day at Wembley—also Valve Allocation day. All in good spirits. Brundrett delighted at the low naval losses at the first landings when many German submarines had apparently been drawn back to French coast to operate against a landing.

On valve preproduction the general feeling is that the factories are more ready than formerly to take over our valves and that before long we may be able to concentrate on the newer types.

June 8th

Messrs. Harley and Ogg came to talk to us about Top Secret, Secret, Confidential, Restricted, etc. etc. papers. The matter in USA and England seems to be in the greatest confusion and no one here has that courage to pass to industry an interpretation of what we are intended to do. We shall probably carry on as we are using our own gumption.

Touche's auditor (Mr. Barrow) has had a bad time with our accounts. Expansion of the work and deterioration of available personnel leave us very short handed and things have almost reached a breakdown. We must get help somehow.

June 9th

Set up Lab. Accounts Committee, Kersey/Cornwell/Humphreys, to overlook the financial structure of the place and to ensure co-ordination where this is necessary and to see that the parts of the system are kept up to date and properly operated.

Lunched with Mr. Bosanquet of Ministry of Agriculture on how to foster research and development on agricultural machinery.

Meeting with Admiralty (Captain Gladstone) on Bowler, Trumper etc. Farren and Edwards are accepting the main load during Marris's illness which may continue for some time.

June 10th

Gossling is getting on with his valve breakage and packing problem. Tests will be made here on the lamp 'jotters' [joggers?] to try and deal with the problem on fundamentals. We may have an RAE and a GPO research man to work here on it.

Mrs. Marris sends none too good an account of her husband. We fear he is in for a tedious illness and prolonged convalescence.

June 12th

Marris died early this morning—a sad and severe shock to very many both inside and outside the Laboratories. He has been a loyal and wise colleague. Lunched with Sir Harry Railing who had noted the strong points of Marris's character and abilities and appreciated them thoroughly.

Confirmed under the new conditions the arrangements made with Espley and Edwards re. the organisational side of the Electrical Group.

June 13th

Various staff arrangements and responsibilities to be distributed in connection with the Electrical Group. Arranged to take on young James Marris for the period of the war. He is just down from Cambridge.

Cherry is up from TRE with exciting things to record about the use of Oboe in the invasion. It is naturally invaluable for bombing targets at such short distances from England. He has been attending to the equipment on the first operational planes and tuning it up. It is a mixture of GEC and USA designs.

June 14th

A CRT meeting at which our two-colour 'window frustration' was one of the features demonstrated. The meeting went smoothly with plenty to settle.

Seven of us attended Marris's funeral at Golders Green. I have had many letters of sympathy and appreciation of Marris's qualities and a particularly charming one from Captain Gladstone. It has been a poignant time for very many here.

June 15th

A GEC radio co-ordination meeting. Heather away so I presided. A tribute was paid by Sir Harry Railing to Marris's memory.

A somewhat hectic TVC meeting all afternoon with Services Stores members present to finalise the CV nomenclature plan for valves and CRTs. The War Office were in trouble over USA types but the arguments for giving them CV numbers were very strong and although this one point had to be referred the scheme has now virtually gone through. I had a bit of a job in the chair.

June 16th[478]

Daylight raids have started again in the form of automatic bombers without pilots. The first group today operated over SE London but a few stray ones got further afield.

Have had Polgreen and Fielding here on quartz crystals. We are not getting on as fast as we could hope.

The experiments on valve packing and breakage are being pushed, and we have one man from Dollis Hill and one from RAE. The vibration gear used for lamps is proving very useful and work is being carried on in the life test laboratories. Sir Stanley Angwin evidently means that things shall be pushed hard.

June 19th

Preproduction goes on relatively smoothly. There were 2400 valves in the week.

Lunched with Chelioti and discussed Schoenberg's arguments with him supporting the removal of valve research from Wembley and Chelioti's able reply in draft. Promised him a memo. on MOV and Wembley's contribution to new valve developments. Whilst we were lunching there was a heavy crump nearby (actually Goodge Street, Tottenham Court Road) from a pilotless bomb plane. These dominate conversation at the moment. People about the streets are quite unconcerned but people tend to have less than their quota of sleep and in South London some neighbourhoods are in the direct path of the machines. The Germans are inventing stories—the extravagance of which is a measure of their demoralisation.

Meeting at Head Office in afternoon on Marconi, EMI, GEC, MOV arrangements, and correspondence between Mr. Schoenberg and Admiral Grant. The time was spent in trying to detect the policy behind the EMI and MWT proposals.

June 20th

The various valves under development are gradually proceeding on their course. We have made twenty of the large micropups E 1496, [and] the CV 192 for TRE with somewhat better results; it looks as if this could be expected to be a production proposition.

Viewing our research contracts generally there seems to be no outstanding liveliness just now but everyone is working very hard nevertheless.

Dr. Sloane has received his notification to take up his post as professor at Beirut, Syria, and will probably leave us in the late summer if transport to the Near East can be found.

Ryde had some Americans here interested in FIDO; they seem to be copying our efforts at fog clearance and have started with a live burner scheme in the Aleutian Islands.

James's work on measurement of crystals at 10 and 3 cm is making progress and is only just in time to save our reputation as a research establishment. BTH make good progress with their extensive craftsmen facilities.

June 21st[479]

Marris's son James came to be taken on for work of national importance. He is just down from Cambridge.

A CVD policy meeting—mostly sparring and some leg pulling between departments. Met Mr. Gestetner afterwards with Wright (DSR) and Brundrett. He is taking on the job on behalf of the Admiralty (DSR) of formulating general plans and conditions for post-war development contracts. I think DSR wants to shelter behind him as an industrialist in putting their schemes to the Treasury.

I hope we are nearly through with our financial analyses for 1943 – 44. There have been last minute adjustments owing to misconceptions in our allocations in LSO which will raise our overhead probably some 10 points from the last expected figure of 117%.

June 22nd

A day on lamps and household appliances. Bligh has had intimation that the spur link plan is to be dropped. This is one more, and practically the last, of MOS schemes to find a premature resting place in the cemetery of projects discarded or jockeyed out of the hands of industry.

The 'Doodle Bug' flying bomb has been keeping up a desultory supply of incidents. We don't hear what is happening at the point where they are being attacked but a good deal is happening if one can believe rumours.

June 23rd[480]

Had Dr. Grimes here from Greenock Torpedo Factory to try to hasten Bowler. Hawes is doing well with it but is almost single handed. We discussed plans for getting over the immediate bottleneck—it should soon pass when the next two or three sample sets get to Greenock.

Dr. Roxbee-Cox of Power Jets Ltd. came to clinch arrangements re. Reeman who remains 'on loan' to them. His account of his new state supported Limited Company was impressive.

Mr. Cheel and Mr. Caldwell came to clinch matters as regards Donaldson's projected transfer to Watson Electro-Medical. We fixed the end of September and I expect the arrangement will now be confirmed—if Donaldson is still willing.

June 26th

Dudding is back from another successful contact with Coventry on Quality Control at which Mr. Malcom expressed much appreciation of his work. They have had particular success in the Condenser Dept.

CV 221 (Harris) blocks are still cracking and have put us behind hand in our promised delivery schedule. The T & R box is at present the main bottleneck.

No end of a discussion this afternoon on GEC policy vis-a-vis MOV on research and development. Chelioti had written an able summary of the main issues. A.H. Railing in the chair and all the GEC leaders on valve and radio plus F. Smith. Mr. Schoenberg is evidently making a strong bid for setting up an independent MOV research organisation but the meeting was unsympathetic. F. Smith considered his loyalty to MOV required him to urge the claim of separation.

June 27th

To Spon Street, Coventry, for the first Radio Works-Research meeting. It was a very useful meeting for informing both sides of the other's activities, and gave us an opportunity of seeing Spon Street Works. This meeting should help to close the ranks and give added strength.

We saw GR in course of production. Espley reminds me that this is just the set needed for shooting at the Doodle Bugs and when the authorities realise it they may work up a flap for crash production.

Fairbairn returned to England yesterday and Heller is expected later in the week. Their tour has been successful. I understand that while all went well and whilst all over there appreciated the set and what it does the Americans are likely to make their own alternative rather than adopt this one.

June 28th

The Doodle Bugs still keep up their attentions and people North of the Thames are getting used to them. But South of the River the nuisance is more troublesome and gets people down. The rumour is that only 10% get to London proper. When one is heard coming people are getting into the habit of diving under the table to avoid danger of [from] glass breakage. At Wembley we step quickly into corridors. Visits from Mr. Gestetner on behalf of Admiralty to see our resources for research and development contracts, post-war. He saw Bell, Megaw, Waldram and others. He is an interesting man to discuss things with, and I think he appreciates the essentials of the Wembley organisation.

MOV receiving valve meeting in afternoon which emphasised the reality of GEC contentions at the meeting at Head Office on June 26th. The items were all connected with scientific service, techniques etc. and were very little to do with specific receiving valve designs alleged to be made at Wembley—except for special war valves.

D.H. Donaldson was finally fixed up to go to Watsons Electro-Medical as assistant to Mr. Caldwell. This arrangement was our recommendation on Mr. Gamage's request that we would put forward a suggestion from here.

June 29th

After many adjustments and reconciliations with the auditors the Lab. overhead comes out at 133% instead of 117% as previously thought. We are not very proud of the handling of our financial affairs this year. Humphreys has straightened out many irregularities and has done a great work.

A surprise visit from Pollard of MOS on a highly secret mission to get names of people or concerns in Germany who must be gripped the moment the German Armies crack and we get access to Germany and to the Occupied Countries. I am glad we are preparing for all eventualities.

A TVC meeting this afternoon at which the Army could not agree to include the 'International' range in the CV nomenclature—so having gone most of the way towards a complete scheme agreed between the three Services we have to be prepared to leave this blemish.

The war goes well. Cherbourg has fallen this week. The Russians have badly defeated the whole German Central Armies before Minsk.

June 30th

Had a nasty letter from DTM Dept. Admiralty complaining of our handling of Bowler as if we were responsible for its conception. The list of indictments are unfair and most inconsiderate in view of Marris's recent death. The complaints really lie at the door of Admiralty establishments.

Meeting at Head Office on Alumicide and difficulty in making progress with it. It was very much of an indignation meeting at the autocratic and rude behaviour of Mr. McAuly Gracie of the Ministry of Food. Good technical reports have been received from Canada and Australia. Our trouble is that this kind of chemical manufacture and the whole chemical and pestiferous [pesticide?] field is outside GEC knowledge and experience.

A diamond die technical committee to consider mainly the USA coning machine. A useful discussion at which Mr. Smeaton of BTH advanced his 'long channel' die as giving three times the life of the standard short channel.

A valve research meeting yesterday, Megaw and Willshaw attending. It was supposed to be to formulate a programme of magnetron research. But it covered the whole field in a useful way and indicated the objectives of very short pulses, millimetre waves and the like. We contributed magnetron requirements of fundamentals but the Services would not consider that as research because it was directed to the further perfecting of existing valves! It sounds heretical to us but it is after all only a matter of nomenclature. The Services naturally think in terms of new devices leaving all the difficult spade work to anyone but themselves!

July 3rd

Preproduction a record at 2872 valves—which includes some 400 CV 258s. This is very good considering that most of our people have been suffering from want of sleep due to Doodle Bugs and not a few have had homes damaged and relations in trouble.

Had here Air Chief Marshal Sir Frederick Bowhill and Air Commodore Brackley of Transport Command. We showed them Fido and other things including the 300-channel set and found them greatly interested.

July 4th

CVD committees on crystals and triodes—very dull on the whole. The thick wet weather has left us and fewer flying bombs are getting through, though there are continuous warnings in town.

July 5th

Have had to consider a demand from the Factory Department making XL legally a factory—in parts. This seems to me nothing but officious and vexatious.

A CVD meeting on magnetrons—brighter than yesterday's. Our high powered 10 cm magnetrons are at last getting into the clear and progress is being made. A limited number of bombs come over town every day. They seem to concentrate on the lunch hour when people are about!

July 6th

The breakage of valves in transit investigation continues under pressure from Angwin. Through Gossling we seem to be providing the ideas and the programme of work.

Ramsay reports good success with hydrogen thyratrons—600 hours life at 100 A 3000 [c/s] frequency, one microsecond [pulses]. Report from ADRDE is that it works at 1/30 microsecond. There is, as usual, strong opposition from Dr. Truscott's department to this work on the ground that the device could not be produced at the present time. I am constantly having to fight the policy of squashing research at the outset and keeping ourselves in ignorance in certain fields for fear [an] awkward production situation might result.

July 7th

Jenkins has a ciné-theodolite equipment to fit with a small discharge lamp for millisecond exposures—the job looks promising.

A Research Services meeting to review with the Managing Directors the research for the Services (activities of the past month) and back to Wembley to a 'Bowler' meeting where all the past week's dissatisfactions and difficulties were mostly ironed out. We are to give Bowler preference here and at Greenock which means that Farren will have to transfer mainly from Trumper at Greenock.

July 8th

A meeting with an informal Laboratory committee on education to implement a scheme we expect to launch to give our student assistants the benefit of regular coaching by members of the staff at standard fees per hour. We hope this will benefit things all round for it will cost the Laboratories between £1500 and £2000 a year with our present large number of candidates. We shall regard the first year as a trial; there seems no doubt we can afford it and the general intellectual whipping up will do no harm.

No Doodle Bugs today—reason not known.

Gossling's report on the theory of valve packing circulated to Angwin's committee is an excellent and businesslike document and I gather met with no little approbation.

July 10th[481]

Trufitt on his visit on Saturday had a good look round with Warren, Bell, Le Rossignol and Gossling and was particularly intrigued with the E 1457—95—96 series.

Lunched with Sir Harry to talk over a number of matters including gas turbines and Power Jets Ltd. and GEC policy. He has asked me for a memorandum outlining a possible policy for action.

Spent two hours in the cellar at Sanctuary Buildings, with DTM [Director of Torpedoes and Mines] presiding and about 25 officers, to go over Bowler, Trumper, Dealer, Bidder and Fisher. We had a good talk and clearing out discussion and confirmed the present trends and priorities and I am to see what extra help can be got on Trumper.

Heller is back today—a good job.

July 11th[482]

The Informal Technical Committee (discharge lamps) met here today after five years. Davis (BTH) Aldington (Siemens) and RL people. Main subject was the *locus standi*, if any, of the ITC. It was agreed that as a meeting of research men on lamp research or early development it had a function. Old committee washed out and its tasks of detailed standardisation, which belong to ELMA, discontinued. Instead an Informal Research Meeting to be held with its proper task of exchanging experience and concentrating attention on those things which may one day need to be made uniform.

Squadron Leader St. John came to talk over the lively Director of Torpedoes and Mines meeting yesterday. We have found a way to add to Trumper personnel. The RAF group are very loyal and we are able to help each other.

Espley has got inside information re. gunnery against flying bombs which will be useful at our meeting with Sir Harry Railing tomorrow.

July 12th

Ryde and Espley gave us an interesting account at our meeting on ways of counteracting the flying bomb. I left before the end, to join the GEC education meeting where we addressed ourselves to many vexed aspects of the subject and the impact upon it of industrial condition[s].

After a lull in Doodle Bugs several have been over today—another in Kenton.

July 13th[483]

A morning with Cobden Turner on oscilloscopes, ultra HF [UHF] measuring instruments, rectifiers, Alumicide and jewels. On the latter he has been enthused by Leeds into somewhat wild thoughts regarding post-war expectations. Leeds and [Cobden] Turner have considerable possibilities when allowed to react on each other. The plans for oscilloscopes were matured and Tomlin got a forward move on his instrumental development facilities at Stockport. GEC Board Meeting. Same result as last year but 20% up in turnover.

Sir Harry encourages research work on gas black and welding electrodes, but is not enthusiastic on copper–carbon brushes. This gives Cameron and Hawkins a lead.

July 14th[484]

Quite a number of workers here had their houses made uninhabitable through flying bombs. The glass workers seem particularly hit.

Consultative Board Meeting with annual statement of accounts and other papers. Chelioti does not seem keen on post-war preproduction but the matter will be argued out. There is never shortage of money or worry about expense (within reason) these days—it is only a question of how to get things done without labour or materials.

Saw the ciné [kine] theodolite by Jenkins and group before despatch—a nice neat job.

Lang Brown wants principle agreed that we establish here after war work on camouflage for the Admiralty. We concur but I doubt the Treasury and jealousy of Service establishments.

July 15th

We have fixed up 'tutorial' arrangements for our Student and Technical Assistants by which all shall have a chance of being coached by a member of the Scientific Staff. Tutorial fees will be paid to people who will take on the job seriously. It may cost £1500 for a year and it will be tried for one year.

July 17th

The Russians are nearing East Prussia and travelling fast. Things are slow in Normandy specially round Caen—weather against all air action. We have nearly reached Florence. Flying bombs fewer the last few days.

Discussed emergency arrangements in event of an 'incident' at the Laboratories under the new day time conditions.

A Lab. Stewards meeting in which subjects roamed from local workshops to the quality of buns. The latter engendered more heat than the former.

Valve preproduction still keeps up although there are a good many women away or leaving due to Blitz.

July 18th[485]

Edwards has spent two days at Greenock (Torpedo Factory) on Bowler and Trumper. The steps we took seem to have eased the position but there is a strong indication that they want to loosen our position and responsibility in respect of Trumper—and we shall take the corresponding diplomatic steps to counter such move.

Spent all morning studying Lab. accountancy system at first meeting of accounts committee (C.C.P., O.W.H., J.C.C., G.K.). We have arranged definite monthly control checks to ensure that the accounts are up to date and correct.

Attended with Ryde a meeting on Flying Bombs at the Petroleum Warfare Dept. Sir H. Hartley in the chair. We heard all about the bombs and the counter measures being worked at but it seemed to us that we have been invited as an audience to listen rather than to speak. Ryde put forward the idea of grapnel and towing away from inhabited areas.

July 19th

We are arranging to take on work for Salford on jewels for instruments by giving [to] Leeds Mason's undivided time and making other arrangements for die mounting which Mason does at present.

A GEC CVD meeting with Griffiths presiding—also allocations. Nothing outstanding.

July 20th[486]

Discussed CEM metal—last year's results—with Williams and Price. Made a loss as the manufacture is all 'odds and ends'. Heavy Alloy has made a compensating margin.

Spent my turn spotting on roof. Had two 'duck' warnings. One exploded at Park Royal—the other apparently about Highgate. The Doodle Bugs are coming over in batches apparently in order to saturate the defences.

An MOV transmitting valve meeting—interesting and lively.

July 21st

A busy night and morning with a fair number of bombs over—one at East Lane beyond North Wembley station and one more at Kenton. The morning 'bugs' all travelling across London SE to NW.

Ward and another from TRE to study CV 208 and causes of too many returned alleged faulty valves. We shall hear the conclusions anon.

Lunched with F. Smith (MOV) and discussed the pros and cons of preproduction groups—and also the GEC, EMI, MOV relationships and Schoenberg's reasons why the handling of valve research by GEC at Wembley cannot be successful—notwithstanding all the experience of 20 years plus the five years of war.

A TVC meeting, long and rather argumentative. Exciting war news of German Army revolt and effort to kill Hitler. We shall see if this is merely an incident or a precursor of more immediate collapse—but it cannot be encouraging to German soldiers or public.

July 22nd

I gather that the CV 208 has not a very good reputation due to rejects in service. Half of these are apparently not justified, the other half are cathode faults which must be rectified at once.

July 24th[487]

Preproduction still up to 2200 which is good evidence that the flying bombs have not affected our production. This is not because plenty of people have not been bombed and some evacuated but because the remainder have determined that the output shall not fall.

Chirnside reports seance with Mr. Macauley Gracie of the pest department of the Ministry of Food! I am looking forward to Mr. Cobden Turner's own recitation. He, Chirnside and Briscoe went to find out why the Ministry of Food treat the GEC plus Professor Briscoe as some special kind of insects.

We are told by Counsel that we have a good case with which to go to the Appeal on the question of tax exemption for the classes expenses of our student assistants.

Lunched with Gamage and discussed staff who could go abroad to GEC branches.

July 25th

A quiet day from bombs—meetings here on 'butterfly circuits' and discharge lamp auxiliaries.

July 26th

A CVD meeting at which the two-colour 'window frustrating' tube came up for discussion. The trial against 'window' in a practical test showed what a very thorough disturbance it occasioned.

Mr. Davis of ASE working on RCM was here with Gavin. He finds our CV 155s invaluable for his portable jamming equipments and considers the Admiralty has done a first class job with them.

Have agreed that we ought to submit a regular report to MOV on the research work in hand and being undertaken so that no charge can lie against us that we do not give a proper account of our stewardship.

July 27th[488]

Dr. Black here on visit from USA. A lot of valve interest at the present time concerns miniatures in which these Labs. are not so interested but the magnetrons and the like are still a top priority.

Leedham lunched with me as I wanted a lead as to the outlook of MAP on post-war collaboration with industrial research concerns. I got such a revelation as I never expected of post-war plans and recent Cabinet decisions on principles to govern the set up of things. There is apparently to be no sparing of funds to maintain proper research and development facilities in the country and tremendous Service research establishments are visualised. I heard much more in confidence which must not be written down. I was reinforced in my inclination to maintain facilities for carrying on at Wembley on valves and CRTs to meet what one can see that the Services will need.

July 28th

A meeting at HO on education in the GEC where we completed most of the main principles.

The meeting with the scientific staff on the 'tutor' scheme for assisting student assistants went off well according to Gavin who was in the chair. It looks as if the scheme will be a real success.

July 29th

News this morning of R.H. Fowler's death—we are all very grieved. He was an ideal scientific consultant.

We seem to have settled up the 'War Dogs' problem by referring Captain Garle to the Harvest Saver and Implement Co. Ltd. who specialise in electrical protection from animals.

July 31st

Our valve preproduction is about a record again in spite of holidays and Blitz. The week-end nights have had some nasty incidents—four in Wembley district.

Examination results are coming out—we have seven B.Sc. successes so far.

Attended meeting at Head Office to consider GEC, EMI, MOV relationships. There was no inclination towards any radical change of selling policy for valves but the possibility of a compromise arrangement was considered if other and more fundamental irregularities in the situation are rectified.

August 1st[489]

Discussion with Heather and Humphreys on the marrying up of M, X and Y products with the Advisory Board machinery. We can anyway start with a proper instrument committee if Salford are willing. Afternoon discussion continued with addition of Settle and decided, after full review, to proceed with build up of Induction Heating development on the lines of our memorandum—which may involve an allocation of £14 000 per annum.

August 2nd

A TVC meeting this morning and a CVD policy this afternoon. Both interesting and not without some spice of naughtiness—particularly over the USA achievements of Mr. Carlton-Dyer. It was agreed that Bell could be sent by CVD to the USA on a visit—also Broadway of EMI.

We seem to be beginning to overrun France to the South. The Russians are nearly into East Prussia. Churchill's speech today was very optimistic—but he does not forecast a let-up in the enemy's missiles over London.

August 3rd[490]

Espley has been having adventures over GR. Final enigmas at Crowkerne [Crewkerne] Aerodrome, and limit value tests at Coventry where the 20 prototypes are going through quickly. I fear one cause of trouble may be in non-uniform fixing of copper radiators to the anodes of CV 155 valves. But the bad cases may be rare.

Espley, Cherry and Farren had a general talk with Hannam-Clark and his engineers at Coventry on techniques developed during the war and their possible influence upon the carrier [telephony] field.

Last night was a disturbing one with Jerry bugs. One fell within striking distance of XL and sucked out windows and doors at the front. Work not affected but Claudgen were rather badly damaged. Lumifax but little so.

Had an hour with CRT group examining L. Jacob's work—revivification by barium bombardment, and electron microscope up to 15000 x magnification with simple demountable tube. Jacob's devices and Sharp's on measurement of decay and 'storage' tubes show good work.

Research Services meeting with A.H.R., L.G. and T.W.H.—useful and full of interest this time.

Willshaw tells me that our high power magnetron has beaten the field and is being tried out both by ASE (EMI) and TRE. I suppose something will eventually be made by BTH on the 'Birmingham' design, but Willshaw does not appear to be very impressed by their prospects.

A full 'Bowler' meeting all afternoon with Captain Gladstone presiding. Captain Richards of Greenock suggested that these meetings be discontinued but nearly everyone at the meeting disagreed—so they will continue as hitherto at Wembley—which seems to me thoroughly appreciated.

August 5th

Willshaw's account of the magnetron position shows a good outlook on all fronts. The large type of S band has successful samples with ASE and

TRE. It is about 60% of the original output target. The CV 108 troubles are not all with the valves. The chief troubles for us are rejects before delivery and Le Rossignol and Willshaw will arrange for Balls to give first priority to our processing. The CV 192 is now out of its troubles but Willshaw has designed a better version which may or may not be exchanged for the CV 192. After the deep waters we have passed through they will be reluctant to do other than leave well alone. The (preplumbed) CV 214 is going well and the E 1511 which is to increase output from 40 to 200 megawatts [kilowatts] makes promising progress. In the K band we have two valves one of them with 8% efficiency and looking well. Altogether a good report.

August 8th

We seem to be breaking German resistance in Normandy and the position is tense. Flying bombs have eased up noticeably over the week end—but many people are leaving London and it will be surprising if our preproduction does not take a turn downwards. Last week for the first time it was noticeably down (10%).

Had internal meeting on Witton rectifier affairs and came to the conclusion that some £10 000 to £12 000 per annum needs to be contemplated with overheads to tackle their development programme. The next thing is to see Gallizia and talk it over with him.

A visit from Dr. Crampton of ASE—one of the long standing members—to talk over a multi-channel set for the Navy on the basis of the RAF equipment.

August 9th

Education again at HO at which Humphreys and I were astonished to realise the strong prejudice in favour of heavy engineers as the basis of good salesmen. There was obvious and ominous silence when I—backed by Chelioti and Humphreys—urged the need of a few scientifically trained men as a lever to the machinery and switchgear people.

Had an interesting talk with Railing on his Presidential address to the IEE. I think he is going to make a good thing of it.

August 10th

Had remonstrated with Sir George Gee (MOS) at GR being sent into action for the first time without Espley, or someone equivalent, to make

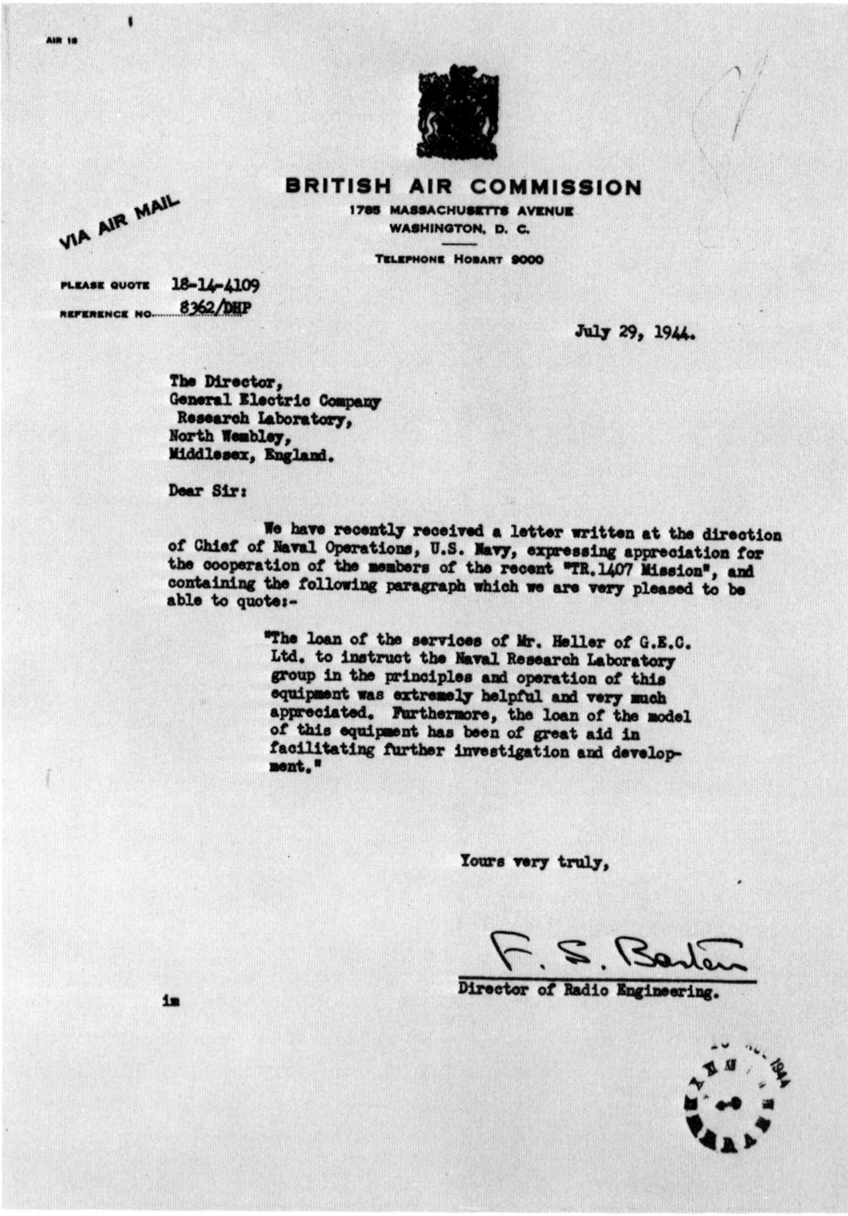

AIR 10

VIA AIR MAIL

BRITISH AIR COMMISSION
1785 MASSACHUSETTS AVENUE
WASHINGTON, D. C.

TELEPHONE HOBART 9000

PLEASE QUOTE REFERENCE NO. 18-14-4109
8362/DHP

July 29, 1944.

The Director,
General Electric Company
Research Laboratory,
North Wembley,
Middlesex, England.

Dear Sir:

We have recently received a letter written at the direction of Chief of Naval Operations, U.S. Navy, expressing appreciation for the cooperation of the members of the recent "TR.1407 Mission", and containing the following paragraph which we are very pleased to be able to quote:-

"The loan of the services of Mr. Heller of G.E.C. Ltd. to instruct the Naval Research Laboratory group in the principles and operation of this equipment was extremely helpful and very much appreciated. Furthermore, the loan of the model of this equipment has been of great aid in facilitating further investigation and development."

Yours very truly,

F. S. Barton

Director of Radio Engineering.

im

Fig.54 Letter dated 29 July 1944, from F.S. Barton, British Air Commission, Washington, on visit by D.M. Heller

sure of its proper functioning. Gray [Grey] of MOS rang up to have a fuller story. He appeared to understand our contention but one does not know the result nor have we heard how GR is functioning.

Clayton gave me his story of work on aerials for 10, 3 and 1 cm. A very good and interesting account.

Had the attached very nice [extract from a] letter from the US Navy—Chief of Operations. This is the first bouquet from this quarter—and not, I believe, a common occurrence (Fig. 54).

Visited XL for a good look round of all groups, I don't know what we should do without this spare space or what we shall do after the war.

August 11th

Espley has an interesting suggestion for combining acoustic and short wave 'waveguide' technique for a system of non-metallic mine detection. A very 'hot' subject and an interesting idea.

A splendid piece of work has just gone through to Appleton's committee by Ryde and his wife on the attenuation of centimetre waves by rain, hail and clouds—a continuation after three years of his pioneer work on the same subject down to 3 cm. There is no doubt that Ryde's work will for long be the classic on this subject.

Peters writes me to say that the new collaboration agreed between GEC and EMI circuit people has broken down on EMI side due to cold feet at high level! What a pity these things cannot be left to the intermediate levels, we should then have real sensible collaboration.

A Consultative Board meeting at which discussions are always lively and useful nowadays.

August 14th[491]

Holidays—but I hope not the Blitz—have brought our preproduction down to 1600. The CV 208 is in trouble and radical steps are being taken to help it onto the lines again.

Dr. James gave me a demonstration of their 'butterfly' circuit work which, with the new valve to match, makes an attractive job. As it means a modification of the CV 90 and so a new valve type I expect it will be violently opposed by ISVPC and have to fight for its life.

The bogy of 'Exercise EXE' has descended and we are requested to consider the steps necessary for meeting any emergency which may arise. The schedule of valves to be made does not agree with the allocations which we had been told to prepare for and there are many points to be made clear.

A meeting called by Dudding in the afternoon to consider oxide coated cathode theory and research—a discussion on a good level.

August 15th

The second 'Laboratory Accounts' meeting. We are not yet going properly with our checks to see that all sections are up to date—and it is difficult to strengthen departments when existing people on whom we lean fall away from us and go into half time or start families. Miss Lock is taking on regular 'sales materials' bookings and other changes we hope will bring the ideal method nearer.

Had a scare letter from DEE Admiralty suggesting looking for 'host' firms to carry on with some at least of production in the event of emergencies arising.

A CVD (triodes) meeting this afternoon. James's valves for butterfly circuits were the main interest to GEC. Troubles over a Gramophone [Company] VM tube, which illustrates that we are not the only people who have devastating teething troubles in preproduction.

August 16th

The landing on south coast of France was yesterday morning. It appears to be almost unopposed so far.

Some of us are living in the atmosphere of the possible evacuation of the Laboratories and incidentally of London as well. I saw Cobden Turner today and he agreed, subject to Admiralty approval, that we could have a floor of one of his factories for part of the Laboratories to move to in the event of the emergency. Have contacted Brundrett to put pressure on DNO dept. to keep hands off it.

A meeting at Head Office on thermostats at which all seemed desirous of amicable co-operation.

Macfadyen tells me of a use at Loch Long on 'Bowler', of his 'talking down the beam' equipment, for 18 months now on the shelf. It is quite a good piece of gear and one would like it to come into its own.

Mr. McLennan (auditor from MAP) came to pass our accounts finally. The overhead of 133% is agreed. O.W.H. as usual had things at his finger ends.

August 17th

Gossling's work on the packing of valves for overseas transport is bearing fruit. They have put up the 'drop' of the NT 98 from 2 ft to 8 ft and the CRTs are also good for a 5 ft drop.

The CVD meeting yesterday on magnetrons went I understand smoothly and generally to our credit. Our largest (10 cm) magnetron is successful and Willshaw is being pressed for samples—therein lies the rub!

A long visit from Lonsdale and Waddington on the programme for gas turbines. Jolley and I discussed a possible compromise programme consisting of energetic work upon the axial compressor—(which is a major component of the gas turbine as well as a normal F & C requirement) and a quite small 100 or 200 kW gas turbine and compressor unit suitable for studying the combined gas turbine programme as a first step and incidentally as a possible prime mover for ancillary electric supplies on aircraft.

August 18th

To Leamington Service camouflage establishment to meet Waldram and Commander Young-Bateman. Their new water model arrangement using diminished image of a large model ship for horizon viewing is excellent.

August 19th

There has been a landslide in France and the German armies holding us back, from Caen southwards, seem to have crumpled up and to be at our mercy. These have been thrilling days and it looks as if Paris would soon be passed or occupied.

This is holiday season and many are away but a number are staying at home. There is feeling of tenseness as between events in France and enemy 'revenge' tactics actual and possible.

August 21st

Preproduction is back again to top level in spite of general disorganisation of flying bombs and holidays.

Farren reports deep water expeditions in submarine from Greenock to test out 'Bowler' in rough water with Fry. They got useful measurements on the effect of rough water on the strength of back echoes. Dudding had successful meeting with senior management at Coventry (25 or 30) under Mr. Malcolm's chairmanship, to acquaint result of Quality Control work at lower levels in the factory. I understand a letter of appreciation is on its way.

Espley has heard that on its first reasonable target shooting, GR got three hits out of four shots.

Lunched with Chelioti and discussed dispersion from the London area. Briefly he is 'agin' it. Glass and Lamps cannot be moved, certainly not Glass, therefore what is the use of assuming Lamps and Valves can be carried on elsewhere—which depend on glass which is assumed not to be available! The glass problem certainly affects the question of Wembley moving its preproduction.

August 22nd

Meeting at Head Office re. metal rectifiers and the Witton desire to market a complete range. The position was explored, and relations with Westinghouse discussed and I think everyone was clear as to GEC intentions.

August 23rd

Smithells was with me last night. His British Aluminium scheme is going strong and he is seeking temporary premises for making a start in.

Two bombs within a mile of the Laboratories at 8.00 this morning—one damaged the glass again at XL.

Lunched at BSI Savoy luncheon to USA and Canadian screw thread delegation. It reminded me of a similar gathering at the Savoy in 1916–17 on fuse standardisation, when Churchill gave the speech of welcome.

An MOV receiving valve meeting in afternoon.

Paris is liberated today by the French patriots—great excitement.

August 24th[492]

An Osram Advisory Committee meeting with a 'post-war' slant on things in spite of the fact that people are being urged to prepare for the virtual evacuation of London.

Had a chat with Appleton at the Club. He is very appreciative of the way that Ryde 'delivers the goods' in the matter of short wave atmospheric transmission and attenuation—referring of course to his second report just sent in. He also asked whether Megaw could be sent to represent Appleton's transmission committee at a conference on short wave transmission, in the USA. Another feather in our cap which is very pleasant and good for the place.

Ran over to Fraser & Chalmers to see Caplan and his set up. A very enjoyable and intimate visit. He is doing very well there.

August 25th[493]

Went with Dudding to be talked to on the subject of general dispersal in view of rocket projectiles putting London out of action. There are a windy lot of people at high levels but I rather gather that politicians (including the Cabinet), having been pretty heavily caned over the flying bombs, are not going to be found unprepared a second time. So they mean to 'plan' in such a way as to make it appear that they knew about the danger and took steps accordingly. I believe the PM is also in the racket. Well we may be in for something and we are going through the motions of planning to meet the eventuality should it arise but certainly not before it arises. We have got the Admiralty to reserve 20 000 sq ft of floor space at Chestergate Mill (Cobden Turner) and I am enquiring in other quarters too for possible space.

The war situation certainly does not look like backing the 'windy'. Romania has joined the Allies and Bulgaria is following suit—half of France is liberated and the Germans are suffering terrific losses all round. The landslide gathers force.

To the Physical Society Council in the afternoon where a strong move was made towards reconciliation of the Society with the Institute of Physics. Jevons had threatened to have to hand over secretarial duties to another so that he could lecture for the Imperial College.

August 26th[494]

Bell tells me that Clegg called here on his way back to the USA. He is handling post-war radio for aviation and is most interested in our latest valves and projects.

Gossling is most incensed over the apathy of Service departments in tackling the valve breakage in transit trouble. He finds that everyone leans on him and he cannot see a way to rid himself of the responsibility, to take up other things including Donaldson's fuse work. He is certainly steering the breakage and packing problem along the right lines and I don't wonder they lean on him.

August 28th

Not a few difficulties with personnel replacements on preproduction, capable people leave and others of equal ability do not materialise to take their place. Preproduction 1950.

A TVC meeting this afternoon at which we discussed prospective meeting with USA delegation coming over to discuss valve standardisation.

Lunched with Mr. Gamage and discussed Fraser & Chalmers affairs, Lab. finance (post-war), Government research in aeronautics (post-war).

August 29th[495]

An *ad hoc* meeting at Wembley on (*a*) street lighting, (*b*) aerodrome lighting, (*c*) cinema studio lighting. J.Y. Fletcher was present and we settled a number of questions of policy which will enable people to plan and think along the right lines.

Lunched with Prain and heard his amazing account of the 150 000 dollars paid by USA Government for the automatic diamond coning machine and the assessment of 45 000 dollars to England for our 'share'. The control therefore wants our patents which are incorporated in our back piercing and other machines sent to USA so that we may claim a quid pro quo! Prain is off to USA again. There is an awkward position on quartz crystal supply because the Brazilians have all the raw material and propose to cut the crystals themselves and compel other countries to buy the finished crystals from them. He wants us to undertake research energetically for a substitute. We must get together a meeting at Wembley to see what can be done.

August 30th

A full meeting with BTH and Siemens on discharge lamps at which many points were settled but there was a considerable wrangle over the suggested non-interchangeability of the starter socket with that used in USA for lamps below 80 watts. This was only tolerated as a temporary measure.

We have agreed internally to undertake, if asked, a research on substitute or artificial quartz for crystals.

Considerable requests are coming in for Heavy Alloy from aircraft firms—for ailerons and other compact balancing arrangements. An enquiry today involved some £40 000.

August 31st

A CVD (GEC) meeting—sparsely attended, for travelling these days is very bad. Stories of people taking eight and ten hours to do the journey from Malvern to London. Flying bomb warnings made people 20 minutes late from town. Thus carrying on is gradually more and more difficult. At the allocations meeting the X band magnetrons and T & R cells were most in the limelight.

Attended overseas meeting at English Speaking Union and met various people including an Indian, Mr. Pundit, introduced by A.V. Hill. Communications from and to India are almost non-existent. As far as I can gather this is just lack of imagination.

Brundrett (re. camouflage research post-war) is not satisfied for Admiralty Research Laboratories to be sniffy of partnership with us in research.

September 1st[496]

The RP 60 scheme has now a hope of getting its power absorbed and may turn out to be practicable.

A Research Services meeting with more subjects to discuss than I can remember. FIDO—on the basis of a 450 ft runway is now clear and Ryde and Cooper will be getting out their report. There is an expectation of saving of 2 to 1, i.e. 250 tons of oil per hour.

Diamond die group has increased in output per worker per hour and we ought to be well ahead in spite of workers being on holiday and blitzed. I presented the Group with much pleasure with £5 last month when they beat Simpsons. This month is better still.

Gallizia came today to discuss his programme of development on rectifiers. It looks as if it might cover work of (total) £12 000 per annum which means about £7000 or £8000 to Witton.

September 2nd

I am discussing with various people the sort of work they may best do after the war. Douglas will have a freelance commission to attach himself to different groups as expediency dictates. For the first year or two he will attach himself to discharge tubes. Bell is off to USA on Monday—left his last instructions today. Donaldson also is preparing to leave at end of month to join Watsons.

Williams has been to Armstrong Vickers re. Heavy Alloy which is definitely prescribed for the aileron balance weights. They had seriously contemplated using gold. So we are apparently for it.

Selenium research with Salford is entering on a new phase because Salford will suspend sales till technical basis is established. L.A. Thomas is the chief figure in this at our end.

Edwards has been to Greenock. He tells me that Farren is not having too easy a job to hold the GEC position of responsibility. Situation needs watching.

So ends the fifth year of the war. The collapse of the enemy in France has brought our armies to the Belgian frontier. Flying bombs have ceased for the past two days and the Pas de Calais with most of the bases will be in our hands very soon and even if they try from Belgium it doesn't look as if it could be for long. Lyons is now in our hands and in Italy the Eighth Army has broken the Gothic Line and is into the Lombardy Plain. The weather is generally very bad and the Air Forces are doing but a tithe of what they could if the weather were clear. But the outlook is more reassuring.

4th September 1944—Sixth Year of the War

Brussels, Calais, Boulogne have fallen. It looks as if the enemy was going to make his next stand at the German frontier. The position in Holland is at present uncertain. I guess he means to hold it—but I may be wrong.

Valve preproduction 10% below the objective total number of 2315 at 2124.

Met Heather, Macqueen and Marriott on a scheme by Angwin and Appleton for radio co-ordination. It is a knotty matter on account of the competitive nature of industry. We favour a co-ordinating body provided the Services do not try to run the show.

Ramsay reports continued success with hydrogen thyratrons. There are now three sizes on the stocks: 200 kW, 500 kW and 2.5 mw. One of the secrets is to make the paths between electrodes very short.

Dudding's quality control work from August 1943 to August 1944 includes 30 public meetings contacting 1500 people and 16 study group meetings.

September 5th

Spent on peace-time post-war projects! But dispersion plans are still the order of the day. No Doodle Bugs for five days and the Pas de Calais and part of Holland in our hands.

September 6th

A CRT meeting but uneventful. This subject is getting straightened out into smooth paths and the meetings will be shortened in future.

Warren has been preparing draft of the first report to MOV Management. We went through it together and recast it in briefer form.

Deputation from Ferranti—Mr. Bass (Director) also Works Manager and a chief assistant all the way from Manchester to talk over emergency dispersion of CV 221 to their works if the word should be given so to do. No difficulty apparently in making a suitable arrangement.

Went down to see Jacob's electron microscope first trial apparatus—a well arranged first experiment showing that the idea is likely to work. We will now plan out a more complete experimental gear and perhaps locate it at the Exhibition [Building].

September 7th

The results of the flying bomb effort for the three months are given—8000 bombs in 80 days—an average of 75% brought down before reaching London! Pretty good. Attached is our own flying bomb record (Fig. 55).

As I leave on holiday there is an atmosphere of optimism that the war will be over sooner than I personally think it will. Anyway everyone is very cheery.

I shall have to come back for a day in a week's time to a TVC meeting with the USA valve standardisation meeting.

September 25th

Back from two weeks' leave. The main happenings apart from general steady progress in the past fortnight are that flying bombs are now very scarce and that all emergency dispersion plans are cancelled. We always felt they would be and that the scare has as much political as actual basis. The speed of events in France and Belgium has caused a tendency to ease off pressure in a number of directions and some of our preproduction may be cut down. But Holland is not falling so easily as Belgium and the big airborne landing has been having a rough time to hold its gains.

The CV 208 magnetron is probably to die out.

Reimbursements of money to the Services for 1942—43 and 1943—44 are now being made by us after going through seas of trouble in getting the figures accurate. The amounts are approximately £27 800.

September 26th

Various staff changes: Donaldson now goes to Watsons Electro-Medical. Dr. Ransley has decided to join Smithells in his new Research organisation for British Aluminium. Knox is leaving to join Napiers. Very probably

RESEARCH LABORATORIES OF THE GENERAL ELECTRIC COMPANY, LTD.

15th September,1944.

A.R.P.

Summary of Flying Bombs Attacks from

16th June to 2nd September,1944.

The following summary of Roof Post action here during daytime working hours has been prepared by Mr. L.G. Clifford, our Chief Spotter, and it is thought that it may interest the staff as a whole.

During the above period the roof post was manned continuously by one spotter during working hours (totalling 627 hours) and by two spotters and a senior man during each "Alert" period (totalling 142 hours 22 minutes). This represents 902 man-hours of duty on the roof.

Telephoned information enabled the senior man to plot the movements of flying bombs on a large squared map and thus to assist spotters in their observations. During this period the courses of 513 flying bombs were plotted. The "Emergency" signal was sounded 46 times (representing a total of approx. 1¾ hours interruption of work). Because of cloud conditions only 11 bombs could be identified visually and of these 6 were seen to dive down to explode.

The longest "Alert" period was 9 hours 58 minutes on 30th June. The busiest day was 5th July, when 31 "divers" were plotted. The "hottest" period of duty was 24th August when, in a 16 minutes duty, 6 bombs were plotted, 2 of which were seen, and 3 emergency signals were given. The most exciting incident was probably that on 5th July when a flying bomb circled the area 1½ times above clouds before diving.

The above duties were carried out by 23 spotters and 22 seniors.

Out of normal working hours fire guards continued the plotting and observing from their roof post by day and night and gave the emergency signals for late workers and night staff. The first flying bomb over this area (midnight 15th June) was reported by fire guards on roof duty and its then novel characteristics correctly described and logged.

N.L. HARRIS.

Dr. Paterson.

I think you may be interested to have a copy of this for your records.

N.L.H.

Fig.55 Summary, by N.L. Harris, of flying bomb incidents between 16 June and 2 September 1944

September 27th

A CVD magnetron meeting. All our lines of work seem to be yielding successful results and Willshaw gave a good account of it all. I was particularly glad that our high power S band magnetron has come out on top and is the only valve in the running. We had to muzzle into this development in just the same way as into the X band magnetron, in face of heavy bias by the CVD management towards the activities of the Birmingham group.

Had Hecht to lunch with me with a long discussion on post-war research and the relative shares of Government Establishments and of Industrial Establishments in the research and development work. I think (or hope) it was clear to both of us that a compromise arrangement is desirable.

September 28th

Mr. Quance came to put me wise as to the manufacturers' views on the proposed new fused plug and socket which official personalities (Beard, Kennedy, Forbes, Jackson [Forbes-Jackson?] and others) are demanding of BSI for standardisation. The situation boiling up for the meeting next week is ominous.

Attended my long delayed TVC meeting with USA representatives. The hurricane on the East coasts of America had caused the delay. They had come over with the clear intention, by compromise, to get common standardisation and nomenclature established. So by 4 pm we had gone a long way towards what looked like agreement.

J.S. Thompson (F & C Group, and on loan to Farnborough) has decided to leave us for the Government service and to join up at Farnborough. Jolley will feel this a serious loss.

September 29th

Discussed various issues with Heather—setting up an Advisory Board Committee for instrument development, situation re. F & C research, GEC radio position and the like.

Fixed with Dr. Ricardo to lunch with Jolley and me over gas turbine research.

TVC meeting with large numbers of submissions for use of valves. The Services are very inclined to use USA valves indiscriminately, so making serious difficulties ahead, for there will be no dollars when Lease – Lend comes to an end.

September 30th

Mr. Aldis came to get us to take up again a compact source for one of his high power signalling lamps. I think we shall do it, as it links up with the Admiralty establishments.

October 2nd – 3rd[497-498]

By car with Espley, Bligh, Gavin (Megaw following) to Coventry for Works – RL meeting on radio. Peters presided. A two hours talk in morning on all broad problems. GR project very much absorbs Spon Street facilities and goes slowly owing to backward deliveries from Witton and elsewhere. Coventry are being kept busy on 1403 with the 'lightweight set' being required in disappointingly small quantities. MAP find (which might have been obvious earlier) that the squadrons know the 1403 inside out and don't want to change. It is also in 'shadow' factory production. The 1407 (multi-channel set) makes normal progress.

Went to Witton to see Garrard and development facilities there and the Helmore job and fire control. Found Dunham and Hall at it with the 10 ton gun mounting.

On to Rugby for diamond die technical committee at Control headquarters tomorrow. Spending night with Smithells. Was glad to see Campbell so well and much like his old self.

October 4th[499]

We have to set up proper technique for photographic timing of Sashalite and such like sources for post-war developments. Winch is handling. He reports that his mechanical mortar ranging gear when taken to Italy was bombed before it had a chance of being tried in earnest.

Dunham reports the conference on the RP 60 job yesterday concluded that the scheme, as it can be further developed, looks promising enough and people concerned are not entirely depressed.

Ramsay reports good progress with hydrogen thyratrons. The small type has run to 800 hours and the large one, after a few failures, has now started on a career. There is considerable interest in the Services.

October 5th

Spent an hour with Jesty and Jacob at XL examining their television (colour) experiments. They may not lead to new equipment but their immediate object is to help us to appraise the value of colour television and to find our bearings in the techniques.

Called on Dr. Clark at Unilever House at his request to establish a liaison on high frequency heating applied to their food preparation processes.

A restart of the BSI Street Lighting Committee at which I presided. The meeting was nearly unanimous in voting for a specification rather than a code of practice. I got Walsh appointed deputy chairman.

Sir Harry Railing's Presidential Address at the IEE. A success I think. All spoke well of it.

October 6th

A Research Services meeting. T.W.H. away. Nothing special. Discussed the implications of Leedham's communication that there will in future be no 'blanket' research contracts only contracts for specific objectives. I have written to him to point out that this mainly favours development rather than research. There is also a strong directive from MAP that they mean to levy royalties on use by manufacturers of designs developed under contracts from MAP. They are clearly over-reaching themselves.

Lunched with F. Smith and talked over MOV matters. I feel that the MOV direction is now thoroughly mesmerised by Mr.Schoenberg.

Presided at a full meeting of the Electrical Industry Committee on plugs and sockets. The HMOW [Office of Works] – IEE Committee under Beard has insisted on introducing a new standard non-interchangeable with existing standards—and contrary to past policy of the BSI. All the bureaucrats were there: Kennedy, Forbes Jackson, supply engineers etc. They all supported the scheme with the knowledge that it would change a rapidly improving situation in interchangeability of plugs into one of chaos for the next 20 years—and that on the lightest of pretexts. The manufacturers took the other view supported by a few, including I am glad to say Mr. Hobson of the Central Board who has taken Archibald Page's place.

The final touch of comic opera was supplied by IMEA led by Mr. Hoadley who came to press for a BSI standard specification for the Wylex plug—but said that if the new proposed plug used a flat pin, no matter if it was not interchangeable with their Wylex type, the IMEA would no longer ask for a standard spec. for the latter. It seemed to be the statesmanship of Bedlam.

October 7th

The rapid advances of a month ago have ceased and hard and slow progress is the order of the day. Holland is terribly hit and the Germans are flooding and destroying everything.

October 9th[500]

Valve preproduction keeps up to about 2300 valves per week but we shall hear on Wednesday at our meeting with Brundrett and Truscott what valves are likely to be cut down in numbers.

Lunched with Sir Harry and discussed MAP intentions to charge industrial concerns a royalty on designs or patents which emerge from development contracts. He agrees with me that this is quite unacceptable and we must not enter into such a contract.

October 10th[501-503]

A Glass Advisory Committee. Partridge carries his work and his liaison well. The pot position seems critical—short life and no replacements immediately in view.

In afternoon Chelioti, Gant, Smith, Cosgrove and Marden all came to confer with us on preproduction group policy. B.P.D., R. le R., B.S.G., G.W.W., J.H.P., L.C.J., J.W.R. and C.C.P. from the Labs. A statesmanlike and helpful talk with a very fair definition of principles which should govern post-war policy. Three groups to remain—lamps, valves and CRTs.

October 11th

Dudding had last week the most gratifying of all his quality control meetings. Chelioti called a meeting for him and Jennett to address all the responsible managers and departmental staff of the Osram factories. His address was received with expressions of conviction and of desire to introduce QC on a considerable scale into their departments. This will open a new era in our GEC statistical compaign.

A CVD-GEC meeting this morning attended by Brundrett, Lewis and others showed a tendency to change from the urge for large production of extensive projects to a small production of a limited number of examples of new equipment. It was deduced that this would result in a demand for the sort of limited production or preproduction facilities which exist at a place like Wembley.

At the afternoon preproduction and allocations meeting it was shown that our preproduction of valves would drop to 25% of present level about January, and the best ways and means for telescoping our effort were discussed. Mr. Brundrett urged the importance of maintaining legitimate preproduction facilities for supply of samples and the like—since these would be needed also post-war. This is a big question of policy and needs to be taken up officially with us if it is to be implemented. The cutting down of activities in this way alters our overheads and bare costs for preproduced valves and will involve a readjustment of prices of valves—and probably of research contracts.

I gathered that the Ministries have immense stocks of valves distributed over the world in order to meet all eventualities. The concentration of Germans within Germany which is taking place releases large quantities of equipment from these dumps—and has caused a radical cutting down of demand.

October 12th

Gossling attended a Fuse meeting yesterday in which prejudice and warped facts fostered by Mr. Rangaby in favour of his particular design of rugged valve were rather well exposed. We do not think the hopes being built upon Mr. Rangaby's efforts will be fulfilled.

I had a nasty fall backwards last evening on a slippery floor outside my office and damaged my left shoulder; hope it will soon recover.

Gavin says his RCM work has fallen flat and we shall use his people on various jobs under his supervision. One urgent matter is to push forward with the wide band transmission problem which Espley particularly is fostering.

It does not look as if the shrinkage of preproduction need cause an adjustment of our overhead this financial year—but it will probably go from 150% to 170% on April 1st.

October 13th

A CB meeting at which MOV relationships and preproduction were the chief topics.

The USA standardisation delegation came to look over the Laboratories. A very charming trio who gave us to think that Wembley was one of the research establishments in England which must not be missed.

The diamond die group efficiency keeps up and for the second time we have come out top of the production groups in unit efficiency value.

October 16th[504]

The atmosphere of shrinkage pervades valve preproduction. Pearce's estimate is 50% by February.

The trials of Heavy Alloy for tank projectiles has not so far shown advantage, but they are to be tried in the larger sizes which we are now making (3 in shell).

I visited Marconi valve factory on Saturday. They have built up a considerable production but mostly in high frequency valves for radar and special war purposes.

Discussed with A.H.R., C.C.G., Gallizia and R. Le R. at Head Office the policy of making or not glass mercury rectifiers. It appears that we might go up to 50 amperes with hot cathode types which would cover half the range up to the smallest pumpless type. Dudding and Jennett had a good day with Hannam Clark at Coventry on quality control on Friday—and the ideas catch on.

October 17th

We have a regular 'accounts' monthly meeting now to make sure that all parts of the accounts system are keeping up to date. Things are getting better but it is almost impossible to get hold of people with any appreciable experience. So we have to make do with half timers who will follow the system rigidly and leave the few seniors to do the real accountancy. It has been a good move to pass the whole invoice system to LSO [the Laboratory Stewards Office].

Annual meeting of BSI. We have appointed Lord Woolton as President—which should be a very good move.

October 18th

The USA standardisation party came again and were much interested in a number of things. Mr. Ritter of Corning has been in touch with Chirnside and Partridge and has told us what we want to know about silicones. A welcome contact.

Gossling reports that valve packing in the USA appears to be as defective as it is here and the party were greatly interested in our studies. He was at a tropical and packing exhibition yesterday. This war appears to be teaching people a lot about tropicalisation.

A receiving valve meeting at Hammersmith with plenty to discuss. It is clearly the policy to rely no more on supplies of valves from the USA—mainly because Lease–Lend may stop at any moment.

A CRT meeting at which it appeared that all present desired a continuance of the morning and afternoon meetings. I was able to have a word with Mr. Hecht about RAE and its ways. Its trouble is that it does not know as much as we do but has to pose as a judge and criticiser of our recommendations.

Dudding and Jennett's Quality Control brochure is at last published by GEC and is about to be distributed.

October 19th

Discussed television optical developments with Espley and Francis. There appear to be possibilities in new projector arrangements. It will be well to try them out.

I find that the letter from Admiralty Contracts Department is not supported by DSR. Cornock and I have designed a reply to Mr. Fitch (DNC) which I shall now send to Brundrett first.

Nearly all day on Household Appliances at HO. Gavin came and was introduced to the subject and the 'personnalia'.

October 20th[505]

Visit from British Aluminium Company (and C.J.S.) re. anodised wire for electrical purposes. We are looking into it but the alumina film has not a high electric strength.

Megaw's clearances for USA seem now to be in order and he expects to leave next week.

October 21st

Appleton's propagation meeting are urging work on the 5 mm band in view of removes [news] from USA; Megaw was asked for Wembley to start up a valve (also Oxford and Bristol). So we are to have a go at it.

Willshaw has beaten the field over the high power S band magnetrons. It is two years since the proposal was launched and we now have 2 megawatts peak with half the mean power asked for and see our way to design cathodes for the full mean power. I understand BTH are giving over their efforts and Birmingham are having no luck.

The proof of my 21st celebrations address on the history of the Laboratories has now come along but it is slow work getting things through the printers these days.

We are going to sell and distribute Dudding and Jennett's Quality Control brochure from the Labs.

October 23rd

The war is progressing slowly in Holland and West Germany and faster in East Prussia. USA has invaded the Philippines.

A preproduction valve consultation with MOV shows that they will not be able to relieve us of several of our valves for many months, which will at least enable us to ease off our groups in a gradual way. The number of valves last week was 2200. CV 240, CV 257 and E 1495 are coming along.

Gavin wants some of his radio team to transfer with him to heating and appliances. The men voted without exception to give up radio.

Lunched with Gamage and had useful reaction from him on the household appliance work and on the general research programme. His desire to know what is the proposed distribution of our research effort over the various fields is welcome and his criticism will be equally so.

October 24th

Helmore wants us to handle the whole 'Fish' which he will locate at Heston. King-Brewster's transference to Head Office for the Edmunson's street lighting job is problematical on account of his commitments to prior urgency war jobs. Herwald may be able to tackle the Z Dept. [street lighting] job as a temporary measure.

A CVD policy meeting. Half an hour wasted on the question of bringing over a Dutchman, wife, children and nurse to do some research on luminescent powders. It is difficult to hold one's tongue! Standardisation affecting post-war valve policy was to the fore and a special committee is to meet on the subject.

October 25th

A further visit from USA valve delegation which has some further points to look into.

A long TVC meeting at which means of collaboration with USA 'JAN' committee was outlined. Arrangements for continuance of TVC and a permanent office in London were canvassed. My suggestion to continue it in the Admiralty was favoured.

Telfer tells me of general remarks passed in Government circles that MOV 'have far and away the best research organisation for valves in the country and do not seem to be able to take adequate advantage of it'—a sentiment which rings a good many bells, but which I must not promulgate! I am wondering if Telfer will go back to MOV.

TVC appointed me as representative on the special Service Valve Standard-Line-of-Valves Committee to meet the manufacturers.

October 26th

Gossling had a trying meeting on valve packing in which certain Government officers wriggled against taking action on the packaging required by the tests. He seems to have to fight the Services battle for them—which of course is not surprising except to those who think that Government Establishments are manned with a different breed of human nature from that to which we are accustomed in industry.

Our discharge lamp work—high pressure and compact source—is expanding: (*a*) for street lighting, (*b*) cinema projection (*c*) Admiralty requirements. Francis has an interesting account to give.

An OAC meeting this morning with a non-war feel about it.

A rocket bomb this morning hit two trains near London Bridge at 8.30—a piece of bad luck.

October 27th[506]

Transport Command Liberty plane pilot came yesterday with an SOS that we would repair a special USA device for heating the plane, due to leave at once with very high level passengers. I don't know who sent them to us but they came to the right place. Foxon handled it.

Made formal presentation to Puleston, who is leaving today, in my office. One of the 1922 comers. A very attractive little ceremony. Also received Lewis (glassblower) resignation, to join Smithells in two months time.

A searchlight panel meeting at which Cameron made the major contributions. I had thought that the panel might go out of action but we all felt that a meeting once a year to exchange experiences would be an advantage. I shall be writing to Couch for guidance.

A 'Bowler' meeting all afternoon went smoothly under Gladstone's chairmanship. The parties seem to have settled down at Greenock under the latest directive from DTM.

Many questions are arising re. size of post-war groups and even more difficult—how to find people at the present time to help with ordinary industrial problems.

October 28th

Primary Batteries have been going through some troublesome times. CIME [Chief Inspector of Mechanical Engineering] are dead against the new HT inert battery and unsatisfactory tropical tests have given them a stick to beat with. The Services succeed in playing off one manufacturer against another and will continue to do so until the makers can establish a prior loyalty amongst themselves.

Winch and Hill have an attractive portable photometer of precision type for coloured light sources. It is a fundamental piece of measuring of a universal type complete with accurate voltage regulator.

The war continues in its 'tough' stage. Holland is having to be fought for and Antwerp is not yet usable by us—so transport difficulties hinder progress. Russia is beginning on East Prussia but there is no German collapse yet.

October 30th[507]

The CV 208 magnetron is still in dire trouble with flashing on test indicative of gas present. The CV 215 is out of its difficulties under Ramsay's plan, and the CV 221 (TR cell) has yielded to Breadner's efforts.

Saw Pinkham and Marchant about publishing articles which will give due credit to the GEC Research Laboratories' effort. Not too easy. It looks as if I should have to write up an article on our war effort.

The GEC is to set up a Works–Sales Instrument Committee for collaboration with Salford and in connection with the Advisory Board scheme. Asked Gavin to act as focus for the Labs.

October 31st

Heavy Alloy is passing through an interesting phase; for certain specimen shapes Williams and Price are finding that pressing of the powder is not necessary and consequently one heating can be eliminated. This cheapens the process greatly.

Drs. Pierce and Hagstone from the Bell Labs. are here on a visit—magnetrons and local oscillators. They are full of good will and desire to collaborate and to carry on the happy atmosphere which we have always had with them.

Had Dr. I. Jones here of the Ebbw Vale Colliery Co. to study our organisation and layout. He was very interesting and seemed to me to have quite unnecessary organisational and psychological difficulties to contend with. Cornwell and I enjoyed the talk.

Called at USA Embassy on Mr. Houghton—an official of the Corning Glass Co. Sowed the seed of collaboration between the GEC and the new Corning–Dow Co. founded to develop silicones. They are favourably disposed to us as the result of many good contacts in the past but there may be prior affiliations.

November 1st[508]

Dr. James, as I expected, is not attracted by the 'forward applications' post at MOV. I hope Cosgrove may be attracted by one of the other names I have put forward to him.

Jolley has in hand interesting work on an 'impactometer'. We have various jobs involving shock, e.g. devices dropped on the water, projectiles, transport crates etc. which will find it useful.

Saw Brundrett with Hughes to try to get an office for TVC in town. The effort succeeded beyond my hopes. Brundrett will arrange one at DSR and Hughes's assistant Webber will be stationed there. Brundrett also told me of an inter-Service meeting yesterday (attended by all but Leedham's department) at which it was agreed in principle to continue post-war the inter-Service valve arrangements as at present, i.e. CVD, Preproduction, TVC and ISC (standardisation). This means that Wembley should make allowance for staff and facilities.

Lunched with Ricardo and Jolley to bring our possible liaison with Power Jets Ltd. a stage nearer. They are not without their difficulties partly owing to the location of the Power Jets establishment five miles from Leicester in dreary open country. I hear that RAE will be transferred ultimately to Cardington in Lincolnshire. Also not a bright prospect for the staff.

November 2nd

Have had a general rampage on crates and packing cases piling up in the open and in the corridors until the situation has become hopeless. It is probably a question of having courage to scrap.

Ryde's report on USW atmospheric characteristics has run into 250 copies and new problems requiring solution pile up on him. He tells me that right down to K/Z [?] wavelengths what practical data there is agrees with his predictions.

H.G. Jenkins is to go to USA on lamps and as one of a party of two GEC and two BTH people. I appreciate Chelioti's invitation to an RL man.

Ward of TRE speaks very warmly to Willshaw of the good work done by our CV 208 X band magnetron. It handled the X band operational doings of the invasion and was for critical months the only reliable valve of the type. Probably the H_2S service was its principal field. It has been the victim of certain more recent troubles and will soon be superseded but nothing can deprive it of the credit when it stood alone—Birmingham having had bad luck in failing to produce a usable 3 cm valve. The BTH CV 108 also had bad luck. Willshaw feels that his 'muscling in' on the job in 1943 was justified and worth while.

The trial on first eight prototypes of the 1407 multi-channel set is now in progress at aerodromes. It seems to be popular but we await conclusion of the full tests before presuming that all's well. A meeting at MAP on the 600-channel successor to the 1407 indicated that a number of substantial advances are to be made and we hope Works will not grudge Wembley tackling a functional prototype.

November 3rd

Gossling's valve chart (which describes it from cradle to grave and even further) is becoming an accepted standard having been canonised by recognition in the USA valve development schedules and again in this country in the regularly issued statements on the various CVD valves under development. When Gossling and I first got this chart out and circulated it we felt it was good but feared ridicule—it has come into its own now.

Ramsay is back from two days at Malvern and RRDE on hydrogen thyratrons. They are rapidly coming into favour because they will stand a higher repetition frequency and have generally a better margin of safety. Four sizes are now projected: (*a*) a small one copying a USA valve, (*b*) virtually our present E 1463, (*d*) our very large one—nearly ready and (*c*) one between (*b*) and (*d*) in size. Our CV 115 is needed in the full numbers originally asked for. It now seems out of the wood so far as our immediate setback is concerned.

There was a meeting here yesterday with Mr. Haliday from Coventry on their condenser manufacture. They wish us to establish a regular group here to handle their research and development with them. It looks as if Macfadyen might take it on as his main job and responsibility.

November 4th

Today we said goodbye to George Woodward, founder member of the staff and in charge of the Woodshop and minor internal building. A presentation in the morning in the Woodshop and lunch and theatre party to follow, attended by all foremen and some ten others. He is 70, and has given everything he could to building up the place.

November 6th[509-510]

I wish one heard better reports of RAE radio administration standards. No one seems to have any other opinion than that they are intriguing for more control by RAE and have not staff of sufficient ability to secure influence by brains rather than by man-made authority.

Important meeting (Maqueen, Bligh and others) with Mr. Jeffcock (MAP civil aviation section) on future policy for radio equipment. The GEC nailed its colours to the mast and said they intended to be in *all* electrical equipment of aircraft even to the point of accepting the job of co-ordinating the products of other industrial contractors for equipment.

Sir George Preece and Mr. Montgomery of Institution of Mechanical Engineers came to vet the Labs. as a suitable training place for candidates for I.Mech.E. They were shown round and took much trouble in questioning apprentices.

November 7th[511]

CVD meetings respectively on crystal valves and triodes. Ryde had a good record to give on our high burn out crystals which BTH have not duplicated with the consequence that GEC crystals are the only ones which give really high burn out. These are more widely in demand than was anticipated.

Had lunch with Rogers and Colebrook (NPL) re. calling together again my original IEE Radio Interference Committee to endeavour to get the medicos to co-operate in suppression and also of the high frequency induction heating question vis-a-vis radio interference.

November 8th

The regular CVD meeting on magnetrons. Our CV 208 came in for criticism. The Marconi copy of it is free from the flashing and noise troubles which our valves suffer from. We are hard after the causes but meanwhile the Services don't like the valve. Apart from this Willshaw had several successful forward steps to report, particularly with the high power S band valve.

Called on Hecht, with Hughes, to consult him about persons for TVC representation in USA. We want a series of some six people who will go over there for three months apiece.

November 9th[512]

An informal research meeting—BTH and Siemens on discharge lamps at which we ventilated the proposal for 150 and 250 watt quartz lamps for street lighting and discussed future arrangements for research exchanges.

The discussion showed the need for such exchanges of view in the early stages.

Spoke with O'Kane re. a possible job at Head Office and T.W.H. will probably see him.

November 10th

A Consultative Board meeting at which there are now good discussions on general questions of Company policy.

To lunch at Savoy with G.C., Mr. Sporberg and Mr. Warren BTH re. future valve agreements. It appears that rearrangements in USA under the [anti-] trust laws will dissolve the RCA arrangement with the other big electrical concerns and also leave GE to vest their patents solely in BTH. A BTH – MOV or GEC agreement on valves may result, similar to the GEC – BTH lamp agreement. The upshot of all this may be important.

November 11th

Now that sapphire can be obtained in this country on a reasonable scale many people are seeing uses for it, notably the BBC for cutting discs, for gramophone needles, for replacement for agate in balances etc. Leeds finds it very interesting.

Our diamond production figure of merit is keeping slightly above that of Simpson's and I hope we shall not slump back again. We suffer much from absences of key people. The diamond powder making is still a key to most of our jewel work.

There is an urge for a small regular supply of Ramsay's large thyratron for several uses both inside and outside the laboratories.

Bell is back from the USA.

November 13th

Valve preproduction 2035. Bell back but with nothing dramatic to report from the USA. Machin attended to urge the provision of 250 E 1480 [E 1280?] hydrogen thyratrons not only for the supply of samples outside but for the running of our own life tests on large and high recurrence rate magnetrons.

Met Sir Harry and Mr. Lonsdale re. programme of post-war research at Fraser & Chalmers. Our general scheme was accepted with the first and foremost attention to steam turbine problems and second priority to axial blower. The 100 kW gas turbine was not to be constructed yet but regarded as the fourth priority matter for the present. Personnel matters also touched upon.

Lunched with A.H.R. and discussed mainly [the] discussion with Mr. Sporberg on valves.

November 14th

Discussed with Bligh and Macfadyen the development of our engineering service group in the Labs.—dealing mainly with electronic and control devices in general engineering.

Our plastic mine detection work (Espley and George) is having a sticky passage. The first combined sound and HF scheme is too difficult, and a better one by Espley and George based on frequency modulation is holding the field. The Services are glad of this, and the practical making of the trial equipment will be handled by them.

Air Commodore Helmore came to see films of the dropping of his FISA fish and to see the control equipment got out by Foot. He was most appreciative of the latter and the trouble and workmanship expended.

November 15th

I hear through Dudding that our intervention with Mr. Hecht over RAE and CRTs culminating in a meeting of all concerned at Hammersmith has much improved the atmosphere. Jesty has done a good job behind the scenes in this.

Megaw has arrived in USA.

A Royal Society of Arts lecture on HF heating by Hartshorn at which I took the chair—a brisk discussion.

November 16th

The 3 cm magnetron CV 214 is following the evil ways of the CV 208—flashing, and all concerned are baffled. Production is going to be cast to the winds and the whole effort will go into an experimental programme to find clues.

A BSI meeting with USA and SBAC delegates on aircraft standardisation. A luncheon at the Dorchester to follow. I was glad to see SBAC and BSI getting together. There is going to be an Aeronautics Industry Committee.

November 17th[513]

A TVC committee at which my plan for an office at the Admiralty for centralising TVC and standardising matters was received with damping second thoughts—partly owing to jealousy by MAP (DCD) of this work being centred at Admiralty. All had agreed at the meeting and Lewis at TRE had supported Brundrett's proposal.

Lunched with Fortescue to talk over manning of the teaching staff at Universities. They are considering salaries for Readers of £750 to £1000 and for Lecturers of £500 to £800.

A Carbon Committee (GEC) at which policy was discussed—a useful talk but the Four Ashes experiment to see whether that plant can produce good carbons has not got a result yet.

November 18th

Nothing dramatic in the war. The weather for the past three months has been a bad hindrance to operations and everyone is reconciled to another six months at least. The V2 rocket missile is being a nuisance in NE London and Essex. We have not yet got across North of the Rhine in Holland.

November 20th

Preproduction—several valves in trouble, CV 215, CV 208, CV 240, and little in the way of clues. Flashing and poisoning of cathodes are the main troubles.

Lunched with Chelioti and discussed amongst other things salary at the lower levels. On account of dearth of technical staff, Works have been compelled to offer higher rates so that scales for technical staff now tend to be higher than for commercial staff.

An Advisory Board meeting—buoyant and informative.

November 21st

A talk by Bell on his visit to USA dealing with transmitting valves (with the exception of magnetrons). One has the impression that they adopt most difficult techniques of manufacture and put up with a wastage in manufacture which would not be tolerated here. It is the reverse of what one expected of the USA. MOV management joined with us in the talk.

November 22nd[514]

A CVD (GEC) meeting which usually stirs us up. The easing off of preproduction seems to have whittled down to about one fifth reduction by March, i.e. about 100 people.

Had long and interesting talk with Brundrett on post-war contracts, position of other firms, terms and conditions for such contracts. He visualised some £10 000 per annum for valve contract and £5000 extra for preproduction facilities. He apparently found BTH the best collaborators after Wembley. His remarks about some others are best not recorded, but to me, knowing the individuals in authority, they are understandable. He told me that Goodeve would be leaving the Admiralty for some other job. He also hinted at a surprising situation in which one well known firm was finding itself.

Lastly, certain developments proceeding in USA, of which I only knew vaguely, are too secret and too amazing to be put in writing.

November 23rd

An Osram Advisory Committee meeting at which the position of Philips was discussed. Philips have not made themselves very acceptable to the GE of America and that is liable to have repercussions on their acceptability to BTH.

The diamond die control are mad about electric arc drilling of diamonds which happens to be our patent but which the Americans (Bureau of Standards) have brought to a measure of success. They only seem interested in the USA effort and not in ours.

November 24th

The FIDO report for 450 ft runways is out and Ryde, Cooper and Strachan have done a really good job. The saving of petrol is about 2 to 1 if the forecast turns out right. Copies are being sent to the Petroleum Warfare Dept.—I expect the proposal to be cold shouldered, but we shall see that it is not just buried.

Had long talk with Jolley and Reeman about future conditions and work at Fraser & Chalmers. They are ready to start up with a good heart as soon as the agreed personnel readjustment is made. Reeman will be wanted at Power Jets Ltd. for some months yet.

November 27th

Preproduction down to 1700—the worst week for many a day—mainly due to technical troubles and a certain amount of personnel trouble. To

XL with Jesty and Espley to see Jesty's group's first colour television tryout gear. It was at too low a frequency from a flicker point of view so colour fringes were exaggerated. But it showed a real grasp of the field.

A long talk with Maqueen, Goldsmith and Sanderson to accentuate the GEC work and position in layer batteries and to decide priorities vis-a-vis high tension inert blocks for tropical use and Admiralty cells. It appears to me that we have our techniques and products well under way.

November 28th

Lunch with Mr. [John Logie] Baird and to Sydenham (with Espley and Jesty) to see his colour television demonstration using two luminescent powders of different colours coated onto the sides of a disc of mica. The idea is simple and free from a rotating colour disc.

November 29th

A day on household appliances at Head Office. The Birmingham Works are very ill placed from a labour point of view and are being caught on the wrong foot. They hardly seem able to contribute anything.

November 30th

Gossling's group's valve breakage report is finished and is a substantial document. Gardiner is coming into the matter to handle its practical aspect through MOV. Gossling may still have some policy to attend to.

A verbal message from Witton that they want to back out of the Helmore 10 ft searchlight development contract because Southworth is wanted for reconstituting his Traction Dept. I don't know yet what will be the upshot.

A War Savings celebration dance at Town Hall where the Research Labs. having topped the list of contributions from concerns of over 300 employees received a very fine cup to hold for three months. Mrs. Paterson and Cornwell were proud to see Mrs. Johnson go up to receive it and its presence in the Labs. should be a stimulation for the new cycle beginning next week.

December 1st[515-517]

The CV 79 and CV 89 (the 6 cm little magnetrons) used in the military communication sets have been doing well, but a sudden trouble has appeared. Hitherto the Lab.-made valves—in our opinion poor quality—have been in satisfactory use. The MOV-made product has just come into use and troublesome noise in operation is appearing. The MOV valves have always seemed to us better than ours and the trouble is one of marrying valve and set.

A Lamp Committee yesterday, BTH, Siemens, GEC, passed off reasonably well—rather discursive I believe.

James reports continued interest in Warren's and his small triode on the part of the Establishment—8 cm upwards. It seems to have struck the RCA visitors from USA.

Had Oliphant, Ackers [Akers] and Perrin of 'Tube Alloys' here today for three or four hours talk. Oliphant is back for a few weeks from his work in the USA with the research team there. It reminded us of the first days when we were initiated into Radio Location secrets. This is vaster in its significance and possibilities.

Oliphant having viewed these Laboratories in retrospect during his twelve months in the USA emitted the opinion that 'Wembley did more than any other research establishment to win the war and if it had not been for Wembley the whole radio (and other) efforts would have crashed'. A handsome and quite unsolicited testimonial but an over simplification!

December 2nd

A long talk with Espley, Jesty and Dudding about internal collaboration on CRTs. and television and size of post-war groups.

December 4th[518]

The war is sticky in more senses than one and the big W. German north and south rivers now in spate form serious barriers. Rocket bombs are coming over and hitting NE London and suburbs rather severely. One burst during the weekend over the Thames opposite the IEE and broke windows badly I hear.

Preproduction still low at 1719 due to CV 208 and CV 215 being out of action.

Had to settle with Gavin about the P 55 Admiralty set for which Johnson & Matthey can do little towards the rhodium plating for which they seem to have lost the art.

A long talk with Le Rossignol, Gossling, Bell and Ryde over the Oliphant job now known as TA. We mean to have a go at it but it means segregating some good men for the job. Shall be seeing Mr. Akers of DSIR on Thursday about a contract for the experimental work.

Cornock has had a busy time with our appeal against the Patent Office ruling against our having a patent for a luminous material. Burrell did a good job for us and Cornock is sanguine that we shall win the appeal.

December 5th[519]

Partridge and Manners are getting interesting results both in factory and Lab., the latter on the glass sheathing investigations, the former on tank blocks at Lemington.

Willshaw is looking after a magnetron meeting here—have not heard upshot.

December 6th

A meeting at Head Office (Heather and others) on organisation of GEC instrument activities under Advisory Board machinery. Gavin came with me. A very useful and unanimous talk and agreement to organise a regular committee on the subject.

Lunched with F. Smith, MOV, to discuss current confidential issues—post-war contracts amongst them and also the form of our research reports to MOV Board—which always raises the subject of Mr. Shoenberg and all his attributes.

A CVD policy meeting where CW magnetrons for RCM work were the most interesting matter, chiefly S band for the Pacific war.

It does not look as if my scheme for lodging TVC at the Admiralty is going to help since Telfer of MOS has declared it impossible for him to centre his standardisation work there. I shall have to tell Brundrett.

Campbell has come back to us for a visit—mainly to discuss his work with Francis on valve noise.

December 7th[520]

The CV 79 and 89 difficulty with the Army sets wanted for the front is acute as only 5% of the stock of MOV made valves are usable. Butement came to see me to see if we could possibly raise 50 or 100 Wembley made type to carry on with. It is not clear yet what is the trouble. Had meeting with Willshaw, Douglas and Cheeseman and we may be able to make or find a certain number.

Mr. Akers and Mr. Block came to carry the TA matter a stage further. We discussed contract arrangement and patent clauses for an hour on the basis of £10 000 per annum. The visit then turned into a meeting to consider some technical details [raised] by Oliphant.

December 8th

Francis showed me the new 250 W compact source lamp—it looks simple and attractive and a good engineering job.

Captain Gladstone and many other Naval brethren from Torpedo School came for a Bowler, Trumper meeting. Had a preliminary meeting with Gladstone, Grime [Grimes], Edwards and Farren to complain of lack of equipment in the Greenock radio laboratory. DTM dept. did not seem very resourceful in supplying the needful chasers we are accustomed to from the other Ministries. Gladstone is leaving for sea I am sorry to say. He is a good man to work with.

A Z Dept. policy meeting—J.Y. Fletcher present.

Willshaw has been to Witley who have disappointed him by putting his large 10 cm magnetron amongst the 'may be used.' All interest at the moment is with magnetrons for CW loads.

We hear indirectly that EMI are cutting off some of their Government work in August and are tailing it off now 'because they are the one firm for maintaining the reputation of British television'. I wonder if they have made application to the authorities on this score.

December 9th[521]

Willshaw, Clayton and Lamont have finished their IEE paper on HF measurements in the centimetre range. It remains to see how to handle it from the security point of view.

December 11th

Valve preproduction 1883—not yet up to the requirements which are 2150. The trouble with CV 208 magnetrons is definitely with the cathode system. We have yet to find in what particular.

I have the 'instructions to proceed' with the TA work from DSIR. I think Boyland will have to be the first person to be transferred to this work. I hope we can arrange for Tremlett to take on Boyland's job. Ransley had been approached by Non Ferrous Metals Research Association to join them for the purpose and asked my advice. I could not advise him to leave (*a*) Wembley (*b*) British Aluminium (in both of which places are to be found the resources wanted) in order to join a concern with very few resources.

December 12th

An aviation meeting at Head Office to consider official proposals for post-war aviation equipment. The competition for development facilities at Wembley and Coventry is heavy and both establishments are conscious of the heavy expenditure involved in keeping up a staff for it. Television and communications work take first priority in GEC programmes and will be expensive anyway, so we shall have to enter upon the civil aviation problem with discretion.

December 13th[522]

The P 55 set (Gavin) for the Admiralty is still in a bad way and at a meeting yesterday we were asked to forget Johnson & Matthey's rhodium plating and try our original suggestion of silver tubing. Marconi seem to be trying to 'get in' on every valve they can. Having tried their own ideas on an alternative CV 214 (high power 3 cm) they have come to us for information re. our improved design. We are going to ask in exchange for a supply of blocks to the better design so that we can execute our contract also in these.

A TVC meeting—intense and somewhat hectic as usual. We had politely to defy the CVD Technical Committee on which four of our own members sit by disagreeing with their objection to our suspending temporarily the VX system of nomenclature so far as the CV equivalents are concerned. There were other tender points but somehow always when MAP are in a minority of one.

Had Leedham to lunch with me and talked over aspects of the above especially the sacrosanct nature of the CV series. He wants a main policy meeting called to discuss it which he can attend. He tells me that the 'Civilian Scientists' tend to throw their weight about too much and expect to run the technical side of the war right up to the top levels.

To MOV for a transmitting valve meeting—useful but uneventful.

December 14th

An *ad hoc* meeting under Brundrett's chairmanship to lay down requirements for a post-war range of receiving valves to suit the Services and form basis for a civilian range. An interesting meeting and surprising unanimity amongst the Services.

To Ministry of Labour (Elect. Eng. Advisory Committee) to hear and advise regarding future engagement of personnel orders. It had been agreed that electrical engineers should be dealt with nationally and that after the European war every man should be free to leave his job if he wishes to—with one or two exceptions for National reasons.

December 15th[523]

A Research Services meeting at which I introduced Sir Harry and Mr. Gamage to Tube Alloys—in a vague form. Agreement to charge no profit on post-war research contracts. L.G. disagrees idea of taking them out in joint GEC – MOV names. Helmore largest searchlight contract is to lose Southworth and if possible is to be put into abeyance. I am to see Helmore first.

Lunched with French Government standardisation representative as guests of BSI to get his reaction and help him regarding post-war international standardisation.

December 16th[524]

Jenkins and Breadner have produced a luminescent lamp of compact form made out of two concentric cylinders of glass fluted to form enough length for 30 watts. I saw one alight today.

December 18th

The war in France has been going heavily but there is a sharp wake-up today by what appears to be an all out attack by the enemy on the American position.

In valve preproduction the trouble with the CV 208 seems to have been the temperature of cooking the magnesium – nickel core material of the cathode system. We appear always to have cooked at too low a temperature.

The magnetron group under Wright is now moving over to the 'new' NE room which Pearce has operated for the past two months. This will leave the Lamp Room free for the lamp preproduction group as soon as it can be formed again.

We are notified that the Helmover (fish) job is now a major priority with an urgent staff requirement. Our part is mainly the radio control gear which Foot, under Bligh, has been handling.

G.M. Wright (superintendent at ASE) came to look over our USW work—SW propagation—modulation, and millimetre wavelength valves. The latter are particularly of interest in view of the proved strong oxygen absorbtion band at about ? mm. I was glad to get G.M.W. here, and discussed post-war USW with him. The present rate of expenditure is about £10 000 per annum.

December 19th

Espley has gone down sick with slight heart trouble. The cases of this kind which we are having are disturbing—and seem to point to low reserve of resistance. There is a discussion at IEE this evening which Espley was to have opened. Edwards will take his place.

Jesty repeated his demonstration of future CRT shapes—and a strong bias towards a 'square' screen.

CVD 'triode' meeting at Admiralty—no incidents.

December 20th

Le Rossignol and Miss Lock brought in their six months valve preproduction returns (April—Sept.). In spite of price reductions it shows a margin of £100 000 including a reduced overhead figure to 150%. I should not have been surprised at a margin of £30 000 but did not expect so much. However we know that the present six months cannot do much more than strike even and we also shall reduce certain prices from January 1st. The S band magnetrons have given the largest margin and next come the CV 55 and the CV 90. On the other hand we have been recently having a bad time with CV 215, CV 208. One cause of the surplus is the absence of purchasing of sundry apparatus. We dare not reduce valve prices for continuing types because the rub comes when groups have to be wound up, with a falling output and workers to be paid till they can be found other jobs.

A CVD magnetron meeting—which was fairly lively. K band types were pressed but the BTH had not much to report. Some criticism was made that Wembley did not throw up their effort to make normal K band and switch over to the Rising Sun type altogether. The answer is that Willshaw wants to establish and extend his theoretical groundwork by using his S and X band design data extrapolated to the K range and so correlate this with the Rising Sun structure and secure a basis for this in his sequence.

An MOV receiving valve meeting—cathode coatings and ceramics were the most live subjects—but the whole meeting was full of interest.

The German break through in France seems a major affair and we are able to be given no news.

December 22nd

A meeting which Partridge had with Mr. Butement yesterday concerned insulated leads for components. We put forward various proposals including the *B9G* valve base. We think there should be an opportunity of manufacture of suitable leads after the war.

We close at 12.0 today for Christmas. It will be a good thing for everyone to have a break. The enemy break through on the American front is not the cheeriest news but some people find it better than the slogging away on fixed lines which was getting monotonous and heart breaking.

December 27th

Examined personnel of all prospective research groups with Humphreys. After reducing juniors the total number of Student Assistants come to 100—far too many if they are to remain permanent members of the Lab. staff. So there is a problem to solve.

Chelioti, Gant and Woodward called to see the new discharge lamp and gave a blessing to its continuance.

December 28th

An OAC Meeting. The reluctance of the GE of America to collaborate and the difficulty of pushing up production of discharge lamps were the main features.

A TVC meeting all afternoon to consider a TVC secretarial office to handle standardisation, collaboration with USA, and to function generally in TVC matters.

December 29th

ASE are pressing us for completion of our high power S band magnetron by March 31st. This brings this valve into prominence again and it must be a collision of priorities between high power S and X band magnetrons. The K band strapped vane type appears very good with 20 to 25% efficiency.

Sanderson reports continued good results using wax in carbon making to eliminate flow lines and cracks. Four repeats have been made but we are not yet allowing ourselves an attitude beyond sober hopefulness.

A very interesting visit from Mr. Mills and Mr. Brighty from MAP to ask for advice on the measure of wastage to be tolerated in valve manufacture. They gave us a useful talk on 'change over' manpower policy particularly in relation to the rights of manufacturers to choose who they will keep when a condition of 'redundancy' arises in their works. The Board of Trade came up for some poignant probing.

December 30th

Saturday—this ends the year which has been wonderful in achieving what seemed to be impossibilities. People are tired through the demands which have been made on them—nervous, emotional as well as physical. It is in their home life that people have felt the controls and restrictions, the privations and the burden of work and frustration—and the organised quarrelling which is to start up politically after the war is making people ask whether nothing will teach them the better way through the forgetting of past antagonisms.

1945

January 1st

To Coventry (Spon Street) to a meeting with the Works' radio engineers. Bligh, Edwards, James, Gavin came. I am sure these meetings do good though there is little enough post-war effort being put into things as yet.

January 2nd[525]

Selenium rectifiers at Salford (Chestergate) have been looking up considerably under Mr. Underwood's stimulation. He is leaning on Wembley for all technical guidance and has for the present reduced the thing to a preproduction unit where processes can really be controlled. Rooksby and Thomas have apparently been doing a really good job.

In the chair at a BSI Engineering Industry Committee for approving Certification Mark policy (of the Mark Committee). Everything went through as we hoped.

January 3rd

Oliphant came yesterday about TA and had a talk with the others. It appears that GEC is to be asked to attend to the valve and rectifier equipment, control panels etc. Two to three million pounds—provided we can get Witton to play. I am seeing Oliphant tomorrow.

CVD–GEC meetings all day at Wembley. Brundrett discussed with me his own position at the Admiralty in relation to the future. I was glad of the talk. The chief matters discussed at the meeting were: (*a*) hydrogen thyratrons; (*b*) CV 12 modulators—of which there is a sudden shortage (We are the only makers at present so Douglas is called on to restrict CV 215s. The latter are still in trouble and we are leaving [learning?]

much which will be of help in future types.); (*c*) high priority of CW magnetrons which we are being asked to graft into an already distended programme in which the high power X band and the high power S band magnetron are already urgent.

January 4th

Discussed with representatives of Design Office the organisation of the section and 'conditions'. I hope the talk will help the dovetailing of this dept. into the Lab. structure.

Saw Oliphant and Akers re. TA—and arranged for him to see Sir Harry and Garrard. The interest GEC are asked to take will need all the salesmanship which Oliphant possesses to stir up interest in such a specialist project.

January 5th

A.G. Pearce has received a well deserved MBE. That is the seventh decoration for the Labs.

A propagation meeting here today—Megaw still in USA but I gather that as usual British work on short wave propagation is used by the USA people but no recognition given! But we shall hear all about it when Megaw gets back.

Flying and rocket bombs are giving NE London and S Essex none too good a time of it. Their pitch runs NE to Chelmsford.

I have word from George Wilson that he will be joining up with us again post-war.

January 8th

Megaw is back from USA and attended our valve preproduction meeting today (with B.S.G., J.B. and W.E.W.). He has some useful contributions on tuneable magnetrons and on cathodes. His experiences with the conference on short wave propagation were, as usual, that the Americans ignored all evidence of the British contribution to the art and were not guiltless of serious plagiarism.

The production figures are badly down. Lunched with Sir Harry Railing and expected Oliphant to join us at 12.15 but he didn't turn up. His meeting with Garrard on Friday seems to me to have been futile because he did not talk over the control gear which was the main subject he was to elaborate to Garrard. All we are wanting is a 'cover' figure for O. [Oliphant] to give the Treasury.

January 9th[526]

Glass Advisory Committee which seemed to me to be a very good review and criticism of the various issues before the Works—refractories, pots, hard glass, etc., etc. Partridge ventilated the various possibilities for glass-metal electrical lead-in connections.

Lunched with Dr. Slade, Chirnside and Snow re. the possibility of ICI being interested in Alumicide. A further meeting will take place. They also discussed with us the possibility of exciting atomic resonances in chemical insecticide and suggest further collaboration. I fear that DDT and an ICI improvement on it will put dust insecticides such as Alumicide in the shade.

Discussion with Macqueen, L.D.G. and R.W.W.S. on the need to proceed with the tropical inert battery for MOS. We shall have to get a man living in Birmingham district to increase our strength on battery work.

January 10th

Discussion with Mr. Thompson (Diamond Dies) re. specifications and USA activities. The USA report on diamond powders and the best method of making them is 'funny to silly'. Our efficiency figure for production has beaten all records for December and we have now led the field for several months.

A CRT meeting—large but not exciting. G. Wilson called—only to tell me that he wants to withdraw his letter to me in which he gave me his decision to join up with us again.

January 11th

Discussed with Williams his group work and the insufficiency of staff and space for the metallurgical programme ahead. Ransley and Ivor Jenkins are a serious loss to contemplate. We are wondering if Ivor Jenkins would return!

FIDO report will be discussed at PWD [Petroleum Warfare Dept.] shortly. Sir Donald Banks says 'as an after-war possibility'.

Brundrett writes helpfully regarding publication of certain aspects of high frequency technique, which were recently secret. We have several papers in preparation and have registered our place in the IEE queue.

A conjoint work with ICI on atom resonances and chemical reactions is permissible so far as CVD (Admiralty) is concerned.

January 12th[527]

Jennett, Jesty, Miss Keen and Lamp Works had a successful meeting with RAE over specifications for CRTs and had a good opportunity to demonstrate the Quality Control factor. We hope a long standing irritant is now liquidated.

Lamp Works metal 'problems' are to the fore. They are working under heavy pressure and processes are liable to go wrong. There is then no way out but a long detective job to define what had gone wrong—usually something quite elementary.

At my Consultative Board meeting today we had out the RL/Government Research issue vis-a-vis MOV. Chelioti has this issue—with others—under discussion with L.G. [Gamage] and I hope the CB discussion was of some use.

A British Association organised symposium at the Royal Institution on Research and Industry. But as usual the main speakers were not industrial—except Lord MacGowan; Blackett made a distressing appeal for salvation by materialism. The usual clap trap of the materialistic scientist.

January 13th[528]

High Frequency heating is all embracing and covers it seems every industry and yields interesting problems. The latest are connected with rubber treatment. Gavin tells me that RCM is not so dead as appeared it might be a few months ago and we have still a team working on it.

January 15th

The war is waking up a bit—the Russians have started a winter campaign in Poland and East Prussia and seem to be doing good. Budapest is almost occupied—the German salient which looked so dangerous a fortnight ago is dissolving—the Americans are well on to Luzon (Philippines) and Burma is going well—Mandalay within our reach. But the German push in Belgium has shown the USA that Germany is by no means down and out. Plenty of rocket bombs in NE London and S Essex.

Valve preproduction is still slumping badly and we are plagued by technical troubles with four or five of the valves.

Willshaw reports a thorough exchange of view and programme on V (K/2) band valves and systems. G.M. Wright (ASE) has given Oxford the same problem as ourselves, to produce a V band system for Admiralty signals.

Willshaw also went to TRE and found them urgent on both high power S and X band magnetrons.

There is a large demand for Dudding and Jennett's quality control pamphlet which is being distributed from here.

I have finished the first approximation to our after-war laboratory organisation, but have not yet costed it. I fear it will turn out fairly expensive to judge by the staffs which are being asked for by the groups—but my first step is to see what would be the cost if we attempted to provide what everyone considers reasonable facilities for the programme ahead of them.

January 16th

Had visit from Mr. Underwood of Chestergate Mill (Salford Instruments) who asked our collaboration in the field of waveguide engineering associated with measuring and testing gear. He is prepared to secure certain expert staff to assist. We gave him encouragement but the final policy regarding the continuance or otherwise of the several branch establishments at Salford seems undecided.

Visited XL and spent an hour with Shaylor and Mrs. Brown examining the CV 114 and 221 T & R boxes group. Technically it is a first class and well run show and is surmounting the many difficulties of this type of valve.

January 17th

Had to see Leeds about handling a situation in which the husband of one of his staff is threatening him. Got Cornock to help and he thinks the matter is clearing up.

A long TVC meeting and agenda only half done. A long CVD policy meeting in the afternoon when we thrashed out magnetron research priorities which have been upset by the urge for a CW S band magnetron.

The Russians seem to be well away with it towards Germany.

January 18th

Household Appliances meeting at Head Office. Dr. Black has spent the day here. He brings the usual news from USA and difficulty of getting them to realise the British effort. My impressions are that in Radar and Valves the USA and British results are say 50/50 in their value but the USA effort is from five to ten times the size of the British: the USA effort seems to be much less efficient in useful output per man than the British. The reason why is an interesting speculation.

January 19th

Research Services meeting with Railing and Gamage. The latter mentioned the GEC – MOV issue in relation to post-war research contracts and noted how the undue exaltation of the GEC tended to affect loyalties between MOV and GEC staffs. I hope this will result in agreement for post-war contracts in the joint names. The conditions regarding patents attaching to development contracts are to be unified throughout the Company through the Patent Dept.

Gavin still has a residuum of RCM jammer development. He is completing a 90 cm model with what remains of his group. Hecht speaks of giving us a 10 cm model to develop and the query is whether we have staff left to handle it.

The Helmore searchlight contract and how the GEC can recede from it is still a difficulty as Helmore is most unwilling to agree to anything but a slowing down of the work. I shall see that Witton tackles its own difficulties with Helmore and the Ministry, for I hardly know what is sincere and what is hollow in the representations made.

The TA party has returned from a visit to BTH to meet their men who have examined the USA plant. Most information is on the engineering side.

January 20th

A conference at MAP led by Helmore has asked for the making of normal torpedoes dirigible. The success with the 'fish' has inspired Helmore. We are hard pressed for staff but Bligh and Foot seem keen on doing it and I am agreeing provided our undertaking is defintely limited to the first functional model.

January 22nd

The Russians are into Silesia. We shall see what they are able to overrun there but presumably the Germans will have to save it somehow.

Valve preproduction is beginning to climb out of its depression—at least we hope so. The performance of our X band magnetrons in operational sets may yet set us back. Tests have been awaited for some time.

An unveiling ceremony of a plaque of Lord Hirst took place today at Hirst Hall when Mrs. Gamage, Sir Harry and Chelioti were the speakers.

January 23rd[529]

Lunched with Mr. Matthey of John & Matthey.

Took steps to settle our expansion of Rooksby's X-Ray section which is becoming very crowded. Partridge is willing to help with some space. It is typical of the difficulties we shall be in in almost every group.

Visit from Dr. Wooldridge of the Bell Labs. to see about Espley and radar.

January 24th[530]

Went carefully over Winch's section with him—group membership and programme. He has been rather unlucky during the war in developing in a masterly way a number of equipments which were pioneer of their kind, but as most of them were for MOS they were never pushed to operational adoption. So he has never had the satisfaction of having a project adopted—through no fault whatever of his own or of his colleagues Hill, Macfadyen and Michaelis. His present project 'Mortimer' for ranging electrically on mortars is an excellent job but we know in advance that it has only a 1 in 10 chance of adoption.

Gossling and Bell have returned from two days with Dr. C.C. Garrard in Witton helping in the preparation of an estimate for Tube Alloys plant. They were greatly impressed by Garrard's ability to tackle such an estimate of some two million personally and out of his head.

Ryde gave a few of us a very interesting talk on nuclear and atomic physics up to date. We want a second sitting.

An MOV transmitting valve meeting. Bell told us that Mr.England from TRE had indicated the extent of his requirement for CV 257 and 259 valves for a project. Some 450 valves I believe.

January 25th

Jesty has had some good meetings on CRTs with MAP and Hammersmith on methods and measurement and specification—and (in the GEC) the electronic scanner meets with interested approval.

January 26th

To Salford (Chestergate and Warrington) and to have a talk with Cobden Turner, temporarily out of hospital. The Works were impressive in the equipment being turned out. All phases of the work were discussed, but particularly measuring instruments in view of the inaugural Instrument Committee meeting next week.

The Jewel Works will now use our own boule for making instrument jewels instead of putting it all into stock and using Swiss material. We may now get some reliable data.

January 29th

The sick preproduction valves are not out of hospital yet and are far from convalescent—so our self respect still at a low ebb.

Prolonged cold weather is rendering the fuel situation serious and further economies are being demanded. The extraordinary Russian advances into East Germany are most encouraging but everyone seems to take these things much for granted, or else wishes that the Russians did not appear to be doing it at all.

A meeting at Head Office to forward the policy of an exchange arrangement with BTH on valves. Much will depend upon EMI who take, as usual, an unco-operative attitude.

January 30th

Visit from Mr. Chapman, Managing Director of Jessop, who spent all day here studying the place in view of the intended expansions of his own nucleus research organisation under Oliver. He expressed a desire that we should use his facilities for experimental melts of special steel or any other collaboration which seems to us desirable without any obligation on our part. A most charming and sincere gesture.

At the CVD Admiralty meeting today Brundrett passed me a note to say there is an RAF staff demand that the CW magnetron shall be given overall priority and everything to give way to it. Have warned Willshaw that it is likely to come up tomorrow at the CVD magnetron meeting.

January 31st

Inaugural meeting of the Instrument Committee—part of the Advisory Board series. It was a useful general review of Salford activities, present and prospective. Gavin and I attended and N. Hill took the minutes.

As a result of my 'prayer wheel' showing the post-war Lab. grouping, Humphreys and I proceeded with our second approximation to the size and cost of the post-war Laboratory. Personnel (total) came to some 900, overheads to 135% and cost, after deducting expected receipts from outside, to, say, £325 000 p.a. This has now to be examined in detail under each group and its cost.

February 1st[531]

Examined with Espley the large Astec [Asdic?] generator for Fairlie which Gledhill has been working on—also our first TTS (transient test set) made by Espley's group and of which we are expecting great things in the future.

Council of the IEE has chosen me for the year's recipient of the Faraday Medal.

February 2nd

A TVC meeting at which we finished our memo on the extended functions of the committee to meet the requirements of collaboration on valve standardisation and nomenclature etc. with the USA. This will now go to OPTEC (Radio Board) for approval.

Megaw gave a three hour talk on his USA visit—propagation and magnetrons.

February 3rd

Francis tells me that his visit with others to ASE to test out our CV 191 and 214 magnetrons (about which we had had serious doubts) turned out much better than was feared. All the 191s in the operational equipment passed muster which releases some 150 valves awaiting the verdict. This is for the moment the most urgent valve. The CV 214 is not so good and has flashing trouble and remains a problem yet to be mastered.

February 5th[532]

The latest CV 215 preproduction hydrogen thyratrons seem to pass muster all right and a batch of 120 have been passed through. This puzzle seems to be on the way to being resolved.

Had an hour's conference with Espley and Pearce (preproduction groups) over the post-war allocation of the large preproduction laboratory at the East end [of the Laboratories]. As we could see no likelihood of this laboratory being freed from preproduction anyway for a year or two it had to be agreed that television should get established in such space as Espley now has in the main Labs. plus the hutments at the back. We hope this can include the television hut if Coventry don't need it. I agreed that, as a long distance policy, the Lab. in question was more suited geographically to circuit rather than valve work.

Lunched with F. Smith (MOV) and discussed the eternal subject of EMI jealousy of the GEC acquiring prestige through its valve research.

Dr. Halbertsma (Philips, Holland) just arrived from the centre of the war area spent the night with us. They have a grim situation and he showed signs of having been through it. He seemed ravenous and said that hunger became a disease.

February 6th

James reports desire by Spon Street for vacuum insulators prescribed by TRE for tropicalised leading in connections. Partridge was able to show samples already developed.

A diamond die technical meeting at Wembley at which we discussed for one and a half hours the weight in carats for the die specification. Dudding put up a jolly good analysis of large selections of stone in order to decide on minimum weight of stones permissible for the different holes.

Hawkins turned up for the first time after several month's illness.

February 7th

Discussed with Bligh and Humphreys the knotty problem of junior technical assistants of whom some are being called up and the question arises how they are to be replaced, for this is a dead-end occupation for young fellows who are not going in for research. The three alternatives are skilled craftsmen who are not to be got; articled apprentices who must be properly trained and of whom we fear to have too many; and thirdly we may get Myers to train some young girls and try to make craftsmen of them. If we can do it, it should be an interesting experiment.

A Diamond Die Advisory Committee and lunch after with Prain. We took the first step towards forming a trade association.

To the BEAMA to discuss increased grants to ERA. There was support provided ERA will keep to their fundamental background work and not enter on the field of development and design.

February 8th

Quality Control makes continuous progress and Dudding and Jennett are constantly called in to advise and help both in factory and technical college. MOV are prosecuting its application with certain valves where control charts are now being instituted. Miss Keen tackles factory jobs single handed.

Gavin reports a P 55 receiver meeting with the Admiralty and the many GEC Depts. concerned. Not a good situation for Johnson Matthey have thrown in their hand over chromium plating and the whole job will probably have to be recast. Various schemes were favoured for adopting James's non-sliding contact (butterfly) circuit. So we await final instructions.

Commander Tomlinson (security officer) called to insist that on account of TA we must maintain our security precautions. Generally over the country these are being relaxed.

Farren has undertaken some trials on 'Dealer' with Squadron Leader St. John, and the device operated satisfactorily in the vertical as well as in the horizontal plane. He felt that results were successful and promising.

February 9th

Discussed Marconi valve matters with Bell who reported visit from N.R. Davis and Green from MWT over valves wanted for peace-time television. They want considerably higher powers for transmission purposes.

Interesting visit from Mr. Essington Lewis of Broken Hill, Australia. Williams mainly took him in hand! Mr. Gamage went round the metallurgical end of the Labs. with us.

ERA lunch at Savoy when Sir Harry proposed the toast. Fleming was not very sparkling in his reply.

February 10th

A regular attack is in progress at the British end of the line in Holland—the Americans are also active so we may see some active advance in an effort to take advantage of the Russian pressure which is now up to the Oder and nearly up to Stettin and 45 miles from Berlin.

February 12th

Our valve preproduction troubles are 50% cleared but plenty of problems are left. Some types like CV 90, CV 155 and CV 214 are finishing off so Le Rossignol and I do some hasty scrutiny from time to time to see if we think we are going to lose money. Our personnel drops by about six per week, which is not far off what we want.

A TVC meeting and lunch with Sir Harry. He is naturally concerned to be sure that we cannot tackle 'TA' without very special provision of facilities by the Government.

February 13th

Examined Lab. accounts at regular meeting with G.K., O.W.H. and J.C.C. Some of the processes for keeping research contract accounts require tidying up but otherwise the whole system is much better up to date.

Helmore's fish, Jolley reports, is in a bad mess. The engine has been fitted by Rolls Royce and other unco-ordinated things done to it, and when put in the water it had no buoyancy and sank! I told Jolley that the responsibility for the project could not be accepted by the GEC and whilst we were prepared to continue to act as consultants to Helmore, he must hold the responsibility and direct the project. In the early stages 18 months ago Helmore cold shouldered Jolley out of the job when the latter tried to help him.

An Illumination Engineering Society meeting at which I introduced our Study report on the lighting of dwellings and schools. Not a particularly inspiring discussion. Halbertsma attended and spoke about conditions in Holland.

February 14th

E.R. Davis of Kodak Research called for a talk on laboratory organisation. He is modifying his and introducing our feature of 'Leading Staff'. I had to insist that there is a distinction between Group Leaders—who are part of the paper system of organisation and executive status—and Leading Staff whose office is on a higher plane altogether and seeks to recognise and organise the imponderables of intellectual and moral leadership and responsibility for the spiritual values in the place.

A CVD (GEC) meeting and allocations meeting. Brundrett away. It went well. Atmosphere was good. We discussed general aims and ways of shrinking the preproduction effort—how to decide the types of valve to be perpetuated into the post-war period.

February 15th

An informal Research Meeting—Oakley in the chair. We discussed hot cathode luminescent lamps for normal fittings and showed Jenkins and Breadner's lamp. Also other matters from a research angle. I hope we are getting together on research matters more as a team. Chirnside and I lunched with Slade (ICI) and afterwards had meeting at ICI on Alumicide. Found ICI had somehow received a sample of our material and had been reproducing it. Ours was admittedly best as an insecticide but the degree is to be decided by further liaison between us. I think a good meeting—our case is not a very strong one.

February 16th[533]

Discussed with Simms the art of 'glass processing' as represented by what Breadner and he have done and written about during the day. Dudding and I agree that his idea of emphasis upon this should be supported by having group meetings here of those specially interested in this art.

A Research Services meeting at Head Office where we agreed that Heavy Alloy should be handled and pushed by Mr. Buckell in the aeronautical field, and he will distribute literature etc. about it.

February 17th

Very heavy bombing of Germany—particularly Dresden—and a successful meeting of the 'Big Three' in the Crimea make us feel that things are still moving fast. The weather in the West and in Italy is the worst ever.

February 19th

Chelioti lunched with me. He is full of his experiences in running 'works production committees' and shop stewards' relationships.

An Advisory Board Meeting at which Macqueen gave a glowing account of possible radio business to be done. It appears that existing police radio requires replacing with shorter wave equipment.

February 20th

The new 'Fuse' Committee of CVD is going to cover all sorts of weapons which may contain valves—a considerable list. So Gossling and

R.O. Jenkins are likely to find themselves busy again notwithstanding the fact that we are now to be consultants. It appears also that the GEC design of rugged valve is the one which new contractors will have to make, and not the much pushed 'Rangaby' type.

Attended conference organised by 'Research Group No. 1' on remuneration of Scientific and Engineering Staffs. A useful discussion but not with very radical ideas.

February 21st

A CRT meeting with Mr. Hecht in the chair—not very eventful. An MOV meeting (receiving valves) in the afternoon also not very eventful. When things are uneventful it usually means that solid work is being done.

February 22nd

An Osram Advisory Committee meeting where general policy was discussed of Government price approval after the war. The BOT [Board of Trade] desires to work through trade associations and there is no longer any licence needed for export.

Called in at the English Speaking Union to meet scientific delegations from abroad.

February 23rd

Breakfasted with Smithells at his new Lab. in Chesham, a very nice converted hotel in a park by Gerrards Cross golf course. It is like early days at Bushy Park (NPL) or Bawdsey.

Discussed with Gavin the feebleness of the GEC high frequency heating effort particularly on the production side. The Wembley contribution is by no means adequate too.

Jolley reports a test of the 'Fish' before an Air Marshal who will report to the Air Ministry staff. The only part which worked was apparently the electrical control. The engine failed and the buoyancy was negative due to water leaks. One is sorry for the flop and Helmore's ability to carry off situations may serve this one, but he clearly wants someone to 'manage' things with him—which Jolley offered to do 18 months ago but was cold shouldered off the job. It is in fact somewhat off our line of country.

February 26th

There is a general offensive to clear the Germans out of the ground west of the Rhine and it appears to be going well.

Bligh reports continued collaboration with BBC and partially for our efforts [?]. We must take it up actively with Head Office and Spon Street.

Cornock returned from the Courts today delighted (as we were too) at our winning our zinc–beryllium silicate patent case against the Patent Office opinion that they could not grant our application.

February 27th

Visited Johnson & Matthey Research Labs. and factories at the Exhibition Grounds and saw a number of interesting processes. They seem to have many contacts with different parts of the GEC.

Discussed with Ransley his easing off of crystal production work in order to push on with his researches before he transfers to British Aluminium.

Had an unaccountable cable from Corning USA asking us to send over a man to receive information 'as requested' on glass–metal soldering techniques! It appears that officials at the MOS have been presuming on our co-operation after the event! Very awkward.

February 28th

A discussion with Bligh and others on whether we should do work on crystals of the ammonium [aluminium] phosphate type as a substitute for quartz and later got into conversation with C.S. Wright who has plans for one of his establishments to do work on the subject. I left our meeting to consider matters further and try to make a recommendation.

Meeting with Heather, Macqueen and Marriott to decide GEC policy vis-a-vis an approach by Appleton (DSIR) to the radio industry to form a Board to consider co-operation in radio research. It is going to be a thorny path probably. I am contacting Appleton with the object of getting his unofficial guidance.

A CVD policy committee this afternoon at which publication of war work in radio was discussed. The first release is somewhat circumscribed—no radar and no magnetrons. Lewis (TRE) had some hard things to say to me about their high power X band magnetron and the fact that we did not take a sufficiently critical view of what their needs would turn out to be—chiefly on grounds of frequency pulling. He is coming down to give us a talking to.

March 1st[534]

The Corning contact has moved some steps further. Gray [Grey?] of MOS agrees to the MOS umbrella for the visit and Chirnside agrees to go to the USA. Chelioti also agrees. Having now apparently made the path clear the difficulties will probably begin.

Interviewed the chief WRN [WREN?] with Humphreys and Gardiner about having girls from them for mechanical laboratory work. The visit looked as if it were to prove quite useful in our future recruitment of suitable women for Lab. work.

An FM equipment meeting with Macqueen and others, to decide our policy on FM vis-a-vis BBC and also collaborative research in the manufacturers' organisation. The discussion centred round a demonstration in the upstairs conference room.

Attended a meeting of the BOT sponsored committee on Art in Industry at Claridges—Sir Thomas Barlow, Leslie [Gamage] and Lord Woolton spoke and the speeches improved in that sequence. They started by preaching to the heathen and ended by the heathen doing the preaching.

March 2nd

Visit by Dr. Slade and Snow from ICI to discuss HF applied to chemical reactions—an interesting talk and visit. Ryde came to conclusion there was not much to be done as yet.

March 5th

Valve preproduction was 1848 last week but due to drop. CV 90 finishes this month, CV 155 is finished. CV 52 and CV 58 are accumulating stocks before fading out.

Lunched with F. Smith, MOV. They feel that their loyalty to the war effort has put them at some disadvantage to certain 'main chance' competitors.

A meeting with Settle, Bligh, Humphreys and Gavin on the programme and planning connected with HF heating. We want Settle's dept. to have a technical administrative group like industrial heating—we want special 'process' jobs to be for Wembley to investigate and advise Head Office on.

Cherry has taken a lectureship in Manchester. Bad show.

March 6th

Mass X-ray examination of all staff on the site is in progress. I attended today.

Meeting of General Board of BSI at which we approved the report of the Mark Committee on certification marks. Followed by lunch with Lord Woolton presiding and speaking—very good.

March 7th[535]

Megaw does not find that changes in the chairmanship of the propagation committee are helpful in regard to our share in the work and there are influences in ASE which seem to be getting 'exclusive' in their outlook. I have suggested that Megaw asks Landale if he wants us to drop out of this work.

TRE are now coming into the USW millimetre wavelength work with us and Oxford.

Dr. Kellaway—director of the Wellcome Foundation—visited us today. Le R. mainly looked after him. He is recently from Australia and a very delightful fellow. He seemed to appreciate what he saw of the place.

Spent afternoon at special meeting for forming a trade association for Diamond Dies. We made good progress—Clements in the chair. It is to be on the modern 'industry' basis: makers and users.

There has been a great push to the Rhine during the past week and Cologne has fallen. It looks as if the whole west side of the Rhine would now fall to us. The Russians will soon be in Stettin and have made an amazing grasp of a long stretch of the Baltic coast. Burma also goes well.

March 8th

Meeting last evening at IEE on Frequency Modulation. Bligh and his group gave a demostration and scored points.

Coales was here yesterday discussing with Willshaw the future demands for valves. He needs higher power X and high power K band—both new problems. He expressed the view that they could never rely on internal valve development and must have this sort of thing done by collaboration. I must try to get an estimate of what this sort of thing will cost us supplying 100 to 200 samples of new types. It is a new problem.

March 9th

The USA Third Army is across the Rhine. The frequency of rocket bombs over the London area is going up. The sooner the Germans are 100 miles east of the Rhine the better.

A Consultative Board Meeting at which we discussed MOV receiving valve situation and the usual jibes from the EMI people alleged backwardness. Also ventilated subject of salaries of professional engineers and scientists.

A Bowler Meeting with Captain Boyle presiding. The device is more vulnerable to counter measures and more particular as regards technique than the Staff had hoped so large numbers are not likely to be needed.

I ventilated the post-war cost of the Laboratories which to handle our prospective programme is about double pre-war (gross) and about 75% pre-war (nett). I am not surprised. [refers to Consultative Board Meeting above].

March 12th

The war still goes well on most fronts. We have the Rhine (West Bank) down to the Moselle.

A TVC meeting which ended with a brief discussion of the peace time future of TVC and its functions. It will be an interesting discussion at our next meeting. Lunched with Sir Harry and found coincidence of view on most fronts, e.g. high frequency engineering, USA visit to Corning. There are vacancies for a number of people at Head Office and it is not easy always to see candidates.

There seems to be a strong come-back over the Code of Practice Committee's proposal for a non-interchangeable plug, which was rather unwillingly supported by the BSI Industry Committee. A stormy joint meeting of the Study Committee (Beard) and the Code of Practice Committee was equally divided on a proposal to up grade the existing 5 amp plug and discard the idea of having a fuse in the plug.

The CV 12 (mercury vapour valve) has been behaving promisingly as a grid controlled rectifier and Ramsay says there is much interest in it from a number of quarters.

March 13th

Trumper is coming much to the fore as Bowler recedes on to the shelf and Farren reports conversations which point to pressure being put on us to get on with our anti-submarine weapons. The discarding of Bowler will be a blow to Coventry.

Our Lab. Accounts are further advanced this financial year than they have been for very many years past. An accounts meeting with G.K., O.W.H. and J.C.C. this morning left us guessing as to the whereabouts of past balances which must be somewhere out of sight.

Arranged with Dunham about his after-war group which we propose to centre in Electrotechnics bringing Michaelis into partnership.

Gavin has been asked to apply for a good job with the Ministry of Education (Inspection) but he tells me today that he has decided to stay here—for which I am deeply thankful. He is one who is in a few years should be carrying a substantial load of responsibility here.

Ryde urges an increase of effort on silicon crystals in order to match that of the BTH Co. It is difficult to respond to every urge for more competitive effort.

March 14th[536-537]

Called on Sir W. Palmer (BOT) who wanted my personal view on an Anglo-American pool of radio patents. Such an idea has our support.

A Television meeting at Head Office at which the Hankey report was reviewed. We agreed with it in the main, in fact it follows our evidence closely—but it is not clear how they propose to stimulate industry to work on 'links' and transmission problems.

Carbons made at Four Ashes are not yet giving the good result expected and we must find out why. They seem uniform but the arc flares back.

Willshaw is expecting a trying magnetron CVD meeting. The high power X band valves are not behaving as they should.

March 15th

The CVD meeting yesterday on magnetrons passed quietly but I imagine that is only because the whole issue is to be raised in a special meeting here on Tuesday.

A household appliances meeting at Head Office—very lively through objections by Research Laboratories to certain Head Office technical policies. We seem to be fairly well forward on designs and agreed models of new apparatus.

March 16th[538]

The TA party are back from a visit to Met. Vick to discuss the electrical side of the project. There is some ridiculous American security ban on transmitting information and not much transpired.

7422
TELEPHONE: MAYFAIR ~~9020~~.
Extn..............

Our Reference SB63238/RDC7/B
Your Reference ....

MINISTRY OF AIRCRAFT PRODUCTION,
RDC7
STRATTON STREET,
BERKELEY STREET,
W.1.

SECRET

14th March, 1945.

Dear Dr. Paterson

Development of C.W. Magnetron Transmitter.

I understand from conversations which have taken place between Mr. Catt and Mr. Gavin, that undertaking this development work has necessitated some diversion of engineering effort from other projects. I am writing to assure you that this work is regarded as being of the highest importance by the Air Staff and that any reasonable inconvenience occasioned by the transfer of Engineers is more than justifed by the urgency of the work involved.

I should like to take this opportunity of saying how much I appreciate the very helpful and co-operative attitude of Mr. Gavin and his group of engineers, I find the combination of this ~~[illegible]~~ with their technical ability and enthusiasm a source of recurring satisfaction and a contributory factor to the smooth running of the work handled by this Directorate.

Yours sincerely,

[signature]

The Director,
Research Laboratories of the
General Electric Co, Ltd.,
WEMBLEY,
Mddsx.

16 MAR

Fig.56 Letter dated 14 March 1945 from Mr. Hecht, MAP, to Clifford Paterson on work of M.R. Gavin and his team

Our diamond die production efficiency figure continues to top the list but we and Simpson are very close.

Glass bushings for tropicalisation were discussed with Mr. Starr of TRE yesterday and I think his visit will give better direction to our effort. Breadner and Simms are working actively on the development.

A nice letter from Hecht expressing appreciation for Gavin and his group's work on RCM and urging the CW magnetron project for which valves are already over here from USA (Fig. 56).

A Research Services meeting with Sir Harry away. A useful review and talk over many things including good intermediate level personnel in the GEC.

Had Coales of ASE to lunch. He was wanting to ensure that the high power S band magnetron needed by the trial EMI set shall be available. It is to undergo sea trials.

A Carbon Committee. The Four Ashes trials are at present inconclusive.

March 18th

Dr. Landale called to look into the magnetron position here and the priorities. He expects to be here for the main priorities meeting on Tuesday.

March 19th

Valve preproduction figures sailed up to 2150 owing to certain techniques being more successful. In the CV 215 valves we are still replacing previous duds and shall be doing so for some time. We must debit our preproduction account this year to the amount of the cost of the dud 215s.

Discussed Watsons Electro-Medical programmes and cost thereof with Mr Caldwell. It looks as if they must allow at least £3000 for the first year.

I am trying to find convenient space for all the people who are needing post-war research accommodation.

Dr. Bard of FBI called to talk over industrial research in relation to the Research Committee. I tried to give him a picture of Industrial Research Establishments within industry with Research Association assisting with general problems—and not vice versa.

Ramsay emphasises coming pressure for hydrogen thyratrons from TRE. We are getting on with our programme and we cannot see our way to try to increase staff to meet an enhanced request unless we have evidence that the equipment is a war winner.

March 20th[539]

Discussed with T.W.H. [Heather] the points for our forthcoming talk with Appleton on radio development co-ordination.

An instrument (GEC Salford) meeting and then a heart to heart talk with Dr. Lewis of TRE on the needs of TRE in magnetrons and the shortcomings of Wembley in fulfilling them. R. Le R., E.M., H.W.B.G., J.B. and W.E.W. joined in. We are in a weak position because the high power X band valve has got into a very sticky spot and we could give Lewis no immediate assurance of the supply of samples they could use. This was followed by a general priority meeting—Lewis, G.M. Wright, Landale, Griffiths, Hecht. An hour's talk boiled the matter down to the need for three instrument maker craftsmen and how we could get them here on loan. TRE will have a shot at it.

March 21st[540-541]

Dr. Whitehead of ERA called to talk over their housing difficulties and the plan (supported by Captain Donaldson) to take temporary premises near Leatherhead. I thought the scheme worth the risk.

Lunched with G.C., J.Y.F., F.W., Ryle, Marden and Balmain to talk over pressed glass making in relation to illumination developments. The case for proceeding with the making of pressed glassware was of course established.

MOV transmitting valve meeting—everyone in good spirits although some fields of needed production are very backward.

Waldram read his paper on photometric properties of the atmosphere at a joint meeting of the Royal Meteorological Society and the IES. A very good paper and exposition. Sir George Simpson, Good, myself and others spoke in the discussion.

March 22nd

The American armies are wiping out all Germans west of the Rhine. It appears to be a complete rout. The Russians have nearly cleaned up all Germans east of the Oder—all this before the campaigning season has properly begun.

An Osram Advisory meeting this morning—lunch with the editors of the electrical Press and a good talk with Appleton and Heather on general collaboration in the electrical industry and how to achieve it. Appleton indicated his plan for a Radio Research Station as a part of DSIR; industry to be brought in with some sort of machinery.

March 23rd

A TVC meeting on policy from 10.0 to 1.0. Discussed future place and organisation of TVC and its interconnection with CVD after the war. I was at pains to know what definite machinery would be fixed up for industrial collaboration, particularly in connection with a prospected valve development, research and preproduction establishment. I find there is reticence on both sides to development of a practical liaison scheme.

A conference in Chelioti's office on receiving valve progress and prospects—Warren, Bell and James from Wembley attended also the management of MOV plus J.Y.F. [J.Y. Fletcher].

Peters is very ill.

March 24th

A message from Helmore via Foot that the 'Fish' has functioned very well on trial. It is now buoyant and travels fast and manoeuvres with our control gear.

I am seeing various people who are not likely to be permanent so that they can be planning well ahead of their time of leaving.

March 26th[542]

Since Saturday there has been a general advance over the Rhine and beyond it and another stage seems opening up. V2 rocket and also flying bombs arrive over here at fairly frequent intervals. Kenton, Neasden and Willesden have had several visitations lately.

Preproduction is on its feet again—that is to say technical snags have virtually cleared up. Certain new types CV 273, CV 259 have very limited demands at present although we are proceeding with their tooling.

Peters is still very ill and causing much anxiety.

March 27th[543]

Partridge reports progress and enthusiasm by the Ministry in connection with our 'condenser bushings' for tropicalised equipment. We have now an order for these.

Ramsay reports that he finds molybdenum protective coating on his hydrogen thyratron electrodes increases their life by ten times or more and makes them into really respectable valves. This makes an increase in our effort on these valves very tempting but in view of other commitments I fear we must keep going substantially at our present pace.

Mr. Matthey came to see over the Labs. today with Mr. McDonald. They are appreciative visitors.

A Watson Electro-Medical liaison meeting to talk over Mr. Caldwell's proposals for a condenser HV build-up X-ray equipment in place of a transformer type of gear. His rotating switch and the functional model is an excellent piece of pioneer work. A most interesting occasion.

The news in the evening papers is sensational but it may be an exaggeration to claim that there is no longer any German opposition.

We are at last beginning to take Kovar seriously and Partridge and Williams are making some melts with a view to establishing a uniform and controlled product. Already we are admitting probable advantages. Years ago this alloy, so widely used in USA, was pressed but no one here could admit any place for it. I fear even research staff suffer from the failings of some factory people whom we condemn for lack of enterprise.

March 28th

A CVD (GEC) meeting and preproduction in the afternoon. Brundrett was hurried, and worried over his Admiralty personnel questions which he says are acute owing to difficulty in finding sufficient technical people to handle all that has to be done here and abroad. It was not a very suitable occasion to urge that CVD is required, and appears to require much more of our facilities than Brundrett suggested to me for the post-war regime. I must get him to face up to this question because our facilities must be devoted as soon as possible to the purposes which peace time requires.

March 29th

We are breaking off for Easter with everyone feeling that the European war is rapidly collapsing—Montgomery starting to overrun the rest of Holland and North Germany. I am reminded of the period which started in August or September 1918 and ended on November 11th 1918!

Today we had a meeting (Machin and group) to review the Admiralty valve life test installation. It is a very satisfactory piece of work and is now fully loaded.

Cornock had success yesterday in his share of a further meeting with MAP (Mr. Pickles) and the Radio Equipment Manufacturers re. conditions for development contracts for post-war equipment. He stood out for reasonable dignified conditions and opposed the oppressive attitude of the Civil Service. He had to lead the protestants and his example and tact won a substantially satisfactory and agreed solution to outstanding points.

April 3rd[544]

Back after Easter break—everyone cheery owing to war news. The rocket bombs have ceased to come down over London and 'Southern England' and the Allies are 100 miles east of the Rhine and closing in on Holland or at least cutting Holland off from Germany.

Had long discussion after the valve preproduction meeting re. the partial disbanding of groups and the allocation of certain members to permanent jobs here. Humphreys is the main link and stimulation.

Caplan came to report an accident to one of the Nechell's turbines—he thinks it may be due to fatigue failure in coupling of main shaft.

April 4th[545]

Foot reports high level demonstration of Helmover—Sinclair, Portal and Sorley all coming to Staines and everything depends on Foot's radio control gear.

CRT meeting. Oatley asked if Wembley would continue to provide service to the Experimental Establishments for CRT experimental work in view of the fact the EMI (Schoenberg) has asked to be relieved of this function. Hecht seemed to be disturbed at the EMI attitude.

Information came through that there was to be a high level inspection of FISA Helmover 'fish', at Staines reservoir. I offered to go and was glad I did for the whole thing and especially Foot's control gear was a great success. Both a launch and the 'fish' were steered by radio which was not subject to jamming. Sir Archibald Sinclair, Sir Charles Portal (Chief of Air Staff), Air Marshal Sorley, Sir Stafford Cripps, Leedham, Helmore, Jolley and others were present. Rolls Royce had fitted the engine and steering mechanisms and Wembley made the radio control. Bligh could not be there—he was in charge of the group. At the conclusion Mr. Hives (Rolls Royce) was asked if he could put in hand 100 sets and GEC was wanted to supply the radio equipment. I don't know if Spon Street will welcome the order. The AI Mark 8 generator is used, the rest of the gear is relatively simple.

April 5th[546]

Megaw reports further progress with ASE vis-a-vis his share in the propagation work under USW and Appleton's Committee. The chief cause of trouble lies not with us but with the desire of ASE not to be overridden in the running of the programme by MOS.

Attended paper at IEE by Humphreys, Rycroft and Connell on process heating. It was very well received.

April 6th[547-548]

Visit by Mr. Bentham and Admiral Perrin who are consultants to the Admiralty on general ship design and construction and are on the look out for places where research and new ideas can come from. They seemed to be much interested and stimulated.

Had call from Preston (NPL) re. the affairs of the International Illumination Commission—the officers are beginning to think about its reconstitution again.

April 9th

The progress into Germany is maintained. Bremen, Hanover and Vienna are about to fall. The main opposition is in Holland which is now nearly cut off.

Valve preproduction is 1395. The weekly figure demanded is now 1440.

The analysis of our post-war financial position has passed another stage. The groups asked for by their leaders to meet the programmes ahead cost about £345 000 nett or as near as possible double the 1938 – 39 level, i.e. seven years increment of development. Salaries – wages and materials appear to be about 35% up.

Visit from Admiral Sir F. Wake Walker, Third Sea Lord, to see what we are like and the sort of things we do. I think he was duly impressed by the variety and scope of our research.

To our surprise the three instrument makers wanted for high power X band work have arrived—two from Union Works!

April 10th

A meeting with Radio Works at Spon Street. All were there from Wembley but we missed Peters who, though better, will be away for some time. I thought it was the best meeting we had had there.

April 11th

A TVC meeting with heated discussion on Security definitions 'Top Secret', 'Secret', 'Confidential', 'Restricted'and 'Unclassified'. It seems an odd subject to be spending time on with the war daily speeding to a close.

A CVD policy committee—a short one and uneventful.

April 12th

The first IEE Radio Interference Meeting—Chairman C.C.P.—I had contemplated this with some anxiety remembering the difficulties of keeping a smooth course with its predecessor ten years ago. Today's meeting went excellently and we got things shaped up so that the course of action seems clear. I succeeded in taking a rapid move and amalgamating the IEE and the BSI organisations before they knew where they were. I think Angwin was pleased and no one expected things to progress so fast.

Ackerley's IEE paper on lighting—a good attendance. [See Ref 12]

April 13th

Humphreys has been cleaning up my post-war cost figures for the various groups and I shall show them to the Consultative Board tomorrow in an interim way.

April 14th

The country has had a shock—President Roosevelt has died suddenly. This seems to us to be a very very great loss—more for winning the peace than the war.

A Consultative Board meeting and lunch with F. Smith when I am afraid Schoenberg was the chief topic. The continual assaults by him on the GEC for not agreeing to research being removed from Wembley to MOV are getting wearing.

We have some £250 000 of Government research contracts to complete. Probably £150 000 of them will be finished by March 1946.

A Z Dept. conference in the afternoon. J.Y.F. present.

April 16th[549]

We are constantly warned that there is stiff fighting ahead but the armies still advance rapidly and both Western and Eastern Allies are about 45 miles from Berlin. Germans in Holland are cut off but are starting to flood the country again. There is going to be a devilish situation there.

Our X band preproduction seems to be on the rails now. Willshaw reports high power X band (about which there was such a flap) is progressing—but it will take time. The high power S band now has a rival from Boot, who is working at BTH. The CW high power magnetron is making rapid progress.

Have spent much of the day on analysis of prospective Laboratory costs with Humphreys. There are so many aspects to be presented in order that others may get the right picture of our activities. Attended first of series of lectures to the staff in the library by Dudding on statistical outlook on data.

April 17th[550]

Dudding reports request from MAP that the valve industry should organise itself on quality control. MOV have told their competitors that they have now amended their attitude (the result of Jennett and Miss Keen's activities at MOV) and are prepared to back the Ministry. This is a great advance and encourages us all that our efforts which are so acceptable outside the GEC also make progress with our own people.

Visited XL with Gavin to see the 'de-tearing' experiment [?]—most interesting. Also all the equipment for process heating.

P Good came for a talk on BSI policy generally.

Sir Alexander Korda and two others spent the afternoon on consideration with us of cinema studio equipment of the future.

Discussed and decided re. a thanksgiving service on V day—whenever that may be declared. Eisenhower says it will not be until all 'pockets' in the West have been cleared up.

April 18th[551]

Mr. Edwards of Pye, Cambridge, called to ask for experimental work to produce short wave valves suitable for television links. We showed him in confidence Bell's CV 288 which Algar is about to preproduce. Also the CV 257 and the CV 273 (Warren and James's little valve). Edwards was most complimentary regarding these and all the war-produced valves at Wembley—and expressed the view that nearly all the worthwhile war valves had come from the Wembley/MOV party.

A large MOV receiving valve meeting—Davis and Howe attending for the first time in their new capacities at the Works. Cosgrove suggested having future meetings monthly.

MAP is to ask us to make 'Kix' getter pellets for the whole valve industry on this side, their negotiations with the Kemet people to start a factory having broken down. We have no answer yet to our request for their licence terms.

April 19th[552]

A scrutiny with Jesty and Jacob of the equipment programme for the necessarily elaborate testing gear for CRTs. It will pay for itself in the end but it constitutes a heavy lot of work and rather expensive. The CRT work for MAP is expected to continue into post-war and we require to be properly equipped for it.

A visit from Sutton of Bristol to discuss with me his plans for the new Services valve research and development establishment. He is planning for ten principal SOs [scientific officers] and a staff of about 150. We were very frank with each other and agreed that there would be competition between his show and the industrial concerns with development contracts. We also discussed the three blemishes in our past contacts, i.e. (*a*) the first velocity modulator and disc seals which were copied by him with Wembley's help and passed to EMI, (*b*) the Skiatron which Wembley would not exactly copy from Bristol, (*c*) the CV 57 which Sutton declared he could put right in spite of MOV and GEC very active work in progress on it. This caused much ill feeling.

Mr. Guthrie (Scottish Electricity Board) and Mr. Nielson from Witton came for discussion on salmon counting. The problem seems to depend on the reliability of underwater electrical means of guiding the fish into the proper channel where their number can be determined. Winch, Hill, Chirnside and Little joined in the talk.

April 20th

A general assault is now in progress on Berlin. The Russians are nearest but the British and Americans are up to the Elbe and have a bridgehead over it. Resistance at the north German ports Bremen and Hamburg is still vigorous.

Lunched with Heather in his office to discuss various things—finances, industrial heating and other matters.

An industrial heating meeting all afternoon, resuming the regular Advisory Board series of liaison meetings. It showed great activity in furnace and in induction heating. Equipment is coming along slowly and all are very short of staff. Mr. Gamage was present.

We get new minor contracts from time to time from the Services and there are constant consultations as to how to make ends meet.

April 21st

A meeting yesterday on 'butterfly' circuits and valves at which Wright (Bristol) reported on USA progress. Warren says that their circuit work is good but that they have no special valves being designed for the circuits but have to use standard valves.

Saw Machin about taking over 'Education Officer' duties in place of Griffiths.

April 23rd[553]

Dr. Crampton (ASE) called to talk again over a new multi-channel project; he had some strong views upon collaboration after the war.

A TVC meeting at which post-war organisation in valve affairs was discussed at length.

Lunched with Gamage who agreed to the policy of joint GEC – MOV valve research contracts subject to entire Wembley responsibility. He also agreed on the desirability of a uniform 60% overhead for factories on Research allocations, Head Office bearing the rest.

Mr. Block of DSIR (Patents) called and we had a heavy but friendly argument on the virtues of an arbitration clause, he dissenting.

April 24th[554]

Berlin is partially in Russian hands and progress everywhere is good except in the North where it is up against real opposition.

An instrument committee at Head Office and a CVD meeting on triodes at the Admiralty—very dull.

Kirk[e] of BBC came for a talk and also emphasised the importance of post-war collaboration with them and the GPO on 'television links' and the like.

April 25th[555]

A radio co-ordination meeting at Head Office—main topic the MAP choice of firms to take on development contracts for civil aviation radio.

GEC is not satisfied to be left out of the centimetre field in which they were the pioneers. GEC has been treated hardly, by doing the major pioneer work in centimetre radar (AIS, Oboe with GR later) and seeing the production work on these mainly handled elsewhere or cancelled. We are to see if we can get the outstanding microwave project led in our direction.

It is a blow to Spon Street that GR has been entirely cancelled. The progress of the war has overtaken it.

News comes that Peters has had a relapse, the old mysterious symptoms appearing again.

Had a discussion at the Labs. and with Picken later on the position of the proposed new Interservice Valve Dept. and Research Laboratory. Shall ask Brundrett to talk it over with us.

Lecture by Dr. Van der Pol at the IEE. He looks as if he also has 'been through it'.

April 26th

Stevens and Waldram had a good day at Denham film studio and found not only Sir Alexander Korda there but senior representatives of the Rank interests. This was a very satisfactory fusing of objectives and all things were fixed up as we should wish.

Willshaw attended CVD magnetron meeting. The position badly needs more rapid progress on our part. We have now got our three extra instrument makers.

Ryde reports that BTH have drawn ahead of us in their facilities for development of silicon crystal assemblies. I cannot see that we are justified in taking strength from any of our other groups, not [nor] in swelling our personnel still further on CVD even if we could get the men.

A useful Osram Advisory Committee meeting this morning where general policies were discussed.

Kelvin Lecture by Appleton—the first exposé of radio location—also I was presented by Sir Harry (President) with the Faraday Medal.

News still good. The opposition in Italy seems to be crumbling—we have Bremen—and Berlin is surrounded by the Russians. The 'Southern Stronghold' is being attacked by the Americans.

April 27th[556-557]

Discussed with Bell difficulties of fitting in the TR [TA] valve tetrode into our valve development programme. Something must be done about it. We want to see about liberating Wright from the magnetron preproduction.

We have produced a plan for a 150 ft tower for television link. It is just as well to have it prepared.

To Bradford to renew acquaintance with the radio work and personnel there. Had most interesting time with Moxom [Moxham] on the advances of his testing techniques and of automatic proving of characteristics of complicated radio assemblies.

May 1st

Many rumours over the week end of unconditional surrender by the Germans but nothing has materialised. The Russians have practically taken Berlin but it looks as though the Germans are wanting the city to have as much destruction as possible. If Nazis are not to run Germany they don't want it to be possible for any succeeding Germans to do so either. Meanwhile there is complete collapse of the enemy in Italy and the Burma fighting indicates that we shall take Rangoon before the rains. The British are doing well.

Spent yesterday (Monday) going over the various 'Mill' factories run by Salford—very impressive—and had a diamond die meeting at the Midland Hotel in the evening. All members are getting on with the electrical drilling method. Wembley has entirely stopped drilling fine dies by the tapping needle process. We can now drill a fine hole in between one to two hours.

May 2nd

Hitler is supposed to be dead but few except '*The Times*' are prepared to believe it.

Had talk with Southworth and Dunham about ways and means of designing and making the amplifiers for the GU [RP?] 60 fire control project. We called on Mr. Cheel and got it fixed up that Union Works would do so. Apparently a happy arrangement.

Met Lonsdale at Head Office to hear from him an embarrassing story of a visit to the Continent, and Admiralty contract to Power Jets for a 7000 kW gas turbine. I don't see how this will end up—but it must be left for the moment.

May 3rd

The whole German army in Italy has surrendered—about a million men. Hamburg and Rangoon have fallen and Berlin also. Everything now depends on the submarine situation and Norway and Denmark.

The opportunity of getting the TA valve designed and tried is up against a workshop bottleneck. It will move slowly I fear but we shall see if MOV can help.

Spent the morning with Heather, Enticknap and Humphreys reviewing the Research Laboratories post-war budget. This is now in good shape for scrutiny and will come up at the Consultative Board in a week's time.

Bowler is now definitely dropped and Trumper has been reduced to something of much greater simplicity. So Cherry hopes that he can get away before long to take up his teaching job in Manchester.

May 4th

Call from Wing Commander Mealing to talk over aerodrome lighting and the bad position of British industry in the matter in view of USA aggression in export trade.

Lumifax meeting. Our position in powders improves. The phosphates are well under way in manufacture and BTH and Siemens are also banking on it as a cheaper production article.

The VE day code word has gone round so that notice may be sent in advance to establishments which must make pre-arrangements for the one and a half day's holiday. We are into Denmark but the German navy elements seem inclined to fight on.

May 5th

Discussions on TA. Oliphant is back in this country and Garrard is willing to take on the design of part of the plant. We must have a meeting with Garrard, Oliphant and Akers.

May 7th—5.15 pm

The weekend has brought news of surrenders and successes and the end was obviously near. Today we hear that cease fire has been ordered by Hitler's successor to all forces—submarines are to cease operations and return to base. The great announcement of VE day has been expected momentarily but has not yet come through. If it comes through this evening we shall close for Tuesday and Wednesday.

Aug 12. 45

Valve Pre Production

G.E.C. Research Laboratories.

Total of valves delivered to April 30th 1945 - 303,848

Types preproduced - 40

Types supplied in fairly large experimental quantities - 8

Totals of major types individually:-

CV58/258	48,766
All magnetrons	36,869
NT99.	28,487
CV52.	27,539
CV8.	25,232
CV55.	23,785
CV90.	23,347
NGT8 & CV12.	20,899
VT90.	15,704
NT100.	10,902
CV114/221.	14,309

A.G.Pearce.

Fig.57 Note by A.G. Pearce on wartime preproduction of valves in the Laboratories

May 8th and 9th

Public holidays and rejoicing.

We have now passed into the post-war era and this five and a half years Diary comes to an end. At our preproduction meeting Pearce announced that we had made 295 138 valves [but see attached note (Fig. 57)].

Notes

1939

September 1st to 4th

(1) E 1046 (VT 90): the 'micropup', a small transmitting valve, initially for pulsed operation, at up to 600 MHz. It was the first in a series of evolving designs.

September 4th to 8th

(2) Bawdsey: the pre-war radar research establishment on the east coast of England which at about this time moved to Dundee and Perth (Ref. 2, pp. 22 onwards).

(3) Signal School: an Admiralty establishment at Portsmouth which worked on radar and communications.

(4) 'ZZ': derivation not known; possibly Paterson's personal code word for work associated with the highly secret radar projects.

(5) Fogg and Atkinson: Bawdsey staff. Fogg had been a member of the staff of the Laboratories in about 1936 and 1937.

(6) Waldram and Stevenson [Stephenson?]: members of the Illumination Group in the Laboratories; they were not themselves working on CRTs.

(7) Heavy Alloy: see Ref. 1, pp.345 – 347.
Elmet metal: a tungsten-based contact material, see Ref. 1, p.347, third paragraph.

September 12th

(8) CVD: see Appendix 1.

September 13th

(9) National (or Central) Register: a scheme for allocating qualified and skilled manpower to the Services, Government Establishments and industry. There are frequent references in the Diary to the effects of this on the Laboratories.

(10) C.S. (later Sir Charles) Wright: Director of Scientific Research and Experiment, Admiralty (1934–46), (Ref. 3, Vol. 7). Usually referred to as DSR.

(11) M.L. (later Sir Mark) Oliphant F.R.S., Professor of Physics, Birmingham University; Governor of South Australia, 1971–76, (see *Who's Who*, 1989).

(12) R.A. (later Sir Robert) Watson-Watt: founder of radar work in the UK; formerly Superintendent of Bawdsey; at this time Director of Communications Development, Air Ministry; became Scientific Adviser to the Air Council in February, 1940; Deputy Chairman, Radio Board of the War Cabinet, 1943–46 (Ref. 3, Vol. 7).

(13) H.J. Gough: Director of Scientific Research, War Office 1938, Director-General Scientific Research and Development, Ministry of Supply 1942–45. (Ref. 3, Vol. 6). Previously Superintendent of Engineering Department of NPL.

(14) Bragg: W.H. or W.L.; more likely W.L. (Sir Laurence), Cavendish Professor of Physics, Cambridge University, because the Cavendish Laboratory had been brought into work on radar.

(15) E.V. (later Sir Edward) Appleton: Secretary to Department of Scientific and Industrial Research. Member of the reconstituted Committee for the Scientific Study of Air Defence.

(16) W. Ure: Head of Transmitting Division, Signal School. A member and deputy chairman of the meetings between GEC and Signal School from which CVD sprang.

(17) G. Shearing: Chief Scientist, Signal School.

(18) Dr. E.T. (later Sir Edward) Paris: Head of Army Radar Group, Bawdsey, in 1937 having been seconded from ADEE. Deputy Director of Scientific Research, War Office, 1938; DDSR, MOS, 1939; Controller of Physical Research, MOS 1941; and of Signals Development, 1942.

(19) Prof. J.D. (later Sir John) Cockcroft: Jacksonian Professor of Natural Philosophy, Cambridge. Seconded to Ministry of Supply in 1939 as Deputy Director of Research. In 1941 became Chief Superintendent of ADEE (later ADRDE) and in 1943 additionally Chief Superintendent of SEE (later SRDE) (see Ref. 10).

(20) C.C.P.: C.C. Paterson.

September 15th

(21) Dr. F.S. Hawkins and Dr. H.K. Cameron: members of the staff of the Laboratories.

September 18th

(22) Max Railing: Vice-Chairman and Joint Managing Director of GEC; see Ref. 3, Vol. 4.

(23) Colonel Disney: not identified.

(24) E.W. Hall was seconded from the Laboratories to the new valve factory; in about 1949 he returned to the Stanmore Laboratory (then a branch of RL) as Laboratory Engineer.

September 19th[25]

(25) Admiral Sir James Somerville: Admiralty, Assistant to the Third Sea Lord with special reference to RDF.

September 20th

(26) B. S. Gossling: a senior member of the Laboratories' staff particularly concerned with valve research and CVD liaison. He had been on the staff of the Signal School before joining the laboratories in 1919.

(27) Shadow factory: a satellite factory set up in another location to manufacture war material, sometimes managed by staff from the parent company and sometimes by another company with their assistance.

(28) Claude General: (also known as Claudgen) a GEC subsidiary which manufactured illuminated advertising displays.

September 27th

(29) I.S.: Dr. I. Schoenberg, Director of Research of EMI, (GEC and EMI were joint owners of the M–O Valve Company, EMI having purchased their share

from Marconi. Valve research and development were undertaken in the GEC Wembley Laboratories for ultimate production at M–O V. This led to an uneasy relationship, only resolved when GEC bought the EMI share and became sole owners).

September 29th

(30) L.S. Harley: Member of the staff of DCD, Air Ministry, later Ministry of Aircraft Production, concerned with thermionic valve production. Later in British Purchasing Mission, Washington.

(31) Lewis: likely to be Dr. W.B. Lewis, a Cavendish Laboratory, Cambridge physicist who became Deputy Chief Superintendent, AMRE, later TRE.

(32) P.E. Pollard: ADEE, led the early work on GL radar sets; was in charge of Fire Control work until ADRDE moved to Malvern.

(33) J.F. Coales: Signal School, Admiralty, responsible for work on communications and radar at 600MHz and above; later F.R.S. and IEE President, 1971–72.

September 30th

(34) Sir Henry Tizard: Chairman, Committee for Scientific Study of Air Defence which initiated work on radar; Rector, Imperial College. Chairman Aeronautical Council 1933–43; Member of Council Ministry of Aircraft Production and additional member Air Council 1943–45 (Ref. 3, Vol. 5).

(35) Dr. E. Friedlander: a distinguished electrical engineer from Germany (IEE Faraday Medallist, 1978) who joined the Research Laboratories in 1938 as a consultant on power transmission systems. In 1944 he moved to GEC Witton Works.

(36) Dr. M. Pirani: inventor of the Pirani Gauge who worked at the Laboratories in the 1930s.

(37) Professor I.M. (later Sir Ian) Heilbron: Professor of Organic Chemistry, Imperial College, London; later one of three full time scientific advisers to the Minister of Production.

October 6th

(38) G. Chelioti: Manager Osram Lamp Works, later a GEC director.

October 7th

(39) Probably Dr. E.G. Hill, head of infra-red group at Admiralty Research Laboratory.

(40) General A.C. Fuller: Director of Engineering and Signals Equipment Ministry of Supply (Ref. 3, Vol. 6).

October 9th

(41) Dr. E.G. Bowen: One of the first workers on radar at Orfordness and Bawdsey; by this time particularly concerned with airborne radar (see his book: Ref.2).

October 10th

(42) RDF: an early synonym for radar.

(43) DCD: Director(ate) of Communications (primarily radar) Development, Air Ministry.

October 12th

(44) Sir Frank Gill: Chairman of Standard Telephones and Cables (STC) and President of the Institution of Electrical Engineers in 1922–23.

October 24th

(45) H.P. Wells: Manager of GEC Telephone Works, Coventry; retired in 1944.

October 30th

(46) Leighton Buzzard: an RAF signals unit.

November 10th

(47) Approved for RDF: security clearance given for them to be told details of radar developments which were a closely guarded secret.

November 13th

(48) E 1024: a metal–glass valve (also known as VT 114).

November 15th

(49) Christchurch: Air Defence Experimental Establishment (ADEE), an Army establishment.

November 17th

(50) Farnborough order on GEC Radio Works — for airborne VHF radio; see Ref. 1 p. 173.

November 20th

(51) Sir Philip Joubert de la Ferté: Assistant Chief of the Air Staff with special responsibility for application of radar in the RAF. In 1941 became Air Officer Commanding-in-Chief, Coastal Command.

November 23rd and 24th

(52) M.R.: Max Railing.

December 12th

(53) R.H. (later Sir Ralph) Fowler F.R.S.: Plummer Professor of Applied Mathematics, Cambridge; a consultant to the Laboratories; member of the Ordnance Board, 1939; UK representative in Canada, 1940 (see Ref. 2 p. 156); on return joined an Admiralty research section (Ref. 14).

December 15th

(54) A.P. Rowe: Chief Superintendent of AMRE, later called TRE, from 1938 until the end of the war (see Ref. 4).

December 22nd

(55) Dr. A.G. Touch: second-in-command to Bowen in his airborne radar team (see references to him in Ref. 2).

(56) R. Hanbury Brown: another member of Bowen's airborne radar team (see references to him in Ref. 2).

1940

January 6th

(57) F. (later Sir Frederick) Brundrett: Assistant DSR, Admiralty, formerly at Signal School (Ref. 3, Vol. 7). He was largely responsible for control of CVD on behalf of C.S. Wright. Also said to have been active on recruitment and allocation of technical staff.

January 8th and 9th

(58) T.W. Heather: a GEC Director.

January 15th

(59) J.Y.F.: J.Y. Fletcher, a GEC Director.

January 16th

(60) A.I.: Airborne Interception radar.

(61) A 25: work at Wembley on AI at 25 cm, later 10 cm, wavelength; also given the codename 25 I in the Laboratories.

(62) Wing Commander R.G. Hart: (later Air Vice-Marshal Sir Raymond Hart) formerly Signals Officer No. 11 Group, RAF; had been posted to Bawdsey to develop training of RAF personnel in radar techniques; by this time at Fighter Command.

January 17th

(63) Rooksby paper; on 'Application of X-ray techniques to industrial laboratory problems.'

January 30th

(64) R.W. Paul: a retired director of the Cambridge Instrument Company; founder of the Paul Instrument Fund.

(65) B.P.D.: B.P. Dudding, a senior colleague of Paterson, one of the founder members of the Laboratories.

February 1st

(66) C.F. (later Sir Charles) Goodeve: Reader in Physical Chemistry University College, London; Deputy Director Department of Miscellaneous Weapons Development, Admiralty, 1940 – 42; Assistant and later Deputy Controller for Research and Development, Admiralty, 1942 – 45 (Ref. 3, Vol. 7).

February 2nd

(67) B.S.G.: B.S. Gossling.

February 5th

(68) P.P. Eckersley: Chief Engineer, BBC. 1923 – 29, at this time a consultant.

(69) W.E. Bragg: W.H. or W.L; more likely W.L. from Cambridge.

February 7th

(70) H. Larnder: in charge of Operational Research Section, RAF Fighter Command, Stanmore; previously on Bawdsey staff (Ref. 4, pp. 52, 55, 83).

February 23rd

(71) Dr. C.J. Smithells: joined the Research Laboratories in 1919 and led their metallurgy work; left in 1938 to become General Manager of Lodge Plugs, Rugby.

February 28th

(72) E 960 (VT 58): a metal – glass triode valve, alternative to the Admiralty silica valve NT 57.

March 4th

(73) B. (later Sir Ben) Lockspeiser: Assistant Director of Scientific Research, Air Ministry, 1939; Deputy Director of Scientific Research, Armaments, MAP, 1941; Director of Scientific Research, MAP, 1943 (*Who's Who*, 1989).

March 6th

(74) Captain B.R. Willett: Captain, Signal School 1939; Captain Superintendent, Admiralty Signals Establishment, 1941–43; Director of Naval Air Radio, Admiralty and Ministry of Aircraft Production, 1944–45 (Ref. 3, Vol. 6).

(75) F Jervis-Smith: a member of the Laboratories' staff, in charge of valve life-testing; later Deputy-Secretary of the IEE.

March 12th

(76) Autodyne receiver: a radio receiver in which one valve doubles as local oscillator and frequency changer for incoming signals.

March 16th

(77) C – in – C: Air Marshal Sir Hugh Dowding: Commander-in-Chief, RAF Fighter Command.

March 30th

(78) W.J. Picken: Marconi's Wireless Telegraph Company; joined the CVD organisation in 1943 and became its Secretary in 1945.

April 3rd

(79) Sir George Lee: Chairman of the Radio Research Board since 1928; Engineer-in-Chief, GPO, 1932–39; President, IEE, 1937; Director of Communications Research and Development, Air Ministry, 1940 (Ref. 3, Vol. 6).

April 10th to 12th

(80) Accounts: preproduction of valves and developments for the Services were to complicate the Laboratories' previously simple accounting system. In the past they had almost entirely charged (mainly to GEC units) for research work (i.e. labour and overheads). Now they were to charge for goods made and supplied.

(81) Staff: at the outbreak of war the Laboratories' staff consisted of some 200 graduate research workers, supported by about 350 other staff. Total numbers were to rise to more than 1500 of whom about 700 were engaged on preproduction.

April 15th

(82) Lord Hirst: Chairman and Managing Director of GEC.

April 29th

(83) Professor Lindemann, later Lord Cherwell: scientific adviser to Winston Churchill.

(84) First Lord of the Admiralty: Winston Churchill.

April 30th

(85) G.C.: George Chelioti.

May 1st

(86) ITC: Informal Technical Committee between lamp manaufacturers.

May 3rd

(87) A number of men were loaned to the Royal Aircraft Establishment by GEC, including J. Reeman (see Ref. 1). At least some of them worked on gas turbine research and development.

(88) H.E. Wimperis: Director of Scientific Research, Air Ministry, 1925–37.

May 7th

(89) Consultative Board: a group of senior GEC managers as advisers to Paterson on Laboratory policy (see Ref. 1, p. 3).

May 9th

(90) A.C. Bartlett: a distinguished communications theoretician. he was an early member of the staff of the Laboratories; he left to join the Royal Aircraft Establishment, Farnborough, in the 1930s. He was transferred with other RAE staff to TRE, Swanage, in 1940. After the war he rejoined GEC at the Stanmore Laboratories (see Ref. 1, pp. 25, 26).

May 14th

(91) A.C.W. Aldis: originator of the Aldis signalling lamp used by the Royal Navy.

May 21st

(92) AMRE, later TRE: the Air Ministry/Ministry of Aircraft Production radar research establishment which had a large influx of university scientists in 1939 and 1940. It grew from the Bawdsey radar research establishment and was located at Swanage and later at Malvern.

May 23rd

(93) L.D. Goldsmith: joined the Laboratories as a chemist in 1919 and was now a senior man at GEC Battery Works at Witton, Birmingham.

(94) R.W.W. Sanderson: group leader of the battery group in the Laboratories.

May 24th

(95) G.C.M.: G.C. Marris; and D.C.E.: D.C. Espley, the two senior members of the Laboratory staff on telecommunications and television.

May 31st

(96) E.H. Rayner: formerly Superintendent of Electricity Division of the National Physical Laboratory.

June 6th

(97) J.W.R.: J.W. Ryde, probably the most eminent individual member of the staff of the Laboratories; like Paterson, he was elected a Fellow of the Royal Society.

June 7th

(98) Professor R.A. Whiddington: Professor of Physics, Leeds University; Department of Scientific Research, Admiralty, 1940–42; Deputy Director of Scientific Research, Ministry of Supply, 1942–45 (Ref. 3, Vol. 6). Secretary CVD, CVD Policy Committee, CVD Technical Committee.

(99) T.W.H.: T.W. Heather.
C.W.: Christopher Wilson, a GEC Director, who issued the original invitation to Paterson to set up the Laboratories (see Ref. 1, pp. 1, 3).

June 10th

(100) Dr. T. Walmsley: Air Ministry and MAP (Deputy DCD), worked with CVD organisation; Chairman of its RDF Committee. Was very active but little found about his background; may have been seconded from GPO.

(101) G. Bradfield: Secretary of the CVD CRT Committee; a group leader at TRE. previously at Air Ministry.

June 21st

(102) Getters: substances contained within the sealed off envelopes of evacuated valves which, when heated by high-frequency induction current, clean up residual gases.

June 25th

(103) Azides: compounds of nitrogen and hydrogen with metals, used in the manufacture of thermionic valves.

June 28th

(104) Megaw's magnetron: in retrospect this entry records a major event — the first output from the prototype operational magnetron E 1189 — see Ref. 1, pp. 125–129 (Megaw's report written in 1945 gives the date as 29th June).

(105) Norway: probably Neville Shute Norway, later to be a well known novelist. He joined the RNVR and was posted to the Admiralty 'Miscellaneous Weapons Department' in 1940.

June 29th

(106) E 1161 (E 1046B): the first enlargement of the E 1046 'micropup' triode valve.

July 2nd

(107) Sashalite: trade name for a photographic flash bulb.

July 5th

(108) Salford Electrical Instruments manufactured magnetic (nickel – iron) powders for use in telecommunication cable loading coils. The Laboratories provided scientific support when problems in manufacture were encountered.

July 9th

(109) J.B.: J. Bell; the senior man on transmitting valves in the Laboratories.

July 23rd

(110) It has not proved possible to elucidate the first two sentences. There is a later reference (25 April 1942) to the flat glass-to-metal technique used for the sextant bubble graticule.

(111) Lamplough: there is a further reference to Lamplough on August 22nd 1940. There were two Lamploughs at RAE at about this time: S.L. Lamplough (field of work unknown — may have been an administrator) and F.E. Lamplough (field: applied optics). The latter seems more likely.

July 24th

(112) Sir F.E. Smith: Sir Frank Smith, Controller of Telecommunications Equipment, Ministry of Aircraft Production; (also Director of Instrument Production MOS 1939 – 42, Controller of Bearings Production MOS 1939 – 43). Scientific Advisory Council MOS 1941 – 47 (see Ref. 3, Vol. 6).

August 6th

(113) Professor C.D. (later Sir Charles) Ellis: Professor of Physics, Kings College London; seconded to MOS; took over direction of the group on proximity fuses in January 1942; Scientific Adviser to the Army Council 1943 – 46; member of Advisory Council on Scientific Research and Technical Development, Ministry of Supply, 1943 – 46 (Ref. 3, Vol. 7).

(114) R. Le R.: R. Le Rossignol, the senior man on the Laboratories' staff concerned with the design of valves.

August 7th

(115) All from EMI.

August 9th and 10th

(116) Acorn: a type of valve for use up to 300 MHz, made by RCA (USA), shaped like an acorn, with connecting pins in a ring round its circumference.

August 12th

(117) P.I. Dee: a physicist from the Cavendish Laboratory, Cambridge, who led the TRE work on airborne radar (see references to his contributions in Ref. 4).

(118) H.W.B. Skinner: a physicist from Bristol University who joined TRE and made major contributions to microwave radar (see Ref. 3, Vol. 5, and Ref. 4). In 1940 he was in charge of microwave development at TRE.

August 16th

(119) The Tizard Mission which took a cavity magnetron and other developments to America (see Ref. 2, chapters 10 and 11).

August 29th

(120) G.M. Wright: Signal School; a pre-1914 – 18 War associate of Marconi, RNVR during that war, at MWT Co between the wars, joined Admiralty in 1939 as a Senior Scientific Officer; rose to be Chief Scientist, ASE; returned to Marconi after the war and became Chief Engineer.

August 30th

(121) Dr. F.A. Vick: Ministry of Supply, concerned with proximity fuse development; pre-war lecturer in physics at University College London; later Professor of Physics at Keele University.

(122) Proximity device (ROF): a remotely operated fuse being developed by Pye, on behalf of Ministry of Supply, for which GEC were to develop valves and batteries; it was superseded by 'P.F.', a proximity fuse (see Ref. 10).

September 3rd

(123) Major Sandys: possibly Duncan Sandys, later a Cabinet Minister.

(124) Air Marshal R.H. (later Sir Richard) Peck: Assistant Chief of the Air Staff; later Deputy Chief of the Air Staff.

September 18th

(125) USW contract: a programme of research and development for the Admiralty, started in 1936, on the production, propagation, and reception of radio waves at wavelengths below 6 cm.

September 20th

(126) Might be Oliver Lyttleton, later Lord Chandos, but he appears to have been President of the Board of Trade in 1940–41.

September 27th

(127) This appears to be the first initiation into major projects on Radio Countermeasures which the Laboratories were to undertake throughout the war (see Ref. 1, pp. 377–382). The work had the code name 'A.B.' within the Laboratories.

(128) Air Commodore Lywood: Deputy Director of Signals, Air Ministry; (Director 1941; AOC 26 Group 1942: see Ref. 3, Vol. 5).

October 17th

(129) E 1166 (also known as E 1046D): the second enlargement of the 'micropup' triode valve.

October 21st

(130) Group Captain Blandy: Head of the RAF Y Service which intercepted German radio signals; Deputy Director of Intelligence, RAF.

October 26th

(131) E.B. Moullin: Reader in Electrical Engineering at Oxford University, seconded to Signal School in 1940; later joined the Research Laboratories of Metropolitan Vickers at Trafford Park for the remainder of the war; President of the IEE 1952–53.

October 29th

(132) Haslemere: Headquarters of the Admiralty Signal Establishment (formerly Signal School) from late 1940 onwards.

October 31st

(133) Captain A. Frazer-Nash (who had designed, built and driven racing and 'grand touring' motor cars) was principal and Mr. Whittaker was Chief Engineer of Nash & Thompson, an engineering company which produced gun turrets for aircraft; they were to work with the Laboratories on infra-red detection systems and to provide the scanners, to TRE requirements, for centimetric AI.

November 8th

(134) Wing Commander Helmore: a freelance project instigator in the RAF, well known as the commentator at the pre-war RAF displays at Hendon. He developed a close working relationship with L.B.W. Jolley and his colleagues in the Laboratories on a number of projects.

November 11th

(135) F.O. Brown: not traced.

November 13th

(136) Witton Kramer Works, GEC Witton: founded by J.B. Kramer (who was still active); specialised in applications of electromagnets.

November 18th

(137) T.L. Eckersley F.R.S.: Head of the Propagation Section of the Marconi Research Laboratory at Baddow. The section, led by Eckersley, was taken over by the Air Ministry in 1940, and in 1942 Eckersley became Chief Scientific Adviser to the Inter-Service Ionosphere Bureau established at Baddow.

November 25th

(138) Clifford: a member of the workshop staff of the Laboratories; an officer in the Air Training Corps and a keen aircraft spotter.

November 27th

(139) IEE National Register Committee: probably the IEE National Service Committee (see Ref. 5, p. 149).

December 3rd

(140) Sir Noel Ashbridge: Controller of Engineering, BBC; IEE President, 1941 – 42.

December 5th

(141) Lord Beaverbrook: Minister of Aircraft Production at that time.

December 9th

(142) Dr. Bruyne: Dr. N.A. de Bruyne, Managing Director of Aero Research. He had worked at the Laboratories for a short time in the early 1920s.

1941

January 1st

(143) A.H. Reeves: formerly of STC, at this time at TRE; the inventor of pulse code modulation (PCM) and of the precision bombing system, Oboe.

(144) E 1231 (later CV 52): a small concentric triode valve for operation above 600 MHz.

January 8th

(145) This is the first reference in the Diary to the problems of 'mode change' in the original cavity magnetrons. These were to continue until resolved by the introduction, by Dr. Sayers of the Birmingham University team, of 'strapping' of appropriate segments.

January 23rd

(146) E 1061: not identified; it could be a typing error for E 1161.

January 25th

(147) Rooksby and Jackson paper: 'Some applications of radiography to the electrical industry'.

January 31st

(148) Dr. D.N. Truscott had joined Government service, having previously been in industry. He worked at the Air Ministry, Harrogate, later MAP, and was concerned with valve production.

(149) Shaw, Oldham, Lancashire: the shadow factory for valve production previously mentioned.

February 3rd

(150) E 1155 (NT 100, CV 44): a tetrode radar modulator valve.

February 5th

(151) The development was almost certainly of proximity fuses.

(152) DRP: Director of Radio Production, MAP.

February 19th

(153) Thyratron: originally a trade name for a mercury-vapour filled triode valve; later applied to any gas-filled triode.

February 27th

(154) IR: infra-red (the first mention in the diary that the Nash & Thompson device worked at infra-red).

(155) O.W.H.: O.W. (later Sir Olliver) Humphreys, who was to take an increasing role in the management of the Laboratories as the war progressed.

March 5th

(156) H.T. Cobden Turner: Managing Director of Salford Electrical Instruments, a GEC subsidiary.

March 11th

(157) TVC: Interservices Technical Valve Committee.

March 15th

(158) Air Chief Marshal Sir John Salmond; Director of Armament Production, MAP.

(159) Air Marshal Sir Roderic Hill: Director General of Research and Development, MAP, 1940–41.

(160) Balfour: A Balfour, MP?

March 19th

(161) ADEE: Air Defence Experimental Establishment (later Air Defence Research and Development Establishment)—a research and development establishment primarily for the Army.

(162) Dr. D.H. Black: Superintendent ADEE, Christchurch, in 1939, having previously been in industry; he was a UK liaison officer in Washington.

March 22nd

(163) A copy of Paterson's reply to Rowe's letter dated 20th March has been found and included, since it is one of the very few Paterson letters from this period which has been traced.

PERSONAL
A.P. Rowe, Esq., 24th March, 1941.
T.R.E. Worth Matravers,
Swanage.

My dear Rowe,

How nice of you to write me a line of mutual congratulation. I nearly did the same to you when Marris returned here with the news that AI had gone up in the air—But he was rather conscious of the long road yet to be travelled and suggested to me that such a letter from this end might give the impression that we thought the time had come for the shouting! But, remembering the talks and heart searchings we had together, I do feel very thankful indeed to all concerned at both our establishments that rivalries and all similar instincts have been allowed to take only a very secondary place in the joint effort to produce what was needed.

Your sentiment, hoping that the present high degree of co-operation will continue, will be loyally acted upon by us here. Long may it last.

Though AI still has a long journey before being a practical service equipment, we have the consolation that T.R.E. work in applying its electrical side to Naval uses has already borne fruit.

I still have your note of February 18th before me asking if I am coming down for another talk. I want to – for there is a good deal of interest to discuss. I have to be in Bournemouth (Parkstone) on April 9th, and had hoped to look in at Signal School and A.D.E.E. on the 10th. If I came on to Swanage would you be available on Good Friday or the day previous (Thursday)?

We have a large number of investigations and developments here mainly for the three Ministries and covering a very varied range – from drilling 10 μ holes in diamond dies to equipping planes with Heaven knows what. At the moment every one of them seems to be going exceptionally well! By the time I see you they will probably be doing the reverse.

In valves, without interfering with our responsibilities for evolving new types required for C.V.D. we have tackled the job of getting into production in the Laboratories with from 6 to 8 of the new types on the basis of 600 to 1000 of each type, and all with unskilled labour. We are nearly there, and I am feeling inordinately proud of accomplishing this in a very short time and without the assistance either of the Professional Productionist or of the exponents of Scientific Management.

I wil try to muster a decent measure of humility—before coming to you.

What a war, and how magnificent that we are able to have a real share in the fight—thanks partly to you!!

Yours ever,

March 26th

(164) E 1190 (CV 55): milli-micropup valve, with an oxide coated cathode, for CW operation.

(165) E1209 (CV7): a small gas-filled triode valve.

March 27th

(166) H.H.: Lord (Hugo) Hirst.

April 1st

(167) Carrol Wilson: see references to him in Ref. 2.

April 4th

(168) Spon Street: the GEC Radio and Television Works in Coventry; it was an offshoot from the much larger GEC Telephone Works at Stoke, Coventry. During the war it had satellite factories, particularly in Bradford.

April 7th

(169) W.C.M. Couch: Assistant Director of Engineering, Admiralty, 1937 – 41; Deputy Director, 1941.

April 10th

(170) GL: Gun Laying, i.e. anti-aircraft gunnery radar.

April 15th

(171) Professor K. Bainbridge: a senior member of the staff of the MIT Radiation Laboratory (see Ref. 2).

April 23rd

(172) Sir Andrew Duncan: Minister of Supply (Ref. 3, Vol. 5).

April 25th

(173) George Wilson: pre-war head of the Illumination Group in the Laboratories who left on the outbreak of war, presumably for conscientious reasons.

April 28th

(174) This suggests that the Frazer – Nash IR system was an active one.

(175) J.R. Beard, C.B.E.: Senior Partner in Merz and McLellan, IEE President, 1940 – 41. Member of the Engineering Advisory Committee of the War Cabinet, 1941 – 45.

May 1st

(176) Bradford: a satellite works of GEC Coventry set up for production of radio equipment for the Services.

May 2nd

(177) Lord Falmouth: Member of Advisory Council DSIR 1936–41, Fuel Research Board 1936–41, Head of a Division in Research and Engineering Department Ministry of Home Security 1941–45. (Ref. 3, Vol. 6).

(178) Sir Arthur Fleming C.B.E.: Director of Research and Education, Metropolitan Vickers; IEE President 1938–39.

May 20th

(179) E 1232 (CV 92): a triode for 600 MHz, 80–100 kW pulse power output.

(180) E 1248 (CV 8): a diode switch for 200 MHz.

May 22nd

(181) N.L.H.: N L Harris, a member of the staff of the Laboratories.

May 27th

(182) Pye: a radio and television manufacturing company in Cambridge.

May 29th

(183) Dr. Kahn: from GEC Witton.

June 5th

(184) A.G.P.: A.G. Pearce, a member of the staff of the Laboratories who took a leading role in the production at the Laboratories of valves developed there.

(185) M.B.: M. Benjamin, an eminent younger member of the staff of the Laboratories who worked on vacuum physics, thermionic emission and receiving valves.

June 7th

(186) Bankes [Banks?]: under A.J. Young was one of the first Marconi people to work on valves, including magnetrons. Later he moved to English Electric Valve Company in charge of research.

June 9th

(187) NC 11: a cathode-ray tube for naval use.

(188) T & R working: the connection of a radar transmitter and a radar receiver to the same aerial.

June 23rd

(189) W.A. Stanier: President of the Institution of Mechanical Engineers, 1941; Chief Engineer, LMS Railway. In September 1942, he became one of three full time scientific advisers to the Minister of Production.

June 25th

(190) NT 93 (E 1161) and NT 97: triode valves in the micropup series.

June 27th

(191) Boart: a finely, randomly, crystallised form of diamond, used as a cutting agent (see Ref. 1, p. 393).

June 30th

(192) NC 12: a cathode-ray tube for naval use.

July 2nd

(193) Dr. P. Dunsheath O.B.E.: Director and Chief Engineer of W.T. Henley Ltd., cable manufacturers; Chairman of the IEE Committee on the Organisation of Post-War Electrical Research; IEE President, 1945 – 46.

July 8th

(194) E 1239: a high power derivative of the micropup.

July 12th

(195) E 1191 (CV 12): a gas-filled triode radar modulator valve.

July 28th

(196) A.R.: Dr. A.H. (later Sir Harry) Railing; Director and General Manager, GEC, at this time.

(197) L.G.: Leslie Gamage; Director and General Manager, GEC, at this time.

August 1st

(198) Air Commodore Leedham: Director of Radio Production MAP 1939 – 42, Director of Radio Research and Development MAP 1943 – 45 (Ref. 3, Vol. 4).

(199) In April 1941 the Government had set up an Engineering Advisory Committee under Lord Hankey. Paterson was one of four members drawn from the IEE out of the Committee's total membership of nine (see Appendix 5 and Ref. 5, p. 148).

August 5th

(200) An internal meeting, which Paterson usually attended, was held every Monday morning to review progress and problems on production of valves and other devices in the Laboratories.

August 8th

(201) Sir John Madsen: Professor of Electrical Engineering, Sydney University; Member of Council for Scientific and Industrial Research, Australia (Ref. 3, Vol. 6).

August 14th

(202) W.H.W.: W.H. Woodward, a GEC – Osram manager.

August 18th

(203) NT 98 (E 1189): A 10 cm cavity magnetron, primarily for naval use.

August 22nd

(204) ELSIE (SLC): Searchlight Control Radar.

August 25th

(205) J.S.V. Walton: Admiralty, a member of the CVD staff; later in British Joint Services Mission in Washington.

August 26th

(206) M.M. MacQueen: Manager of Radio and Television Department, GEC Head Office.

(207) R.L. Prain: Controller, Diamond Die (and Machine Tool) Control, Ministry of Supply (based in Rugby); apparently seconded from a company, Selection Trust.

September 1st

(208) W.A.S. Butement: an Australian, was an early member of the Bawdsey Army radar team. He led the team which built the prototype CD (200MHz) radar. Earlier he had been a scientist in the Army Signals Establishment.

(209) A W Ross: Admiralty Signal School, who had been engaged on a programme of antenna and propagation research.

(210) Dr. A.O. Rankine: attached at this time to Dr. Walmsley. Since 1937 he had been Chief Physicist of the Anglo-Iranian Oil Company, before which he had been a Professor of Physics at Imperial College, London. He was seconded for special work for the Admiralty and MAP and with the Petroleum Warfare Department. Later he was in charge of an attempt to find a method of fog clearance on airfields. A system, FIDO, which burned 100 000 gallons of petrol an hour, was used in the latter part of the war. The Laboratories explored an alternative scheme.

September 4th

(211) E 1267: a cavity magnetron for 10.7 cm wavelength, 100 – 200 kW pulse output.

(212) Percy Good: Deputy Director, British Standards Institution (BSI); he became Director in 1942 and was President of the IEE in 1947 – 48

(213) J R Beard: see note for April 28th 1941.

(214) T G N Haldane: another power engineer, a partner in Merz & McLellan; President of the IEE 1948 – 49.

(It is not apparent what brought these three men together to the Laboratories)

October 2nd[215]

(215) R. W. Sutton: an Admiralty scientist who had joined from industry just before the war. He was engaged on valve and other device development. At this time he was in charge of the Admiralty Signal Establishment Extension at Bristol University. He was ultimately Superintendent, Services Electronics Research Laboratory, Baldock. (*Who's Who*, 1989).

October 6th

(216) W.H. Peters: Manager, GEC Radio Works, Coventry. This was an offshoot of GEC Telephone Works, set up for production of radio and television receivers, and now for electronic equipment for the Services.

October 9th

(217) Farnborough: the Royal Aircraft Establishment.

(218) Mr W. (later Sir William) Farren: Superintendent of RAE.

October 10th

(219) Elton: A member of Dr. Truscott's staff on valve production.

(220) J.T.C. Moore-Brabazon (later Lord Brabazon), Minister of Aircraft Production.

October 15th

(221) Ryde's report was indeed to prove a classic (see Ref. 1, p. 382).

October 20th

(222) Chelioti *et al.*: senior staff of GEC – Osram Lamp Works.

October 25th

(223) E 1273 (CV 58): an axial diode valve of which, with its variant CV 258, nearly 50 000 had been made in the Laboratories by the end of the war.

(224) E 1271 (CV 57): a tetrode radar modulator valve.

October 27th

(225) Skiatron: a cathode-ray tube in which the electron beam varies the transparency of the screen, which is illuminated from behind so that its image is projected onto a screen.

November 6th

(226) Polgreen: Chief Engineer of Salford Electrical Instruments (SEI), a GEC subsidiary.

November 11th

(227) E 1148 (VR 135): a small glass triode valve.

November 17th

(228) Advisory Board: a GEC co-ordinating committee which regularly reviewed the technical programmes of units of the Company, including their relationships with the Laboratories (see Ref. 1, p. 18).

November 18th

(229) Levy and West: the principals in Levy – West Laboratories, experienced in the manufacture of phosphors, with whom GEC formed a joint company, Lumifax (see Ref. 1, p.89).

November 24th

(230) A twin-dish helical scanner, developed by the Laboratories for AI, was converted for the first experiments on the bomber navigational aid, H_2S, and the first flight trials were carried out by Dr. B.J. O'Kane, a member of the Laboratories' staff, who had been working at Christchurch on AI and who was then seconded to TRE for the rest of the war.

November 25th

(231) E 1001: a small glass triode for 300 MHz.

November 26th

(232) Oboe: a precision bombing system, invented at TRE; the Laboratories were to work on Mark II, a microwave version.

November 28th

(233) Hankey Committee: Lord Hankey was Chairman of the Government's Scientific Advisory Committee as well as the Engineering Advisory Committee referred to in note 199 to August 1st, 1941. He was Minister without Porfolio in the War Cabinet in 1939–40; Chancellor of the Duchy of Lancaster, 1940–41; Paymaster-General, 1941–42. He was also Chairman of the Inter-Departmental Committee on Further Education and Training, and Chairman of the Committee on Television in 1943 (Ref. 3, Vol. 6).

December 8th

(234) The Dunsheath memorandum was from an IEE Committee (see Ref. 5, p. 152).

December 15th

(235) Dr. Milner: this entry remains a mystery.

December 18th

(236) R.E. Robinson: a senior manager in GEC Telephone Works, Coventry; later General Manager of this Works and a GEC Director.

December 29th

(237) Dr. P.B. Moon: Birmingham University (Oliphant's group) (*Who's Who* 1989)

December 30th

(238) Ackerley: Chief Lighting Engineer, GEC Head Office: a President of the Illuminating Engineering Society.

1942

January 1st

(239) Walsh: a member of the staff of the NPL; a world authority on photometry.

(240) Sir Arthur Fleming: see note for May 2nd 1940; a member of the Government Engineering Advisory Committee.

January 14th

(241) M.J. Railing had been Vice-Chairman and Joint Managing Director of GEC. On his death Dr. A.H. Railing became Vice-Chairman and joint General Manager with Mr. L.C. Gamage.

January 19th

(242) Lord Hankey was Chairman of a Technical Personnel Committee set up by the Ministry of Labour (Ref. 5, p. 149).

January 26th

(243) Strachan was the GEC Constructional Engineer; his office was on the GEC Wembley estate, next to the Laboratories.

February 4th

(244) J.M. Kennedy: a partner in Kennedy & Donkin, consulting engineers; IEE President 1935–36.

February 10th

(245) Central Register: see Ref. 5, p. 149.

February 12th

(246) Dr. B.G. Dickins: Head of Operational Research Section, Headquarters Bomber Command (see *Who's Who*, 1989).

February 17th

(247) Dr. Randall had been a pre-war member of the staff of the Laboratories working on fluorescent lamps and powders (see Ref. 1, pp. 49 and 87–89).

February 20th

(248) E. Yeoman Robinson: Manager of Cosmos Valve Company, a subsidiary of AEI.

(249) E 1335 (CV61): a midget, high gain, pentode valve.

February 24th

(250) West Drayton: Admiralty Research Laboratory (ARL).

(251) N.J. McAinsh: Chief Engineer, GEC Radio Works, Coventry.

February 28th

(252) E 1210: an S band glass envelope magnetron.

(253) Sir Archibald Sinclair: Secretary of State for Air.

(254) Sir Charles Portal: Chief of the Air Staff.

March 6th

(255) Chesham Hivac Factory; a factory which made thermionic valves, particularly subminiature types for fuses. Hivac was a small independent company.

March 7th

(256) Pitkin: a member of the staff of the Laboratories, seconded to Salford Electrical Instruments.

March 12th

(257) Presumably between Head Office Departments and Coventry and Bradford Works.

March 17th

(258) The raid which captured parts of a German radar from a station at Bruneval on the French coast (see Ref. 6, pp.305–321 etc.).

March 23rd

(259) Dr. C.G. (later Sir Charles) Darwin: Director of NPL since 1938; seconded to Washington in 1941 as first Director of what became the British Central Scientific Office; in 1942 became first Scientific Adviser to the War Office.

March 26th

(260) Refers to Engineering Advisory Committee.

March 27th

(261) Rosen: Managing Director of Ultra Electronics; member of Radio Manufacturers Association, apparently concerned with manpower allocation.

(262) Smythe: a member of the staff of the Illumination Group in the Laboratories.

April 4th

(263) Miss Parslow also ran the Laboratories' canteen.

April 24th

(264) Kersey: the Laboratories' cashier.

April 29th

(265) H.G. Hughes: Head of the valve division in the Admiralty Signal Establishment; secretary for some years of the CVD – GEC liaison committee.

April 30th

(266) F.W.: F. Winstanley, a GEC Director with responsibilities in the lighting field.

May 1st

(267) Research Services: the first mention of a series of meetings between Paterson and other senior GEC managers to discuss work in the Laboratories. 'Services' appears to have the connotation of work for the Armed Services.

(268) Viscount Margesson is first mentioned as a GEC Director in the GEC annual report for the year ending 31st March 1942. See also entry for 12th March 1942.

May 4th

(269) Sir William Bragg had been Director of the Royal Institution since 1923 and had died on 12th March 1942 (see Ref. 3, Vol. 4).

May 5th

(270) Mackenzie: a senior Laboratory electrician with long service.

May 9th

(271) Premiums: prizes awarded each year by the Institution of Electrical Engineers for the best published papers in the year.

May 13th

(272) Enticknap: became joint GEC Company Secretary in the year ending 31st March 1943.

(273) '4X': not elucidated. The project was subsequently known as GR and was an integration of radar and predictor.

(274) Southworth: a senior member of the staff of GEC Witton.

May 19th

(275) Y Group: an RAF unit which monitored enemy radio transmissions.

(276) Transmission trials: the 'USW' contract held by the Laboratories, from the Admiralty, (out of which indirectly grew CVD) was an attempt to provide communication between ships at frequencies which would not go over the horizon. It was to continue through and after the war at higher and higher frequencies.

(277) Castle was a senior man in the GEC traffic signalling company located on the GEC Wembley estate. Chapple was a senior member of the GEC Patent Department attached to the Laboratories.

May 21st

(278) Mandrel: a radio noise generating jammer for use against enemy radars (see Ref. 1, pp. 380, 381).

May 23rd

(279) RX: apparently a 25 cm radar project.

May 26th

(280) H.E. (later Sir Harry) Ricardo: Chairman and Technical Director of Ricardo and Co. Ltd., consulting engineers. President of the Institution of Mechanical Engineers 1944–45.

May 30th

(281) 'Young' Stewart: possibly because he was an ex-member of the staff of the Laboratories!

June 3rd

(282) Miss (later Dame) Caroline Haslitt: a member of many organisations particularly those concerned with electricity and engineering. (see Ref. 3, Vol. 5).

(283) J.A. Ratcliffe: a radio physicist from the Cavendish Laboratory, Cambridge, who had senior responsibility in TRE for introduction into Service use and training on newly developed equipment (see Ref. 5, p. 235).

June 4th

(284) Would be W.L. Bragg.

June 6th

(285) US naval victory: presumably the battle of June 3rd in the Midway Islands.

(286) Following the British raid on the Bruneval radar station on the French coast, TRE and ADRDE were moved hastily from Swanage and Christchurch to Malvern.

June 9th

(287) T. Woodget: the foreman electrician who had joined the Laboratories in 1921.

June 16th

(288) Lord Hirst's proposal to provide funds for a charitable trust.

June 17th

(289) SALI: Sound and Light Indicator?; known in the Laboratories as 'Flash-Bang'

(290) E 1368 (CV 90): a disc seal triode used as a receiver local oscillator. The 'Lighthouse' valve was an American development using stacked glass and metal sections (hence its name), said to require assembly in a controlled atmosphere.

June 26th

(291) Professor C.W. (later Sir Charles) Oatley F.R.S. came into Government service in 1939 from the Department of Physics, Kings College, London. He became deputy to Cockcroft at ADRDE and later took charge of its scientific work with the rank of Acting Superintendent. He was a member of the CVD Technical Committee and the GEC-CVD. Liaison Committee. Throughout the war he was a member, and later Chairman, of the CRT Committee.

June 29th

(292) Mr. F.G. Miles was Managing Director (formerly Technical Director) of Phillips & Powis, of Woodley, near Reading. His wife was a senior member of their technical staff; both were shareholders in the company, which developed and manufactured the 'Miles' range of aircraft. Their association with the company is described on pp. 21 to 44 of 'Miles Aircraft since 1925' by Don L. Brown (Putman, 1970) which also contains on pp. 178 – 181 descriptions of their pilotless bomber and auto-pilot projects (see also entry in this Diary for November 9th, 1942).

July 2nd

(293) Dr. and Mrs. Wooster: Cambridge crystallographers.

July 4th

(294) CVD reports: copies of Nos. 6 onwards are retained in the GEC Hirst Research Centre registry.

July 14th

(295) Sir Robert Robertson (see Ref. 3, Vol. 4).

July 17th

(296) Heller's altimeter (see Ref. 1, p. 176).

July 20th

(297) J.S. Highfield: President of the IEE, 1921–22.

July 21st

(298) Z Dept.: GEC Head Office Department dealing with street lighting, airfield lighting and film studio lighting.

July 23rd

(299) Professor A.V. Hill: a distinguished physiologist who had been a member of the Tizard Committee for the Scientific Study of Air Defence.

August 7th

(300) Helmore's confidential work on Staines Reservoir was probably early experiments on his idea for a large guided torpedo, the first trials using a motor-boat.

August 14th

(301) Sir Henry Dale: member of the Scientific Advisory Council of the War Cabinet, 1940–47; Chairman 1942–47; President of the Royal Society,

1940–45: Director of the Laboratories of the Royal Institution, 1942–46 (Ref. 3, Vol. 6).

August 18th

(302) Professor R.G. Emeleus: Professor of Electrical Engineering, Queen's University, Belfast; undertook research on gas discharge phenomena.

August 19th

(303) May be M. Solomon, a GEC Director throughout the war period; alternatively could be H.G. Solomon, a past editor of *Science Abstracts* who retired in 1937.

August 21st

(304) Canal navigation job: a scheme for navigating at night on a ship canal by the use of fluorescent poles on the banks of the canal, illuminated by an ultra-violet lamp on the ship.

August 25th

(305) Nicholson: not traced as a member of staff of the Laboratories; possibly a member of the staff of Fraser & Chalmers.

August 27th

(306) Sir Harold Hartley: Chairman, Fuel Research Board; Director, Gas Light & Coke Company.

(307) Sir J. Greener: not traced.

September 3rd

(308) Sir Robert (later Lord) Renwick: Controller of Communications, Air Ministry, and Controller of Communications Equipment, MAP, 1942–45.

(309) Dr. de Laszlo: on staff of Sir Robert Renwick.

September 4th

(310) Canal lighting: see note (304) for August 21st, 1942.

September 5th

(311) Inter-Service Research Bureau: apparently concerned with development and production of equipment for special operations.

(312) Jolley's job with the Bureau appears to have been on a one-man submarine.

September 8th

(313) Professor Willis Jackson (later Lord Jackson of Burnley); Professor of Electrotechnics, Manchester University, 1938–46 (see Ref. 3, Vol. 6).

(314) Projector: appears to be a code name for an Army communication transmitter.

September 19th

(315) CV 14: type not identified; it could be a typing error for CV 114, an X band T–R cell.

(316) Bright annealing: see Ref. 1, p. 344.

September 23rd

(317) Professor G.P. (later Sir George) Thomson: Professor of Physics, Imperial College of Science; Chairman and Secretary of the first British committee on atomic energy, the Maud Committee, 1940–41; later Scientific Liaison Officer, Ottawa; Deputy Chairman of the Radio Board and Scientific Adviser to the Air Ministry, 1943–44 (see Ref. 3, Vol. 7 and Ref. 14).

September 24th–26th

(318) Flare job: for more details of Wembley work on flares see Ref. 1, p. 103.

September 28th

(319) CV 93: type not identified; NT 93?

(320) CV 97: type not identified; NT 97?

September 30th

(321) Mr. Allcock: from Callenders Cables; he was also to be Chairman of the Inter-Services Valve Production Committee.

October 1st

(322) Dudding and Jennett: see Ref. 1, pp. 398 – 400.

October 7th

(323) IEE post-war planning research committee: see Ref. 5, p. 152.

October 8th

(324) Jeffcock: not traced.

(325) Fog dispersal on runways: the GEC proposal involved air jets at intervals on each side of a runway, those on one side being heated. An alternative scheme involved the burning of petrol.

October 9th

(326) Air Vice Marshal Linnell: not identified.

October 10th

(327) Sapphires: see Ref. 1, pp. 391, 392.

October 12th

(328) Mr. Geoffrey Lloyd: Minister of Petroleum Warfare.

(329) Sir Donald Banks: Director-General Petroleum Warfare Department, 1940 – 45 (see Ref. 3, Vol. 7).

(330) Mr. Simon Marks (knighted in 1944 and later Lord Marks of Broughton): Chairman of Marks & Spencer who among other public duties was an honorary adviser to the Minister of Petroleum Warfare.

October 16th

(331) Admiral Grant: Chairman and Managing Director of Marconi's Wireless Telegraph Co. Ltd.

October 22nd

(332) Air Marshal Sir Wilfred Freeman: had been Air Member for Research and Development and later Vice Chief of the Air Staff.

October 26th

(333) Oboe: Oboe Mark 1 on a 1.5 metres wavelength.

October 27th

(334) CV 79 and CV 89: glass magnetrons for communications at 6 cm wavelength.

November 2nd

(335) Pacific: peaceful or the geographical area? More likely to be the former.

November 23rd

(336) Sir Cecil Kisch: Director General Petroleum Department 1939–42: Special duties Petroleum Warfare Department 1943–46; seconded from the India office where he had been Assistant and then Under Secretary of State (see Ref. 3, Vol. 6).

November 25th

(337) Wireless Set No. 10: an Army communications system used during the invasion of Europe (see Ref. 7).

(338) 300-channel set: a VHF multi-channel radio transmitter – receiver for use in aircraft (see Ref. 1, pp. 174–175).

November 28th

(339) Dr. T.R. Merton: Treasurer of the Royal Society; one of three scientific advisers to the Minister of Production.

December 8th

(340) Professor Patrick (later Lord) Blackett: Langworthy Professor of Physics, University of Manchester; a member of the original Tizard Committee; in 1941 was Scientific Adviser to Commander in Chief Anti-Aircraft Command; later he was Head of Operational Research in Coastal Command and then Scientific Adviser to the C-in-C Coastal Command before becoming Director of Operational Research at the Admiralty in December 1941.

December 11th

(341) Dr. Irving Langmuir: a distinguished senior member of the staff of the laboratories of General Electric (USA).

December 14th

(342) Mr. Graham (later Sir Graham) Cunningham: Chief Executive and Controller General of Munitions Production Ministry of Supply 1941 – 46 (Ref. 3, Vol. 7).

(343) Leigh Light: a light fitted to aircraft to assist in seeking enemy submarines (see Ref. 2, p. 113).

December 22nd

(344) Dr. Suits: Research Laboratories, General Electric (USA).

(345) Professor G.I. (later Sir George) Taylor FRS: Yarrow Research Professor, Cambridge University; an authority on the mechanics of fluids and solids and their applications. A member of the Home Office Civil Defence Committee and the Explosives Committee of the MOS (Ref. 13).

(346) Air Marshal Sir Ralph Sorley: Assistant Chief of the Air Staff (Technical Requirements) 1941; Controller of Research and Development Ministry of Aircraft Production 1943 – 45 (Ref. 3, Vol. 7).

1943

January 22nd

(347) Officers from RAF Technical Training Command, Reading. This visit and subsequent collaboration were to lead to GEC involvement in torpedo guidance over many years (see Ref. 1, pp. 390 – 391).

(348) HMS *Vernon*: the Admiralty Mining and Torpedo Establishment.

January 25th

(349) Wing Commander St. John was responsible for getting Squadron Leader Cropper, Flight Lieutenant (later Squadron Leader) Griffith and other RAF personnel seconded to Wembley to work with Research Laboratories' staff on torpedo guidance.

January 26th

(350) Lord Selborne: Minister of Economic Warfare 1942–45 (Ref. 3, Vol. 7).

January 29th

(351) Rangaby: of Hivac Valve Company, later to work at SERL.

January 30th

(352) Dr. D.H. Priest: TRE; took part in the raid on the Bruneval radar station.

February 8th

(353) Air Marshal Babington: Technical Training Command.

(354) Professor J.A. (later Sir John) Carroll: Professor of Natural Philosophy University of Aberdeen; later Deputy for Research and Development to the Controller of the Navy and then Scientific Adviser to the Board of Admiralty and Chief Scientist Royal Navy (Ref. 3, Vol. 7).

February 11th

(355) Sapphire boules: see Ref. 1, pp. 391–392.

February 12th

(356) Consultative Board: see Ref. 1, pp. 34–35.

February 13th

(357) Induction heating: see Ref. 1, pp. 318–322.

February 16th

(358) Griffith: Flight Lieutenant G.W.S. Griffith who was to join the Laboratory staff at the end of the war but to work on telecommunications.

February 17th

(359) 'RG': not identified.

February 19th

(360) Colonel W.E. Phillips: see R. Watson Watt: 'Three Steps to Victory', (Odhams Press, 1957), p. 380.

February 22nd

(361) Claudgen: a GEC subsidiary principally concerned pre-war with manufacture and sale of illuminated advertising signs. It undertook production for the Laboratories on a number of projects.

(362) du Parc: du Parq? (there was a Lord Justice of Appeal du Parq from 1938 to 1946).

February 27th

(363) BM 313: A Birmingham University magnetron for 3 cm operation.

(364) CV 57: there was nevertheless a serious problem in the field with these valves due to the conditions in operational equipment, and until a solution was found an emergency delivery service of replacement valves had to be set up.

March 1st

(365) Sir Alexander Ramsay: Director, Engineering and Allied Employers National Federation (Ref. 3, Vol. 6).

(366) Ben Smith: Parliamentary Secretary, Ministry of Aircraft production, 1942 (Ref. 3, Vol. 6).

March 5th

(367) E.S. Shire: ex-Cavendish Laboratory; was involved in erection of early radar stations and in setting up the Army Training School at Petersham where

he was Deputy Superintendent. He worked on a 50 cm GL set and at Malvern was the group leader in charge of all GL work.

March 6th

(368) Laboratory Stewards; see Ref. 1, pp. 14–15.

March 12th

(369) G.M. Wells: a member of the staff of the Laboratories.

March 13th

(370) Dunwich: on the East Coast of England between Aldeburgh and Southwold.

March 22nd

(371) CV 55 etc: see Appendix 2.

(372) Mr. Borosenko: not identified.

March 23rd

(373) Post-war television: see Ref. 1, p. 189.

March 24th

(374) Dr. B.V. Bowden: later Principal of UMIST. Became Lord Bowden of Chesterfield and was a Minister in a Labour Government.

April 2nd

(375) Gramophone Company: EMI.

April 6th

(376) Dr. Constant: Head of Engine Department, RAE, 1941 (see Ref. 3, Vol. 6).

April 8th

(377) Leading Staff: the one senior scientific staff grade in the Laboratories (see Ref. 1, p. 13–14).

April 12th

(378) Phantom submarine problem: effect of temperature distribution in sea?

April 28th

(379) OPTEC committee: a committee of the Radio Board.

May 1st

(380) For an account of the work on diamond dies see Ref. 1 pp. 392–395.

May 3rd

(381) 'Anti-netting' research: netting was a system of manual setting of the frequencies of a group of receivers and transmitters used by the Army with non-crystal controlled equipment. Narrow frequency spacings and drift led to problems, and later in the war crystal-controlled equipment was introduced. Anti-netting research was presumably directed towards this end.

May 6th

(382) The mobile workshops and drawing offices were most useful but were embarrassingly difficult to return after the war; the last of them had to be returned unofficially at dead of night and parked at the end of a row of similar vehicles at RAF Uxbridge.

May 10th

(383) Reliance Telephones: a GEC subsidiary whose principal business was the supply, installation and maintenance of private telephone exchanges.

May 14th

(384) Air Vice-Marshal Sir Victor Tait: Director of Radar and Director General of Signals, Air Ministry, 1942–45 (*Who's Who*, 1989).

May 18th

(385) 'the Cabinet influence plus the Clarendon': probably Lord Cherwell and Dr. R.V. Jones; see the latter's book 'Most Secret War' (Ref. 6).

May 19th

(386) Sir Stanley Angwin: Engineer-in-Chief of the Post Office; President of the IEE, 1943–44.

May 21st

(387) Defford: the RAF Station in Worcestershire to which flight trials moved from Christchurch when TRE moved from Swanage to Malvern. A number of members of the staff of the Laboratories worked there, and flew from there, particularly G.W. Edwards and later E.C. Cherry.

May 25th

(388) CV 114; a T R cell, i.e. a gas-filled cell which enabled a radar transmitter and receiver to be connected to a common aerial.

May 26th

(389) Dr. J.H.E. Griffiths: Oxford University; Secretary of CVD from 1943 until 1945.

(390) N.F.S. Hecht: Assistant Director, DCD, MAP.

May 28th

(391) Grocer: see Ref. 1, p. 381.

(392) Perrin: W.G. Perring, Deputy Director, Research and Experiment, RAE (later Director, RAE).

(393) Breadner: Foreman glass blower in the Laboratories.

May 29th

(394) Flight trials of the precision bombing system, Oboe Mark 2, used one tracking station on the Laboratories' roof and one on Bredon Hill in Worcestershire.

June 2nd

(395) Dr. Hadfield: probably Dr. W.H. Hatfield F.R.S., Director of Research, Thomas Firth & John Brown Ltd.; a member of the FBI Industrial Research Committee.

June 5th

(396) D.J. Jones: see Ref. 1. p. 341.

June 7th

(397) The hut at the back of the Laboratories used, before the war, for testing production television receivers (since they could only be finally tested in the London area) was taken over for valve preproduction.

June 8th

(398) The TR 1407 airborne transmitter – receiver (see Ref. 1, pp. 174 – 175).

June 10th

(399) Probably the halophosphate phosphors (see Ref. 1, p. 89).

June 15th

(400) Lumifax: this and other entries in the Diary put a later date on the establishment of Lumifax than had been suggested by contributors to Reference 1. The Diary, written at the time, must be taken as the correct version.

June 22nd

(401) Night Illumination Chart: a guide to the illumination of ground objects to be expected on clear cloudless nights, as a function of date and time, to assist in assessing suitability of date and time for particular tasks (not included as an illustration but a copy is in the original diary in the IEE Archives). See Biographical Memoirs of Fellows of The Royal Society, Volume 8, 1962, p. 114.

June 28th

(402) Mr. Ashley Cooper: a Director of GEC.

June 29th

(403) CV 108: a magnetron of Birmingham University design.

July 7th

(404) Sir Geoffrey Burton: Adviser to Governor of Central Provinces, India? (Ref. 3, Vol. 7).

(405) Major General Arcedeckne-Butler: Deputy Director General, Ministry of Supply (Ref. 3, Vol. 5).

July 12th

(406) L.H.B. Knox: a pre-war member of GEC Radio Works staff, responsible for final testing of television receivers in the 'Television Hut' at the back of the Laboratories. On the outbreak of war he joined Espley's team and was responsible for manufacture of prototype equipment. At the same time he was something of a 'thorn in the flesh' of the Laboratory management because he was an active member of the Association of Scientific Workers.

July 22nd

(407) Air Vice Marshal Baker: from Technical Training Command.

July 29th

(408) Dr. Barron: not identified.

August 6th

(409) Student Assistants: see Ref, 1, p. 15.

August 9th

(410) Pirelli Works, Milan: of some interest because in the UK GEC had an interest in a joint company, Pirelli General Cables.

(411) D.A. Oliver: of William Jessop & Sons Ltd. Worked with Megaw on high field magnets. A pre-war member of the staff of the Laboratories (see Ref. 1, p. 183).

(412) Professor Sucksmith: Sheffield University; worked on magnetic materials with Megaw and Oliver.

August 21st

(413) Marriott: manager of GEC Head Office Valve Department.

August 27th

(414) Squadron Leader Cropper remained with the Laboratories when demobilised at the end of the war and led, for many years, the work on torpedo guidance, first at Wembley and later at Stanmore (see Ref. 1, p. 391).

September 2nd

(415) RPC: probably Radio Production Control(ler).

September 21st

(416) RI: the Royal Institution.

September 27th

(417) Dr. R.: Dr. Railing.

October 5th

(418) Dr. Wolfenden (later Lord Wolfenden): Headmaster of Uppingham School.

October 13th

(419) N.R. Campbell: a distinguished physicist who was one of the first members of the staff of the Laboratories. For many years he undertook the task of drafting patent applications.

October 21st

(420) Dr. (later Sir Robert) Cockburn, at that time in charge of countermeasures work at TRE.

October 22nd

(421) *Science Abstracts* (which consisted of Part A, Physics, and Part B, Electrical Engineering) were started as a joint venture between the IEE and the Physical Society in 1899 but the Physical Society, with its small membership, and hence funds, had to withdraw in 1903 and *Science Abstracts* became an IEE publication. The Physical Society continued to make a modest financial contribution to *Science Abstracts* and to be represented on its Committees. The American Physical Society, and later the American Institute of Physics, as major subscribers also influenced policy.

During the 1939–45 war years the management of *Science Abstracts* was particularly concerned with the introduction of the Universal Decimal Classification System (UDC) and with proposals for closer working, and probable ultimate integration, with abstracting services in other scientific and technical fields. Paterson, who had been President of the IEE in 1930–31 and of the Institute of Physics (which was later to merge with the Physical Society) in 1937–39 — and who was attending the Physical Society Council — was clearly involved in these discussions. The UDC was adopted, there was not integration with other abstracting services and *Science Abstracts* was to lead on to the IEE's highly successful INSPEC activities.

October 23rd

(422) Fire control project: this was the introduction to 'RP 60', a project for the remote control of a naval 'pom-pom' anti-aircraft gun in which the windings of DC motors were fed through grid-controlled mercury-arc rectifiers from a DC supply.

October 27th

(423) Oxide coated cathodes: this and other entries in the Diary give a later date for the 'perfection' of this process than other sources; e.g. Ref. 1, pp. 116–117.

November 1st

(424) E 1487 (CV 208): the Wembley design of X band magnetron.

November 16th

(425) O.W.H.: O.W. Humphreys.
J.C.C.: J.C. Cornwell, Chief Laboratory Steward, in charge of all maintenance staff.

H.W.B.G.: H.W.B. Gardiner, in charge of drawing offices and workshops.

November 23rd

(426) Watsons Electromedical: A GEC subsidiary which marketed diagnostic X-ray equipment. The equipment was manufactured in GEC Union Works; both were on the GEC Wembley estate close to the Laboratories.

December 6th

(427) Professor Truman: not identified.

(428) Dr. McGee: a member of the staff of EMI Research Laboratories.

December 8th

(429) Increasing consideration was to be given to moving work from north-west London as the potential threats from the V1 (flying bomb) and V2 (rocket) weapons became clearer.

December 15th

(430) CV 69 and CV 76: high power magnetrons for 10 cm waveband.

December 16th

(431) Dr. Kahn: GEC Witton.

December 23rd

(432) DTM: Directorate of Torpedoes and Mines.

1944

January 4th

(433) CV 221; an improved CV 114 X band T – R cell.

January 5th

(434) E 1440: a smaller version of the CV 90 triode.

January 6th

(435) O.S. Ratsey: ASE, Head of group designing radar transmitters.

January 9th

(436) Nuffield Kremlin: not identified.

(437) Oxford foreign economists: not identified.

(438) A conference, with a very distinguished attendance — including Paterson — was held at Nuffield College, Oxford, in January and April 1944, which led to a publication 'Problems of Scientific and Industrial Research — a Statement'. Paterson was a member of the final drafting committee – see entry for April 15th 1944.

January 10th

(439) Ivor Jenkins: a metallurgist who had joined the Laboratories in 1934 (see Ref. 1, p. 344); he left the Laboratories but returned in 1946 as Head of the Metallurgy Department, becoming Chief Metallurgist in 1952 and leaving again in 1961.

January 13th

(440) K band; centre frequency 24 GHz; see Appendix 4.

January 27th

(441) Confidential History of the Laboratories: as part of the celebrations of the 21st anniversary of the official opening of the Wembley Laboratories Dr. Paterson delivered a discourse on the history of the Laboratories before an audience which included GEC Directors and managers and many members of the Laboratory staff. It was subsequently printed and circulated privately. It is not clear why it was 'Confidential'. A copy has been deposited in the IEE Archives.

January 28th

(442) Lord Suirdale: later Lord Donoughmore, role not identified.

February 7th

(443) Sir George Gee: not traced; at first thought to be a misprint for Sir George Lee, but Sir George Gee is referred to again on April 19th 1944.

February 9th

(444) Air Grocer: an airborne jammer for use against enemy 600 MHz radars (see Ref. 1, pp. 380–381).

February 16th

(445) Glass-to-metal seals: see Ref. 1, pp. 327–329.

February 23rd

(446) E 1496 (CV 240): the largest oxide coated cathode micropup with four times the output of the NT 99.

(447) CV 192 (E 1481): CV 64 type magnetron for higher loading.

March 4th

(448) Dr. Dale Bussell: not identified.

March 9th

(449) Air Commodore Erlanger: possibly Sir Gerard d'Erlanger who was Commanding Officer Air Transport Auxiliary and a member of the Board of British Overseas Airways.

March 28th

(450) Air Commodore Cardell: not traced (Cadell?).

(451) Multi-channel set: see Ref. 1, p. 175.

(452) Fairbairn: a senior engineer at GEC Radio Works, Bradford.

March 30th

(453) Discussion on Physics in Industry Journal: nothing was to happen until 1948 when the *Journal of Scientific Instruments* added to its title *and of Physics in Industry*. In 1950 the *British Journal of Applied Physics* started publication and took over the Physics in Industry role from JSI.

April 1st

(454) E 1457 (CV 257): a disc seal triode, usable as an amplifier between a CV 273 and a CV 288.

April 3rd

(455) Dr. Snow: C.P. Snow, later Lord Snow, physical chemist, novelist and later Government minister, at that time involved in allocation of scientific and technical manpower.

April 4th

(456) Helmover gear: the phrase 'for controlling '?' ' is not understood. The project was the 36 in diameter radio-controlled torpedo. The Laboratories were involved in the development of both the aircraft and the torpedo equipment. The project was known as Helmover and as Fisa. '?' may stand for Fisa if this name was applied to the torpedo.

April 13th

(457) Lord Woolton: Minister of Food, 1940–43, member of the War Cabinet 1943–45 (Ref. 3, Vol. 6).

April 15th

(458) Sir William Larke: Director, British Iron & Steel Federation; Controller Non-ferrous Mineral Development 1942–45 (Ref. 3, Vol. 5). He was a participant in the Nuffield conference.

(459) Sir Charles Bartlett: Managing Director General Motors Ltd. (Ref. 3, Vol. 5). Also at the Nuffield Conference.

(460) Captain Bennett: Fuel Utilisation Research Association.

April 17th

(461) Professor M. Polanyi: Professor of Physical Chemistry in Manchester University. He turned to economic problems in 1939 and in 1944 he wrote a paper on the need for reform in patent law to make more favourable provision for the encouragement of inventiveness ('Patent Reform', *Review of Economic Studies*, **11**, Summer).

April 19th

(462) The equipment of which Witton were making six sets was probably the large Helmore ground searchlight (see entry for May 22nd 1943).

(463) Sproule: said to have been seconded to staff of Air Commodore Leedham for production of GL equipment.

May 9th

(464) CIRES Department: not traced.

May 10th

(465) Mr. S. Cautauld: not traced; Mr. S. Courtauld was Chairman of Courtaulds at this time; he had been at the Nuffield conference.

(466) 'Window' frustration: see Ref. 1, p. 143.

May 15th

(467) CV 215: a hydrogen filled diode.

May 16th

(468) NT 57: a silica envelope valve; a Signal School development. NT 59: type not traced.

(469) E 1495: a 'higher power' valve in the micropup series for CW operation at 600 MHz.

May 17th

(470) Lonsdale: Manager of Fraser & Chalmers, a GEC subsidiary company in the mechanical engineering field. Later he was a GEC Director.

(471) Pochobradsky: not traced.

(472) Butterfly circuits; developed by ASE, BTH and Marconi, Baddow. Distributed radio frequency circuits using butterfly-shaped vanes which could rotate between plates with similarly shaped apertures.

May 19th

(473) This is the first mention in the Diary of what was to become the British Standard 13 A three flat pin plug.

(474) IMEA: possibly the Incorporated Municipal Electrical Association.

May 22nd

(475) Dr. C.J. Smithells: see the entry for February 23rd 1940. In 1944 he became Director of Research of the British Aluminium Company and set up their Research Laboratory at Chalfont St. Giles, Buckinghamshire.

May 24th

(476) Professor D.R. Hartree F.R.S.: Professor of Theoretical Physics, University of Manchester; an authority on numerical analysis; he applied a differential analyser to the study of the behaviour of magnetrons, including the effects of strapping (Ref. 13).

May 25th

(477) Mr. Touche: a principal in the GEC auditors.

June 16th

(478) E. Fielding: the SEI expert on quartz crystals.

June 21st

(479) Mr. Gestetner: Head of Gestetner Duplicators. He worked with his son as a consultant to ASE during the war.

June 23rd

(480) Dr. Roxbee Cox (later Lord Kings Norton): Director of Special Projects, Ministry of Aircraft Production, 1943–44; Chairman and Managing Director, Power Jets (Research & Development) Ltd. 1944–46 (*Who's Who*, 1989).

July 10th

(481) Dr. D.C. Truefitt: MAP staff; served in British Purchasing Commission, Washington.

July 11th

(482) ELMA: Electric Lamp Manufacturers Association.

July 13th

(483) G.M. Tomlin: manager of the R & D laboratories of SEI, he was later to be manager of GEC Telecommunications. He received an award at the end of the war for an invention in the field of proximity fuses.

July 14th

(484) Lang Brown: a senior member of the Admiralty Engineering Department, Bath, concerned with the work of the Laboratories on camouflage and other illumination projects. He had joined DSRE in about 1933.

July 18th

(485) G.K.: G. Kersey, the Laboratories' cashier.

July 20th

(486) CEM metal: not identified.

July 24th

(487) Professor Briscoe: not identified.

July 27th

(488) CVD work was, in fact, to continue at Wembley without interruption after the war.

August 1st

(489) M, X and Y products: products dealt with by a Head Office Department with this name, including, for example, small switchgear.

August 3rd

(490) Hannam Clark: in charge of work on transmission at GEC Telephone Works.

August 14th

(491) Another mention of Government plans to evacuate at least vital industrial production from London to north-west England in the event of intensification of V-weapon bombardment.

August 24th

(492) Caplan: a member of the staff of Fraser & Chalmers.

August 25th

(493) The Institute of Physics and the Physical Society were not to merge until 1960.

August 26th

(494) Clegg: TRE, said to be an expert on 200 MHz receiver design.

August 29th

(495) Research on alternatives to quartz: see Ref. 1, pp. 252–255.

September 1st

(496) Gallizia: from GEC Rectifier Works, Witton.

October 2nd–3rd

(497) 1403 set: not identified.

(498) 10 ton gun mounting: presumably the RP 60 project.

October 5th

(499) Dr. Clark: Dr. C.H. Clarke, Technical Director, Unilever. See also the entry in the Diary for April 15th, 1944.

October 9th

(500) Contract conditions: see Ref. 1, p. 287.

October 10th

(501) G.W.W.: G.W. Warren: responsible for development of a number of types of receiving valve.

(502) J.H.P.: J.H. Partridge; leader of the glass and ceramics group in the Laboratories.

(503) L.C.J.: L.C. Jesty: leader of the CRT group in the Laboratories.

October 16th

(504) C.C.G.: Dr. C.C. Garrard, Manager GEC Witton, a GEC Director.

October 20th

(505) C.J.S.: C.J. Smithells.

October 27th

(506) Puleston: in charge of the Laboratories' stores.

October 30th

(507) Charles Pinkham: manager of GEC Publicity Department.

November 1st

(508) Power Jets Ltd.: nothing is known to have come from this possible liaison. The site in Leicestershire was later to be the site for a number of English Electric (later GEC) units, including the Company's Mechanical Engineering Laboratory.

November 6th

(509) Civil aviation: this intention was not to be fulfilled post-war.

(510) Engineer Vice-Admiral Sir George Preece: a Vice-President of the Institution of Mechanical Engineers.

November 7th

(511) F.M. Colebrook: a senior member of the staff of the Radio Department in the National Physical Laboratory.

November 9th

(512) O'Kane: the job was part of the abortive GEC initiative on electronics for civil aviation. O'Kane was to join International Aeradio and later Marconi, a move which was to bring him back to GEC as a result of the GEC-English Electric merger in the late 1960s.

November 17th

(513) Professor C.L. Fortescue, Professor of Electrical Engineering, Imperial College; IEE President, 1942–43.

November 22nd

(514) Presumably the atomic bomb.

December 1st

(515) Tube Alloys: the code name for British work on the atomic bomb project.

(516) W.A. (later Sir Wallace) Akers: Research Director of ICI, an early member of the Maud Committee on the feasibility of nuclear fission who by this time was Director of the Tube Alloys project.

(517) M.W. Perrin: ex-ICI; deputy to Akers in Tube Alloys.

December 4th

(518) Admiralty P 55 set: not identified.

December 5th

(519) Glass: see Ref. 1, pp. 327–330.

December 7th

(520) Mr. Blok: senior patents officer, Tube Alloys.

December 9th

(521) See Ref. 8.

December 13th

(522) 'Civilian Scientists': see Ref. 4; pp. 83, 85.

December 15th

(523) GEC nuclear work: see Ref. 1, p. 287.

December 16th

(524) The lamp seems to have been far ahead of its time. Lamps of this type were not to be marketed until about the 1970s.

1945

January 2nd

(525) The beginning of the 'kite mark' scheme?

January 9th

(526) L.D.G.: L.D. Goldsmith.
R.W.W.S.: R.W.W. Sanderson.

January 12th

(527) Lord MacGowan: a Director of Imperial Chemical Industries.

January 13th

(528) High frequency heating: see Ref. 1, pp. 318–322.

January 23rd

(529) Dr. Wooldridge: one of the founders of Ramo Wooldridge, later Thomson Ramo Wooldridge, the major American electronics company with whom GEC was to have a joint venture on process control in the 1960s.

January 24th

(530) CV 259 (E 1495): a CW 600 MHz valve in the micropup series.

February 1st

(531) Transient Test Set: an instrument developed by Espley and Cherry to examine the performance of circuits, cables etc., in the time domain, i.e. their response to very short pulses.

February 5th

(532) Dr. N.A. Halbertsma: an authority on lighting who combined an academic career with one in Philips. In 1939 he was appointed Professor extraordinary in illumination technology at the University of Utrecht and in the same year he was appointed Chairman of the Commission Internationale de l'Eclairage — a position which Paterson also held at one time (see Ref. 1, p. 112).

February 16th

(533) Frank Buckell: for many years manager of Aviation Department, GEC Head Office.

March 1st

(534) Sir Thomas Barlow: Director General of Civilian Clothing, 1941–45 (Ref. 3, Vol. 6).

March 7th

(535) S.E.A. Landale: ASE, designed the Radar Type 271, the first centimetric surface vessel search equipment; at this time Superintending Scientist at ASE Witley.

March 14th

(536) Sir William Palmer: Second Secretary, Ministry of Supply, 1939 – 42; Second Secretary, Ministry of Production, 1942 – 44; Principal Industrial Adviser, Board of Trade, 1944 – 46 (Ref. 3, Vol. 6).

(537) Television links and transmission: the Laboratories were to take up work on radio relay links for transmission of television and to develop and supply the London to Birmingham link (see Ref. 1, pp. 194 – 199).

March 16th

(538) Glass bushings: for insulating electrical leads into hermetically sealed containers of electronic equipment.

March 20th

(539) W.E.W.: W.E. Willshaw.
E.M.: E. Megaw.

March 21st

(540) Waldram's paper: see Ref. 11.

(541) Sir George Simpson: former Director of the Meteorological Office (Ref. 3, Vol. 6).

March 26th

(542) CV 273 (VX 3018): a small microwave triode.

March 27th

(543) Kovar: a trade name for an alloy of nickel, cobalt and iron used in glass-to-metal seals.

April 3rd

(544) Nechell's turbines: not identified.

April 4th

(545) Mr E. W. Hives: a Director, later Chairman of Rolls Royce (Ref. 3, Vol. 6).

April 5th

(546) Process heating paper: see Ref. 9.

April 6th

(547) Admiral Perrin: possibly Engineer Rear-Admiral H.H. Perring who retired from the Royal Navy in 1938 but was made C.B.E. in 1945.

(548) Mr. Bentham: not identified.

April 16th

(549) Boot: of Randall and Boot, the inventors of the original cavity magnetron.

April 17th

(550) Sir Alexander Korda: film producer and head of London Films which owned Denham studios.

April 18th

(551) The CV 273, 257 and 288 were to be the basis of the Laboratories' own London – Birmingham television link.

April 19th

(552) The Services Electronic Research Laboratory (SERL), Baldock.

April 23rd

(553) Dr. Crampton: Head of Radio Direction Finding at ASE Haslemere.

April 24th

(554) H.L. Kirke: Head of Research, BBC.

April 25th

(555) Dr. Van der Pol: for many years a senior scientist at Philips, Eindhoven, renowned for his application of mathematics to radio, particularly non-linear phenomena.

April 27th

(556) A 200 ft tower was built and was a Wembley landmark for many years (see Ref. 1, pp. 199–201).

(557) Moxham: later to be the manager of GEC Broad Oak Works Portsmouth.

Appendix 1

The CVD organisation and its relationship with the GEC Research Laboratories

An account of the origins of CVD is given in a GEC Research Laboratories report to DSR, Admiralty, entitled 'C.V.D. – General survey of work February to December 1939', dated 2nd January, 1940. The following is an extract from this report:

> 'The C.V.D. liaison was initiated towards the end of 1938 in order to regularise and extend an informal emergency arrangement, known as U.S.W.A., made in the preceding summer, between D.S.R. and G.E.C. Research Laboratories for the rapid provision of trial valves, other than silica valves, for R.D.F. purposes, the ordinary channels for obtaining such samples having been found to be unsuited to confidential applications.
>
> 'The initials C.V.D. were intended to denote "communication valves development", not as sometimes interpreted "commercial". According to the original terms of reference commercial considerations were to be eliminated as far as possible, and in fact they have been. For example, contractors, such as Metropolitan – Vickers, have embodied auxiliary valves manufactured by their own associates in lieu of G.E.C. valves used in the earlier stages of development.
>
> 'The object of the C.V.D. liaison was to assist D.S.R. to keep continuously in review the needs of all three Services and to ensure that valve development was suitably directed or initiated to meet them.
>
> 'The original demand was for transmitting types, but at an early stage we pressed for and obtained the inclusion of receiving types. At that time, as subsequently, tentative requests for samples of receiving types not yet in production were multiplying.
>
> 'In later months, as the demands on the transmitting side have approached satisfaction, the receiving side has become more and more prominent. There has been a growing need for correlating experimental trials of samples with the possibilities of early production. Further the view has been gaining strength that there is more to be hoped for from improvement of receiver performance than from enhanced transmitter power.

'Finally, there has recently appeared some difference of opinion as to whether the C.V.D. scheme is, or is not, to be considered as confined to the selection or development of valves for R.D.F. purposes. Whatever the original object may have been it would now appear that on the receiving side especially, where valves are not so specialized, some consideration of wider fields of application will be inevitable.'

The report goes on to survey the position on six types of transmitting valve which had been developed and which 'either have been or shortly will be adopted for immediate use'. Some of these valves were for use at frequencies up to 600 MHz and, although the report was to the Admiralty, there are references to their application in GL (Gun laying) presumably for the Army, and to AI (Airborne Interception) and ASV (Air-to-Surface Vessel) for the Royal Air Force. These make it clear that by the date of the report Paterson and his senior colleagues on valve research and development had been brought into discussions on the requirements of such systems. There are also surveys of transmitting valve types which had been developed but discarded and to a number of types of receiving valve.

From this original liaison between GEC and Service Departments, led by the Admiralty, grew the C.V.D. organisation which was responsible for research and development on thermionic valves and other electronic devices to meet the needs of the Services' equipment for radar, navigation, communication and other purposes, which involved a number of companies, university departments, and government establishments. Its development, organisation, and activities and the part played by the Laboratories are themes throughout Paterson's diary.

Confirmation of the origins of C.V.D. comes to light in the minutes of early meetings.

Minutes of the 'first C.V.D. meeting held at Research Laboratories of the General Electric Co. Ltd., Wembley, on 1st February 1939.' show that it was attended only by Messrs. Shearing, Hughes and Ure of HM Signal School and Dr. Paterson and Messrs. Bell, Gavin, Gossling, Megaw and Le Rossignol of GEC.

The minutes of the second C.V.D. meeting (held at Signal School) on 14th February 1939 show that, in addition to representatives of GEC Research Laboratories and Signal School, R.A. Watson-Watt was present.

Commander Willett (Signal School) in his opening remarks on this occasion stated that Signal School had been entrusted with the very important duty of valve development for the three Services. An important feature was that the GEC Research Laboratories had become part and parcel of the development scheme. Meetings (which it was proposed to call C.V.D. meetings) would be held at Signal School and at Wembley, and representatives from both the Air Ministry and War Office would always be welcome.

The third C.V.D. meeting was held at Wembley on 4th May 1939, and DSR, Admiralty, and the Bawdsey Research Station were additionally represented. DCD (Air Ministry) were also represented at the fourth meeting on 1st August 1939. By the fifth meeting, on 29th September 1939, Bawdsey had moved and become the Air Ministry Research Establishment and the Air Defence

Experimental Establishment was additionally represented. A sixth C.V.D. meeting was held at Wembley on 1st December 1939, with similar representation. In parallel with these C.V.D. meetings there were informal 'liaison meetings' at Wembley.

In November 1939 the Admiralty was asked to expand efforts on meeting valve requirements of all three services; by this time other industrial and academic organisations were becoming involved and the bilateral GEC – Government C.V.D. meeting was becoming an anomaly. The GEC – Government meetings were to continue, at about six-weekly intervals, throughout the war, as the GEC – C.V.D. liaison meetings.

A C.V.D. Policy Committee, confined to representatives of Government organisations, was established in November 1939. At first this only had two subcommittees: a Technical Committee (also attended only by Government staff) and the C.V.D.-GEC Liaison Committee. In 1940 a 50 cm committee, a 10 cm committee and a modulation techniques committee were set up. The remits of the group of committees and their titles were to be changed from time to time but the committee structure was to remain in being throughout the war and each individual committee was to meet at about six week invervals. Paterson and staff of the Laboratories were to play a part in substantially all of the subject committees.

Early in 1941 an Inter-Services Technical Valve Committee (ISTVC) was set up to standardise valve types with the object of checking the haphazard introduction of new types of valves. Paterson was appointed first Chairman of this committee and continued as Chairman throughout the war. In June 1941 Paterson joined the C.V.D. Policy Committee in his role of Chairman of ISTVC and continued to be a member throughout the war.

The ambit of C.V.D. ultimately embraced research and development on all valves and related electronic devices for radar, navigational aids, countermeasures, fuses and communications. A parallel series of meetings dealt with research and development on cathode-ray tubes but for much of the war these appear to have been the responsibility of the Air Ministry and then the Ministry of Aircraft Production (MAP).

In addition to research and development, the Laboratories were to manufacture many valves during the war; this 'preproduction' being necessary to get newly developed designs into equipment for trials and operational use before a factory could get into production. The rate of preproduction rose to over 2000 per week and about 300 000 valves were produced in all.

Appendix 2

Valve types referred to in the Diary

Valve types are listed under E numbers used in the Laboratories, inter-Service CV numbers, and in some cases individual Service numbers or other descriptions. Where several type numbers are known for one design, all are given and cross referenced. A bracketed type number indicates that it has not been found in the Diary but identified from other sources, e.g. reports to CVD. Valve descriptions have been obtained from various sources including the Diary and reports to CVD; they are not always unambiguous and should be treated with caution. It should be borne in mind that there must be chances of typing errors which could lead to false identification or failure to identify. Powers are generally peak pulse powers unless stated to be C.W.

Type Number			Description
E	CV	Other	
960		(VT 58)	Metal – glass alternative to silica NT 57, based on ACT 10.
1001			Triode, small, glass, for 1 metre.
1024		(VT 114)	Glass – metal tetrode, 'for MB'.
1046		VT 90	Micropup, for Signal School, also used in AI and ASV.
1061			Not traced.
1148		(VR 135)	Small glass triode oscillator, for RDF and CW.
1149			Glass – metal triode, not traced.
1155	(44)	NT 100	Tetrode modulator, 12 kV, 10 A.
1161	(92)	NT 93(97)	First enlargement of micropup (also known as 1046B).
1166			1046D. Second enlargement of micropup.
1189		NT 98	Magnetron, 10 cm.

1190	55		Enlarged oxide coated milli-micropup — CW.
	155		Enlarged oxide coated milli-micropup — pulsed.
1191	12	NGT8	Gas triode modulator, 15 kV, 200 A, 300 c/s; NGT 8: lower rating.
1198	(38)		Magnetron, 9.1 cm.
1199			Peak-voltmeter diode.
1209	(7)		Gas triode ('small').
1210			Glass magnetron, 'receiving', 9.1 cm.
1231	52		Concentric triode oscillator.
1232	92	NT 99	Oxide coated micropup valve for 80 – 100 kW at 50 cm.
1233			Magnetron, 10 cm, 50 – 100 kW.
1239			Micropup series, largest, oxide coated cathode.
1248	8		Diode, switching, for 150 cm.
1255			Magnetron, 10 cm, 100 – 200 kW.
1256			Magnetron, 9.1 cm, 100 – 200 kW.
1267	(41)		Magnetron, 10.7cm, 100 – 200 kW.
1271	57		Tetrode hold-off for use with gas triodes. Series modulator, parallel operation.
1273	58 258		Axial mixer diode, CV 258: lower grade CV 58.
1274			All concentric triode (transmitter) for <50 cm.
1290			Triode amplifier, anti-microphonic.
1305			Grounded grid triode for 200 MHz.
1324			Magnetron, not traced.
(1325)	56		Magnetron, strapped E 1189.
(1326)	69		Magnetron, strapped E 1198.
1335	61		High gain rugged midget pentode (CV 61 for RNTF).
1336			Rugged miniature, 10 mm diameter, oscillator/receiver.
1342			Magnetron, E 1326 with reduced anode length.
1344			Triode, common grid-earthed grid oscillator, double ended.
(1359)	76		Magnetron, cf. E 1325, 25 kV, 40 A, 2 μS, heavy duty cathode.
1364			Gas triode, rugged, miniature.
(1365)	114		Waveguide T/R cell, 3 cm, brass body, soldered.
1366			Midget valve for Torpedo School.
1368	90		Common grid-earthed anode oscillator (single ended 1344).

(1373)	99		Magnetron, similar to 1325/1359, 8.5 cm.
1374			Concentric valve for up to 2 GHz.
(1379)	79		Magnetron, glass, formerly E 1210DE, 6.6 cm; for MOS.
(1380)	89		Magnetron, glass, formerly E 1210DE, 6.3 cm; for MOS.
(1415)	115		Slotted disc cell (anti-T/R), for 10 GHz.
1440			Smaller version of CV 90, 2 watt cathode.
1457	257		Larger disc seal triode, 5 – 10 kW, pulsed, 3 GHz.
1463			Hydrogen filled gas triode, 160 kW, 2500 pps.
1480			Triode, (1419) without internal feedback.
(1481)	192		CV 64 for higher input.
1487	208		Magnetron, 12 segment, strapped, 10 GHz.
1494	191	(VX 3006)	1487 for higher frequency.
1495	259		'Higher power' CW valve, 10 × CV 55, for 600 MHz.
1496	240		1239 improved and modified for lower frequencies. 4 × NT 99.
(1497)	215		Hydrogen diode for absorbing overswing energy in a network.
1511			High power X band valve, lower frequency than 1487.
1516	221		CV 114 with shorter discharge gap, wider band, gold seals, copper body.
(1531)	214		E 1494 with coaxial to waveguide feed; limited frequency range.
	14		Not traced.
	64	(BM 717)	Magnetron, strapped, S band, Birmingham design.
	93		Not traced, NT 93?
	97		Not traced, NT 97?
	108		Magnetron, 3 cm, Birmingham design.
	204		Not traced.
	273	(VX 3018)	Small common-grid earthed-anode triode.
	288	(VX 3034)	Larger cathode CV 259 for pulsed operation.
		BM 313	Magnetron, Birmingham design.
		NC 11	Cathode-ray tube for Naval use.
		NC 12	Cathode-ray tube for Naval use.
		NT 57	Signal School silica valve.
		NT 59	Not traced.

NT 89	Magnetron, not traced, possibly typing error for NT 98.
VC R97	Cathode-ray tube.
VT 93	Not traced, NT 93?
VT 97	Not traced, NT 97?

Appendix 3

Some Abbreviations used in the Diary

AA	Anti-aircraft
AB	GEC work on RCM Elsewhere, an internal GEC Advisory Board
ADEE	Air Defence Experimental Establishment (later ADRDE)
ADRDE	Air Defence Research and Development Establishment
AI	Airborne Interception radar
AIS	Airborne Interception Radar (short-wave), i.e. on 9 cm
AM	Air Ministry
AMRE	Air Ministry Research Establishment, later TRE
ARL	Admiralty Research Laboratory
ARP	Air Raid Precautions
ASE	Admiralty Signals Establishment, formerly Signal School
ASEE	Admiralty Signals Establishment Extension
ASW	Anti-Submarine Warfare
BEAMA	British Electrical & Allied Manufacturers Association
BEF	British Expeditionary Force
BRS	Building Research Station
BSI	British Standards Institution
CB	GEC Consultative Board for the Research Laboratories
CRF	Compact Range Finder
CRT	Cathode Ray Tube
CVD	Originally Communication Valves Development — a GEC RL project arising out of USW to provide valves for RDF. Later, Co-ordination of Valve Development
CW	Continuous Wave generation, transmisssion or reception
DCD	Director(ate) of Communications (and radar) Development
DEE	Director(ate) of Electrical Engineering (Admiralty)
DGMS	Director General of Munitions Supply.
DNC	Director of Naval Contracts
DRP	Director of Radio (and Radar) Production
DSIR	Department of Scientific and Industrial Research
DSR	Director of Scientific Research
DTM	Director of Torpedoes and Mines

EAW	Electrical Association for Women
ELMA	Electric Lamp Manufacturers Association
EMI	Electrical and Musical Industries Ltd.
ERA	Electrical Research Association
FBI	Federation of British Industries
f.c.	Foot-candle
FM	Frequency modulation
GCI	Ground Controlled Interception
GEC	The General Electric Company Ltd. (now plc)
GL	Anti-aircraft Gun Laying radar
GR	25 cm gunnery radar
HF	High Frequency
HO	GEC Head Office
HV	High voltage
IEE	Institution of Electrical Engineers
IES	Illuminating Engineering Society
IR	Infra-red
ISTVC	Inter-Services Technical Valve Committee
ISVPC	Inter-Services Valve Production Committee
ITC	Informal Technical Committee, between Lamp Manufacturers
ITP	Instruction to proceed
MAP	Ministry of Aircraft Production
MIT	Massachusetts Institute of Technology
MOS	Ministry of Supply
MOV	M – O Valve Company, at that time a joint GEC – EMI company
MWT	Marconi's Wireless Telegraph Company
NPL	National Physical Laboratory
OAC	Osram Advisory Committee
OC	Officer Commanding
PF	Proximity Fuse
QC	Quality Control
RAE	Royal Aircraft Establishment
RC	Radio Communications
RCM	Radio Countermeasures
RDF	Radio Direction Finding: an early code name for radar
RL	Research Laboratories
RMA	Radio Manufacturers Association
RNTF	Royal Naval Torpedo Factory
ROF	Remotely (or Radio?) Operated fuse
RS	Research Services — a meeting between the Director, RL, and GEC Directors, primarily on work for the Services
SEE	Signals Experimental Establishment, late SRDE
SEI	Salford Electrical Instruments, a GEC subsidiary
SRDE	Signals Research and Development Establishment
SS	Admiralty Signal School
STC	Standard Telephone and Cable Company Ltd
SW	Short Wave

TF	Torpedo Factory
TRE	Telecommunications Research Establishment formerly AMRE
TVC	(Interservices) Technical Valve Committee
USW	Ultra Short Wave (Code name for an RL project for the Admiralty)
UV	Ultra-violet
VM	Velocity modulation
VPC	(Inter-Services) Valve Production Committee
XL	RL extension in the Wembley Empire Exhibition Grounds
Z Dept.	A GEC Head Office Department on Lighting

Appendix 4

Frequency and wavelength designations

Designation	Approximate central frequency	Approximate central wavelength
L band	600 MHz	50 cm
S band	3 GHz	10 cm
X band	10 GHz	3 cm
K band	24 GHz	1.25 cm
Q band	40 GHz	0.75 cm
V band	50 GHz	0.6 cm

Appendix 5

Engineering Advisory Committee

The Lord President of the Council, with the approval of the Prime Minister, has appointed an Engineering Advisory Committee under the chairmanship of Lord Hankey to advise the Government upon engineering questions connected with the war effort and in particular:

(1) To consider how the resources of the engineering profession can best be utilised in connection with the war work of Government Departments; and to nominate engineers who might suitably be invited to undertake particular tasks;

(2) To suggest means of improving or supplementing the methods adopted by Government Departments for utilising engineering science for war purposes;

(3) To advise on problems referred to them in connection with the development of new engineering devices; and to advise, as required, on methods of bringing such devices speedily into production and on means of meeting new war requirements in the engineering field;

(4) To examine new ideas or devices in engineering likely to assist the war effort, to test their technical validity, and to bring to the notice of the Government those which appear to merit further consideration by the Department concerned.

The Committee will consist of Rt. Hon. Lord Hankey, G.C.B., G.C.M.G., G.C.V.O., Chancellor of the Duchy of Lancaster (Chairman); Rt. Hon. Viscount Falmouth, M.I.Mech.E. (Vice-chairman); Sir Henry Tizard, K.C.B., F.R.S., F.R.Ae.S., F.Inst.P.; and the following members who have been selected after consultation with the Presidents of the Institutions of Civil, Mechanical, and Electrical Engineers: J R Beard, M.Sc., M.Inst.C.E., M.I.E.E., Dr. A. P. M. Fleming, C.B.E., M.I.Mech.E., M.I.E.E., F.Inst.P., W. T. Halcrow, M.Inst.C.E., B.W. Holman, O.B.E., A.R.S.M., M.I.Mech.E., M.Inst.Chem.E., Dr. C.C. Paterson, O.B.E., M.I.E.E., F.Inst.P., H.R. Ricardo, B.A., F.R.S., A.M.Inst.C.E., M.I.Mech.E., Principal A. Robertson, F.R.S., M.I.Mech.E.

Representatives of the Government research establishments and production departments will be associated with the Committee as required from time to time.

IEE Journal, 1941, **88**, Pt. 1, p.207

References

1 CLAYTON, R., and ALGAR, J.: 'The GEC Research Laboratories 1919–1984' Peter Peregrinus, 1989

2 BOWEN, E.G.: 'Radar days' Adam Hilger, 1987

3 'Who was Who', Vol. 4 (1941–50), Vol. 5 (1951–60), Vol. 6 (1961–70), Vol. 7 (1971–80)

4 ROWE, A.P.: 'One story of radar' Cambridge University Press, 1948

5 READER, W.J.: 'A history of the Institution of Electrical Engineers 1871–1971' (Peter Peregrinus IEE, 1987

6 JONES, R.V.: 'Most secret war' Hamish Hamilton, 1978

7 JAMES, E.G., DIX, J.C., COPE, J.E., ELLIS, C.F., and ANDERSON, E.W.: 'The development of the Wireless Set No.10: an early application of pulse-length modulation', *J.IEE, Pt. 3A, 1947 (13)*

8 CLAYTON, R.J., HOULDIN, J.E., LAMONT, H.R.L., WILLSHAW, W.E.: 'Radio measurements in the decimetre and centimetre wavebands', *J.IEE, Pt. III,* March 1946

9 CONNELL, L.C.J., HUMPHREYS, O.W., RYCROFT, J.L.: 'The place of radiant dielectric and eddy current heating in the process heating field', *J.IEE,* 1945 **92**, and 1946, **93**

10 COCKCROFT, J.E.: 'Memories of radar research', *IEE Proceedings*, 1985, **132A** (6) (Special issue on historical radar)

11 WALDRAM, J.M.: 'Measurement of the photometric properties of the upper atmosphere,' *Trans. IES*, 1945 (10)

12 ACKERLEY, R.O.: 'Some factors affecting the design of electric lighting installations for building interiors,' *J.IEE*, 1945, **92**, Pt. II, pp. 365–384

13 Biographical Memoirs of Fellows of the Royal Society

Index of Names

Index of Subjects

Met Lonsdale at H.O. to hear from him an embarrassing story of a visit to the Continent, and Admiralty contract to Power Jets for a 7000 K.W. gas turbine. I dont see how this will end up - but it must be left for the moment.

May 3rd. The whole German Army in Italy has surrendered - about a million men. Hamburg and Rangoon have fallen and Berlin also. Everything now depends on the submarine situation and Norway and Denmark.

The opportunity of getting the TA valve designed and tried is up against a workshop bottleneck. It will move but slowly I fear but we shall see if M.O.V. can help.

Spent the morning with Heather, Enticknap and Humphreys reviewing the RL post war budget. This is now in good shape for scrutiny and will come up at the C.B. in a week's time.

Bowler is now definitely dropped and Trumper has been reduced to something of much greater simplicity. So Cherry hopes that he can get away before long to take up his teaching job in Manchester.

May 4th. Call from Wing Comm.Mealing to talk over aerodrome lighting and the bad position of British industry in the matter in view of U.S.A. aggression in export trade.

Lumifax meeting. Our position in powders improves. The phosphates are well under way in manufacture and B.T.H. and Siemens are also banking on it as a cheaper production article.

The VE day code word has gone round so that notice may be sent in advance to establishment which must make prearrangements for the 1½ days holiday. We are into Denmark but the German Navy elements seem inclined to fight on.

May 5th. Discussions on T.A. Oliphant is back in this country and Garrard is willing to take on the design of part of the plant. We must have a meeting with Garrard, Oliphant and Akers.

May 7th. 5.15 p.m. The week end has brought news of surrenders and successes and the end was obviously near. Today we hear that cease fire has been ordered by Hitler's successor to all forces - submarines are to cease operations and return to base. The great announcement of VE day has been expected momentarily but has not yet come through. If it comes through this evening we shall close for Tuesday and Wednesday.

May 8th and 9th. Public holidays and rejoicing.

We have now passed into the post war era and this 5½ years diary comes to an end. At our pre-production meeting Pearce announced that we had made 295138 valves.

The last page of the original typescript of the diary.

WAR DIARY

C.C.PATERSON.

September 1st to 4th 1939.

Considered and arranged many staff transferences to work of more urgent importance.

Ryde and Harris	to Klystrons.
Laycock	to receiving valves.
H.G.Jenkins & Bowtell	to CRT screen efficiencies.
Lewer	to radio work Coventry.

Started organisation under RLeR of 1046 valve production at Laboratories.

September 4th to 8th.

Intensification of activity on valve and radio work for Bawdsey and Signal School (ZZ).

Despatch by car of special CRT's to S.Coast for Fogg and Atkinson's application - drivers Waldram and Stevenson.

Heavy alloy bullets made for certain trials.

Negotiations with Mr.Sims for making Elmet Metal at Wembley.

September 11th.

Discussed with Stevens their plans for ARP lighting particularly in regard to IES-BSI activity. Considerable activity on subject at Wembley.

September 12th.

CVD meeting with Signal School. It appears that with Air Ministry moved 'N' and Admiralty 'W' the Research Laboratories is the most convenient fixed point for future liaison between all concerned. It was agreed that meetings must be held in future more frequently, i.e. a full meeting 2 monthly and intermediate meeting sectionalised.

September 13th.

Attended meeting of National Register - dearth of Radio men.

Meeting with all the tonnage in C.S.Wright's office to decide programme of special research to be undertaken at Birmingham by Oliphant with treasury grant.

Watson Watt /